introduction to algebra

4th edition

K. Elayn Martin-Gay
Supplemental Work by Walter Hunter

Pearson
Custom
Publishing

Cover Art: George Herman, "Matisse Dreams"

Excerpts taken from:
Beginning Algebra, Second Edition, by K. Elayn Martin-Gay
Copyright © 1977, 1993 by Prentice-Hall, Inc.
A Pearson Education Company
Upper Saddle River, New Jersey 07458

Copyright © 2000, 1999, 1998, 1997 by Walter Hunter.
All rights reserved.

The information, illustrations, and/or software contained in this book, and regarding the above-mentioned programs, are provided "As Is," without warranty of any kind, express or implied, including without limitation any warranty concerning the accuracy, adequacy, or completeness of such information. Neither the publisher, the authors, nor the copyright holders shall be responsible for any claims attributable to errors, omissions, or other inaccuracies contained in this book. Nor shall they be liable for direct, indirect, special, incidental, or consequential damages arising out of the use of such information or material.

All rights reserved. No part of this book may be reproduced, in any form or by any means, without permission in writing from the publisher.

This special edition published in cooperation with Pearson Custom Publishing.

Printed in the United States of America

10 9 8 7 6 5 4 3 2 1

Please visit our web site at www.pearsoneducation.com

ISBN 0-536-60998-5

BA 992196

PEARSON CUSTOM PUBLISHING
75 Arlington Street, Boston, MA 02116
A Pearson Education Company

CONTENTS

UNIT 1

Signed Numbers S-1
    Homework S-1

Introduction to Variables S-5
    Class Work S-5
    Group Work S-9
    Homework S-11

Simplifying Algebraic Expressions S-17
    Class Work S-17
    Group Work S-19
    Homework S-21

Solving Equations Part I S-23
    Class Work S-23
    Group Work S-25
    Homework S-27

Solving Equations Part II S-29
    Class Work S-29
    Group Work S-31
    Homework S-33

Applications of Linear Equations Part I S-35
    Class Work S-35
    Group Work S-39
    Homework S-41

Applications of Linear Equations Part II (Literal Equations) S-48
    Class Work S-48
    Group Work S-49
    Homework S-51

Percentages S-54
    Class Work S-54
    Group Work S-57
    Homework S-59

Review for Test I S-62

UNIT 2

Inequalities S-66
- Class Work S-66
- Group Work S-69
- Homework S-71

Applications of Inequalities S-73
- Class Work S-73
- Group Work S-75
- Homework S-77

Scatter Plots S-83
- Class Work S-83
- Group Work S-89
- Homework S-91

Interpreting Graphs S-93
- Class Work S-93
- Group Work S-95
- Homework S-97

Graphing Lines by Plotting Points S-99
- Class Work S-99
- Group Work S-101
- Homework S-103

Graphing Lines by Finding the Intercepts S-105
- Class Work S-105
- Group Work S-107
- Homework S-109

Introduction to Slope S-112
- Class Work S-112
- Group Work S-117
- Homework S-119

Slope S-124
- Class Work S-124
- Group Work S-127
- Homework S-129

Applications of Graphs S-133
- Class Work S-133
- Group Work S-141
- Homework S-143

Review for Test II S-149

UNIT 3

Introduction to Exponents and Positive Exponents S-155
    Class Work S-155
    Group Work S-158
    Homework S-159

Negative Exponents and Scientific Notation S-163
    Class Work S-163
    Group Work S-165
    Homework S-167

Properties of Exponents S-170
    Class Work S-170
    Group Work S-173
    Homework S-175

Introduction to Algebraic Fractions S-176
    Class Work S-176
    Group Work S-179
    Homework S-181

Adding and Subtracting Algebraic Fractions S-183
    Class Work S-183
    Group Work S-185
    Homework S-187

Solving Equations with Fractions S-189
    Class Work S-189
    Group Work S-191
    Homework S-193

Ratio and Proportion Problems S-196
    Class Work S-196
    Group Work S-199
    Homework S-201

Review for Test III S-203

UNIT 4

Introduction to Quadratics S-206
    Class Work S-206
    Group Work S-209
    Homework S-211

Applications of the Quadratic Formula S-214
    Class Work S-214
    Group Work S-217
    Homework S-219

Quadratic Applications and Their Graphs S-222
    Class Work S-222
    Group Work S-229
    Homework S-231

Factoring S-235
    Homework S-235

Review for Test IV S-236

Review for Final Exam S-239

Answers to Homework and Review Tests S-248

CONTENTS

Excerpts from *Beginning Algebra,* Second Edition, by K. Elayn Martin-Gay

1 Review of Real Numbers — 1

- 1.1 Symbols and Sets of Numbers / 2
- 1.2 Fractions / 11
- 1.3 Exponents and Order of Operations / 19
- 1.4 Introduction to Variable Expressions and Equations / 25
- 1.5 Adding Real Numbers / 30
- 1.6 Subtracting Real Numbers / 36
- 1.7 Multiplying and Dividing Real Numbers / 41
- 1.8 Properties of Real Numbers / 48
- 1.9 Reading Graphs / 55

Group Activity: Creating and Interpreting Graphs / 62
Highlights / 63
Review / 67
Test / 70

2 Equations, Inequalities, and Problem Solving — 73

- 2.1 Simplifying Algebraic Expressions / 74
- 2.2 The Addition Property of Equality / 81
- 2.3 The Multiplication Property of Equality / 88
- 2.4 Solving Linear Equations / 94
- 2.5 An Introduction to Problem Solving / 104
- 2.6 Formulas and Problem Solving / 112
- 2.7 Percent and Problem Solving / 122
- 2.8 Further Problem Solving / 131
- 2.9 Solving Linear Inequalities / 139

Group Activity: Calculating Price Per Unit / 149
Highlights / 150
Review / 156
Test / 159
Cumulative Review / 160

3 Graphing

3.1	The Rectangular Coordinate System / 164	
3.2	Graphing Linear Equations / 175	
3.3	Intercepts / 185	
3.4	Slope / 195	
3.5	Graphing Linear Inequalities / 208	

Group Activity: Financial Analysis / 217
Highlights / 218
Review / 222
Test / 224
Cumulative Review / 225

4 Exponents and Polynomials

4.1	Exponents / 228
4.2	Adding and Subtracting Polynomials / 238
4.3	Multiplying Polynomials / 246
4.5	Negative Exponents and Scientific Notation / 257
4.6	Division of Polynomials / 266

Group Activity: Making Predictions Based on Historical Data / 273
Highlights / 274
Review / 276
Test / 278
Cumulative Review / 279

5 Factoring Polynomials

5.1	The Greatest Common Factor and Factoring by Grouping / 282
5.2	Factoring Trinomials of the Form $x^2 + bx + c$ / 289
5.4	Factoring Binomials / 304
5.5	Choosing a Factoring Strategy / 309
5.6	Solving Quadratic Equations by Factoring / 314
5.7	Quadratic Equations and Problem Solving / 324

Group Activity: Choosing Among Building Options / 333
Highlights / 334
Review / 337
Test / 339
Cumulative Review / 340

6 Rational Expressions 343

- 6.1 Simplifying Rational Expressions / 344
- 6.2 Multiplying and Dividing Rational Expressions / 351
- 6.3 Adding and Subtracting Rational Expressions with Common Denominators and Least Common Denominator / 356
- 6.4 Adding and Subtracting Rational Expressions with Unlike Denominators / 363

- 6.6 Solving Equations Containing Rational Expressions / 376
- 6.7 Ratio and Proportion / 383
- 6.8 Rational Equations and Problem Solving / 390

  Group Activity: Comparing Formulas for Doses of Medication / 399
  Highlights / 400
  Review / 405
  Test / 407
  Cumulative Review / 408

9 Roots and Radicals

- 9.1 Introduction to Radicals / 498
- 9.2 Simplifying Radicals / 504

10 Solving Quadratic Equations

- 10.3 Solving Quadratic Equations by the Quadratic Formula / 558
- 10.4 Summary of Methods for Solving Quadratic Equations / 564
- 10.6 Graphing Quadratic Equations / 572

Appendix F

Answers to Selected Exercises / A-20

Index I-1

Homework

For questions 1 through 5, write an arithmetic expression for each sentence, then find the person's net worth.

1. I have $80 in my checking account and I owe PECO $100. What is my net worth?

2. I am in debt for $60 and I receive a check for $90. What is my net worth?

3. I am in debt for $210 and I owe M.C.C.C. $80. What is my net worth?

4. I have $60 in my checking account and I owe $50 to ITT. What is my net worth?

5. I have $97 in my checking account and I owe ACME $155. What is my net worth?

6. What is the rule for adding numbers with the same sign?

7. What is the rule for adding numbers with different signs?

8. Perform the operation.

    a. 7 - 5 =

    b. -6 - 7 =

    c. -15 + 6 =

    d. -21 + 36 =

    e. 4 - 11 =

    f. -6 + 15 =

    g. -8 - 21 =

    h. 10 - 15 =

    i. -11 - 21 =

    j. 17 - 28 =

    k. -8 - 10 + 4 =

    l. 6 - 7 - 8 =

    m. -3 + 6 - 11 =

    n. 7 - 11 + 4 - 15 =

9. What is the opposite of a number?

10. How do you find the opposite of a number on your calculator?

11. Use your calculator to perform the operation. Estimate if the answer will be positive or negative <u>before</u> you do the arithmetic.

    a. 4.3 - 8.7 =

    b. -6.5 + 7.21 =

    c. -6.83 + 7.42 =

    d. -5.62 - 8.21 =

    e. 18.3 - 36.5 =

    f. -133.5 + 118.2

12. What is the rule for multiplying or dividing numbers with the same sign?

13. What is the rule for multiplying or dividing numbers with different signs?

14. By what number can you not divide?

15. Perform the operation.

    a. (-6)(7)   b. (-5)(-6)   c. 3(-5)

    d. -21(-3)   e. $\dfrac{18}{-3}$ f. $\dfrac{-35}{7}$

    g. $\dfrac{0}{11}$ h. $\dfrac{11}{0}$ i. $\dfrac{-24}{-12}$

    j. $\dfrac{25}{-5}$ k. 8(-5)   l. 4(-3)(-5)

    m. (-8)(-4)(-3)   n. 9(-3)(-4)(0)(5)

16. Use your calculator to perform the operation. Estimate if the answer will be positive or negative <u>before</u> you do the arithmetic.

    a. (-3.56)(-7.41)   b. -7.83(6.3)

    c. $\dfrac{-13.88}{4.7}$ d. $\dfrac{-6.21}{-3.51}$

17. List the order of operations.

18. Perform the operations.

   a. 2(3-8)

   b. $-7(2) - \dfrac{-20}{5}$

   c. $\dfrac{-12+14}{2}$

   d. $\dfrac{36}{-12} - 4(-8)$

   e. $(-3)^2 + 8(-5)$

   f. $-6 - (-2)^2$

   g. $\dfrac{8(-2) - 5(4)}{(-3)^2}$

   h. $\dfrac{(-5)^2 + 3^2}{2(-17)}$

   i. $\dfrac{3(-5) - 2(-8)}{2}$

19. The bar chart at the right shows the annual profit and loss, in millions of dollars, for National Silver Company.

   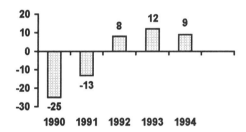

   a. Determine the profit for the five years.

   b. What is the difference between the profit in 1994 and 1990?

   c. What is the difference between the profit in 1994 and 1993?

   d. According to the graph, what does 12 - (-13) mean?

20. The high temperature is for a six-day period in Boise Idaho. The temperatures were 10°F, 8°F, 15°F, 25°F, -10°F, -2°F. Calculate the average daily temperature for the six day period.

Introduction to Variables
Class Work

1. The manager of a 33 Flavors Ice Cream Shop pays $800 per month for fixed expenses such as rent, lights, and wages. They sell ice cream cones for at $1.85 each. It costs $1.40 to serve the ice cream cone.

    A. Without considering fixed expenses of $800 per month, how much profit do they make per cone?

    B. Calculate the monthly profits when they have sold the following number of ice cream cones per month.

CONES	CALCULATION	PROFIT
10,000	10,000 x .45	4,500
15,000	15,000 x .45	67
20,000		
C		

    C. What is the equation that relates profit and the number of cones sold?

    D. Estimate the number of ice cream cones they must sell if they want to make $7,750 a month.

GUESS NUMBER OF CONES	CALCULATION OF PROFIT	PROFIT	TOO HIGH/ TOO LOW

    E. Suppose the expenses increase to $875 a month and they charge $2.10 a cone ($1.40 still goes for ice cream, cone, and napkin). What will be the new equation for their monthly profits?

2. A rental car company, Wrecker, charges $21.95 per day plus 41¢ a mile.

   A. Calculate the cost of renting a car for one day if you drive the following miles:

MILES	CALCULATION	COST
10	21.95 + 41M = Cost	26.05
20		30.15
30		34.25
M		

   B. What is the equation that related cost and number of miles driven?

   C. Another rental company, Limo, charges a flat rate of $39.95 a day with unlimited miles. How many miles would you have to drive to make Limo cost the same as Wrecker?

GUESS NUMBER OF MILES	CALCULATION OF COST	COST	TOO HIGH/ TOO LOW

   D. A third company, Ertz, charges $18.95 a day and 50¢ a mile. What is the formula that calculates the cost of renting a car from Ertz for a day? When is Ertz the same price as Wrecker? (Just set up the equation; do not solve it.)

3. Farmer Nixon has 300 cows. Each cow requires 50 square feet of land. Farmer Nixon wants to put his cows in a rectangular pasture.

A. What is the area of the pasture that Farmer Nixon needs?

B. Complete the table below and find a formula for the perimeter.

W Feet	L Feet	P = 2L + 2W Feet
50		
100		
125		
140		
W		

C. Use the table to find the minimum amount of fencing he needs to enclose his pasture.

Group Work

Today we will look at the cost of renting a car from three different rental companies. Avis will rent us a Ford Taurus for $35 a day and $0.25 a mile. Hertz rents the same model for $25 a day and $0.50 a mile. Budget will rent it to us for a flat daily rate of $55 with no mileage charge.

1. For each company, find a formula for the cost of renting a car for a day. Complete the table below. (Pick values for the variable miles.)

Avis			Hertz		
Miles	Calculation	Cost	Miles	Calculation	Cost
M			M		

2. What are the equations that relate cost and number of miles driven for each company?

3. Which company charges the least for a small number of miles? Large number of miles? Write a sentence explaining your choices.

4. Which company would you rent from if you planned to drive the car 60 miles? 100 miles? 20 miles?

Homework

1. Complete the table for each rectangle.

WIDTH IN.	LENGTH IN.	AREA A = L · W SQ. IN.	PERIMETER P = 2(L+W) IN.
5 in.	6 in.	A = 6 · 5 = 30 in²	P = 2(6+5) = 22 in.
7 in.	8 in.		
9 in.	12 in.		
8 in.		80 in²	
11 in.			60 in.

2. The area of a circle is given by the formula $A = \pi r^2$ ($\pi \approx 3.14$)

    a. One pizza has a radius of 5 inches. Another pizza has a radius of 7 inches. How much larger is the 7 inch pizza than the 5 inch pizza? If the 5 inch pizza costs $5.95 and the 7 inch pizza costs $6.95, which pizza is the better buy? Write a sentence explaining why.

    b. One pizza has a radius of 14 inches. Another pizza has a radius of 16 inches. How much larger is the 16 inch pizza than the 14 inch pizza? If the 14 inch pizza costs $10.95 and the 16 inch pizza costs $11.95, which pizza is the better buy? Write a sentence explaining why.

3. The formula $C = \dfrac{5}{9}(F - 32)$ converts temperature in Fahrenheit to Celsius.

    a. If it is 32° F, find the temperature in Celsius.

    b. If it is -5° F, find the temperature in Celsius.

    c. If it is 98.6° F, find the temperature in Celsius.

4. A server at the gourmet restaurant, Slow Eddies, earns $80 per week in salary and averages $7.50 in tips per table.

   a. Calculate the server's wages when he has served the following number of tables per week.

TABLE	CALCULATION	WAGES
10	7.50(10) + 80	$155
15		
20		
t		

   b. What is the equation that relates wages and number of tables serviced?

   c. If he waited on 36 tables, how much money would he make?

   d. Estimate how many tables he would have to serve if he wanted to make $215 for the week.

   e. If he averaged $8.50 per table, how would the equation change?

   f. If he averaged $7.00 per table and earned $95 in salary, then how would the equation change?

5. Megan's making plans for a summer business. She wants to enter the lawn-mowing business. She can buy a power mower for $160, and she hopes to charge $8 an hour for her work.

   a. Calculate Megan's income for the summer if she has worked the following number of hours.

Hours	Calculation	Income
40		$160
60		
80		
h		

   b. What is the equation that relates income and hours worked?

   c. Estimate the number of hours Megan would have to work if she wants to earn $280 for the summer.

   d. What would the income equation be if she buys a lawn mower for $200 and charges $9 an hour?

6. a. A phone company, Ringer, charges $7.46 per month plus 13¢ a call. Calculate your phone bill if you make the following number of calls per month:

CALLS	CALCULATION	PHONE BILL
10		$8.76
15		
20		
c		

b. What is the equation that relates the phone bill and the number of calls?

c. A second phone company, Busy, charges $6.17 per month plus 17¢ per call. What is the equation for your phone bill if your phone company is Busy?

d. Fill in a table to estimate when Ringer is a cheaper phone company than Busy?

CALLS	RINGER	BUSY

e. For what number of calls is the cost the same for both companies?

7. A 10 cm stick is broken into two pieces. One is placed at a right angle to form an upside down "T" shape. By attaching wires from the ends of the base to the end of the upright piece, a framework for a sail will be formed.

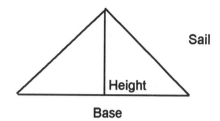

a. Calculate this area of the sail, for the given length of the base.

BASE CM	HEIGHT CM	AREA $\left(A = \frac{1}{2} b \cdot h\right)$ CM²
4	10 - 4 = 6	$\left(\frac{1}{2}\right) \cdot (4)(6) = 12$ CM²
5		
8		
b		

b. What is the equation that relates area and the base of the sail?

c. Why can't the base be 15 cm long?

d. Use the table to determine what base gives the maximum area.

e. Could the base that gives the maximum area be a fractional number?

S-15

8. A farmer has 1800 feet of fencing. She wants to fence her pasture. The pasture borders on a river. How should she do it if she wants to maximize the area of her pasture? Find a formula for the area of her pasture.

Hint: You have to find a formula for L involving W. You do not have to enlarge the table to find the maximum area.

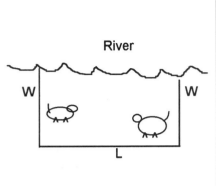

W FT.	L FT.	$A (A = L \cdot W)$ SQ. FT.
300	1200	$1200 \cdot 300 = 360{,}000 \text{ ft}^2$
400		$= 400{,}000 \text{ ft}^2$
500		
450		
W		

Simplifying Algebraic Expressions
Class Work

What are like terms?

How do we add or subtract like terms?

Combining Like Terms

1. $3x + 7 + 2x + 5$
2. $6x - 8 - 11x + 2$
3. $-\dfrac{1}{2}x - \dfrac{3}{4}x + \dfrac{7}{5}$

The Distributive Property

1. $2(3x + 5)$
2. $-(x-8)$
3. $-6(2x + 4)$

Simplify

1. $4x + 2(3x + 8)$
2. $6x - (5x + 7)$
3. $9 - 3(4x + 6)$

Group Work

Simplify

1. $\frac{1}{5}x - \frac{2}{3} + \frac{2}{15}x - 4$

2. $3(2x - 5) - 7x$

3. $4(6-x) - 3(4x + 5)$

4. Describe what is wrong with the following problems.

    a. $6x - 7 + 4x + 8 =$
         $10x + 1 =$
         $11x$

    b. $4(3x + 5) - 9x =$
         $12x + 5 - 9x =$
         $3x + 5$

    c. $5(3x - 8) - 4x$
         $15x - 40 - 4x =$
         $19x - 40$

    d. $4x + 2x = 8x$

    e. $6 - 2(4x + 5) =$
         $4(4x + 5) =$
         $16x + 20$

Homework

1. What are like terms?

2. How are like terms combined?

3. State the distributive property.

4. Simplify

   a. 9x - 4x

   b. 6x + 11x - 13

   c. $\frac{1}{2}x - \frac{3}{5} + \frac{2}{3}x$

   d. 3(2x + 5)

   e. 6 - 4(3x - 8)

   f. 4(2x - 5) + 3(7 - 2x)

   g. 4x + 5(3x + 1)

   h. 9x - 18x + 2(4x - 8)

   i. 4 - 2(3x + 4) - (x - 8)

5. Describe what is wrong with the following problems.

a. $6x + 5x = 30$

b. $5x - 3 + 4x - 8 =$
$9x - 11 =$
$-2x$

c. $6(4x - 5) - 7x =$
$24x - 5 - 7x =$
$17x - 5$

d. $2(5x - 8) - 3(2x - 4) =$
$10x - 16 - 6x - 12 =$
$4x - 28$

e. $4x - (x - 7) =$
$4x - x - 7 =$
$3x - 7$

f. $3(4x - 8) - (3x - 5) =$
$12x - 24 - 3x + 5 =$
$15x - 19$

Solving Equations Part I
Class Work

1. A rental car company, Wrecker, uses the formula C = .41M + 21.95 to calculate the cost, C, of renting a car driven M miles. If your vacation budget allows you to spend $100 for car rental, how far can you drive?

2. Consider the equation 5x - 2 = 13

    Is x = 2 a solution?

    Is x = 3 a solution?

    Is x = 22 a solution?

3. Solve each equation.

    a. x + 8 = -3

    b. x - 2 = 5

    c. 4x = -12

    d. $\dfrac{-2}{5}x = 4$

4. Solve the equation and describe each step..

   a. $4x - 7 = -27$

   b. $\frac{1}{3}x + 4 = 6$

5. Solve each equation.

   a. $-\frac{4}{5}x = 12$

   b. $5x - 11 = 4$

Group Work

1. Describe each step.

	Step	
$\frac{2}{5}x - 8 = 12$		2. Find where the first mistake occurs.
$\frac{2}{5}x - 8 + 8 = 12 + 8$	_____	$6x - 11 = 35$
$\frac{2}{5}x = 20$	_____	$6x - 11 + 11 = 35 - 11$
$\frac{5}{2} * \frac{2}{5}x = \frac{5}{2} * 20$	_____	$6x = 24$
$x = 50$	_____	$x = 4$

3. Solve.

    a. $x - 5 = -2$ b. $3x + 8 = -13$

    c. $\frac{3}{5}x = -18$ d. $5x - 3 = 8$

Homework

1. Describe each step.

   a. $6x + 7 = 31$ Step

       $6x + 7 - 7 = 31 - 7$ _____

       $6x = 24$ _____

       $\dfrac{6x}{6} = \dfrac{24}{6}$ _____

       $x = 4$ _____

   b. $\dfrac{1}{5}x - 4 = 3$ Step

       $\dfrac{1}{5}x - 4 + 4 = 3 + 4$ _____

       $\dfrac{1}{5}x = 7$ _____

       $5\left(\dfrac{1}{5x}\right) = 5 * 7$ _____

       $x = 35$ _____

2. Find where the first mistake occurs.

   a. $3x + 5 = 4$ b. $\dfrac{2}{3}x - 4 = 5$

       $3x + 5 - 5 = 4 + 5$ $\dfrac{2}{3}x - 4 + 4 = 5 + 4$

       $3x = 9$ $\dfrac{2}{3}x = 9$

       $\dfrac{3x}{3} = \dfrac{9}{3}$ $\dfrac{2}{3} * \dfrac{2}{3}x = \dfrac{2}{3} * 9$

       $x = 3$ $x = 6$

3. a. Is $x = 2$ a solution to $4x - 8 = 0$?

   b. Is $x = -3$ a solution to $2x + 5 = -1$?

   c. Is $x = 8$ a solution to $4x - 12 = -16$?

4. Describe the procedure to solve an equation.

5. Solve.

   a. $6x = 18$ b. $-2x = 5$ c. $\dfrac{x}{4} = -10$

   f. $x - 15 = 21$ d. $\dfrac{2}{3}x = 21$ e. $x + 8 = 10$

   g. $6x - 5 = 31$ h. $\dfrac{1}{7}x + 4 = -3$ i. $-3x - 4 = 11$

6. A company uses the equation $V = C - 500t$ to determine the depreciated value V, after t years, of a printing press that originally cost C dollars. If a printing press originally cost $22,000, in how many years will the depreciated value be $10,000?

Solving Equations Part II
Class Work

1. Consider the equation $5x + 6 = 3x - 2$

   Is $x = 4$ a solution?

   Is $x = -4$ a solution?

2. Solve the equation and describe each step.

   a. $8x - 2 = 11x + 7$

   b. $2(4x + 5) - 3x = 24 - 2x$

3. Solve each equation.

   a. $3x + 5 = 4 - 5x$

   b. $3x - 8 = 4(5 - 3x) + 9$

   c. $3(2x + 8) = 6x - 7$

   d. $3(2x + 8) = 8x + 24 - 2x$

4. A company determines that the cost, C, of making x items is $C = 2.2x + 78$ and the revenue, R, is $R = 2.25x$. Find the break even point.

Group Work

1. Describe each step.

	Step
$5(4x - 8) - 13x = 2x + 10$	
$20x - 40 - 13x = 2x + 10$	_____
$7x - 40 = 2x + 10$	_____
$7x - 2x - 40 = 2x - 2x + 10$	_____
$5x - 40 = 10$	_____
$5x - 40 + 40 = 10 + 40$	_____
$5x = 50$	_____
$\dfrac{5x}{5} = \dfrac{50}{5}$	_____
$x = 10$	_____

2. Find where the first mistake occurs.

$3(4x - 5) + 4 = 3x + 47$
$12x - 15 + 4 = 3x + 47$
$12x - 11 = 3x + 47$
$12x - 11 + 11 = 3x + 47 - 11$
$12x = 3x + 36$
$12x - 3x = 3x - 3x + 36$
$9x = 36$
$\dfrac{9x}{9} = \dfrac{36}{9}$
$x = 4$

3. Solve each equation.

   a. $6x - 5 = 8 - 4(2x + 7)$

   b. $5(3 + 2x) = 10x - 11$

Homework

1. Describe each step.

   a. $4(2x + 5) - 11x = 3x - 4$ Step

   $8x + 20 - 11x = 3x - 4$ _____

   $-3x + 20 = 3x - 4$ _____

   $-3x - 3x + 20 = 3x - 3x - 4$ _____

   $-6x + 20 = -4$ _____

   $-6x + 20 - 20 = -4 - 20$ _____

   $-6x = -24$ _____

   $\dfrac{-6x}{-6} = \dfrac{-24}{-6}$ _____

   $x = 4$ _____

   b. $7x + 2 = 2(5x - 3) - 3x$ Step

   $7x + 2 = 10x - 6 - 3x$ _____

   $7x + 2 = 7x - 6$ _____

   $7x - 7x + 2 = 7x - 7x - 6$ _____

   $2 = -6$ _____

   no solution   _____

2. Find where the first mistake occurs.

   a. $2(3x - 8) - 4x = 6 - 7x$

   $6x - 8 - 4x = 6 - 7x$

   $4x - 8 = 6 - 7x$

   $4x + 7x - 8 = 6 - 7x - 7x$

   $11x - 8 = 6$

   $11x - 8 + 8 = 6 + 8$

   $11x = 14$

   $\dfrac{11x}{11} = \dfrac{14}{11}$

   $x = \dfrac{14}{11}$

   b. $4 - (x + 6) = 4x + 7$

   $4 - x - 6 = 4x + 7$

   $-x - 2 = 4x + 7$

   $-x - 4x - 2 = 4x - 4x + 7$

   $-5x - 2 = 7$

   $-5x - 2 + 2 = 7 - 2$

   $-5x = 5$

   $\dfrac{-5x}{-5} = \dfrac{5}{-5}$

   $X = -1$

3. Describe the procedure to solve an equation.

4. When does an equation not have a solution?

5. When does an equation have all real numbers as a solution?

6. Solve each equation.

   a. $6x - 8 = 3x + 7$ b. $-2x + 5 = 7x - 31$ c. $3(2x-5) = 6 + 6x$

   d. $6 - (x+4) = 2x - 5$ e. $4 + 3x = 2(5x + 2) - 7x$ f. $4 - 3(2x + 6) = -(x - 5)$

7. A phone company, Ringer, charges $7.46 per month plus 13 cents a call. A second company, Busy charges $6.17 per month plus 17 cents per call. How many calls does it take for the two companies to charge the same for a month?

Applications of Linear Equations Part II
Class Work

A. Two girls want to enter the lawn-mowing business for the summer. They plan to buy a lawn mower for $180 and they hope to charge $8 an hour.

1. If they work for 20 hours, how much money will they make?

HOURS	CALCULATION	PROFIT
20		
t		

2. What is the equation that relates profit and hours worked?

3. How many hours will they have to work in order to break even?

4. How many hours will they have to work in order to make $780 for the summer?

B. You are offered two very similar jobs selling math textbooks. One pays 8% commission plus $10,000 a year and the other pays 12% commission.

1. If you sell $200,000 for the year, which job would pay you more?

COMPANY A			COMPANY B		
Sales	Calculation	Wages	Sales	Calculation	Wages
200,000			200,000		
S			S		

What are the equations that relate wages and sales for Companies A and B?

2. How much would you have to sell for the two companies to pay you the same amount of money for the year?

C. The women's recommended weight formula from the Cambridge, Mass. HMO/U.S. Death Plan says:

"Give yourself 100 lbs. plus 5 lbs. for every inch over 5 ft. tall."

1. Complete the table:

HEIGHT	CALCULATION	WEIGHT
61 inch		
65 inch		
72 inch		
h inches		

What is the equation that relates weight and height? Simplify the equation.

2. How tall should you be if you weigh 135 lbs.?

3. How tall should you be if you weigh 85 lbs.?

4. How tall should you be if you weigh zero lbs.?

5. Solve for h from the formula in part 1b.

Group Work

A. A company pays $10,000 a year plus 10% commission on sales over $50,000.

1. Complete the table below.

SALES	CALCULATION	WAGES
80,000		
100,000		
S		

What is the equation that relates wages and sales? Simplify the equation.

2. If you want to earn $30,000 a year, how much do you have to sell?

3. Solve for the variable S from the formula in part 1.

B. Rental Car Company, Wrecker, charges 21¢ a mile plus $31 a day.

   1. Complete the table below.

MILES	CALCULATION	COST
20		
M		

   What is the equation that relates cost and miles?

   2. If you spent $44.02 renting a car for the day, how many miles did you drive?

   3. A second company, Limo, charges 11¢ a mile and $42 a day. How many miles do you have to drive for both companies to charge you the same amount of money?

Homework

1. You plan to rent a car for the day, you are told it costs $25.00 for the day and 16¢ a mile.

   a. Complete the table below.

MILES	CALCULATION	COST
20		
m		

   b. What is the equation that relates cost and miles?

   c. If you drive 31 miles, how much will it cost?

   d. If you want to spend $31 on renting a car, how far can you drive?

   e. If you want to spend $65 on renting a car, how far can you drive?

2. A phone company charges 80¢ a call and 15¢ a minute.

   a. Complete the table below.

MINUTES	CALCULATION	COST
10		
m		

   b. What is the equation that reflects cost and minutes?

   c. If you are on the phone for 32 minutes, how much did it cost?

   d. If it costs $1.85, how long were you on the phone?

   e. If it costs $2.50, how long were you on the phone?

3. Phone Company Ringer, charges 8¢ per minute and 50¢ per call while Company Busy, charges 10¢ per minute and 25¢ per call.

   a. Complete the table below.

RINGER			BUSY		
MIN.	CALCULATION	COST	MIN.	CALCULATION	COST
10			10		
m			m		

   b. What are the equations that relate cost and minutes on the phone for the two companies?

   c. How long would you have to talk if the cost of the phone call was the same for both Ringer and Busy?

4. You are offered two very similar jobs selling encyclopedias. One job pays 7.5% commission and $5,000 a year, while the second company pays 5% commission and $9,000 a year.

   a. If you sold $100,000 for the year, which job would pay more?

Company A			Company B		
Sales	Calculation	Wages	Sales	Calculation	Wages
100,000			100,000		
s			s		

   b. What are the equations that relate wages and sales for the two companies?

   c. How much would you have to sell for the two companies to pay you the same amount of money for the year?

5. A job pays 15% commission on sales over $1,000 a week.

   a. Complete the table below.

SALES	CALCULATION	WAGES
1,500		$75
2,000		
s		

   b. What is the equation that relates wages and sales? Simplify the equation.

   c. If you sell $10,000 in a week, how much money will you make?

   d. If you earned $270 in a week, how much did you sell?

   e. If you earned $3,372 in a week, how much did you sell?

6. The men's recommended weight formula from the MONTCO HMO says: "Give yourself 160 lbs. plus 7 lbs. for every inch over 6 ft."

   a. Complete the table.

HEIGHT	CALCULATION	WEIGHT
74 in.		174
78 in.		
h		

   b. What is the equation that relates weight and height? Simplify the equation.

   c. How tall should you be if you weigh 216 lbs.?

   d. How tall should you be if you weigh 220 lbs.?

7. Phone Company Hook, charges 22¢ a call and 10¢ a minute after the first 5 minutes.

   a. Complete the table below.

MINUTES	CALCULATION	COST
2		
5		
7		
12		
m		

   b. What is the equation that relates cost and minutes? Simplify the equation.

   c. If it costs $2.02 for a phone call, how long were you on the phone?

   d. If it costs $3.81 for a phone call, how long were you on the phone?

Application of Linear Equations Part II (Literal Equations)
Class Work

1. An appliance repair store charges $50 for the first hour and $18 an hour for each additional hour.

   a. Complete the table to find the cost of repairing an appliance.

HOURS	CALCULATION	COST
3		
8		
t		

   b. What is the equation that relates cost and hours? Simplify the equation.

   c. If it cost $230, how long did it take?

   d. If it cost $140, how long did it take?

2. Solve each equation for the indicated variable.

   a. $P = 2L + 2W$, solve for W.

   b. $8x + 3y = 16$, solve for y.

Group Work

A. The accompanying bar graph shows the number of registered shareholders (in thousands) of Toxin Company stock at the end of each year between 1986 and 1990. The data in this figure can be modeled by the equation $s = 788 - 28t$, where s is the number of registered shareholders (in thousands) of stock at the end of year t, and t is the number of years since 1985.

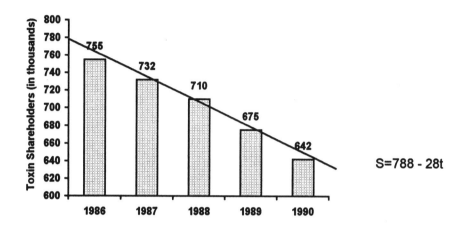

1. When will there be 600,000 Toxin Company shareholders? (hint: s = 600)

2. When will there be 544,000 Toxin Company shareholders? (hint: s = 544)

3. Solve for the variable t in the formula $s = 788 - 28t$.

4. Solve for t in the equation from part b.

Homework

1. The equation $C = \dfrac{5}{9}(F - 32)$ converts temperature in Fahrenheit, F, to Celsius, C.

    a. If the temperature is 72° Fahrenheit, what is the corresponding temperature in Celsius?

    b. If the temperature is 11° Celsius, what is the corresponding temperature in Fahrenheit?

    c. If the temperature is -6° Celsius, what is the corresponding temperature in Fahrenheit?

    d. If the temperature is 63° Celsius, what is the corresponding temperature in Fahrenheit?

    e. Try to solve the equation $C = \dfrac{5}{9}(F - 32)$ for the variable F.

2. From a modest beginning, Q-Mart Stores have risen to become one of the largest retailing chains in America. The accompanying bar graph shows the number of stores at the end of selected years in the period 1983-1991. The figure also shows the graph of s = 118t +146, where s represents the number of stores at the end of year t, and t is the number of years since 1982.

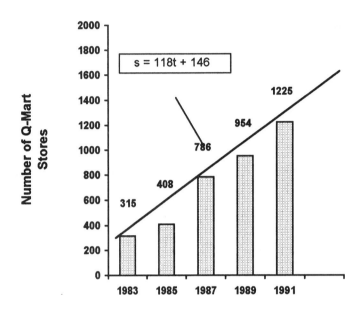

a. According to the equation, when will Q-Mart have 1,500 stores?

b. According to the equation, when will Q-Mart have 1,600 stores?

c. Solve for t in the equation s = 118t + 146.

3. Solve for the indicated variable.

   a. $P = 2L + 2W$, solve for L

   b. $D = rt$, solve for r

   c. $3x - 4y = 12$, solve for y

   d. $6x + 3y = 9$, solve for y

Percentages
Class Work

1. You get 9 out of 12 questions right on a quiz.

    a. What percent did you get right?

    b. If you got 75% of the questions on a test correct and there were 52 questions, how many did you get right?

2. a. Walden Department Store computes retail prices (the price Walden charges its customers for an item) by marking up the wholesale price (the price Walden pays for the item) by 40%. Complete the table. Recall that wholesale price plus markup equals retail price.

WHOLESALE PRICE ($)	MARK UP ($)	RETAIL PRICE ($)
10	0.40(10) = 4	10 + 0.40(10) = 14
20		
40		
w		
		70

   b. What is the equation that relates retail price to wholesale price and markup?

   c. Suppose an item costs $82 at the Walden Department store. How much did Walden pay for the item?

3. If 20% of the people at the State Fair on Saturday had free passes and there were 26,500 free passes used that day, how many people were at the fair?

4. Phillips Hardware Store is having a 20% off sale. What is the sale price of a mower that is usually priced $380?

5. Phillips Hardware Store is having a 20% off sale. A stepladder has a sale price of $44. What was its price before the sale?

   a. Complete the table.

Pre Sale Price	Discount	Sale Price
60		
50		
x		

   b. Set up the appropriate equation and solve it.

6. A math teacher, Dr. Kaczynski, computes a student's grade for the course as follows:

> 10% for homework
> 65% for the average of 4 tests
> 25% for the final exam

a. Compute Bill's grade for the course if he has a 78 on the homework, 81 for his test average and a 79 on the final exam.

b. Suppose Sue has an 82 homework average and a 63 test average. What does Sue have to get on the final exam to get a 70 for the course?

Group Work

1. A grocer purchases a can of fruit juice for $.68. Find the selling price if the markup is 30%.

2. A grocer sells a bottle of mustard for $1.15. Find out how much the grocer pays for the mustard if the markup is 25%.

    a. Complete the table.

WHOLE SALE	MARKUP	RETAIL PRICE
.85		
1.00		
s		

    b. Solve the appropriate equation.

3. Professor Passall computes his grades as follows:

    15% for homework
    55% for the average of 3 tests
    30% for the final exam

    a. Compute Howie's grade if he has a 71 homework average, an 83 test average, and a 68 final exam.

    b. Nancy wants to get an 80 for the course. She has an 82 homework average and a 74 test average. What does Nancy have to get on the final exam to get an 80 for the course?

Homework

1. a. If 98 girls and 87 boys attend a dance, what was the percent of girls that attended the dance?

   b. If 60% of the people at a dance were girls and 125 people attended the dance, how many were girls?

2. Dairy Prince has given each employee a 12% raise.

   a. Complete this table.

Employee	Susan	Joe	Mary	Pat
Old earnings per hour	$3.35	$3.50		
New earnings per hour			$4.48	$4.20

   b. What is the equation that relates new earnings to old?

3. Harry's Hardware Store is having a 32% off sale. Complete the table.

Old Price ($)	Discount ($)	Sales Price ($)
25	0.32(25) = 8	25 - 0.32(25) = 17
35		
45		
x		
		37.40

4. A math teacher, Dr. Pi, computes a student's grade for the course as follows:

   10% for Homework
   65% for the average of the 5 tests
   25% for the Final Exam

   Compute the following students' grades for the course.

   a. George: 78 Homework, 88 for the 5 tests, 71 for the final

   b. Darrel: 87 for Homework, his 5 test scores of: 89, 71, 95, 97, 88, and a 90 for the final

   c. A student, Rachel, in Dr. Pi's class wants an 80 for the course.
      Rachel has an 81 homework average and a 75 test average.
      What does Rachel have to get on the final exam to get an 80 for the course?

5. An English teacher, Dr. Austin, computes a student's grade for the course as follows:

   15% for a research paper
   15% for short essays
   50% for the average of 4 tests
   20% for the final

   a. A student, Harold, in Dr. Austin's class wants a 75 for the course. Harold has an 82 for the research paper, 61 for the short essays and a 71 for the 4 tests. What does Harold need to get on the final exam to get a 75 for the course?

   b. A student, Thomas, in Dr. Austin's class wants a 90 for the course. Thomas has a 78 for the research paper, 80 for the short essays and a 69 for the 4 tests. What does Thomas need to get on the final exam to get a 90 for the course? What do you conclude about Thomas' chances of getting a 90?

Review for Test I

1. Water Witch Well Drillers charge their customers $350.00 to come to the well site and $20.00 per foot to drill a well.

    a. Complete the table.

DEPTH OF THE WELL	CALCULATION	COST ($)
50		
70		
90		
d		

    b. What is the equation that relates cost and depth?

    c. If a person was charged $2,150.00 for a well, how deep is the well?

2. Take-Taxi Co. charges $1.35 immediately upon entering the taxi. The first 3 miles are free, and after that it costs $1.80 per mile.

    a. Complete the table.

MILES	CALCULATION	COST ($)
10		
15		
20		
m		

    b. What is the equation that relates cost and miles?

    c. If it cost $27.50, how far was your ride?

3. Professor Failure computes his grades as follows:

   Test Average: 60%
   Homework: 15%
   Final Exam: 25%

   a. Otto has a test average of 82, homework average of 99, and a final exam score of 71. What is Otto's grade for the course?

   b. Tito has a test average of 71 and a homework average of 76. What does Tito have to get on the final to get a 70 for the course?

4. The equation

   $$G = .022B + .359$$

   relates the price per gallon, G, of gasoline with the price per barrel, B, of crude oil.

   a. Find the price per gallon if the price per barrel of crude oil is $47.50.

   b. Find the price per barrel of crude oil if the price per gallon of gasoline is $1.27.

5. Clancy's Burgers has given a 12% raise to all of its employees. Complete the table.

   a.

OLD SALARY	CALCULATION	NEW SALARY
$1.00 per hr.		
$3.00 per hr.		
x per hr.		

   b. What is the equation that relates old salary to new salary?

   c. If your new salary is $5.25 per hour, what was your old salary?

6. Stats Department Store is having a 20% off sale.

   a. Complete the table.

ORIGINAL PRICE	CALCULATION	SALE PRICE
10.00		
20.00		
p		

   b. What is the equation that relates the original price to the sale price?

   c. If the original price was $81.00, what was the sale price?

7. The net profits and losses for Rose Stores for the years 1990 through 1993 are shown in the graph below.

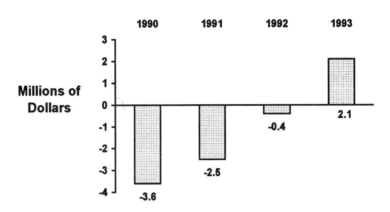

   a. What is the difference between the profit or loss in 1990 and that in 1992?

   b. What is the difference between the profit or loss in 1993 and that in 1991?

8. Simplify $-2(x-3)+2(4-x)$

9. Is -3 a solution of $x^2+6x+9=x+3$

10. Solve   a. $7x-8=-29$ b. $8x-3(4x-5)=-2x-11$

11. A business manager has determined that the cost per unit for a camera is $70 and that the fixed costs per month are $3,500. Find the number of cameras that are produced during a month in which the total cost was $21,000. Use the equation

   $$T = U \cdot N + F,$$

   where T is the total cost, U is the cost per unit, N is the number of units produced, and F is the fixed cost.

12. $P = 2W + 2L$, solve for W.

Inequalities

Class Work

1. Is x = 3 a solution to x + 5 ≥ 1?

   Is x = -2 a solution to x + 5 ≥ 1?

   Is x = -4 a solution to x + 5 ≥ 1?

   Is x = -8 a solution to x + 5 ≥ 1?

2. Use the number line to describe the solutions to an inequality equation.

   a. x ≤ 5

   b. x > -3

   c. 5 ≥ x

   d. -2 < x ≤ 5

3. Solve the inequality and graph the solution on the number line. Describe each step.

   a. 4 − 3x ≥ 22

   b. -6 ≤ 15 + 7x < 50

4. Solve each inequality and graph the solution on the number line.

   a. $4x - 11 > 15x + 25$

   b. $3(2 - 8x) \leq 3x - 5$

   c. $-2 \leq 7 + 2x < 8$

   d. $-8 < 7 - 3x < 31$

5. An English teacher, Bill Shakespeare, computes a student's grade for a course as follows:

   20% for a research paper
   40% for the average of 4 tests
   15% for short essays
   25% for the final exam

   A student, Rosetta Stone (author of *Because a Little Bus Went Ka-Choo*), in Dr. Shakespeare's class wants to get a C for the course. This means that Rosetta's grade must be greater than or equal to 70, but less than 80. Rosetta has an 85 on the research paper, a 68 average for the tests and 62 average on the short essays. What does Rosetta have to get on the final exam to get a C?

Group Work

1. Describe each step.

	Step
$4 - 8x > 19 - 3x$	
$-8x > 15 - 3x$	_____
$-5x > 15$	_____
$x < -3$	_____

2. Find where the first mistake occurs.

   $-6 \leq 4 + 5x \leq 15$

   $-10 \leq 5x \leq 15$

   $-2 \leq x \leq 3$

3. Solve and graph the solutions on the number line.

   a. $8x - 5 \leq 4x + 23$

   b. $-6 \leq 10 - 2x < 22$

Homework

1. Describe each step.

   a. $6 - 3x \leq 30 + x$ Step

       $-3x \leq 24 + x$ _____

       $-4x \geq 24$ _____

       $x \leq -6$ _____

   b. $-4 < 6 - 2x \leq 14$ Step

       $-10 < -2x \leq 8$ _____

       $5 > x \geq -4$ _____

2. Find where the first mistake occurs.

   a. $5 - 7x > 19$
       $-7x > 14$
       $x > -2$

   b. $-15 \leq 4x + 9 < 12$
       $-24 \leq 4x < 12$
       $-6 \leq x < 3$

3. a. Is $x = 5$ a solution to $3x - 4 > 8$?

   b. Is $x = 21$ a solution to $3x - 4 > 8$?

   c. Is $x = 1$ a solution to $3x - 4 > 8$?

4. a. Describe the procedure to solve a simple inequality equation.

   b. Describe the procedure to solve a multiple inequality problem.

5. Solve and graph the solutions on the number line.

   a. $-6x > 54$
   b. $\frac{1}{5}x < 4$
   c. $7 - x \leq 15$

   d. $15 - 2x \geq 21$
   e. $5x - 11 < 2x + 8$
   f. $2(4x - 5) - 11 \leq 41$

   g. $5 < x + 11 \leq 15$
   h. $-6 \leq 5 - 2x \leq 11$
   i. $0 \leq 4x - 8 < 18$

   j. $-7 \leq \frac{1}{3}x + 5 \leq 15$
   k. $-6 < 8 - \frac{2}{3}x < 24$

Applications of Inequalities
Class Work

1. The equation $S = 118t + 146$

   represents the number of stores, S, Q-Mart has opened at the end of t years, where t is the number of years since 1982.

   a. When will Q-Mart have more than 1500 stores?

   b. When will Q-Mart have between 1000 and 1300 stores? (Make sure your interval truly expresses the time described.)

2. Two companies offer you very similar sales positions. Random House will pay you $10,000 a year plus 7% commission on the dollar amount of book sales. Reader Publishing Co. will pay you $8,000 a year plus 11% commission. For what dollar amount of book sales does Random House pay more than Reader Publishing Co.?

3. Pop Bell charges 20¢ per phone call and 11¢ for each minute over 3 minutes.

   a. Find an equation for the cost of making a phone call. Simplify the equation.

   b. How many minutes were you on the phone if the cost was more than $2.50?

   c. How many minutes were you on the phone if the cost was between $3.00 and $4.50?

Group Work

1. Solve and graph.

   a. $-4x < 10$

   b. $4.8 \leq 6x + 7.5 \leq 11.21$

2. Graphic Inc. offers you a sales position. They will pay you $18,000 a year plus 13% commission on sales over $100,000.

   a. Find an equation for your yearly salary. (You may need to make a table.) Simplify the equation.

   b. How much do you have to sell if you want to make more than $50,000 a year?

   c. How much do you have to sell if you want to make between $45,000 and $75,000 a year?

Homework

1. The equation $C = \frac{5}{9}(F - 32)$ converts temperature in Fahrenheit, F, to Celsius, C. For each problem set up an equation and solve it.

   a. If the temperature must be above 8° Celsius, what must the corresponding temperature in Fahrenheit be?

   b. If the temperature is between -2° Celsius and 12° Celsius, what is the corresponding temperature in Fahrenheit?

2. Two companies have offered you very similar jobs selling cars. Company A pays $30,000 a year and Company B pays 8% commission plus $10,000 a year. When will Company A pay more than Company B? Set up an inequality and solve.

3. The accompanying bar graph shows the population growth of the United States since 1980.

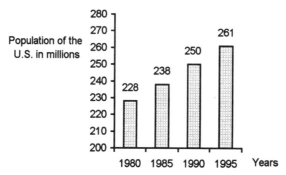

The equation P = 2.31t + 227 can be used to approximate the population of the U.S. In the equation, P represents the population in millions and t is the number of years since 1980.

a. Use the equation to estimate the population in 1989.

b. Use the equation to find when the population will be more than 266,500,000. (Set up an inequality and solve it.)

c. When will the population of the U.S. be between 270 million and 300 million? (Set up an inequality and solve it.)

4. Two girls want to enter the lawn-mowing business for the summer. They plan to buy a power mower for $200 and they hope to charge $8.50 an hour.

   a. Use the table below to find a formula for the girls' profit.

HOURS	CALCULATION	PROFIT
30		
h		

   b. What is the equation that relates profit and hours worked?

   c. Find out how many hours the girls will have to work to make more than $200 for the summer.

   d. Find out how many hours the girls will have to work to break even, but not earn more than $500.

5. You are planning on renting a car. Wrecker charges 50¢ a mile and $28 for a day, while Ertz charges 75¢ a mile and $20 for a day.

   a. Use the table below to help you set up a formula for finding the cost of renting a car for the day.

WRECKER			ERTZ		
MILES	CALCULATION	COST	MILES	CALCULATION	COST
20			20		
M			M		

   b. What are the equations that relate cost and miles for the two companies?

   c. When will Wrecker cost more than Ertz? Set up an inequality and solve it.

6. Two companies have offered you very similar jobs selling appliances. Company A pays 12% commission on sales over $50,000 a year plus $8,000, and Company B pays 5% commission plus $15,000 a year.

   a. Use the table below to find a formula for your wages.

	COMPANY A			COMPANY B	
SALES	CALCULATION	WAGES	SALES	CALCULATION	WAGES
50,000			50,000		
100,000			100,000		
125,000			125,000		
S			S		

   b. What are the equations that relate wages and sales for the two companies? Simplify them.

   c. When does Company A pay more than Company B? Set up an inequality and solve it.

   d. For Company A, why is the formula not valid for S = 20,000?

7. You are trying to decide which long distance phone company to choose. Company WHAT?! charges 15¢ a call and 11¢ a minute after the first 4 minutes. Company Disconnect charges 10¢ a call and 14¢ after the first 5 minutes. Which company will you choose? Why? Find equations for both companies and state some limitations on the variable M by writing inequalities. (Hint: you may have to use a table to find the equation.)

Scatter Plots
Class Work

Sarah has a job that pays her 8% commission plus $10,000 per year. She has computed her possible wages in the table below.

Wages	$11,600	$12,400	$14,000	$15,600	$18,000	$19,600
Sales	$20,000	$30,000	$50,000	$70,000	$100,000	$120,000

Since Sarah is interested in her wages, then we can say that her wages <u>depend</u> on her sales. Wages are called the <u>dependent</u> variable and sales are called the <u>independent</u> variable.

The General Rules:
1. The <u>independent</u> variable, S, goes on the <u>horizontal</u> axis.
2. The <u>dependent</u> variable, W, goes on the <u>vertical</u> axis.
3. The ordered pair is written as (S,W) or (Independent, Dependent)

In each problem below, find the equation and then identify the independent variable, dependent variable, label the axis, and write down the ordered pair.

1. It costs $25 per day plus 16 cents a mile to rent a car.

2. A phone company charges 15 cents a minute plus 80 cents for the call.

3. The equation S = 114t + 146 models the number of Q-Mart stores, S, at the end of the year t, where t is the number of years since 1982.

1. When NASA sends a rocket into space, they monitor the temperature of certain gases. The table below gives a sample of the type of data collected. Time is the independent variable and temperature is the dependent variable. Notice that negative time is used to denote time before lift off.

Time	Temperature	Time	Temperature
-20	-20	10	80
-15	-20	15	170
-10	-10	20	90
-5	0	25	25
0	15	30	5
5	40	35	-15

    a. Graph the data pairs, (time, temperature).

    b. Answer each of the following questions about the temperature of the nose cone of the space shuttle and give the coordinates you used to get the answer. Estimate where needed.

    1) What was the temperature at take off?

    2) What was the temperature 5 minutes into the flight?

    3) What was the temperature 17 minutes into the flight?

    4) At what times was the temperature 90°?

    5) At what times was the temperature 0°?

    6) What was the change in temperature for the first 15 min. of the flight?

    7) What was the change in temperature of the flight between 15 min. and 35 min.?

    8) What was the maximum temperature?

    9) At what times was the temperature rising most quickly?

    10) At what times was the temperature falling most quickly?

2. Animal populations tend to rise and fall in cycles. Suppose the following data shows how squirrel populations in a central Pennsylvania city varied from 1975 to 1984.

Year	75	76	77	78	79	80	81	82	83	84
Population	750	700	520	680	730	650	550	625	780	700

a. Graph the data on graph paper using the two different scales.

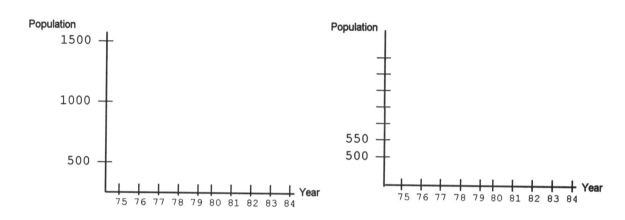

b. Which graph describes the data the best? Why?

Group Work

1. You are given data relating two variables. Choose reasonable scales for axes on a graph and plot the given data. Then write a sentence describing the pattern in the graph and what it says about the relationship between the two variables.

Price per Barrel of Crude Oil, X	Price per Gallon of Gasoline, Y
$ 3.35	$ .38
3.75	.40
4.29	.43
5.80	.57
18.00	.80
21.30	.85
30.00	.98
35.75	1.20
37.05	1.15
42.50	1.35
45.75	1.40
45.75	1.25

2. A 10 cm stick is broken into two pieces. One is placed at a right angle to form an upside down "T" shape. By attaching wires from the ends of the base to the end of the upright piece, a frame work for a sail will be formed.

   a. Calculate the area of the sail, for the given length of the base.

Base b cm	Height h cm	Area $\left(A = \frac{1}{2} b \cdot h \text{ cm}^2\right)$	Point (base, area)
0			
2			
4 cm	10 - 4 = 6 cm	$\frac{1}{2}(4)(6) = 12$ cm	(4, 12)
5			
8			
9			
10			
b			

   b. What is the equation that relates Area to the base?

   c. Use the graph to estimate the maximum area.

Homework

1. You are given data relating two variables. Choose reasonable scales for axes on a graph and plot the given data. Then write a sentence describing the pattern in the graph and what it says about the relationship between the two variables.

   a. **The following table shows the school year and total revenue spent for public elementary and secondary schools.**

School Year	1930	1946	1954	1966	1974	1985	1990
Revenues (millions of dollars)	2	3	8	25	58	137	208

   Source: *Digest of Educational Statistics*.

   b. **The following table shows the number of establishments of Insurance, Agents, Brokers and Service in Montgomery County.**

Year	1990	1991	1992	1993	1994
Number of Establishments	519	523	536	552	554

   Source: *Pennsylvania County Industry Trends 1990-1994*.

2. At a pizza store, Sarah is in charge of scheduling workers. To help make these decisions, Sarah collected the following data on the number of people waiting for pizzas each hour on the hour for three days.

Time	Day 1	Day 2	Day 3	Average
11:00	11	9	8	
12:00	15	18	19	
1:00	18	21	20	
2:00	9	11	13	
3:00	5	4	6	
4:00	6	8	7	
5:00	12	11	15	
6:00	22	4	21	
7:00	15	17	9	
8:00	8	7	7	

   a. Average the values for the different times and plot them on a graph.

   b. Write a paragraph interpreting the data and the graph.

   c. Suppose you were told that on the second day at 6:00 pm awater repair crew blocked your driveway, would you change your interpretation of the graph? How?

S-91

3. Suppose that a baseball player hits a high pop-up straight above home plate. If the bat meets the ball 1.5 meters above the ground and sends it up at a velocity of 30 meters per second, then the height of the ball, in meters, t seconds later is given by the table.

t seconds	0	1	2	3	4	5	6
h meters	1.5	27	42	47	43	29	5

a. Use the table to make a graph. Choose your scales so that you can answer the questions that follow. Be sure to indicate time and height scales along the axes.

b. What point shows the starting height of the ball? Give its coordinates.

c. What point shows height of the ball after 2 seconds? Label this point B and give its coordinates.

d. What point(s) show a height of 20 meters? Label it (or them) C and give the coordinates.

e. What point shows where the ball reaches its maximum height? Label this point D and give its coordinates.

f. What point shows when the ball hits the ground? Label this point E and give its coordinates.

g. When is the ball more than 40 meters above ground?

h. When is the ball exactly 35 meters above the ground? (Be careful!)

i. What is the height of the ball when the time is zero seconds?

j. What is the height of the ball when the time is 4.5 seconds?

k. When is the ball less than 50 meters above the ground?

l. What was the change in height for the first two seconds?

m. What was the change in height between 2 seconds and 4 seconds?

n. What was the change in height between 4 seconds and 6 seconds?

o. At what times was the ball rising most quickly?

p. At what times was the ball falling most quickly?

Interpreting Graphs
Class Work

1. The graph below shows the temperature during a winter day in Chicago, Illinois.

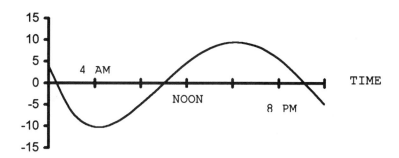

a. What was the temperature at noon?

b. When was the temperature 0°?

c. What was the high temperature for the day?

d. When was the high temperature?

e. What was the low temperature for the day?

f. When was the low temperature for the day?

g. When was the temperature rising?

h. When was the temperature decreasing?

2. The graphs below show the profit for the calculator companies, PA Instruments, and Hewett Luggage.

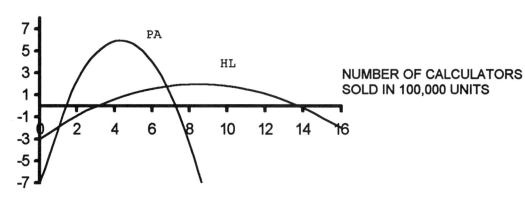

a. How many calculators does Hewett Luggage have to sell to break even?

b. How many calculators does PA Instruments have to sell to break even?

c. For what amount of calculators sold is the profit the same for both companies?

d. How much profit does PA Instruments make (or lose) if they don't sell any calculators?

e. How much profit does Hewett Luggage make (or lose) if they don't sell any calculators?

Group Work

1. The graph below shows the temperature during a winter day in Snowbound Montana.

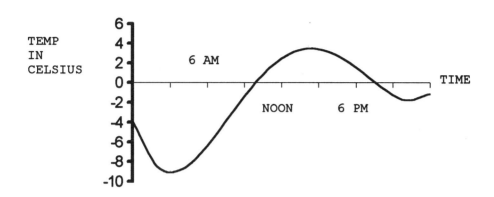

a. What is the temperature at 9 a.m.?

b. When was the temperature 0°

c. What was the high temperature for the day?

d. When was the high temperature for the day?

e. What was the low temperature for the day?

f. When was the low temperature for the day?

g. When was the temperature rising?

h. When was the temperature falling?

2. The graph below shows the profit for two companies, AMATYC and PSMATYC.

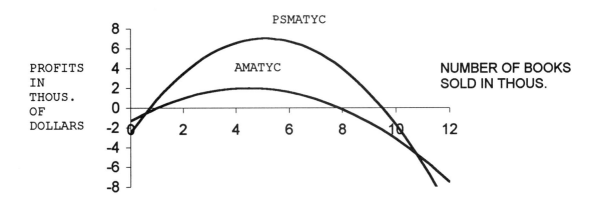

a. How many books does AMATYC have to sell to break even?

b. How many books does PSYMATYC have to sell to break even?

c. For what number of books is the profit the same for both companies?

d. What is the maximum profit for AMATYC?

e. How many books does AMATYC have to sell to maximize its profits?

f. How much profit does PSYMATYC make (or lose) if they don't sell any books?

g. How much profit does AMATYC make (or lose) if they don't sell any books?

h. When are PSYMATYC's profits increasing?

i. When are AMATYC's profits decreasing?

Homework

1. The graph below shows the temperatures during a winter day in Killington, Vermont.

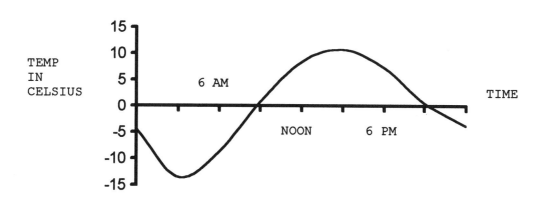

a. What were the high and low temperatures recorded during the day?

b. When were the high and low temperatures recorded?

c. During what time intervals was the temperature above 0° F?

d. During what time intervals was the temperature below 0° F?

e. When was the temperature 10° F?

f. What was the temperature at noon?

g. When was the temperature rising?

h. When was the temperature falling?

2. The graph below shows the fish population of a lake.

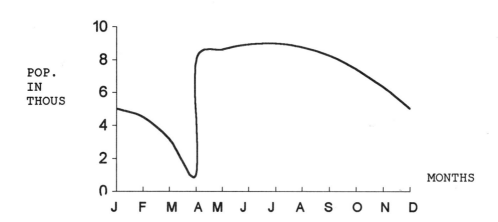

   a. How many fish are there at the beginning of June?

   b. When is the fish population the greatest?

   c. What is the maximum number of fish in the lake?

   d. When is the population of the fish increasing?

   e. When is the population of the fish decreasing?

   f. What do you think happened on April 1?

3. The graphs below show the cost of making a phone call for two different companies, WRONG # and BUZZ.

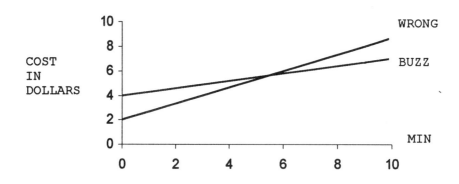

   a. When do the two companies charge the same amount for a phone call?

   b. Which company charges the most per minute?

   c. Which company charges the most just to pick up the phone?

   d. When is Buzz cheaper than Wrong #?

Graphing Lines By Plotting Points

Class Work

1. An appliance repair shop charges an initial fee of $30 plus $15 per hour to fix an appliance.

    a. Complete the table to find the cost of repairing an appliance.

Hours	Calculation	Costs
3		
5		
10		
t		

    b. What is the equation that relates cost and hours?

    c. Use the results in part a. to graph the equation in part b. Choose an appropriate scale and only graph the portion that makes sense to the problem. Label the axes.

2. Solve: $y = 3x - 4$

    Is $x = 2$, $y = -1$ a solution?
    Is $x = 5$, $y = 11$ a solution?
    Is $x = -2$, $y = -10$ a solution?
    Is $x = 3$, $y = 2$ a solution?

3. Represent all of the solutions by graphing a line in the Cartesian Coordinate system

   a. $y = 3x - 4$

   b. $6x - 4y = 18$

4. Graph the line by finding three points.

   a. $y = \dfrac{-1}{2}x + 5$

   b. $6x + 5y = 35$

Group Work

1. A sales position pays 12% commission plus $200 per week.

   a. Complete the table below.

Sales	Calculation	Wages
500	.12 of 500 + 200	$260
1000	.12 of 1000 + 200	$310
2000	.12 of 2000 + 200	420
s		

   b. What is the equation that relates wages and sales?

   $W = .12S + 200$

   c. Use the results in part a. to graph the equation in part b. Choose an appropriate scale and only graph the portion that makes sense to the problem. Label the axes.

   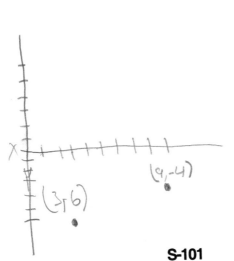

2. Graph the line by finding three points.

   a. $y = \dfrac{2}{3}x - 7$ b. $8x - 3y = 24$

   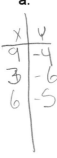

S-101

Homework

1. An internet provider, Wild Web, charges $15 per month plus 50 cents an hour.

   a. Complete the table below.

Hours	Calculation	Cost
20		
50		
100		
h		

   b. What is the equation that relates cost and hours?

   c. Use the results in part a. to graph the equation in part b. Choose an appropriate scale and only graph the portion that makes sense to the problem. Label the axes.

2. A car rental company, 4Wheels, charges 35 cents a mile plus $15 a day.

   a. Complete the table below. (You must pick values for hours.)

Hours	Calculation	Cost
h		

   b. What is the equation that relates cost and hours?

   c. Use the results in part a. to graph the equation in part b. Choose an appropriate scale and only graph the portion that makes sense to the problem. Label the axes.

3. Find the mistake. Write a sentence describing what the student did wrong.

Graph y = 2x – 5

Let x = 3	Let x = 5	Let x = -3
y = 2(3)–5	y = 2(5)–5	y = 2(-3)-5
y = -1	y = 5	y = -11
(3, -1)	(5, 5)	(-3, -11)

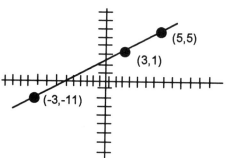

4. Solve y = 2x + 7

Is x = 2, y = 11 a solution?
Is x = 5, y = 15 a solution?
Is x = -1, y = 5 a solution?
Is x = 7, y = 21 a solution?

5. Graph the line by finding three points.

    a.   y = 2x + 7

    b. $y = \dfrac{-1}{2}x + 5$

    c.   3x – 8y = 32

    d.   6x + 4y = 18

Graphing Lines by Finding the Intercepts

Class Work

1. Sally started a lawn-mowing business for the summer. She bought a lawn mower for $200 and she charges $5 an hour. The equation that relates profit and hours worked is

   $$p = 5h - 200$$

   a. How many hours does she have to work to break even?

   b. How much money will she make if she doesn't work any hours?

   c. Graph the line $p = 5h - 200$ by plotting the points obtained in parts a. and b. Choose an appropriate scale and only graph the portion that makes sense to the problem. Label the axes.

2. Graph the line by finding the x and y intercepts. Choose an appropriate scale and label the axes.

   a. $y = \dfrac{2}{3}x - 5$ b. $.03x - 4y = 16$

3. Vertical and horizontal lines.

4. Graph the lines.

   a. y = 18   b. x = -4   c. 28x - .13y = 15

Group Work

1. Johnny Lift runs a snow clearing business for the winter. He bought a snow blower for $310 and charges $10 per hour. The equation that relates profit and hours is

    $$p = 10h - 310$$

    a. How many hours does he have to work to break even?

    b. How much money will he make if he doesn't work any hours?

    c. Graph the line $p = 10h - 310$ by plotting the points obtained in parts a. and b. Choose an appropriate scale and only graph the portion that makes sense to the problem. Label the axes.

2. Graph the line. If possible, find the x and y intercepts. Choose an appropriate scale and label the axes.

    a. $.02x - 15y = 8$ b. $y = 7$

Homework

1. Joe Skuppy runs a pool cleaning service. The equipment costs $1095 and he charges $15 an hour. The equation that relates profit and hours is

    $$p = 15h - 1095$$

    a. How many hours does he have to work to break even?

    b. How much money will he make if he doesn't work any hours?

    c. Graph the line p = 15h – 1095 by plotting the points obtained in parts a. and b. Choose an appropriate scale and only graph the portion that makes sense to the problem. Label the axes.

2. Find the mistake. Write a sentence describing what the student did wrong.
    Graph 3x – 2y = 6

    Let x = 0   Let y = 0
    3(0) – 2y = 6   3x – 2(0) = 6
       – 2y = 6   3x = 6
         y = -3   x = 2

    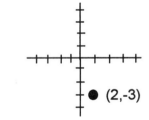
    (2,-3)

3. a. How do you find the x intercept?

    b. How do you find the y intercept?

4. Graph the line by finding the x and y intercepts. Choose an appropriate scale and label the axes.

   a. $y = \dfrac{3}{5}x + 11$

   b. $11x + 1241y = 100$

   c. $3x - .02y = 12$

   d. $y = \dfrac{-3}{4}x - 5$

5. a. Write an equation for a horizontal line?

   b. Write an equation for a vertical line?

6. Graph the equation.

   a. $y = -3$ b. $x = 6$

   c. $x = \dfrac{-1}{2}$ d. $y = 8$

8. Graph the line $y = 0.2x + 70$ by finding the x and y intercepts. Use the scale from the two graphs below. Which graph is steeper? Should one graph be steeper than the other graph? Why or why not? (You will need to use graph paper.)

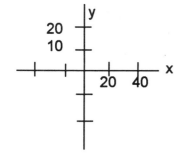 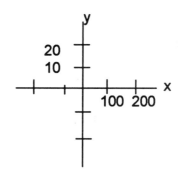

Introduction to Slope
Class Work

1. A rental car company, Wrecker, charges $21.95 per day plus 41 cents a mile.

   a. Calculate the cost of renting a car for one day.

Miles	Calculation	Cost
10	.41*10 + 21.95	26.05
20	.41*20 + 21.95	30.15
30	.41*30 + 21.95	34.25
M	.41*M + 21.95	C

   b. The graph of the equation C = .41M + 21.95 is given below.

   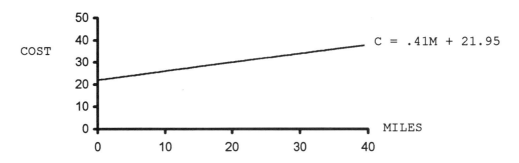

   c. What does $\dfrac{34.25 - 26.05}{30 - 10}$ mean?

   d. Suppose Wrecker starts charging 61 cents a mile, graph the new cost equation C = .61M + 21.95 using the axis in part b.

2. Which grows faster: Hybrid A corn seedlings, which grow 14.6 centimeters in 15 days, or HybridB, which grow 11.2 centimeters in 12 days?

3. What does the highway sign mean?

Average Rate of Change and Percent Change

4. Using the information in the table:

   a. Make a graph of year versus average time spent by women in a supermarket. Make sure you label your axis.

Year	Time Spent in Supermarket
1991	32
1996	78.3

   b. Calculate the average rate of change. (Be sure to include the correct units.)

   c. Calculate the percent increase in the time spent in the supermarket.
   $$\text{Percent Change} = \left(\frac{\text{New} - \text{Old}}{\text{Old}}\right) \times 100$$

   d. What is the slope of the line through the two points?

5. Using the information in the table:

   a. Make a graph of year versus U.S. trade deficit. Make sure you label your axes.

Year	U.S. Trade Deficit in Billions of Dollars
1983	.70
1984	1.20
1985	1.25
1986	1.45
1987	1.60
1988	1.20

   b. Calculate the average rate of change between 1983 and 1987 (include the correct units). Write a sentence describing your results.

   c. Calculate the average rate of change between 1987 and 1988 (include the correct units). Write a sentence describing your results.

   d. What is the percent change between 1983 and 1987. Write a sentence describing your results.

6. The graph and the table give the mens average Math SAT scores.

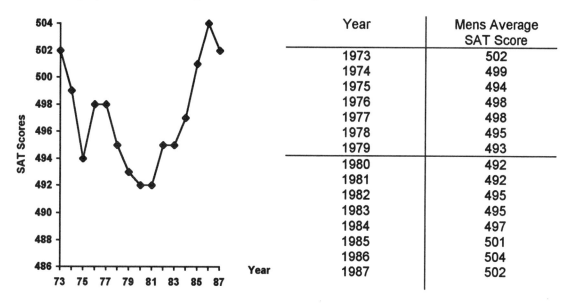

Year	Mens Average SAT Score
1973	502
1974	499
1975	494
1976	498
1977	498
1978	495
1979	493
1980	492
1981	492
1982	495
1983	495
1984	497
1985	501
1986	504
1987	502

a. According to the graph, what was the longest time period for declining SAT scores?

b. According to the graph, over what year did the SAT scores decrease the most?

c. Use the table to find the average rate of change of your answer in part b.

d. According to the graph, over what year did the SAT scores increase the most?

e. Use the table to find the average rate of change of your answer in part d..

f. According to the graph over what years did the SAT scores remain constant?

Group Work

Use the table below to make a very accurate graph of year versus women's Math SAT scores. Make sure you label your axes and choose a correct scale.

Year	1973	1974	1975	1976	1977	1978	1979	1980	1981	1982	1983	1984	1985	1986	1987
SAT Score	463	461	461	450	447	445	444	444	444	444	446	451	449	450	452

1. According to the graph, over what year did the SAT scores decline the most?

2. Use the table to find the average rate of change of your answer in part 1.

3. Use the table to find the percent decrease of your answer in part 1.

4. According to the table, over what year did the SAT scores increase the most?

5. Use the table to find the average rate of change of your answer in part 4.

6. Use the table to find the percent increase of your answer in part 4.

7. According to the graph, over what time period did women's math SAT scores remain constant?

Homework

Compute ratios to answer the following questions.

Example: Carl runs 100 meters in 9.8 seconds and Anthony runs 200 meters in 19.7 seconds. Who has the faster average speed?

Carl: $\dfrac{\text{Meters}}{\text{Seconds}} = \dfrac{100}{9.8} = 10.2$ meters per second

Anthony: $\dfrac{\text{Meters}}{\text{Seconds}} = \dfrac{200}{19.7} = 10.15$ meters per second

Carl has the faster average speed.

1. Acme sells a dozen grade A eggs for $1.89. The Fresh Stop sells 18 eggs for $2.68. Which store is the better buy?

2. Which is steeper: the car ramp for U-Haul rental company, which rises 5 feet for every horizontal distance of 9 feet, or my son Darrel's toy garage which rises 3 inches for every horizontal distance of 8 inches.

Example: Given the table to the right, calculate the average rate of change and the percent change of the number of employees at hotels, camps and other lodging places in Montgomery County.

Year	Number of Employees at Hotels, Camps and Other Lodgings in Montgomery County
1990	3,554
1994	2,285

Source: Pennsylvania County Industry Trends 1990-94

Average rate of change = $\dfrac{2{,}285 - 3{,}554}{1994 - 1990}$

= -317.25 per year

Percent change = $\dfrac{2{,}285 - 3{,}554}{3{,}554}$

= -35.7%

3. Use the table to answer the following questions.

   a. Make a graph of year versus number of employees. Make sure you label your axis.

Year	Number of Employees in Chemical and Allied Products in Montgomery County
1990	13,329
1991	14,486
1992	15,132
1993	17,355
1994	18,072

Source: Pennsylvania County Industry Trends 1994-95

   b. Calculate the average rate of change between 1990 and 1994. (Include the correct units.)

   c. Calculate the average rate of change between 1993 and 1994. (Include the correct units.)

   d. Use parts b. or c. to estimate the numbers of employees in 1995. Write a sentence that describes your logic.

   e. What is the percent change between the years 1990 and 1994?

   f. Write a topic sentence summarizing what you think is the central idea to be drawn from this data.

4. Use the table to answer the following questions.

   a. Make a graph of year versus number of newspapers.

Year	Number of Daily Newspapers in the U.S.
1980	1,745
1985	1,676
1990	1,611
1992	1,570

Source: 95 Statistical Yearbook by the United Nations Educational, Scientific and Cultural Organization

   b. Calculate the average rate of change between the years 1980 and 1992 (include the correct units).

   c. Calculate the average rate of change between the years 1990 and 1992 (include the correct units).

   d. What is the percent change between the years 1990 and 1992?

   e. Write a topic sentence summarizing what you think is the central idea to be drawn from this data.

5. Use the table at the right to make a very accurate graph of year versus men's median age of first marriage. Make sure you label your axis and choose a correct scale.

Year	Men's Median age of First Marriage
1890	26.1
1900	25.9
1910	25.1
1920	24.6
1930	24.3
1940	24.3
1950	22.8
1960	22.8
1970	23.2
1980	24.7
1990	26.1

a. According to the graph, over what 10 year period was there the greatest decrease in men's median age of first marriage?

b. Use the table to find the average rate of change over the 10 year period for your answer in part a.

c. According to the graph, over what 10 year period did the greatest increase in men's median age of first marriage occur? Why are you not sure you have the correct answer?

d. Use the table to find the average rate of change over the 10 year period for your answer in part c. Use the table to find the average rate of change over the 10 year period in men's median age of first marriage for your "other" answer in part c.

e. According to the graph, over what 10 year period was there no change in men's median age of first marriage?

f. Use slope to predict the men's median age of first marriage in the year 2000. Write a sentence explaining what you did.

S-122

6. Below is an approximation of the costs of doctors' bills and Medicare between 1963 and 1979.

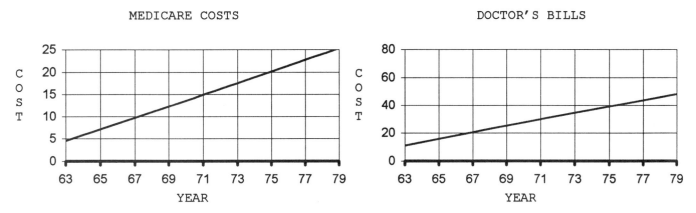

a. Which *appears* to have been growing at a faster rate: doctors bills or Medicare costs? Why?

b. Which actually grew at a faster rate and how can you tell?

8

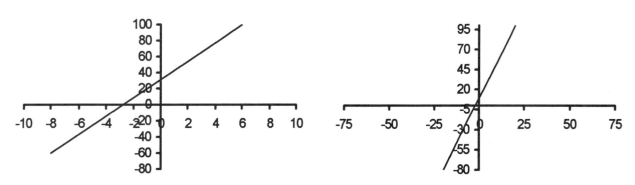

a. Which line appears to have the greater slope?

b. Estimate the slope of both lines by finding two points from both lines.

S-123

Slope

Class Work

Definition: a. $M = \dfrac{Change\ in\ Y}{Change\ in\ X}$

b. $M = \dfrac{Y_1 - Y_0}{X_1 - X_0}$ (X_0, Y_0) and $(X_1$ and $Y_1)$ are two points on the line.

Graph the line that passes through the two points and find the slope of the line.

1. (6, -2) and (5, 7) 2. (8, 2) and (4, -12)

3. (2, 4) and (2, -5) 4. (7, -3) and (-2, -3)

The equation y = mx + b is called the slope intercept equation. The coefficient of x, m, is the slope of the line and (0, b) is the y intercept.

5. Identify the slope and y-intercept of the equation of the line. Graph the line.

   a. $y = -\dfrac{3}{4}x - 5$

   b. $y = 4x + 5$

Group Work

1. Find the mistake in the problems below. Write a sentence describing the mistake.

    a. (6, 4) and (-7, 2)
    $$M = \frac{6-(-7)}{4-2}$$
    $$M = \frac{13}{2}$$

    b. (-3, 5) and (2, 6)
    $$M = \frac{5-6}{2-(-3)}$$
    $$M = \frac{-1}{5}$$

    c. (4, -2) and (-5, 1)
    $$M = \frac{-2-1}{4-5}$$
    $$M = \frac{-3}{-1}$$

2. Graph the line that passes through the two points and find the slope of the line.

    a. (-1, 5) and (3, -2)

    b. (-2, 4) and (5, 4)

Homework

1. Find the slope of the line.

   a.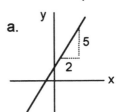

   b.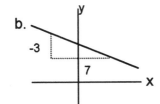

   c.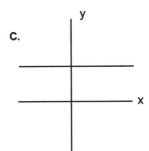

2. Find the mistake in the problems below. Write a sentence describing the mistake.

   a. (-6, 1) and (-2, 3)
   $$M = \frac{1-(-2)}{3-(-6)}$$

   b. (5, -1) and (3, 7)
   $$M = \frac{5-3}{-1-7}$$

3. Graph the line that passes through the two points and find the slope of the line.

   a. (8, -1) and (-2, 4)

   b. (3, 5) and (7, 5)

c. (-3, 1) and (6, 5)   d. (6, -2) and (6, 4)

e. (9, 6) and (-1, -2)

4. Identify the slope and y intercept of the equation of the line.

   a. $y = 7x + 5$ b. $Y = \dfrac{-2}{3}x - 2$

Application of Graphs
Class Work

1. Two girls want to enter the lawn mowing business for the summer. They buy a lawn mower for $400 and plan to charge $8.75 an hour.

    a. Write an equation for the amount of profit they plan to make in terms of the number of hours they work.

    b. Graph the equation. Label your axes and use an appropriate scale. Only graph the portion that is relevant to the problem.

    c. What does the y intercept mean? What does it mean in terms of the problem?

    d. What does the x intercept mean? What does it mean in terms of the problem?

    e. What is the slope? What does it mean in terms of the problem?

    f. How much profit will they make if they work 200 hours for the summer?

    g. Use the equation to find out how many hours they have to work to make $500.

2. Two companies, ACME and EMAC, offer you very similar sales positions. ACME pays $25,000 a year while EMAC pays $10,000 a year plus 10% commission.

   a. Write an equation for your yearly wages from ACME.

   Write an equation for your yearly wages from EMAC.

   b. Graph both equations on the same set of axes. Label your axes and choose an appropriate scale. Only graph the portion that is relevant to the problem.

   c. Find where the two lines intersect. Label this point.

   d. Use the graph to find when ACME pays more than EMAC.

   e. Use the graph to find when EMAC pays more than ACME.

   f. What do the y intercepts mean in terms of the problem?

   g. What does the slope of each line mean in terms of the problem?

3. You are going to rent a car for a day. You have two choices, Wrecker Car Rental and Caddie Car Rental. Wrecker charges $20 a day and 75¢ a mile, while Caddie charges $27 a day and 40¢ a mile.

    a. Write an equation for the cost of renting a car from Wrecker.

    Write an equation for the cost of renting a car from Caddie.

    b. Graph both equations on the same set of axes. Label each axis and choose an appropriate scale. Only graph the portion that is relevant to the problem.

    c. Find where the two lines intersect. Label the point.

    d. Use the graph to find when Wrecker costs more than Caddie.

    e. Use the graph to find when Caddie costs more than Wrecker.

    f. What do the y-intercepts mean in terms of the problems?

    g. What does the slope of each line mean in terms of the problem?

4. A phone company, DADBELL, charges 75¢ a phone call and 15¢ per minute after the first three minutes.

   a. Write an equation for the cost of a phone call.

   b. What is the cost of a phone call 2 minutes long?

   c. What is the cost of a phone call 1 minute long?

   d. What is the cost of a phone call 10 minutes long?

   e. When is the formula in part a. not valid?

   f. Graph the cost of a phone call. Make sure you include the first three minutes in your graph. Label the axis and choose an appropriate scale.

Group Work

Two companies offer you a similar sales position. Brisk Inc. will pay you $40,000 a year while Bipton Tea Co. will pay you $18,000 plus 8.8% commission.

a. Write an equation for your yearly wages from Brisk Inc.

   Write an equation for your yearly wages from Bipton Tea Co.

b. Graph both equations on the same set of axes. Label your axes and choose an appropriate scale. Only graph the portion that is relevant to the problem.

c. Find where the two lines intersect. Label this point. (Do not use the graph to estimate where they intersect.)

d. Use the graph to find when Brisk Inc. pays more than Bipton Tea Co.

e. Use the graph to find when Bipton Tea Co. pays more than Brisk Inc.

f. What do the y intercepts mean in terms of the problem?

g. What does the slope of each line mean in terms of the problem?

Homework

1. Darrel plants sunflower seedlings, each 4 inches tall. With plenty of water and sunlight, they will grow approximately 1.7 inches a day.

    a. Write an equation for the height of the plants in terms of the number of days since they were planted. (If you don't recognize the equation, form a table.)

    b. Graph the equation. Label you axes and use an appropriate scale. Only graph the portion that is relevant to the problem.

    c. How tall is the sunflower after two weeks?

    d. How tall is the sunflower after two months?

    e. How tall is the sunflower after four months? Why does this not make sense? How can your graph reflect this idea?

    f. Estimate from your graph how long it will take before the sunflower is 5 feet tall.

    g. Use the equation to find out how long it will take before the sunflower is 5 feet tall.

2. You are going to rent a car for a day. You have two choices, Lemon Car Rental and Limo Car Rental. Lemon Car Rental will rent you the car for $35 a day and unlimited mileage, while Limo Car Rental will rent you the car for $15 a day plus 80¢ a mile.

   a. Write an equation for the cost of renting a car from Lemon Car Rental. Write an equation for the cost of renting from Limo Car Rental. (If you don't recognize the equation, form a table.)

   b. Graph both equations on the same set of axes. Label your axes and choose an appropriate scale. Only graph the portion that is relevant to the problem.

   c. Use the equation to find the point where the two lines intersect. Label this point on the graph.

   d. Use the graph to find when Lemon Car Rental is more expensive then Limo Car Rental.

   e. Use the graph to find when Limo Car Rental is more expensive.

   f. What do the y intercepts mean?

   g. What does the slope of each line mean?

3. You have a choice of two phone companies, Ringer and Buzz. Ringer charges 50¢ a phone call and 18¢ a minute. Buzz charges 25¢ a phone call and 27¢ a minute.

   a. Write an equation for the cost of making a phone call using Ringer. Write an equation for the cost of making a phone call using Buzz. (If you don't recognize the equations, form a table.)

   b. Graph both equations on the same set of axes. Label your axes and choose an appropriate scale. Only graph the portion that is relevant to the problem.

   c. Use the equations to find the point where the two lines intersect. Label this point on the graph.

   d. Use the graph to find when Ringer is more expensive than Buzz.

   e. Use the graph to find when Buzz is more expensive then Ringer.

   f. What do the y intercepts mean in terms of the problem?

   g. What does the slope of each line mean in terms of the problem?

4. Two companies offer you very similar sales positions, RATCO and EXPLOITCO. RATCO will pay $5,000 a year and 8% commission while EXPLOITCO will pay $10,000 a year and 4% commission.

   a. Write an equation for your yearly wages from RATCO. Write an equation for your yearly wages from EXPLOITCO.

   b. Graph both equations on the same set of axes. Label your axes and choose an appropriate scale. Only graph the portion that is relevant to the problem.

   c. Use the equations to find where the two lines intersect. Label this point on the graph.

   d. Use the graph to find when RATCO will pay more than EXPLOITCO.

   e. Use the graph to find when EXPLOITCO will pay more than RATCO.

   f. What do the y intercepts mean in terms of the problem?

   g. What does the slope of each line mean in terms of the problem?

5. The phone company, Hook, charges 50¢ a call and 20¢ per minute after the first three minutes.

   a. Write an equation for the cost of a phone call. (You may need to make a table.)

   b. When is the formula in part a. *not* valid?

   c. What is the cost of a phone call for the first three minutes?

   d. Graph the cost of a phone call. Make sure you include the first three minutes in your graph. Label the axes and choose an appropriate scale.

6. A company, Mathematics, Inc., will pay you $10,000 a year plus 10% commission on sales over $50,000.

   a. Write an equation for your yearly salary. (You may need to make a table.)

   b. When is the formula in part a. not valid?

   c. What is your pay if you sell under $50,000 worth of merchandise?

   d. Graph your salary for the year. Make sure you include the possibility of selling less than $50,000 worth of merchandise. Label the axes and choose an appropriate scale.

Review for Test II

1. The following table shows the temperature, in Celsius, of a T.V. dinner while it is in the freezer, taken out to defrost, cooked in the oven and served at the table. The food was removed from the freezer at t = 0 min.

Time in Minutes	Temperature Celsius
-15	-10°
0	-10°
10	-5°
20	3°
25	23°
30	45°
35	90°
50	125°
60	160°
65	90°
70	55°
75	40°

    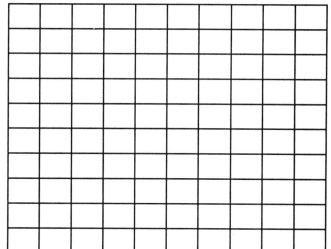

    a. Make a graph of temperature versus time. Make sure you label your axes and choose an appropriate scale.

    b. When do you think the T.V. dinner was placed in the oven?

    c. When do you think the T.V. dinner was taken out of the oven?

    d. From the graph, when was the T.V. dinner 100°?

2. Graph the lines

    a. y = -4

    b. x = 7

3. Find the x and y intercepts and graph the following lines. Label your axes and choose an appropriate scale.

a. $y = .1x + 210$

b. $76x + 2y = 4$

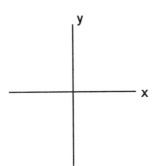
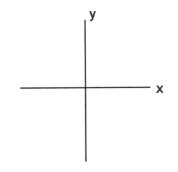

c. $4x + 7y = 20$

d. $y = \frac{4}{3}x + 8$

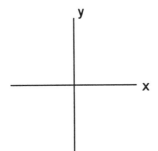
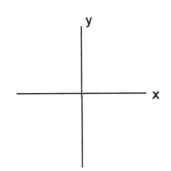

4. SITUATION: The following diagram shows the temperature in Frostburg, Maryland for a typical December day. The temperature is a function of the time of day.

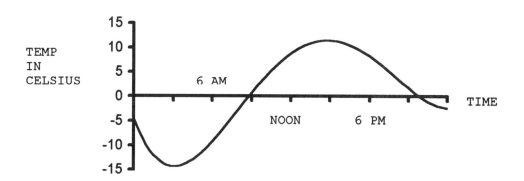

For each of the following questions, list the coordinates of the point(s) that give the answer. Then write your answer in a sentence. Estimate where necessary.

a. What was the temperature at 4 a.m.?

b. What was the temperature at 10:30 a.m.?

c. When was the temperature -2°C?

d. When was the temperature 3°C?

e. What was the maximum temperature and when did it occur?

f. What was the minimum temperature and when did it occur?

g. When was the temperature rising?

h. When was the temperature falling?

5. Use the table to answer the following questions.

   a. Find the average rate of change of corn produced in Nebraska between 1980 and 1984.

Year	Millions of Tons of Corn Produced in Nebraska
1980	62.4
1984	78.5

   b. Find the percent increase of corn produced in Nebraska between 1980 and 1984.

   c. Find the slope of the line created by the data.

6. Solve.

   a. $8x - 3 \leq 7$
   b. $4 - 2x \geq -2$
   c. $-3 \leq 7x - 15 < 10$

7. An electric company has a fleet of trucks. The annual operating cost per truck is
   $$c = 0.26m + 3100$$
   where m is the number of miles traveled in a year.

   a. How many miles can a truck travel in a year and still cost less than $4000?

   b. How many miles can a truck travel in a year and still cost between $3500 and $5000?

8. Given the equation, find the slope and y intercept.

   a. $y = -3x + 7$

   b. $y = .15x - 6$

9. Find the slope of the line containing the two points.

   a. (5, -8) and (-3, 5)

   b. (2, 6) and (-9, 6)

10. P.M.A. offers you a sales position. They pay 10% commission plus $20,000 a year. The equation that models this situation is $W = .1S + 20000$. W is your wages for the year and S is the dollar amount that you sold for P.M.A.

    a. What is the slope of the line? What does the slope mean in terms of the problem?

    b. If you don't sell anything for P.M.A., how much money will you make for the year? If you were to graph this point, what would it be called?

11. You want to rent a car for a day. Limo will charge you $32.00 for the day, while Lemmon will charge you $19 for the day plus 26¢ a mile.

   a. Write an equation for the cost of renting a car from Limo. Write an equation for the cost of renting a car from Lemmon.

   b. Graph both equations on the same set of axes. Label your axes and choose an appropriate scale. Only graph the portion that is relevant to the problem.

   c. Use the equations to find where the two lines intersect. Label this point on the graph.

   d. Use the graph to find when Limo is more expensive than Lemmon.

   e. Use the graph to find when Lemmon is more expensive than Limo.

   f. What do the y-intercepts mean in terms of renting a car?

   g. What is the slope of each line and explain what the means in terms of the problem.

Introduction to Exponents and Positive Exponents
Class Work

1. What is the definition of an exponent?

2. Complete the following using a calculator.

   a. $(-6)^2$ b. $(-8.2)^4$ c. 7^0 d. -6^2

   e. -8.2^4 f. 8.6^0 g. $(-3.1)^3$ h. -3.1^3

Compound Interest

For many transactions interest is added to the principal, the amount invested, at regular time intervals so that the interest itself earns interest. Examples of accounts that use compound interest are: savings accounts, certificates of deposit, savings bonds and money markets.

3. Assuming that the interest is compounded monthly the formula below computes how much money will be in your account at sometime in the future.

   $$FV = P(1 + i)^n$$
   where   FV is Future Value
   P is Amount Invested
   i is Interest rate per month
   n is the number of times compounded

   a. A couple invests $2000 at an annual interest rate of 12%. How much money will they have after 10 years?

   b. How much money should you invest at an annual interest rate of 6%, if you want $10,000 in 20 years?

4. a. The population of Shanghai was 10,820,000 in 1974. If the population increased by 1% each year, complete the table below.

YEAR	CALCULATION	POPULATION
1974		10,820,000
1975		
1976		
1977		
1978		
n		

b. What is the equation that relates Shanghai's population and the year?

c. Use the equation to find Shanghai's population in 1985.

5. a. When a ball is dropped it may bounce, but not up to its original height. Call the ratio of the rebound height to the original height the "resiliency" of the ball for whatever surface on which it is being dropped. A ball is dropped from two meters and for that surface its resiliency is .6. Complete the table below.

BOUNCE	CALCULATION	HEIGHT OF BOUNCE
1st		
2nd		
3rd		
4th		
B		

b. What is the equation that relates the height of the bounce to the number of times the ball has bounced?

c. Use the equation to find the height of the 15th bounce.

Group Work
Positive Exponents

1. There are 2,000 bacteria present initially in a culture. The culture grows at 13% each day.

    a. Complete the table.

TIME	CALCULATION	NUMBER OF BACTERIA
Initial Day		2000
1 Day Later		
2 Days Later		
3 Days Later		
4 Days Later		
n Days Later		

    b. What is the equation that relates number of bacteria to time?

    c. Use the equation in 1b to find how many bacteria there will be in

    1) 15 days 2) 20 days

    d. Use the table below to find out when there will be 16,000 bacteria.

TIME (GUESS)	NUMBER OF BACTERIA	TOO LOW/TOO HIGH

Homework
Positive Exponents

1. Use a calculator to compute the following.

   For example: $8^5 = 32768$, 8 $\boxed{y^x}$ 5 $\boxed{=}$

   a. 6^5 b. 7.82^3 c. 8.51^4

2. Use a calculator to compute the following. First decide if the answers should be positive or negative.

   a. $(-6)^2$ b. $(-7.4)^6$ c. $(-3.2)^4$

   d. $(-8.1)^2$ e. $(-6.8)^4$ f. $(-2.4)^8$

   g. All of the exponents in problems a. - f. were _____.

   h. Use question g. to complete the sentence:

   All bases raised to an _____ exponent yield answers that are _____.

3. Use a calculator to compute the following. First decide if the answers should be positive or negative.

   a. -8.2^2 b. -6.7^4 c. -2.5^2

   d. -6^3 e. $(-6)^3$ f. -7^5

   g. $(-7)^5$

   h. Look at questions d - g, do you need ( ) for these questions? Why? or Why not?

4. There are 1000 bacteria present initially in a culture. The culture grows at 5% each day.

   a. Complete the table.

TIME	CALCULATION	NUMBER OF BACTERIA
Initial day		1000
1 day later	1000(1.05)	1050
2 days later		
3 days later		
4 days later		
n days later		

   b. What is the equation that relates number of bacteria to time?

   c. Use the equation in 4b. to find how many bacteria will there be
   (Give to the nearest whole number)
   1) 20 days later   2) 40 days later

   d. Use the table below to find when there will be 2000 bacteria.

TIME (GUESS)	NUMBER OF BACTERIA	TOO HIGH/TOO LOW

5. The population of Farmington in 1980 was 1,052,000. The population is decreasing by 3.5% each year.

   a. Complete the table.

YEAR	CALCULATION	POPULATION
1980		1,052,000
1981	1,052,000(.965)	
1982		
1983		
1984		
n Years after 1980		

   b. What is the equation that relates population to time?

   c. Use the equation from 5b. to find the population of Farmington.
   (Give to the nearest whole number.)
   1) in 1990 2) in 1996

   d. Use the table below to find when there will be 900,000 people living in Farmington.

YEAR (GUESS)	POPULATION	TOO HIGH/TOO LOW

6. Assuming that the interest is compounded monthly the formula below computes how much money will be in your account at sometime in the future.

$$FV = P(1 + i)^n$$

where: FV is Future Value
P is the amount invested
i is the interest per period
n is the number of times compounded

a. Find the Future Value of a $1,000 deposit if the annual rate is 8% compounded monthly for 20 years. (Hint: $i = .08/12$, $n = 12 \cdot 20$)

b. Find the Future Value of a $3,000 deposit if the annual rate is 7.5% compounded monthly for 10 years. (Hint: $i = .075/12$, $n = 12 \cdot 10$)

c. Find the Future Value of a $5,000 deposit if the annual rate is 6% compounded monthly for 10 years. (Hint: $i = .06/12$, $n = 12 \cdot 10$)

d. Determine how much money must be invested today at an annual rate of 6% compounded monthly if the sum of $15,000 is desired fifteen years from now. (Hint: $i = .06/12$, $n = 12 \cdot 15$)

e. Determine how much money must be invested today at an annual rate of 7.5% compounded monthly if the sum of $10,000 is desired 20 years from now. ($i = .075/12$, $n = 240$)

f. Use the table below to find how long it will take for $5,000 to grow to $15,000 if it is invested at 9.5% compounded annually. (Hint: $I = .095/12$)

GUESS N	$5000(1.0079)^n$	TOO HIGH/TOO LOW (COMPARE TO 15,000)
10 years		

Negative Exponents and Scientific Notation
Class Work

1. What does a negative exponent mean?

2. Use a calculator to compute the following.

   a. 8^{-2} b. -6^{-4} c. $(-6)^{-4}$ d. $.08^{-5}$

3. Use the following formula to find the monthly payment of a loan.

$$P = A\left[\frac{i}{1-(1+i)^{-n}}\right]$$

   P is the monthly payment
   A is the amount of the loan
   n is the number of payments
   i is the interest rate per month

   a. Find the monthly payments on a 48-month car loan of $18,000 at 3% annual interest. (Hint $i = .03/12$).

   b. The Clintons can afford a $1,000 monthly payment for a house. They find a loan for 7.2% annual interest rate for 360 months. How expensive of a home can they afford? (Hint: $i = .072/12$)

4. A number in Scientific Notation has the form

$$P \times 10^n$$

where $1 \leq p < 10$ and n is an integer.

a. Write the number in Scientific Notation.

1) 8,200,000

2) 0.000517

3) 0.0028

b. Write the number in decimal form.

1) 7.3×10^6

2) 3.141×10^{-4}

3) 8.14×10^3

c. Use your calculator to compute the following.

1) $(6.3 \times 10^8)(4.2 \times 10^9)$

2) $\dfrac{2.8 \times 10^5}{8.6 \times 10^{-3}}$

d. Light travels at a rate of 1.86×10^5 miles per second. How far does light travel in a day? (There are 8.64×10^4 seconds in a day.)

e. The distance from the sun to the earth is 93,000,000 miles. How long does it take for the light of the sun to reach the earth?

Group Work

1. Compute the following using your calculator

   a. 2^{-4} b. -2^{-4} c. $(-2)^{-4}$

2. Use the following formula to find the monthly payment of a loan.

$$P = A\left[\frac{i}{1-(1+i)^{-n}}\right]$$

   P is the monthly payment
   A is the amount of the loan
   n is the number of payments
   i is the interest rate per month

   a. Find the monthly payments on a 36-month car loan of $7,800 at 4% annual interest. (Hint: i = .04/12)

   b. Eddie Nerder can afford a $250 car payment at 6% annual interest for 36 months. How expensive a car can he afford? (Hint: i = .06/12)

3. The average amount of water flowing past the mouth of the Amazon River is 4.2×10^6 cubic feet per second. How much water flows past in a day?

Homework

1. Compute the following using a calculator.

   a. 8^{-3}
   b. 3.1^{-2}
   c. $\left(\frac{1}{5}\right)^{-4}$
   d. -3^{-4}

   e. $(-3)^{-4}$
   f. -8^{-3}
   g. $(-8)^{-3}$
   h. $.2^{-5}$

2. Compute using a calculator.

   a. $(3.8 \times 10^5)(6.2 \times 10^7)$
   b. $\dfrac{8.7 \times 10^{-3}}{2 \times 10^5}$
   c. $(7.3 \times 10^{-8})(9.3 \times 10^{20})$

3. Use the following formula to find the monthly payments of a loan.

$$\text{Payment of Debts} \qquad P = A\left[\frac{i}{1-(1+i)^{-n}}\right]$$

P is the monthly payment
A is the amount of the loan
n is the number of payments
i is the interest rate per month

a. Find the monthly payments on a 36-month auto loan of $2,500 at 15% annual interest compounded monthly. (Hint: $i = .15/12$)

b. Jackson Pollock can afford a $300 a month car payment at 12% annual interest for 36 months. How expensive a car can he afford? (Hint: $i = .12/12$, find A.)

c. Find the monthly payments of a 30-year home mortgage of $100,000 at 9% annual interest compounded monthly. (Hint: $i = .09/12$, and $n = 30(12)$.)

d. The Switzer's can afford a $1,000 monthly mortgage payment for a house. How large a 15-year mortgage can they afford at 8% annual interest compounded monthly. Hint: $i = .08/12$ and $n = 15(12)$. Suppose they have a 30-year mortgage?

4. Write each number in scientific notation.

   a. 78,000

   b. 0.00000167

   c. 0.00635

   d. 1,160,000

5. Write each number in decimal notation.

   a. 7.86×10^8

   b. 8.673×10^{-10}

   c. 3.3×10^{-2}

   d. 2.032×10^4

6. The distance light travels in 1 year is 9.460×10^{12} kilometers. Write this number in decimal notation.

7. A beam of light travels 9.460×10^{12} kilometers in one year. How far does it travel in 10,000 years?

8. Suppose $1000 is invested at a rate of 9% and compounded monthly. The amount (A) after one year is given by

$$A = (1 \times 10^3)(1.09381)$$

Compute this amount in decimal notation.

Properties of Exponents
Class Work

Property 1. $a^n \cdot a^m = a^{n+m}$

Simplify each expression.

1. $x^5 \cdot x^3$
2. $x^4 \cdot x^2$

Property 2. $(a^n)^m = a^{nm}$

Simplify each expression.

3. $(x^3)^2$
4. $(x^5)^3$

Property 3. $(ab)^n = a^n b^n$

Simplify each expression.

5. $(3x^2)^4$
6. $(5x^2)^3$

Property 4. $a^{-n} = \dfrac{1}{a^n}$

Simplify each expression. Write the expression with positive exponents only.

7. $8x^{-3}$
8. $(2x)^{-4}$

Property 5. $\dfrac{1}{a^{-n}} = a^n$

Simplify each expression. Write the expression with positive exponents only.

9. $\dfrac{1}{4x^{-3}}$
10. $\dfrac{1}{(7x^2)^{-3}}$

Property 6. $\dfrac{a^n}{a^m} = a^{n-m}$

Simplify each expression. Write the expression with positive exponents only.

11. $\dfrac{x^8}{x^3}$ 　　　　　12. $\dfrac{x^2}{x^5}$

Property 7. $a^0 = 1$

Simplify each expression.

13. $8x^0$ 　　　　　14. $(7x^3)^0$

Property 8. $\left(\dfrac{a}{b}\right)^n = \dfrac{a^n}{b^n}$

Simplify each expression.

15. $\left(\dfrac{4}{x^2}\right)^3$ 　　　　　16. $\left(\dfrac{x^2}{3}\right)^2$

Simplify each expression. Write the expression with positive exponents only.

17. $3x^2 x^3$ 　　　　　18. $4(-3x^2)^2$ 　　　　　19. $6x(-2x^2)^3$

20. $\dfrac{8x^5}{x^{-3}}$ 　　　　　21. $\dfrac{4x^{-3}}{8x^{-1}}$ 　　　　　22. $\left(\dfrac{-2}{x^2}\right)^3$

Group Work

1. Find the mistake.

   Simplify each expression. Write the expression with positive exponents only.

   a. $x^3 \cdot x^2 = x^6$

   b. $\dfrac{1}{6x^{-3}} = 6x^3$

2. Simplify each expression. Write the expression with positive exponents only.

   a. $(5x^2)^3$

   b. $(-6x^3)^2$

   c. $\dfrac{9x^5}{3x^2}$

   d. $4x^{-5}$

   e. $\left(\dfrac{3x^3}{5}\right)^2$

   f. $\dfrac{1}{(3x^2)^{-4}}$

Homework

1. State the eight properties of exponents.

2. Find the mistake.

   Simplify each expression. Write the expression with positive exponents only.

   a. $(3x^2)^4 = 81x^6$

   b. $(3x^2)(4x) = 12x^2$

   c. $-7x^{-2} = 7x^2$

3. Simplify each expression. Write the expression with positive exponents only.

   a. $8x^5x^3$

   b. $2x^3x^{-2}$

   c. $(2x^2)^3$

   d. $(5x^{-3})^4$

   e. $6x^{-1}$

   f. $\dfrac{5}{2x^{-3}}$

   g. $\dfrac{9x^5}{18x^2}$

   h. $\dfrac{x^4}{9x^6}$

   i. 9^0

   j. $\left(\dfrac{6}{x^2}\right)^3$

   k. $\left(\dfrac{x^4}{2}\right)^{-3}$

   l. $(7x^5)^{-2}$

Introduction to Algebraic Fractions

Class Work

I. Evaluating an Algebraic Fraction.

1. Suppose the cost of removing p percent of the particulate pollution from the exhaust gases at an industrial site is given by

$$C = \frac{6800p}{100 - p}$$

Find the cost for

a. $p = 75$

b. $p = 85$

c. $p = 95$

d. $p = 100$

II. Reduce to lowest terms.

2. $\dfrac{12y}{8y}$

3. $\dfrac{-5x}{15x^2}$

4. $\dfrac{24x^3}{32x}$

III. Multiply or divide as indicated and reduce to lowest terms.

5. $\dfrac{8x}{5} \cdot \dfrac{15}{16x}$

6. $\dfrac{21x^2}{10} \div \dfrac{7x^3}{15}$

7. $\dfrac{-24x^3}{8} \div \dfrac{3}{5x^3}$

8. $\dfrac{9x^2}{14} \cdot \dfrac{7}{27x^3}$

Group Work

1. Suppose the cost of removing p percent of the particulate pollution from the exhaust gases at an industrial site is given by

$$C = \frac{7200p}{100 - p}$$

   Find the cost for

   a. $p = 85$

   b. $p = 95$

   c. $p = 99$

   d. $p = 100$ (What does this say about removing all of the exhaust gases at the industrial site?)

2. Reduce to lowest term.

   $$\frac{9x^3}{27x}$$

3. Multiply or divide as indicated and reduce to lowest terms.

   a. $\dfrac{16x}{20} \cdot \dfrac{8x^2}{3}$

   b. $\dfrac{-81x}{10} \div \dfrac{9x^3}{5}$

Homework

1. Suppose that for a certain city the cost C of obtaining drinking water with p percent impurities is given by

$$C = \frac{120000}{p} - 1200$$

Find the cost for

   a. $p = 15$

   b. $p = 5$

   c. $p = 1$

   d. $p = 0$ (What does this say about the drinking water in this town?)

2. Suppose the cost C of removing p percent of the particulate pollution from the exhaust gases at an industrial site is given by

$$C = \frac{8300p}{100 - p}$$

Find the cost for

   a. $p = 80$

   b. $p = 90$

   c. $p = 95$

   d. $p = 100$ (What does this say about removing all of the exhaust gases at the industrial site?)

3. Reduce to lowest terms.

   a. $\dfrac{9x}{6x}$ b. $\dfrac{27y}{15y}$ c. $\dfrac{2x^2}{8x^3}$

   d. $\dfrac{6y^3}{15x}$ e. $\dfrac{-2x}{8x^2}$ f. $\dfrac{16x^3}{-32x}$

4. Multiply or divide as indicated and reduce to lowest terms.

   a. $\dfrac{2x}{3} \cdot \dfrac{6}{3x}$ b. $\dfrac{6x}{5} \div \dfrac{3x}{10}$ c. $\dfrac{7}{3y} \div 3y$

   d. $\dfrac{2x^2}{9} \cdot \dfrac{18}{4x}$ e. $\dfrac{3x^2}{4} \cdot \dfrac{16}{12x^3}$ f. $\dfrac{-6x^3}{5} \div \dfrac{18x}{10}$

Adding and Subtracting Algebraic Fractions

Class Work

Combine into single fractions and reduce to lowest terms.

1. $\dfrac{9x}{10} + \dfrac{3x}{5}$

2. $\dfrac{9y}{8} + 3y$

3. $\dfrac{7}{8x} - \dfrac{5}{6x}$

4. $\dfrac{3}{2x} - \dfrac{7}{5}$

5. $\dfrac{2}{9x} + \dfrac{5}{6x^2}$

6. $\dfrac{9}{5x} - \dfrac{1}{15x^3}$

Group Work

Describe the mistake:

1. $\dfrac{6x}{5} + \dfrac{3x}{2} = \dfrac{9x}{7}$

2. $\dfrac{3x}{8} + \dfrac{5}{2x} =$

   $\left(\dfrac{3x}{8}\right)\left(\dfrac{x}{x}\right) + \left(\dfrac{5}{2x}\right)\left(\dfrac{4x}{4x}\right) =$

   $\dfrac{3x^2}{8x} + \dfrac{20x}{8x} =$

   $\dfrac{3x^2 + 20x}{8x}$

Describe each step.

3. $\dfrac{3}{4y^2} - \dfrac{5}{6y^3} =$

   $\left(\dfrac{3}{4y^2}\right)\left(\dfrac{3y}{3y}\right) - \left(\dfrac{5}{6y^3}\right)\left(\dfrac{2}{2}\right) =$ _____

   $\dfrac{9y}{12y^3} - \dfrac{10}{12y^3} =$ _____

   $\dfrac{9y - 10}{12y^3}$ _____

Combine into single fractions and reduce to lowest terms.

4. $6x + \dfrac{3x}{5}$

5. $\dfrac{7}{4x^3} - \dfrac{9}{10x}$

S-185

Homework

Describe the mistake.

1. $$\frac{2x}{9} + \frac{5}{6x} =$$
    $$\left(\frac{2x}{9}\right)\left(\frac{2x}{2x}\right) + \left(\frac{5}{6x}\right)\left(\frac{3}{3}\right) =$$
    $$\frac{4x^2}{18x} + \frac{15}{18x} =$$
    $$\frac{19x^2}{18x} =$$
    $$\frac{19x}{18}$$

2. $$\frac{7x}{5} + \frac{2x}{3} = \frac{9x}{8}$$

Describe each step.

3. $$\frac{11}{2y} + \frac{8}{3y^2} =$$
    $$\left(\frac{11}{2y}\right)\left(\frac{3y}{3y}\right) + \left(\frac{8}{3y^2}\right)\left(\frac{2}{2}\right) =$$
    $$\frac{33y}{6y^2} + \frac{16}{6y^2} =$$
    $$\frac{33y + 16}{6y^2}$$

Combine into single fractions and reduce to lowest terms.

4. $$\frac{3}{5x} - \frac{2}{5x}$$

5. $$\frac{3}{5x} - \frac{2}{3}$$

6. $\dfrac{4m}{3}+\dfrac{m}{7}$

7. $\dfrac{2}{3}-\dfrac{3}{4y}$

8. $\dfrac{5}{8m^3}-\dfrac{1}{12m}$

9. $\dfrac{5}{3y}+\dfrac{3}{4y^2}$

10. $\dfrac{2}{9n^2}-\dfrac{5}{12n^4}$

11. $\dfrac{3}{2x^2}+\dfrac{4}{3x}$

12. $\dfrac{x^2}{4}-\dfrac{x}{3}$

13. $\dfrac{1}{5x^3}+7$

Solving Equations with Fractions

Class Work

Solve each equation and describe each step.

1. $\dfrac{3x}{5} + \dfrac{7}{2} = \dfrac{x}{10}$

2. $\dfrac{4}{x} - \dfrac{4}{3} = \dfrac{8}{5x}$

Solve each equation.

3. $\dfrac{2x}{5} = \dfrac{x+2}{6}$

4. $\dfrac{y-4}{2} - \dfrac{y-3}{9} = \dfrac{5}{18}$

Group Work

1. Describe each step.

$$\frac{5x}{8} - \frac{3}{16} = \frac{7x}{24} \qquad \text{Step}$$

$$\left(\frac{5x}{8}\right)\left(\frac{48}{1}\right) - \left(\frac{3}{16}\right)\left(\frac{48}{1}\right) = \left(\frac{7x}{24}\right)\left(\frac{48}{1}\right) \qquad \rule{4cm}{0.4pt}$$

$$(5x)(6) - (3)(3) = (7x)(2) \qquad \rule{4cm}{0.4pt}$$

$$30x - 9 = 14x \qquad \rule{4cm}{0.4pt}$$

$$-9 = -16x \qquad \rule{4cm}{0.4pt}$$

$$\frac{9}{16} = x \qquad \rule{4cm}{0.4pt}$$

2. Find the mistake.

$$\frac{x+5}{4} - \frac{x-3}{6} = \frac{7}{12}$$

$$\left(\frac{12}{1}\right)\left(\frac{x+5}{4}\right) - \left(\frac{12}{1}\right)\left(\frac{x-3}{6}\right) = \left(\frac{12}{1}\right)\left(\frac{7}{12}\right)$$

$$3(x+5) - 2(x-3) = 7$$

$$3x + 15 - 2x - 6 = 7$$

$$x + 9 = 7$$

$$x = -2$$

Solve each equation.

3. $\dfrac{6}{5} + \dfrac{8x}{15} = \dfrac{7x}{3}$

4. $\dfrac{x-2}{2} - \dfrac{3x+5}{4} = \dfrac{3}{8}$

Homework

Describe each step.

1.
$$\frac{5x}{9} - \frac{7}{36} = \frac{11x}{6}$$ Step

$$\left(\frac{5x}{9}\right)\left(\frac{36}{1}\right) - \left(\frac{7}{36}\right)\left(\frac{36}{1}\right) = \left(\frac{11x}{6}\right)\left(\frac{36}{1}\right)$$ _____

$$(5x)(4) - 7 = (11x)(6)$$ _____

$$20x - 7 = 66x$$ _____

$$-7 = 46x$$ _____

$$\frac{-7}{46} = x$$ _____

2.
$$\frac{2x+5}{8} - \frac{x-3}{2} = \frac{7}{4}$$ Step

$$\left(\frac{2x+5}{8}\right)\left(\frac{8}{1}\right) - \left(\frac{x-3}{2}\right)\left(\frac{8}{1}\right) = \left(\frac{7}{4}\right)\left(\frac{8}{1}\right)$$ _____

$$2x + 5 - 4(x - 3) = (7)(2)$$ _____

$$2x + 5 - 4x + 12 = 14$$ _____

$$-2x + 17 = 14$$ _____

$$-2x = -3$$ _____

$$x = \frac{3}{2}$$ _____

Find the mistake.

3.
$$\frac{2}{x} - \frac{7}{5} = \frac{6}{15}$$
$$\left(\frac{2}{x}\right)\left(\frac{15x}{1}\right) - \left(\frac{7}{5}\right)\left(\frac{15x}{1}\right) = \frac{6}{15}\left(\frac{15x}{1}\right)$$
$$30x - 35x = 6$$
$$-5x = 6$$
$$x = \frac{-6}{5}$$

4.
$$\frac{3x-2}{2} - \frac{x-5}{4} = \frac{7}{12}$$
$$\left(\frac{12}{1}\right)\left(\frac{3x-2}{2}\right) - \frac{12}{1}\left(\frac{x-5}{4}\right) = \left(\frac{12}{1}\right)\left(\frac{7}{12}\right)$$
$$6(3x-2) - 3(x-5) = 7$$
$$18x - 12 - 3x - 15 = 7$$
$$15x - 27 = 7$$
$$15x = 34$$
$$x = \frac{34}{15}$$

Solve each equation.

5. $\dfrac{5}{18} - \dfrac{4x}{3} = \dfrac{11}{6}$

6. $\dfrac{2x}{21} + \dfrac{4}{7} = \dfrac{5x}{3}$

7. $\dfrac{8}{5x} + \dfrac{4}{3} = \dfrac{1}{15}$

8. $\dfrac{5}{12} - \dfrac{1}{3x} = \dfrac{7}{6}$

9. $2x - \dfrac{7}{5} = \dfrac{3x}{10}$

10. $\dfrac{9}{8} + 3x = \dfrac{5x}{12}$

11. $\dfrac{x+1}{3} - \dfrac{3x+2}{2} = \dfrac{7}{6}$

12. $\dfrac{3x+5}{4} - \dfrac{x-2}{20} = \dfrac{9}{5}$

Ratio and Proportion Problems
Class Work

Ratios

1. Which is the better buy?

    8 oz. of jelly for $1.59 or 12 oz. of jelly for $1.80?

2. If a line rises 8 units for every 2 horizontal units, what is the slope of the line?

3. A ratio is:

Proportions

4. A proportion is:

Solve:

5. $\dfrac{x}{5} = \dfrac{7}{15}$

6. $\dfrac{6}{x} = \dfrac{5}{7}$

7. To establish the number of fish in a lake, 30 fish are caught, tagged, and released. Later 70 fish are caught and 14 are found to have been tagged. Estimate the number of fish in the lake.

8. Two people put their money together to buy lottery tickets. The first person put in $15 and the second person put in $25. If they won 2.4 million dollars, how much does each person win?

9. It is a rainy Saturday morning and your 5-year-old wants to make clay. The ingredients for clay are:

> 1½ cups salt
> 4 cups flour
> 1½ cups water

If you only have 1/3 cup of salt, how much flour should you use?

Group Work

1. Three people pool their money to buy lottery tickets. The first person put in $20, the second put in $30 and the third put in $35. If they won 7.8 million dollars, how much did each person win?

2. If your monthly mortgage payment is $5.955 per every $1,000, how much do you pay per month if your mortgage is $120,000?

Homework

1. Which is the better buy?

   75 ounces of laundry detergent for $2.10 or
   90 ounces of laundry detergent for $2.70

2. Which is the better buy?

   10 ounces of tuna for $1.09 or
   15 ounces of tuna for $1.61

3. To estimate the number of bears in a forest, 8 are caught, tagged, and released. Later 9 bears are caught and 2 are found to have been tagged. Estimate the number of bears in the forest.

4. If a car can go 110 miles on 5 gallons of gas, how far can it go on 12 gallons of gas?

5. To estimate the number of people in Reading, population 27,000, who have no life insurance, 160 people were polled, and 18 said they had no heath insurance. Estimate the number of people in Reading that don't have any life insurance.

6. A laser printer can print 7 pages every 2 minutes. How long will it take to print 81 pages?

7. Three people pool their money to buy lottery tickets. The first person put in $20, the second put in $24, and the third put in $31. If they won 12.6 million dollars, how much does each person win?

8. Solve for x:

   a. $\dfrac{6}{x} = \dfrac{9}{4}$

   b. $\dfrac{7x}{2} = \dfrac{6}{5}$

Review for Test III

1. Simplify. Write with positive exponents.

   a. $x^5 \cdot x^2$
   b. $\dfrac{x^3}{x^5}$
   c. y^0

   d. $\left(x^5\right)^2$
   e. $\left(2x^2\right)^3$
   f. $x^{-7}x^2$

   g. $\dfrac{x^2}{x^{-5}}$
   h. $\dfrac{9x^{-2}}{x^3}$

2. Compute using a calculator.

   a. 2.5^{-3}
   b. -3.6^4
   c. $(-3.6)^4$

3. Compute using a calculator.

   a. $\left(2.5 \times 10^5\right)\left(8.6 \times 10^7\right)$
   b. $\left(8.1 \times 10^{-3}\right)\left(6.2 \times 10^{-11}\right)$

4. Write in Scientific Notation.

   a. 8300000
   b. 0.000614

5. Write in decimal notation.

   a. 4.2×10^5
   b. 3.1×10^{-3}

6. Perform the indicated operation.

   a. $\dfrac{4x^5}{3y^2} \cdot \dfrac{9x}{2y^3}$

   b. $\dfrac{4x^2}{7} \div 14x^5$

   c. $\dfrac{5}{2x} - \dfrac{7}{3x^2}$

   d. $\dfrac{11}{5x} + \dfrac{1}{3}$

7. Solve.

   a. $\dfrac{x}{3} + 9 = \dfrac{3x-3}{2}$

   b. $\dfrac{x}{4} - \dfrac{x-3}{2} = 2$

8. There are 500 bacteria in a culture. The culture grows at 7% a day. Complete the table.

   a.

DAY	CALCULATION	POPULATION
0		
1		
2		
3		
n		

   b. What is the equation that relates population to day?

   c. Find how many bacteria there are after 30 days.

   d. Find how many bacteria there are after 60 days.

9. Use the formula:

   Payment of Debts $\quad P = A\left[\dfrac{i}{1-(1+i)^{-n}}\right]$

   P is the monthly payment

   A is the amount of the loan

   n is the number of payments

   i is the interest rate per month

   a. Find the monthly payments of a 30 year mortgage of 9% annual interest on a $200,000 home.

   b. Find out what price car you can buy if you can afford a $300 car payment for 36 months at 4.5% annual interest.

Introduction to Quadratics
Class Work

I. Evaluating Quadratics

William Tell shoots an arrow straight up with an initial velocity of 160 feet per second. The height (in feet) of the arrow is given by the equation

$$h = -16t^2 + 160t,$$

where t is the number of seconds the arrow is in the air.
Find the height of the arrow for

a. $t = 2$

b. $t = 5$

c. $t = 8$

d. $t = 10$

e. According to the calculations above, when will the arrow reach its maximum height?

f. According to the calculations above, when will the arrow hit the ground?

g. Graph the points obtained in a through d.

II. Combining Like Terms

   1. What are like terms?

   2. How do we combine like terms?

   3. Simplify.

      a. $-8x^2 + 2x - 8 - 6x^2 + 10x + 2 =$

      b. $2(3x^2 + 5x - 10) - 4(x^2 - 6x + 3) =$

      c. $5(2x^2 + 5) - (8x^2 + 2x - 4) =$

      d. The equation for profit is, Profit = Revenue − Cost. If the revenue equation for a company is

      $R = -5x^2 + 17x$

      and the cost equation for a company is

      $C = 3x^2 - 27x + 40.$

      Find the equation for profit.

III. Multiplying Two Binomials (FOIL)

1. What is a binomial?

2. Multiply.

   a. $(x+5)(x+3)$

   b. $(x-6)(x+2)$

   c. $(x-7)(x-5)$

   d. $(2x+5)(3x-8)$

   e. $(3x+4)^2$

Group Work

Triple Bubble Gum Company makes gum. The cost of making x thousand pieces of gum a day is

$$C = x^2 - 19x + 48$$

and the revenue from selling x thousand pieces of gum a day is

$$R = -x^2 + 9x$$

1. Find the profit equation for Triple Bubble Gum Co.
   (Profit = Revenue – Cost)

2. Find the profit for

   a. $x = 2$

   b. $x = 3$

   c. $x = 7$

   d. $x = 10$

   e. $x = 12$

   f. When will the company make no profit? (The break even point)

   g. When will the company make the most profit?

   h. Graph the points a through e.

Homework

1. A rocket is launched from the top of a cliff with an initial velocity of 256 feet per second. The height (in feet) of the rocket is given by the equation

    $$h = -16t^2 + 256t + 80,$$

    where t is the number of seconds the rocket is in the air.
    Find the height of the rocket for

    a. $t = 0$

    b. $t = 3$

    c. $t = 8$

    d. $t = 10$

    e. $t = 16$

    f. $t = 17$

    g. According to the calculations above, when will the rocket reach its maximum height?

    h. According to the calculations above, estimate when the rocket hits the ground?

    i. Graph the points a through f

2. What are like terms?

3. How do you combine like terms?

4. Simplify

   a. $-2x^2 + 8x - 10 - 2x^2 + 4x + 7$

   b. $3(2x^2 + 5x - 8) - 2(5x^2 + 6x - 2)$

   c. $(-16x^2 + 8x - 5) - (3x^2 - 5x + 2)$

   d. $3(-2x^2 + 6) + 2(3x^2 - 9x)$

   e. $2(3x^2 - 8x - 5) + 3(-4x^2 - 3x + 8)$

   f. $(-3x^2 + 4x - 10) - (8x^2 - 2x + 5)$

5. The equation for profit is, Profit = Revenue − Cost. If the revenue equation for a company is

   $$R = -3x^2 + 8x$$

   and the cost equation for a company is

   $$C = x^2 - 9x + 72.$$

   Find the equation for profit.

6. Multiply.

   a. (x+2)(x+3)   b. (x+4)(x-3)

   c. (x-7)(x-4)   d. (x+8)(x-8)

   e. (x+9)(x-9)   f. (3x-1)(2x+3)

   g. $(x+3)^2$ h. $(x-5)^2$

Application of the Quadratic Formula
Class Work

1. Suppose you are standing on top of the 1600 feet tall Rears Tower and you drop a rock. The height of the rock from the ground is given by the equation

$$h = -16t^2 + 1600,$$

   h is in feet and t is in seconds. How long until the rock hits the ground?

2. The revenue generated by selling x items is given by
$$R = 280x - 0.4x^2,$$
and the cost of making x items is given by
$$C = 5000 + 0.6x^2.$$

   a. Find the profit function.

   b. How many items must be sold (and made) if a profit of $439 is to be generated?

3. Suppose you want to enclose a rectangle with a 40 inch string and you use a wall of the room for one side of the rectangle.

   a. Make a table and find the formula for the area of the rectangle.

   WALL

   W | | W
   L

WIDTH	LENGTH	AREA

   b. Find the dimensions of the rectangle if the area is 65 square inches.

4. A 10 cm stick is broken into two pieces. One is placed at a right angle to form an upside down "T" shape. By attaching wires from the ends of the base to the end of the upright piece, a framework for a sail will be formed.

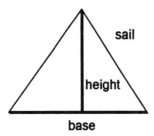

a. Complete the table below.

BASE	HEIGHT	AREA $\left(A = \frac{1}{2}bh\right)$
0	10-0 = 10	$\frac{1}{2}(0 \cdot 10) = 0$ sq cm
2	10-2 = 8	$\frac{1}{2}(2 \cdot 8) = 8$ sq cm
4		
6		
8		
10		
b		

b. Simplify the formula for Area.

c. What should the base of the sail be if the area must be 10 sq cm?

Group Work

1. King Co. makes refrigerators. The cost of making x refrigerators per month is
   $$C = .3x^2 - 100x + 5250 \text{ dollars.}$$
   The revenue from selling x refrigerators per month is
   $$R = -.7x^2 + 800x \text{ dollars.}$$

   a. Find the equation for profit. (P = R - C)

   b. Find how many refrigerators King Co. would have to sell if they want to have a profit of $100,000.

Homework

1. Quadratic Formula

   a. State the formula.

   b. Write a sentence describing what the formula is used for.

2. A rancher has 360 yards of fence to enclose a rectangular pasture. If the pasture should be 8000 square yards in area, what should the dimensions of the pasture be. (Hint: You may need a table to find the equation.)

WIDTH	LENGTH	AREA

3. A flair is fired from the bridge of the Titanic. The height in feet of the flair is given by the equation where t is measured in seconds
$$h = -16t^2 + 140t + 40.$$
How long is the flare in the air?

4. The total cost (in dollars) for a company to manufacture and sell x items per week is
$$C = .6x^2 - 140x + 6200$$
while the revenue (in dollars) by selling all x items is
$$R = -.4x^2 + 130x$$
How many items must be sold to obtain a weekly profit of $10,000?

5. The table below represents the percent of 18-25 year olds who reported using marijuana in the past 30 days.

Year	1974	1976	1977	1979	1982	1985	1988	1990	1991	1992	1993
% of Teenagers	34.2	35	38.7	46.9	40.4	36.3	27.9	24.6	24.5	22.7	22.9

Source: Digest of Educational Statistics.

The quadratic equation $P = -.12t^2 + 1.46t + 35.5$ can be used to approximate the percent of teenagers who reported using marijuana in the past 30 days. P is the percentage and t is the number of years since 1974. Use the equation to:

a. Find the percentage of teenagers who used marijuana in the past 30 days for the year 1997.

b. When will the percentage of teenagers who used marijuana in the past 30 days be 10%?

6. The total cost (in dollars) for a company to manufacture and sell x items per week is
$$C = 60x + 300,$$
while the revenue (in dollars) from selling all x items is
$$R = 100x - 0.5x^2.$$
How many items must be sold to obtain a weekly profit of $300?

Quadratic Applications and Their Graphs
Class Work

1. Earl Black makes tea bags. The cost of making x million tea bags per month is

    $$C = x^2 - 38x + 400$$

    The revenue from selling x million tea bags per month is

    $$R = -x^2 + 78x$$

    C and R are in thousands of dollars.

    a. Find the equation for profit.

    b. Graph the profit equation. Explain what the vertex, x and p intercepts mean in terms of making tea bags. Make sure you label the axis and use an appropriate scale.

    c. Suppose Earl Black needs to make $500,000 in profit (P = 500). Graph this line on the graph above and find out where the line intersects the graph. Explain what the answers mean.

2. An angry algebra student stands at the top of a 250 foot cliff and throws his algebra book upward with a velocity of 46 feet per second. The height of the book above the floor of the canyon t seconds after the book was thrown is given by

$$h = -16t^2 + 46t + 250 \text{ feet}$$

Graph the equation. Explain what the vertex, t and h intercepts mean in terms of the book being thrown off the cliff. Why should there only be one t intercept? Make sure you label the axes and choose an appropriate scale.

3. Ms. Piggie wants to enclose two adjacent chicken coops of equal size against the hen house wall. She has 66 feet of chicken-wire fencing and would like the chicken coop to be as large as possible.

   a. Find the formula for the area of the chicken coops.

WIDTH	LENGTH	AREA

   b. Graph the formula in part a. Explain what the vertex, w-intercept and A-intercept mean in terms of Ms. Piggie's problem.

   c. Kermit comes along and tells Ms. Piggie that the chicken coops should only have an area of 360 square feet. Represent this idea in the graph of part b) by graphing the line A = 360 and find the dimensions of the chicken coop.

Group Work

1. Soul Shoe Co. makes x thousand shoes per week. The cost of making x thousand shoes per week is

   $$C = .3x^2 - 11x + 13$$

   and the revenue from selling x thousand shoes per week is

   $$R = -.7x^2 + 7x.$$

   C and R are in thousand dollar units.

   a. Find the equation for profit. (P = R – C)

   b. Graph the equation for profit. Explain what the vertex, x-intercepts, and P-intercept mean in terms of making shoes. Label your axes and choose an appropriate scale.

   c. Suppose Soul Shoe Co. needs to make $50,000 in profit for the week (P = 50). Graph this line on the graph above and find where the line intersects the graph. Explain what the answers mean in terms of making shoes.

Homework

1. Write a sentence describing the vertex of the graph of the quadratic.

2. What is the formula for the x coordinate of the vertex?

3. Write a sentence describing how to find the y coordinate of the vertex.

4. List the important points you need to find to graph a quadratic.

5. Mr. Snyder makes pretzels. The cost of making x thousands of pretzels per week is
$$C = .4x^2 - 28x + 87.$$
The revenue from selling x thousands of pretzels per week is
$$R = -.6x^2 + 8x.$$
C and R are in thousands of dollars.

   a. Find the equation for profit. (Profit = Revenue - Cost)

   b. Graph the profit equation. Explain what the vertex, x-intercepts, p-intercept mean in terms of making pretzels. Make sure you label the axes and use an appropriate scale.

   c. Suppose Mr. Snyder needs to make $40,000 in profits (P = 40). Graph this line on the graph above and find out where the line intersects the graph. Explain what the answers mean.

6. Dr. Dunkin makes doughnuts. The cost of making x thousands of doughnuts a day is

$$C = 4x^2 - 62x + 35$$

and the revenue from selling x doughnuts a day is

$$R = -3x^2 + 15x.$$

C and R are in thousands of dollars.

a. Find the equation for profit. (Profit = Revenue - Cost).

b. Graph the profit equation. Explain what the vertex, x-intercepts, p-intercept mean in terms of making doughnuts. Make sure you label the axes and use an appropriate scale.

c. Suppose Dr. Dunkin needs to make $80,000 a day in profit (P = 80). Graph this line on the graph above and find out where the line intersects the graph. Explain what the answers mean.

7. The Star Spangle Banner Fireworks Company launches its Liberty Bell rocket every Labor Day (the company is a little confused). The height of the rocket is given by the equation

$$h = -16t^2 + 200t \text{ feet}$$

a. Graph the equation. Explain what the vertex, t-intercepts and h-intercept mean in terms of the Liberty Bell rocket. Make sure you label the axes and choose an appropriate scale.

b. Suppose the Liberty Bell rocket must be ignited 300 feet above the ground. Graph this requirement on the graph above. Find where the line intersects the graph and explain what the answers mean.

8. A farmer has 1800 feet of fencing. She wants to fence a rectangular pasture. The pasture borders on a river.

   a. Find an equation for the area of the pasture.

   RIVER

WIDTH	LENGTH	AREA

   b. Graph the equation for area. Explain what the vertex, W-intercepts, A-intercept mean in terms of the area of the pasture.

   c. What are the dimensions of the pasture that gives the maximum area?

Factoring Homework

1. What does it mean for an algebraic expression to be factored?

2. Which expression is factored?
   a. $2x^2 + 8x$ or $2x(x + 4)$
   b. $(x - 3)(x + 1)$ or $x^2 - 2x - 3$

3. Describe the mistake.
   a. $6x^2 + 3x = 3x(2x)$ b. $x^2 - 9 = x - 3$

4. Factor

   a. $6x + 9$ b. $14x - 7$ c. $-10x + 50$

   d. $8x^2 - 6x$ e. $27x^2 - 18x$ f. $15x^3 - 5x^2$

   g. $x^2 + 7x + 6$ h. $x^2 - 8x + 15$ i. $x^2 - 6x + 9$

   j. $x^2 - 3x - 18$ k. $x^2 + 13x + 30$ l. $x^2 - 15x + 50$

   m. $x^2 - 100$ n. $x^2 - 36$ o. $x^2 - 81$

Review for Test IV

1. Titanic Yacht Company makes yachts. The cost of making x yachts per month is

    $$C = x^2 - 32x + 12$$

    and the revenue from selling x yachts per month is

    $$R = -x^2 + 8x.$$

    C and R are in thousands of dollars.

    a. Find the equation for profit. (Profit = Revenue - Cost)

    b. Graph the profit equation. Explain what the vertex, x-intercepts and p-intercept mean in terms of making yachts. Make sure you label the axes and use an appropriate scale.

    c. Suppose the Titanic Yacht Company needs $35,000 in profits (P = 35). Graph this line on the graph above and find out where the line intersects the graph. Explain what the answers mean.

2. A rabbit breeder has 80 meters of chicken wire. He wants to form two rectangular hutches for his rabbits. One side of the hutch will be against a wall. (See the diagram below.)

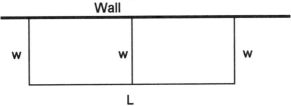

a. Find a formula for the area of the rabbit hutches. (Make a table.)

b. Graph the equation for area. Explain what the vertex, w-intercepts, a-intercept and vertex mean in terms of the area of the rabbit hutch.

c. Suppose he reads in *Breeding Rabbits Magazine* that he should have an area of 477 feet. (A = 477) Represent this idea in the graph of part b. Find where the line intersects the graph and explain what the solutions represent.

3. Simplify. $3(x^2+4x-8)-2(3x^2-5x+3)$

4. Multiply.

   a. $(3x+5)(x-8)$ b. $(x-5)^2$

5. Factor

   a. $4x^2+8x$ b. $x^2+7x-18$

6. Solve. $x^2-8x+12=0$

Review for Final Exam
MAT 011

Name: _____

I. Find an algebraic equation for each situation. (A table may be helpful.)

1. Two students want to open up a gift wrapping store in the mall. They spend $800 on wrapping paper, bows and ribbons, and they plan on charging $1.25 per package. Find an equation for the amount of money they will earn.

2. There are 400 bacteria growing in a culture. The culture grows at 8% a day. Find a formula for the number of bacteria *n* days later.

3. A farmer wants to build a rectangular pigsty off of his barn. He has 150 feet of fencing. Find a formula for the area of the pigsty.

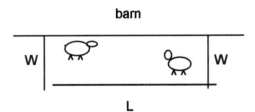

II. Solve each problem.

4. The following equation represents the average time of people commuting to work on public transportation. T is the time it takes to get to work and D is the distance from home to work.

   $T = 2.9D - .3$ (T is in minutes, and D is in miles)

   a. Find out how long it will take you to get to work if you live 15 miles from work.

   b. Find out how far away you live from work if it takes you 30 minutes to get to work.

5. A math teacher, Dr. Kaczynski computes a student's grade for the course as follows:

   10% for homework
   65% for the average of 4 tests
   25% for the final exam

   a. Compute Sue's grade for the course if she has a 72 on the homework; 86 for the test average and a 78 on the final exam.

   b. Suppose Salina has an 82 homework average and a 67 test average. What does Salina have to get on the final exam to get a 70 average for the course?

6. The equation

    $S = 23t + 71$

    represents the number of stores, S, Q-Mart has opened at the end of *t* years, where *t* is the number of years since 1985.

    a. When will Q-Mart have more than 278 stores? (You must set up an inequality.)

    b. When will Q-Mart have between 232 and 416 stores? (You must set up an inequality.)

7. The height of an arrow shot straight up into the air from 6 feet above the ground with an initial velocity of 95 feet per second has the equation:
    $$h = -16t^2 + 95t + 6$$
    where h is the height in feet at *t* seconds.

    a. How high is the arrow at 3 seconds?

    b. When will the arrow be 100 feet above the ground?

8. Use the formula to fund the monthly payment of a loan.

$$P = A\left[\frac{i}{1-(i+1)^{-n}}\right]$$

P is the monthly payment
A is the amount of the loan
n is the number of payments
i is the interest rate per month

a. Find the monthly payments of a 36-month auto loan of $3,000 at 9% annual interest. (Hint: $i = .09/12$)

b. I can only afford $150 a month car payment. How expensive a car can I buy if my loan is for 36 months and the annual interest rate is 15%? (Hint: $i = .15/12$)

III. Graphs

9. Graph the following equation by finding the x and y intercepts. Choose an appropriate scale and label the axis.

$$y = .2x + 41$$

10. Graph: $y = -8$

11. Graph: $x = 3$

12. The graph below shows the temperature during a winter day in Bismarck, North Dakota.

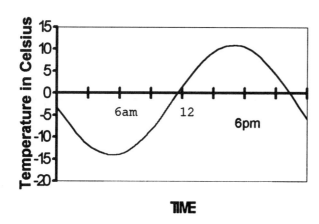

b. When was the temperature 0°?

b. What was the high temperature for the day?

c. When was the temperature rising?

13. You have the choice of two phone companies, Ringer and Buzz. Ringer charges 35¢ a phone call and 10¢ a minute while Buzz charges 15¢ a phone call and 14¢ a minute. The equation for the cost of making a phone call using Ringer is

   C = 10m + 35 cents.

   The equation for the cost of making a phone call using Buzz is

   C = 14m + 15 cents.

   a. Graph both equations on the same set of axes. Label your axes and choose an appropriate scale. Only graph the portion that is relevant to the problem.

   b. Use the equations to find where the two lines intersect. Label this point on the graph.

   c. What is the c intercept of the Ringer equation? What does it mean in terms of the problem?

   d. What is the slope of the Ringer equation? What does it mean in terms of the problem?

14. AIE makes printing presses. The cost of making x presses in a month is

$$C = .5x^2 - 10x + 60$$

and the revenue from selling x printing presses a month is

$$R = -.5x^2 + 14x$$

C and R are in thousands of dollars.

a. Find the equation for profit. (P = R - C)

b. Graph the profit equation. Explain what the vertex, x intercepts and p intercept mean in terms of making printing presses. Make sure your label the axes and use an appropriate scale.

15. The table below gives the population of the U.S. from 1800 to 1810.

YEAR	POPULATION IN MILLIONS
1800	5.3
1810	7.5

    a.   Calculate the percent increase.

    b.   Calculate the average rate of change (slope).

IV. Algebra

16. Solve the equations.

    a. $3x + 7 = 4(2x - 8)$ b. $\dfrac{x}{3} - \dfrac{5}{6} = 2$

17. Factor

    a. $8x^2 + 16x$ b. $x^2 - 4x - 21$

18. Multiply: $(x + 7)^2$

19. Simplify using properties of exponents.

    a. $x^5 \cdot x^2$

    b. $(3x^4)^2$

    c. $\dfrac{x^3}{x^6}$

    d. $\dfrac{9x^{-3}}{x^2}$

20. Perform the operation.

    a. $\dfrac{x^2}{8} \div \dfrac{x^3}{2}$

    b. $\dfrac{x+5}{3} - \dfrac{x-2}{2}$

21. Simplify: $3(x^2 - 4x + 5) - 5(2x^2 + 8x - 4)$

22. Write in Scientific Notation.

    4,300,000

23. Write in Decimal Notation.

    2.4×10^{-4}

24. 3x + 5y = 10, solve for y.

ANSWERS TO HOMEWORK AND REVIEW TESTS

Signed Numbers

1. $-20 2. $30 3. $-290 4. $10 5. $-58 8. a. 2 c. −9 e. −7
g. −29 i. −32 k. −14 m. −8 11. a. −4.4 c. .59 e. −18.2 14. zero
15. a. −42 c. −15 e. −6 g. 0 i. 2 h. −40 m. −96 16. a. 26.3796
c. −2.953 18. a. -10 c. 1 e. −31 g. −4 i. .5 19. a. −9 c. −3
20. 7.67

Introduction to Variables

1.

Width	Length	Area	Perimeter
9 in.	12 in.	108 in²	42 in.
8 in.	10 in.	80 in.²	36. in.
11 in.	12 in.	209 in.²	60 in.

2. a. 75.36 sq. in., 5 inch pizza is cheaper. It costs .0757 per. sq. in.

3. a. 0°C c. 37°C

4.a

Table	Calculation	Wages
15	7.50(15) + 80	$155
20	7.50(20) + 80	$192.5
t	7.50t + 80	W

b. $W = 7.50t + 80$
c. 350
d. 18 tables
e. $W = 8.50t + 80$
f. $W = 7.00t + 95$

5.a

Hours	Calculation	Income
40	8(40) − 160	160
60	8(60) − 160	320
80t	8(80) − 160	480
h	8h − 160	I

b. $I = 8h - 160$
c. 55 hours
d. $I = 9h - 200$

6.a

Calls	Calculation	Phone Bill
10	.13(10) + 7.46	8.76
15	.13(15) + 7.46	9.41
20	.13(20) + 7.46	10.06
C	.13C + 7.46	B_r

b. $B_r = .13C + 7.46$
c. $B_B = .17C + 6.17$
e. about 32

7.a

Base	Height	Area
5	10 − 5	½(5)(5) = 12.5
8	10 − 8	½(8)(2) = 8
b	10 − b	½(b)(10 − b)

b. $A = ½b(10 − 6)$
c. 5

Simplifying Algebraic Expressions

4.
   a. 5x
   b. 17x – 3
   c. $\frac{7}{6}x - \frac{3}{5}$
   d. 6x + 15
   e. -12x + 38
   f. 2x + 1
   g. 19x + 5
   h. -x – 16
   i. -7x + 4

5.
   a. Multiplied the coefficients
   b. Combined unlike terms
   c. Did not multiply the 6 and the –5
   d. -3 * -4 = 12
   e. A negative times a negative is a positive
   f. 12x – 3x = 9x

Solving Equations – Part I

1.
   a. Subtract 7 from both sides.
      Combine like terms.
      Divide both sides by 6.
      Simplify.
   b. Add 4 to both sides.
      Combine like terms.
      Multiply both sides by 5.
      Simplify.

2.
   a. 4 + 5, should be 4 – 5.
   b. Multiply both sides by $\frac{3}{2}$, not $\frac{2}{3}$.

3.
   a. yes
   b. yes
   c. no

5.
   a. x = 3
   b. x = $\frac{-5}{2}$
   c. x = -40
   d. x = $\frac{63}{2}$ or 31.5
   e. x = 2
   f. x = 36
   g. x = 6
   h. x = -49
   i. x = -5

6. The printing press will be worth 10,000 in 24 years.

Solving Equations - Part II

1.
   a. Distributive Property
      Combine like terms
      Subtract 3X from both sides
      Combine like terms
      Subtract 20 from both sides
      Combine like terms
      Divide both sides by –6
      Simplify
   b. Distributive Property
      Combine like terms
      Subtract 7X from both sides
      Combine like terms
      Conclusion

2.
   a. 2 * – 8 = -16
   b. 7 – 2 should be 7 + 2

6.
   a. x = 5
   b. x = 4
   c. no solution

Applications of Linear Equations (Part I)

1. b. C = .16M + 25   c. C = 29.96   d. M = 37.5   e. M = 250
2. b. C = .15m + .80   c. C = 5.60   d. M = 7   e. M = 11.33
3. b. C_R = .08M + .5 C_B = .10M + .25   c. M = 12.5
4. b. C_A = .075S + 5000 C_B = .05S + 9000   S = 160,000
5. b. W = .15S – 150   c. W = 1350   d. S = 2800   e. S = 23,480
6. b. W = 7h – 344   c. h = 80   d. h = 80.57
7. b. C = .1M - .28   c. M = 23   d. M = 40.9

Applications of Linear Equations (Part II)

1. a. 22.2   b. 51.8   c. 21.2   d. 145.4
   e. $F = \dfrac{9C}{5} + 32$ or $\dfrac{9}{5}\left(C + \dfrac{160}{9}\right)$

2. a. t = 11.47, 1994   b. t = 12.3, 1994 or 1995   c. $t = \dfrac{s - 146}{118}$

3. a. $L = \dfrac{P - 2W}{2}$ b. $r = \dfrac{D}{t}$ c. Y = ¾x – 3
   d. $Y = \dfrac{9 - 6x}{3}$ or Y = 3 – 2x

Percentages

1. a. 53%   b. 75 girls
2. a.
Harry	Mary	Pat
3.85	4.00	3.75
4.31	4.48	4.20

   b. N = 1.12 OLD

3.
OLD	Discount	Sales
45	14.4	30.60
X	.32X	.68X
55	17.6	37.40

4. a. 82.75   b. 88.4   c. 92.6
5. a. 90   b. 159

Review for Test I

1. b. C = 20d + 350   c. d = 90
2. b. C = 1.8M – 4.05   c. M = 17.53
3. a. 81.8   b. 64
4. a. 1.40   b. 41.41
5. b. N = 1.125 OLD   c. 4.69
6. b. SP = .8 OP   c. SP = 64.8
7. a. –3.2   b. 4.6
8. -4X + 14
9. Yes

10. a. −3   b. 13   11. 250   12. $W = \dfrac{P - 2L}{2}$

Inequalities

1. a. Subtract 6 from both sides
       Subtract x from both sides
       Divide both sides by −4 and change the direction of the inequality
    b. Subtract 6 from all three parts
       Divide all three parts by −2 and change the directions of the inequalities
2. a. Didn't change the direction of the inequality
    b. Didn't subtract 9 from 12
3. a. yes   b. yes   c. no
5. a. x<−9   b. x<20   c. x≥8
    d. x≤−3   e. x<6.33   f. x≤9
    g. −6<x<4 h. 5.5≥x≥−3 i. 2<x<6.5
    j. −36≤x≤30 k. 21>x>−24

Applications of Inequalities

1. a. 46.4   b. 28.4 < F < 53.6
2. C < 250,000
3. a. 247.79   b. t > 17.09, after 1997
    c. 18.61 < t < 31.60, 1999 < Year < 2001
4. b. h > 47.05, 23.5 ≤ h ≤ 82.35
5. b. C_W = .50M + 28, C_E = .75M + 20   c. M < 32
6. b. W_A = .12(S − 50,000) + 8000, W_B = .05S + 15000   c. S > 185,714
7. C_W = .11(M-4) + .15, C_D = .14(M-5) + .10, M < 10.3

Scatter Plots

2. Averages: 9.3, 17.3, 19.6, 11, 5, 7, 12.6, 15.6, 13.6, 5
3. b. (0, 1.50)   c. (2, 42)   d. (.5, 20) (5.5, 20)   e. (3, 47)   f. (6.5, 0)
    g. 2 < t < 4   h. (1.5, 35)   i. (4.5, 35)   j. 35   k. always   l. 40.5
    m. 8   n. 38   o. between 0 and 1 sec.

Interpreting Graphs

1. a. −14°, 11°   c. 8:30, 9:30   e. 3:00   g. 3 AM to 3 PM
2. a. 8500   c. 9000   e. Jan. to April and July to Dec.
3. a. 5 min 30 sec   b. WRONG #   c. Buzz   d. Min < 5.5

Graphing Lines by Plotting Points

1. a. (Hours,Cost) (20,25) (50,40) (100,65)
    b. C = .50H + 15
2. a. (Hours,Cost)
    b. C = .35M + 15
3. Switched the x and y coordinates when graphing

4. Yes, No, Yes, Yes

Graphing Lines by finding the Intercepts

1. a. (73,0)   b. (0,-1095)
2. Did not graph the intercepts
4. a. (-18.33,0)   (0,11)
    b. (9.09,0)   (0,.0806)
    c. (4,0)   (0,-600)
    d. (-6.66,0)   (0,-5)

Introduction to Slope

1. Acme = $\dfrac{1.89}{12}$, F.S. = $\dfrac{2.68}{18}$
3. b. 1185.75 emp. per yr.   c. 717 emp. per yr.   d. 19000   e. 35.6%
4. b. –14.58 papers per yr.   c. –20.5 papers per yr.   d. –2.5%
5. a. 1940 to 1950   b. -.15   f. 27.5
6. b. Dr. bills: $\dfrac{22.5-11}{5}$ = 2.3 billion per yr., Medicare: $\dfrac{11-4.5}{5}$ = 1.3 billion per yr.

Slope

1. a. $M = \dfrac{5}{2}$ b. $M = \dfrac{-3}{7}$ c. $M = 0$
2. a. should be $\dfrac{1-3}{-6-(-2)}$ b. they did x over y
3. a. M = -.5   b. M = 0   c. M = .8
    d. undefined   e. M = .8
4. a. M = 7, (0,5)   b. M = -2/3, (0,-2)

Applications of Graphs

1. a. h = 1.7d + 4   c. 27.8   d. 106   e. 208   f. 30   g. 33
2. a. Lemmon: C = 35, Limo: C = .80M + 15   c. M = 25, C = 35   d. $0 \leq M < 25$
    e. M > 25   f. Cost if you don't drive anywhere   g. slope is .80 for Limo and 0 for Lemmon
3. a. C_R = .18M + .50, C_B = .27M + .25   c. M = 27, C = 5.36
    d. $0 \leq M < 2.7$ e. M > 2.7   f. initial charge for the phone call
    g.   slope for ringer is .18 and the slope for Buzz is .27
4. a. W_R = .08 S + 5000, W_E = .04 S + 10000   c. S = 125,000, W = 15000
    d. S > 125,000   e. $0 \leq S < 125000$ f. base pay without any sales
    g.   slope for Ratco is .08 and the slope for EXPLOITCO is .04
5. a. C = .20(M-3) + .50   b. M < 3   c. C = .50
6. a. W = .10(S-50000) + 10000   b. S < 50000   c. C = .50, W = 10,000

Review for Test II

1. b. 20   c. 65   d. 45 min. to 62 min.
2. a. horizontal line   b. vertical line
3. a. (0,210) (-2100,0)   b. (0,2)(19,0)   c. (0,2.85)(5,0)
   d. (0,8)(-6,0)
4. a. -12°   b. 5°   c. 10 AM, midnight   g. 3 AM TO 3 PM
5. a. 4.025   b. 25.8%   c. M = 4.025
6. a. $x \le 1.25$ b. $x \le -1$ c. $1.71 \le x 3.57$
7. a. M < 3461   b. 1538 < M < 7307
8. a. M = -3, (0,7)   b. M = .15 (0,-6)
9. a. slope is .1, commission   b. $20,000, y intercept
10. Limo: C = 32, Lemmon: C = .26M + 19   c. M = 50, C = 32   d. $o \le M$ 50
    e. M > 50   f. Rental charge without any miles   g. slope of Limo is 0, Slope of Lemmon is .26

Introduction to Exponents

1. a. 7776
2. a. 36   d. 65.61   g. even   h. even, positive
3. a. −67.24   c. −6.25   e. −216   g. −16807
4. b. $p = 1000(1.05)^n$ c. 2653,7040   d. 14.2
5. b. $p = 1,052,000 (.965)^n$ c. 736,697, 594,912   d. 4.4
6. a. $4926.80   b. $6336.19   c. $9096.98   d. $6112.24   e. $2241.74   f. 12.1 years

Negative Exponents and Scientific Notation

1. a. .00195   d. -.01235   e. .01234   h. 3125
2. a. 2.356×10^{13} b. 4.35×10^{-8}
3. a. $86.66   b. $9032.25   c. $804.62   d. $104,640.59, $136,283.49
4. a. 7.8×10^4 b. 1.67×10^{-6} c. 6.35×10^6
5. a. 786,000,000   b. .0000000008673   c. .033   d. 20,320
6. 9,460,000,000,000
7. 9.460×10^{16} Kilometers
8. $1093.81

Properties of Exponents

3. a. $8x^8$ b. $2x$ c. $8x^6$ d. $\dfrac{625}{x^{12}}$ e. $\dfrac{6}{x}$ f. $\dfrac{5x^3}{2}$

   g. $\dfrac{x^3}{2}$ h. $\dfrac{1}{9x^2}$ i. 1   j. $\dfrac{216}{x^6}$ k. $\dfrac{8}{x^{12}}$ l. $\dfrac{1}{49x^{10}}$

Introduction to Algebraic Fractions

1. a. 6800   b. 22800   c. 118800   d. undefined
2. a. 33200   b. 74700   c. 157700   d. undefined

3. a. $\dfrac{3}{2}$ b. $\dfrac{9}{5}$ c. $\dfrac{1}{4x}$ d. $\dfrac{2y^3}{5x}$ e. $-\dfrac{1}{4x}$ f. $-\dfrac{x^2}{2}$

4. a. $\dfrac{4}{3}$ b. 4 c. $\dfrac{7}{9y^2}$ d. x e. $\dfrac{1}{x}$ f. $-\dfrac{2x^2}{3}$

Adding and Subtracting Fractions

4. $\dfrac{1}{5x}$ 5. $\dfrac{9-10x}{15x}$ 6. $\dfrac{31m}{21}$ 7. $\dfrac{8y-9}{12y}$ 8. $\dfrac{15-2m^2}{24m^3}$

9. $\dfrac{20y+9}{12y^2}$ 10. $\dfrac{8n^2-15}{36n^4}$ 11. $\dfrac{9+8x}{6x^2}$ 12. $\dfrac{3x^2-4x}{12}$ 13. $\dfrac{1+35x^3}{5x^3}$

Solving Equations with Fractions

5. $x = -1.1\overline{6}$ 6. $x = .\overline{36}$ 7. $x = -1.263$ 8. $x = -.\overline{4}$ 9. $x = .8235$
10. $x = -.4355$ 11. $x = -1.5714$ 12. $x = .6429$

Ratio and Proportion Problems

1. 75 ounces is the better buy.
2. 15 ounces is the better buy.
3. The number of bears in the forest is approximately 36.
4. The car can go 264 miles on 12 gallons.
5. The number of people in Reading without medical insurance is approximately 3,038.
6. It will take about 23 minutes to print 81 pages.
7. One person should get 3.36 million, the second person should get 4.032 million and the third person should get 5.208 million.
8. a. 8/3   b. 12/35

Review for Test III

1. a. x^7 b. $\dfrac{1}{x^2}$ c. 1   d. x^{10} e. $8x^6$ f. $\dfrac{1}{x^5}$
    g. x^7 h. $\dfrac{9}{x^5}$

2. a. .064   b. –167.96   c. 167.96
3. a. 2.15×10^{13} b. 5.022×10^{-13}
4. a. 8.3×10^6 b. 6.14×10^{-4}
5. a. 420000   b. .0031
6. a. $\dfrac{6x^6}{y^5}$ b. $\dfrac{2}{49x^3}$ c. $\dfrac{15x-14}{6x^2}$ d. $\dfrac{33+5x}{15x}$
7. a. $x = 9$ b. $x = -2$
8. b. $p = 500(1.07)^n$ c. 3806   d. 28973   9. a. 1609.25   b. 10,085

Introduction to Quadratics

1. a. 80 b. 704 c. 1104 d. 1040 e. 80 f. –192 g. t = 8
   h. 16.3 secs.
4. a. $-4x^2 + 12x - 3$ b. $-4x^2 + 3x - 20$ c. $-19x^2 + 13x - 7$
   d. $-18x + 18$ e. $-6x^2 - 25x + 14$ f. $-11x^2 + 6x - 15$
5. $-4x^2 + 17x - 72$
6. a. $x^2 + 5x + 6$ b. $x^2 + x - 12$ c. $x^2 - 11x + 28$ d. $x^2 - 64$
   e. $x^2 - 81$ f. $6x^2 + 7x - 3$ g. $x^2 + 6x + 9$ h. $x^2 - 10x + 25$

Applications of Quadratic Formula

2. 80 yards by 100 yards   3. 9.026 sec.   4. 90 or 180
5. a. 5.6%   b. 21.875 = t or 1996
6. 60 or 20

Quadratic Applications and Their Graphs

2. $x = \dfrac{-b}{2a}$ 4. x intercepts, y intercepts, vertex
5. a. $P = -x^2 + 36x - 87$ b. vertex (18,237), y intercept (0, -87), x intercepts (2.605, 0)(33.39, 0)   c. (3.964, 40), (32.04, 40)
6. a. $P = -7x^2 + 77x - 35$ b. vertex (5.5, 176.75), y intercept (0, -35), x intercepts (.475, 0)(10.52, 0)   c. (9.217, 80) or (1.78, 80)
7. a. vertex (6.25, 625), y intercept (0, 0), x intercepts (0, 0), (12.5, 0)
   b. (1.75, 300) and (10.75, 300)
8. a. $A = x(1800 - 2x) = -2x^2 + 1800x$ b. vertex (450, 405000) intercepts (0, 0) and (900, 0)   b. 450 by 900

Factoring

2. a. $2x(x + 4)$ b. $(x - 3)(x + 1)$
4. a. $3(2x + 3)$ c. $-10(x - 5)$
   e. $9x(3x - 2)$ g. $(x + 6)(x + 1)$
   i. $(x - 3)(x - 3)$ k. $(x + 10)(x + 3)$
   m. $(x - 10)(x + 10)$ o. $(x - 9)(x + 9)$

Review for Test IV

1. a. $P = 2x^2 + 40x - 12$ b. vertex (10,188) y intercept (0, -12), x intercepts (.305, 0) and (19.70), 0)   c. (1.25, 30)(18.75, 30)
2. a. $A = W(80-3W) = -3W^2 + 80W$ b. vertex (13.3, 533.3), intercepts (0, 0) (26.67, 0)
   c. (9,477)(17.67, 477)
3. $-3x^2 + 22x - 30$
4. a. $3x^2 - 19x - 40$ b. $x^2 - 10x + 25$
5. a. $4x(x+2)$ b. $(x+9)(x-2)$
6. x = 6, 2

Review for Final Exam

1. $S = 1.25p - 800$ 2. $p = 400(1.08)^n$ 3. $A = W(150-2W) = -2W^2 + 150W$
4. a. 43.2 min. b. 10.44 min.
5. a. 82.6   b. 73

6.   a. t = 9, 1994   b. 7 < t < 15 or 1992 < years < 2000
7.   a. 147 ft   b. 1.25 or 4.68
8.   a. P = $95.40   b. A = 4327.09
9.   (0, 41) (-205, 0)
10.   horizontal line
11.   vertical line
12.   a. 11:30 AM and 10:30 PM   b. 10°   c. 6 AM until 5:00 PM
13.   b. (5, 85)   c. (0, 35), cost to place the call   d. slope is 10, cost per minute
14.   a. $P = -x^2 + 24x - 60$ b. vertex (12, 84) x intercepts (2.83, 0) (21.15, 0), y intercept (0, -60)
15.   a. 41.5%   b. .22 million per year
16.   a. x = 7.8   b. x = 8.5
17.   a. 8x(x+2)   b. (x-7)(x+3)
18. $x^2 + 14x + 49$
19.   a. x^7 b. $9x^8$ c. $\dfrac{1}{x^3}$ d. $\dfrac{9}{x^5}$
20.   a. $\dfrac{1}{4x}$ b. $\dfrac{-x+16}{6}$
21. $-7x^2 - 52x + 35$
22. 4.3×10^6
23.   .00024
24. $y = \dfrac{10 - 3x}{5}$

CHAPTER

1

REVIEW OF REAL NUMBERS

SYMBOLS AND SETS OF NUMBERS

FRACTIONS

EXPONENTS AND ORDER OF OPERATIONS

INTRODUCTION TO VARIABLE EXPRESSIONS AND EQUATIONS

ADDING REAL NUMBERS

SUBTRACTING REAL NUMBERS

MULTIPLYING AND DIVIDING REAL NUMBERS

PROPERTIES OF REAL NUMBERS

READING GRAPHS

CREATING AND INTERPRETING GRAPHS

Companies often rely on market research to investigate the types of consumers who buy their products and their competitors' products. One way of gathering this type of data is through the use of surveys. The raw data collected from such surveys can be difficult to interpret without some kind of organization. Graphs serve to organize data visually. They also enable a user to interpret the data represented by the graph quickly.

IN THE CHAPTER GROUP ACTIVITY ON PAGE 62, YOU WILL HAVE THE OPPORTUNITY TO CONDUCT A BRIEF SURVEY ON BREAKFAST CEREAL CONSUMPTION AND CREATE GRAPHS TO REPRESENT THE RESULTS.

The power of mathematics is its flexibility. We apply numbers to almost every aspect of our lives, from an ordinary trip to the grocery store to a rocket launched into space. The power of algebra is its generality. Using letters to represent numbers, we tie together the trip to the grocery store and the launched rocket.

In this chapter we review the basic symbols and words—the language—of arithmetic and introduce using variables in place of numbers. This is our starting place in the study of algebra.

1.1 Symbols and Sets of Numbers

TAPE BA 1.1

OBJECTIVES

1. Identify natural and whole numbers, and picture them on a number line.
2. Define the meaning of the symbols =, ≠, <, >, ≤, and ≥.
3. Translate sentences into mathematical statements.
4. Identify integers, rational numbers, irrational numbers, and real numbers.
5. Find the absolute value of a real number.

1 We begin with a review of the set of natural numbers and the set of whole numbers and how we use symbols to compare these numbers. A **set** is a collection of objects, each of which is called a **member** or **element** of the set. A pair of brace symbols { } encloses the list of elements and is translated as "the set of" or "the set containing."

NATURAL NUMBERS

The set of **natural numbers** is {1, 2, 3, 4, 5, 6, ...}.

WHOLE NUMBERS

The set of **whole numbers** is {0, 1, 2, 3, 4, ...}.

The three dots (an ellipsis) at the end of the list of elements of a set means that the list continues in the same manner indefinitely.

These numbers can be pictured on a **number line.** We will use the number line often to help us visualize objects and relationships. Visualizing mathematical concepts is an important skill and tool, and later we will develop and explore other visualizing tools.

To draw a number line, first draw a line. Choose a point on the line and label it 0. To the right of 0, label any other point 1. Being careful to use the same distance

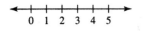

as from 0 to 1, mark off equally spaced distances. Label these points 2, 3, 4, 5, and so on. Since the whole numbers continue indefinitely, it is not possible to show every whole number on the number line. The arrow at the right end of the line indicates that the pattern continues indefinitely.

Picturing whole numbers on a number line helps us to see the order of the numbers. Symbols can be used to describe concisely in writing the order that we see.

The **equal symbol** = means "is equal to."
The symbol ≠ means "is not equal to."

These symbols may be used to form a **mathematical statement.** The statement might be true or it might be false. The two statements below are both true.

$2 = 2$ states that "two is equal to two"

$2 \neq 6$ states that "two is not equal to six"

If two numbers are not equal, then one number is larger than the other. The symbol > means "is greater than." The symbol < means "is less than." For example,

$2 > 0$ states that "two is greater than zero"

$3 < 5$ states that "three is less than five"

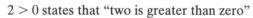

On the number line, we see that a number **to the right of** another number is **larger.** Similarly, a number **to the left of** another number is smaller. For example, 3 is to the left of 5 on the number line, which means that 3 is less than 5, or $3 < 5$. Similarly, 2 is to the right of 0 on the number line, which means 2 is greater than 0, or $2 > 0$. Since 0 is to the left of 2, we can also say that 0 is less than 2, or $0 < 2$.

The symbols ≠, <, and > are called **inequality symbols.**

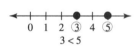

REMINDER Notice that $2 > 0$ has exactly the same meaning as $0 < 2$. Switching the order of the numbers and reversing the "direction of the inequality symbol" does not change the meaning of the statement.

$5 > 3$ has the same meaning as $3 < 5$.

Also notice that, when the statement is true, the inequality arrow points to the smaller number.

EXAMPLE 1 Insert <, >, or = in the space between the paired numbers to make each statement true.

a. 2   3   **b.** 7   4   **c.** 72   27

Solution: **a.** $2 < 3$ since 2 is to the left of 3 on the number line.
b. $7 > 4$ since 7 is to the right of 4 on the number line.
c. $72 > 27$ since 72 is to the right of 27 on the number line.

4 CHAPTER 1 REVIEW OF REAL NUMBERS

Two other symbols are used to compare numbers. The symbol ≤ means "is less than or equal to." The symbol ≥ means "is greater than or equal to." For example,

$7 \leq 10$ states that "seven is less than or equal to ten"

This statement is true since $7 < 10$ is true. If either $7 < 10$ or $7 = 10$ is true, then $7 \leq 10$ is true.

$3 \geq 3$ states that "three is greater than or equal to three"

This statement is true since $3 = 3$ is true. If either $3 > 3$ or $3 = 3$ is true, then $3 \geq 3$ is true.

The statement $6 \geq 10$ is false since neither $6 > 10$ nor $6 = 10$ is true.

The symbols ≤ and ≥ are also called **inequality symbols.**

EXAMPLE 2 Tell whether each statement is true or false.

　　a. $8 \geq 8$　　　b. $8 \leq 8$　　　c. $23 \leq 0$　　　d. $23 \geq 0$

Solution:　a. True, since $8 = 8$ is true.　　　b. True, since $8 = 8$ is true.
　　c. False, since neither $23 < 0$ nor $23 = 0$ is true.　　d. True, since $23 > 0$ is true.

Now, let's use the symbols discussed above to translate sentences into mathematical statements.

EXAMPLE 3 Translate each sentence into a mathematical statement.

　　a. Nine is less than or equal to eleven.
　　b. Eight is greater than one.
　　c. Three is not equal to four.

Solution:　a.

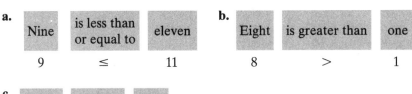

　　c.

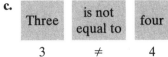

Whole numbers are not sufficient to describe many situations in the real world. For example, quantities smaller than zero must sometimes be represented, such as temperatures less than 0 degrees.

We can picture numbers less than zero on the number line as follows:

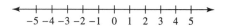

Numbers less than 0 are to the left of 0 and are labeled −1, −2, −3, and so on. A − sign, such as the one in −1, tells us that the number is to the left of 0 on the number line. In words, −1 is read "negative one." A + sign or no sign tells us that a number lies to the right of 0 on the number line. For example, 3 and +3 both mean positive three.

The numbers we have pictured are called the set of **integers.** Integers to the left of 0 are called **negative integers;** integers to the right of 0 are called **positive integers.** The integer 0 is neither positive nor negative.

> **INTEGERS**
> The set of **integers** is $\{\ldots, -3, -2, -1, 0, 1, 2, 3, \ldots\}$.

Notice the ellipses (three dots) to the left and to the right of the list for the integers. This indicates that the positive integers and the negative integers continue indefinitely.

A problem with integers in real-life settings arises when quantities are smaller than some integer but greater than the next smallest integer. On the number line, these quantities may be visualized by points between integers. Some of these quantities between integers can be represented as a quotient of integers. For example,

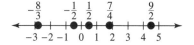

The point on the number line halfway between 0 and 1 can be represented by $\frac{1}{2}$, a quotient of integers.

The point on the number line halfway between 0 and −1 can be represented by $-\frac{1}{2}$. Other quotients of integers and their graphs are shown.

These numbers, each of which can be represented as a quotient of integers, are examples of **rational numbers.** Notice that every integer is also a rational number since each integer can be expressed as a quotient of integers. For example, the integer 5 is also a rational number since $5 = \frac{5}{1}$. In the rational number $\frac{5}{1}$, the top number, 5, is called the numerator and the bottom number, 1, is called the denominator.

> **RATIONAL NUMBERS**
> The set of **rational numbers** is the set of all numbers that can be expressed as a quotient of integers, with denominator not equal to 0.

The set of numbers, each of which corresponds to a point on the number line, is called the set of **real numbers.** One and only one point on the number line corresponds to each real number.

REAL NUMBERS

The set of **real numbers** is the set of all numbers each of which corresponds to a point on the number line.

Some real numbers cannot be expressed as quotients of integers. These numbers are called **irrational numbers** because they cannot be represented by rational numbers. For example, $\sqrt{2}$ and π are irrational numbers.

IRRATIONAL NUMBERS

The set of **irrational numbers** is the set of all real numbers that are not rational numbers. That is, an irrational number is a number that cannot be expressed as a quotient of integers.

Rational numbers and irrational numbers can be written as decimal numbers. The decimal equivalent of a rational number will either terminate or repeat in a pattern. For example, upon dividing we find that

$\dfrac{3}{4} = 0.75$ (decimal number terminates or ends) and

$\dfrac{2}{3} = 0.66666\ldots$ (decimal number repeats in a pattern)

The decimal representation of an irrational number will neither terminate nor repeat. (For further review of decimals, see the appendix.)

On the following number line, we see that real numbers can be positive, negative, or 0. Numbers to the left of 0 are called **negative numbers;** numbers to the right of 0 are called **positive numbers.** Positive and negative numbers are also called **signed numbers.**

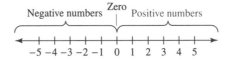

Several different sets of numbers have been discussed in this section. The following diagram shows the relationships among these sets of real numbers.

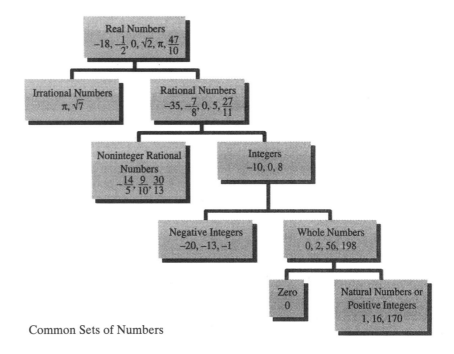

Common Sets of Numbers

EXAMPLE 4 Given the set $\left\{-2, 0, \frac{1}{4}, 112, -3, 11, \sqrt{2}\right\}$, list the numbers in this set that belong to the set of:

   a. Natural numbers **b.** Whole numbers **c.** Integers
   d. Rational numbers **e.** Irrational numbers **f.** Real numbers

Solution: **a.** The natural numbers are 11 and 112.

   b. The whole numbers are 0, 11, and 112.

   c. The integers are -3, -2, 0, 11, and 112.

   d. Recall that integers are rational numbers also. The rational numbers are -3, -2, 0, $\frac{1}{4}$, 11, and 112.

   e. The irrational number is $\sqrt{2}$.

   f. The real numbers are all numbers in the given set.

We can now extend the meaning and use of inequality symbols such as $<$ and $>$ to apply to all real numbers.

ORDER PROPERTY FOR REAL NUMBERS

Given any two real numbers a and b, $a < b$ if a is to the left of b on the number line. Similarly, $a > b$ if a is to the right of b on the number line.

8 CHAPTER 1 REVIEW OF REAL NUMBERS

EXAMPLE 5 Insert <, >, or = in the space between the paired numbers to make each statement true.

    a. −1   0   **b.** 7 $\frac{14}{2}$ **c.** −5   −6

Solution: **a.** −1 < 0 since −1 is to the left of 0 on the number line.

     b. 7 = $\frac{14}{2}$ since $\frac{14}{2}$ simplifies to 7.

     c. −5 > −6 since −5 is to the right of −6 on the number line.

The number line not only gives us a picture of the real numbers, it also helps us visualize the distance between numbers. The distance between a real number a and 0 is given a special name called the **absolute value** of a. "The absolute value of a" is written in symbols as $|a|$.

> **ABSOLUTE VALUE**
>
> The absolute value of a real number a, denoted by $|a|$, is the distance between a and 0 on a number line.

For example, $|3| = 3$ and $|-3| = 3$ since both 3 and −3 are a distance of 3 units from 0 on the number line.

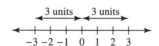

> REMINDER Since $|a|$ is a distance, $|a|$ is always either positive or 0, never negative. That is, **for any real number a, $|a| \geq 0$.**

EXAMPLE 6 Find the absolute value of each number.

     a. $|4|$ **b.** $|-5|$ **c.** $|0|$

Solution: **a.** $|4| = 4$ since 4 is 4 units from 0 on the number line.

     b. $|-5| = 5$ since −5 is 5 units from 0 on the number line.

     c. $|0| = 0$ since 0 is 0 units from 0 on the number line.

EXAMPLE 7 Insert <, >, or = in the appropriate space to make the statement true.

     a. $|0|$ 2   **b.** $|-5|$ 5   **c.** $|-3|$ $|-2|$ **d.** $|5|$ $|6|$ **e.** $|-7|$ $|6|$

Solution:
a. $|0| < 2$ since $|0| = 0$ and $0 < 2$.
b. $|-5| = 5$.
c. $|-3| > |-2|$ since $3 > 2$.
d. $|5| < |6|$ since $5 < 6$.
e. $|-7| > |6|$ since $7 > 6$.

Exercise Set 1.1

Insert $<$, $>$, or $=$ in the space between the paired numbers to make each statement true. See Example 1.

1. 4   10
2. 8   5
3. 7   3
4. 9   15
5. 6.26   6.26
6. 2.13   1.13
7. 0   7
8. 20   0

9. The freezing point of water is 32° Fahrenheit. The boiling point of water is 212° Fahrenheit. Write an inequality statement using $<$ or $>$ comparing the numbers 32 and 212.

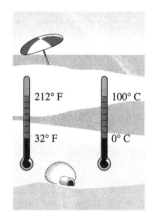

10. The freezing point of water is 0° Celsius. The boiling point of water is 100° Celsius. Write an inequality statement using $<$ or $>$ comparing the numbers 0 and 100.

Are the following statements true or false? See Example 2.

11. $11 \leq 11$
12. $4 \geq 7$
13. $10 > 11$
14. $17 > 16$
15. $3 + 8 \geq 3(8)$
16. $8 \cdot 8 \leq 8 \cdot 7$
17. $7 > 0$
18. $4 < 7$

19. An angle measuring 30° is shown and an angle measuring 45° is shown. Use the inequality symbol $\leq$ or $\geq$ to write a statement comparing the numbers 30 and 45.

20. The sum of the measures of the angles of a triangle is 180°. The sum of the measures of the angles of a parallelogram is 360°. Use the inequality symbol $\leq$ or $\geq$ to write a statement comparing the numbers 360 and 180.

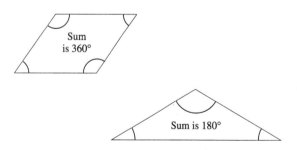

Write each sentence as a mathematical statement. See Example 3.

21. Eight is less than twelve.
22. Fifteen is greater than five.
23. Five is greater than or equal to four.
24. Negative ten is less than or equal to thirty-seven.
25. Fifteen is not equal to negative two.
26. Negative seven is not equal to seven.

The graph below is called a bar graph. This particular graph shows the first three quiz scores for Bill Seggerson in his anatomy class. Each bar represents a different quiz and the height of each bar represents Bill's score for that particular quiz.

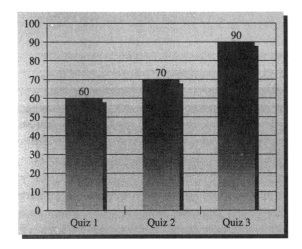

27. What is Bill's highest quiz score?
28. What is Bill's lowest quiz score?
29. Write an inequality statement using ≤ or ≥ comparing the scores for Quiz 2 and Quiz 3.
30. Do you notice any trends shown by this bar graph?

Tell which set or sets each number belongs to: natural numbers, whole numbers, integers, rational numbers, irrational numbers, and real numbers. See Example 4.

31. 0
32. $\frac{1}{4}$
33. -2
34. $-\frac{1}{2}$
35. 6
36. 5
37. $\frac{2}{3}$
38. $\sqrt{3}$

Tell whether each statement is true or false.

39. Every rational number is also an integer.
40. Every negative number is also a rational number.
41. Every natural number is positive.
42. Every rational number is also a real number.
43. 0 is a real number.
44. Every real number is also a rational number.
45. Every whole number is an integer.
46. $\frac{1}{2}$ is an integer.

Insert <, >, or = in the appropriate space to make a true statement. See Examples 5–7.

47. $-10 \quad -100$
48. $-200 \quad -20$
49. $32 \quad 5.2$
50. $7 \quad -7$
51. $\frac{18}{3} < \frac{24}{3}$
52. $\frac{8}{2} = \frac{12}{3}$
53. $-51 \quad -50$
54. $|-20| \quad -200$
55. $|-5| \quad -4$
56. $0 \quad |0|$
57. $|-1| \quad |1|$
58. $\left|\frac{2}{5}\right| \quad \left|-\frac{2}{5}\right|$
59. $|-2| \quad |-3|$
60. $-500 \quad |-50|$
61. $|0| \quad |-8|$
62. $|-12| \quad \frac{24}{2}$

The apparent magnitude of a star is the measure of its brightness as seen by someone on Earth. The smaller the apparent magnitude, the brighter the star. Below, the apparent magnitudes of some stars are listed.

Star	Apparent Magnitude	Star	Apparent Magnitude
Alpha Centauri	0	Spica	0.98
Sirius	-1.46	Rigel	0.12
Vega	0.03	Regulus	1.35
Antares	0.96	Canopus	-0.72
Sun	-26.7	Hadar	0.61

63. The apparent magnitude of the sun is -26.7. The apparent magnitude of the star Alpha Centauri is 0. Write an inequality statement comparing the numbers 0 and -26.7.
64. The apparent magnitude of Antares is 0.96. The apparent magnitude of Spica is 0.98. Write an inequality statement comparing the numbers 0.96 and 0.98.
65. Which is brighter, the sun or Alpha Centauri?
66. Which is dimmer, Antares or Spica?
67. Which star listed is the brightest?
68. Which star listed is the dimmest?

Tell whether each statement is true or false.

69. $5 < 6$ **70.** $7 > 8$
71. $-5 < -6$ **72.** $-7 > -8$
73. $|-5| < |-6|$ **74.** $|-7| > |-8|$
75. $|-5| \geq |5|$ **76.** $|-3| < |0|$
77. $-3 > 2$ **78.** $-5 < 5$
79. $|8| = |-8|$ **80.** $|9| = |-9|$
81. $|0| > |-4|$ **82.** $|0| \leq |0|$

Rewrite the following inequalities so that the inequality symbol points in the opposite direction and the resulting statement has the same meaning as the given one.

83. $25 \geq 20$ **84.** $-13 \leq 13$
85. $0 < 6$ **86.** $5 > 3$
87. $-10 > -12$ **88.** $-4 < -2$

89. In your own words, explain how to find the absolute value of a number.

90. Give an example of a real-life situation that can be described with integers but not with whole numbers.

1.2 FRACTIONS

OBJECTIVES

1. Write fractions in simplest form.
2. Multiply and divide fractions.
3. Add and subtract fractions.

TAPE BA 1.2

A quotient of two integers such as $\frac{2}{9}$ is called a **fraction.** In the fraction $\frac{2}{9}$, the top number, 2, is called the **numerator** and the bottom number, 9, is called the **denominator.**

A fraction may be used to refer to part of a whole. For example, $\frac{2}{9}$ of the circle below is shaded. The denominator 9 tells us how many equal parts the whole circle is divided into and the numerator 2 tells us how many equal parts are shaded.

$\frac{2}{9}$ of the circle is shaded.

To simplify fractions, we can factor the numerator and the denominator. In the statement $3 \cdot 5 = 15$, 3 and 5 are called **factors** and 15 is the **product.** (The raised dot symbol indicates multiplication.)

$$3 \quad \cdot \quad 5 \quad = \quad 15$$
$$\uparrow \qquad \uparrow \qquad \quad \uparrow$$
$$\text{factor} \quad \text{factor} \quad \text{product}$$

12 CHAPTER 1 REVIEW OF REAL NUMBERS

To **factor** 15 means to write it as a product. The number 15 can be factored as $3 \cdot 5$ or as $1 \cdot 15$.

A fraction is said to be **simplified** or in **lowest terms** when the numerator and the denominator have no factors in common other than 1. For example, the fraction $\frac{5}{11}$ is in lowest terms since 5 and 11 have no common factors other than 1.

To help us simplify fractions, we write the numerator and the denominator as a product of **prime numbers.** A prime number is a whole number, other than 1, whose only factors are 1 and itself. The first few prime numbers are

2, 3, 5, 7, 11, 13, 17, 19, 23, 29, and so on.

EXAMPLE 1 Write each of the following numbers as a product of primes.

a. 40 **b.** 63

Solution: **a.** First, write 40 as the product of any two whole numbers.

$$40 = 4 \cdot 10$$

Next, factor each of these numbers. Continue this process until all of the factors are prime numbers.

$$40 = 4 \cdot 10$$
$$= 2 \cdot 2 \cdot 2 \cdot 5$$

All the factors are now prime numbers. Then 40 written as a product of primes is

$$40 = 2 \cdot 2 \cdot 2 \cdot 5$$

b. $63 = 9 \cdot 7$
$$= 3 \cdot 3 \cdot 7$$

To use prime factors to write a fraction in lowest terms, apply the fundamental principle of fractions.

> **FUNDAMENTAL PRINCIPLE OF FRACTIONS**
> If $\frac{a}{b}$ is a fraction and c is a nonzero real number, then
> $$\frac{a \cdot c}{b \cdot c} = \frac{a}{b}$$

EXAMPLE 2 Write each fraction in lowest terms.

a. $\frac{42}{49}$ **b.** $\frac{11}{27}$ **c.** $\frac{88}{20}$

Solution: **a.** Write the numerator and the denominator as products of primes; then apply the fundamental principle to the common factor 7.

$$\frac{42}{49} = \frac{2 \cdot 3 \cdot \boxed{7}}{7 \cdot \boxed{7}} = \frac{2 \cdot 3}{7} = \frac{6}{7}$$

b. $\dfrac{11}{27} = \dfrac{11}{3 \cdot 3 \cdot 3}$

There are no common factors other than 1, so $\frac{11}{27}$ is already in lowest terms.

c. $\dfrac{88}{20} = \dfrac{\boxed{2} \cdot \boxed{2} \cdot 2 \cdot 11}{\boxed{2} \cdot \boxed{2} \cdot 5} = \dfrac{22}{5}$

2 To multiply two fractions, multiply numerator times numerator to obtain the numerator of the product; multiply denominator times denominator to obtain the denominator of the product.

> **MULTIPLYING FRACTIONS**
>
> $$\frac{a}{b} \cdot \frac{c}{d} = \frac{a \cdot c}{b \cdot d}, \quad \text{if } b \neq 0 \text{ and } d \neq 0$$

EXAMPLE 3 Find the product of $\dfrac{2}{15}$ and $\dfrac{5}{13}$. Write the product in lowest terms.

Solution: $\dfrac{2}{15} \cdot \dfrac{5}{13} = \dfrac{2 \cdot 5}{15 \cdot 13}$ Multiply numerators.
Multiply denominators.

Next, simplify the product by dividing the numerator and the denominator by any common factors.

$$= \frac{2 \cdot \boxed{5}}{3 \cdot \boxed{5} \cdot 13}$$

$$= \frac{2}{39}$$

Before dividing fractions, we first define **reciprocals.** Two fractions are reciprocals of each other if their product is 1. For example $\frac{2}{3}$ and $\frac{3}{2}$ are reciprocals since $\frac{2}{3} \cdot \frac{3}{2} = 1$. Also, the reciprocal of 5 is $\frac{1}{5}$ since $5 \cdot \frac{1}{5} = \frac{5}{1} \cdot \frac{1}{5} = 1$.

To divide fractions, multiply the first fraction by the reciprocal of the second fraction.

> **DIVIDING FRACTIONS**
>
> $$\frac{a}{b} \div \frac{c}{d} = \frac{a}{b} \cdot \frac{d}{c}, \quad \text{if } b \neq 0, d \neq 0, \text{ and } c \neq 0$$

14 CHAPTER 1 REVIEW OF REAL NUMBERS

EXAMPLE 4 Find each quotient. Write all answers in lowest terms.

a. $\dfrac{4}{5} \div \dfrac{5}{16}$ b. $\dfrac{7}{10} \div 14$ c. $\dfrac{3}{8} \div \dfrac{3}{10}$

Solution:

a. $\dfrac{4}{5} \div \dfrac{5}{16} = \dfrac{4}{5} \cdot \dfrac{16}{5} = \dfrac{4 \cdot 16}{5 \cdot 5} = \dfrac{64}{25}$

b. $\dfrac{7}{10} \div 14 = \dfrac{7}{10} \div \dfrac{14}{1} = \dfrac{7}{10} \cdot \dfrac{1}{14} = \dfrac{\boxed{7} \cdot 1}{2 \cdot 5 \cdot 2 \cdot \boxed{7}} = \dfrac{1}{20}.$

c. $\dfrac{3}{8} \div \dfrac{3}{10} = \dfrac{3}{8} \cdot \dfrac{10}{3} = \dfrac{\boxed{3} \cdot \boxed{2} \cdot 5}{\boxed{2} \cdot 2 \cdot 2 \cdot \boxed{3}} = \dfrac{5}{4}$

3 To add or subtract fractions with the same denominator, combine numerators and place the sum or difference over the common denominator.

ADDING AND SUBTRACTING FRACTIONS WITH THE SAME DENOMINATOR

$\dfrac{a}{b} + \dfrac{c}{b} = \dfrac{a+c}{b}$,   if $b \neq 0$

$\dfrac{a}{b} - \dfrac{c}{b} = \dfrac{a-c}{b}$,   if $b \neq 0$

EXAMPLE 5 Add or subtract as indicated. Write each result in lowest terms.

a. $\dfrac{2}{7} + \dfrac{4}{7}$ b. $\dfrac{3}{10} + \dfrac{2}{10}$ c. $\dfrac{9}{7} - \dfrac{2}{7}$ d. $\dfrac{5}{3} - \dfrac{1}{3}$

Solution:

a. $\dfrac{2}{7} + \dfrac{4}{7} = \dfrac{2+4}{7} = \dfrac{6}{7}$

b. $\dfrac{3}{10} + \dfrac{2}{10} = \dfrac{3+2}{10} = \dfrac{5}{10} = \dfrac{\boxed{5}}{2 \cdot \boxed{5}} = \dfrac{1}{2}$

c. $\dfrac{9}{7} - \dfrac{2}{7} = \dfrac{9-2}{7} = \dfrac{\boxed{7}}{\boxed{7}} = 1$

d. $\dfrac{5}{3} - \dfrac{1}{3} = \dfrac{5-1}{3} = \dfrac{4}{3}$

To add or subtract fractions without the same denominator, first write the fractions as **equivalent fractions** with a common denominator. Equivalent fractions are fractions that represent the same quantity. For example, $\dfrac{3}{4}$ and $\dfrac{12}{16}$ are equivalent fractions since they represent the same portion of a whole, as the diagram shows. Count the larger squares and the shaded portion is $\frac{3}{4}$. Count the smaller squares and the shaded portion is $\frac{12}{16}$. Thus, $\frac{3}{4} = \frac{12}{16}$.

We can write equivalent fractions by multiplying a given fraction by 1, as shown in the next example. Multiplying a number by 1 does not change the value of the number.

$\frac{3}{4} = \frac{12}{16}$

EXAMPLE 6 Write $\frac{2}{5}$ as an equivalent fraction with a denominator of 20.

Solution: Since $5 \cdot 4 = 20$, multiply the fraction by $\frac{4}{4}$. Multiplying by $\frac{4}{4} = 1$ does not change the value of the fraction.

$$\frac{2}{5} = \frac{2}{5} \cdot \frac{4}{4} = \frac{2 \cdot 4}{5 \cdot 4} = \frac{8}{20}$$

EXAMPLE 7 Add or subtract as indicated. Write each answer in lowest terms.

a. $\frac{2}{5} + \frac{1}{4}$ **b.** $\frac{1}{2} + \frac{17}{22} - \frac{2}{11}$ **c.** $3\frac{1}{6} - 1\frac{11}{12}$

Solution: **a.** Fractions must have a common denominator before they can be added or subtracted. Since 20 is the smallest number that both 5 and 4 divide into evenly, 20 is the **least common denominator.** Write both fractions as equivalent fractions with denominators of 20. Since

$$\frac{2}{5} \cdot \frac{4}{4} = \frac{2 \cdot 4}{5 \cdot 4} = \frac{8}{20} \quad \text{and} \quad \frac{1}{4} \cdot \frac{5}{5} = \frac{1 \cdot 5}{4 \cdot 5} = \frac{5}{20}$$

then

$$\frac{2}{5} + \frac{1}{4} = \frac{8}{20} + \frac{5}{20} = \frac{13}{20}$$

b. The least common denominator for denominators 2, 22, and 11 is 22. First, write each fraction as an equivalent fraction with a denominator of 22. Then add or subtract from left to right.

$$\frac{1}{2} = \frac{1}{2} \cdot \frac{11}{11} = \frac{11}{22}, \quad \frac{17}{22} = \frac{17}{22}, \quad \text{and} \quad \frac{2}{11} = \frac{2}{11} \cdot \frac{2}{2} = \frac{4}{22}$$

Then

$$\frac{1}{2} + \frac{17}{22} - \frac{2}{11} = \frac{11}{22} + \frac{17}{22} - \frac{4}{22} = \frac{24}{22} = \frac{12}{11}$$

c. To find $3\frac{1}{6} - 1\frac{11}{12}$, first rewrite each mixed number as follows:

$$3\frac{1}{6} = 3 + \frac{1}{6} = \frac{18}{6} + \frac{1}{6} = \frac{19}{6}$$

$$1\frac{11}{12} = 1 + \frac{11}{12} = \frac{12}{12} + \frac{11}{12} = \frac{23}{12}$$

Then

$$3\frac{1}{6} - 1\frac{11}{12} = \frac{19}{6} - \frac{23}{12} = \frac{38}{12} - \frac{23}{12} = \frac{15}{12} = \frac{5}{4}$$

EXERCISE SET 1.2

Represent the shaded part of each geometric figure by a fraction.

1.

2.

3.
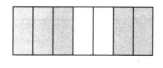

4.
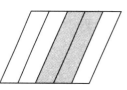

Write each of the following numbers as a product of primes. See Example 1.

5. 20 **6.** 56
7. 75 **8.** 32
9. 45 **10.** 24

Write the following fractions in lowest terms. See Example 2.

11. $\frac{2}{4}$ **12.** $\frac{3}{6}$

13. $\frac{10}{15}$ **14.** $\frac{15}{20}$

15. $\frac{3}{7}$ **16.** $\frac{5}{9}$

17. $\frac{18}{30}$ **18.** $\frac{42}{45}$

Multiply or divide as indicated. Write the answer in lowest terms. See Examples 3 and 4.

19. $\frac{1}{2} \cdot \frac{3}{4}$ **20.** $\frac{10}{6} \cdot \frac{3}{5}$

21. $\frac{2}{3} \cdot \frac{3}{4}$ **22.** $\frac{7}{8} \cdot \frac{3}{21}$

23. $\frac{1}{2} \div \frac{7}{12}$ **24.** $\frac{7}{12} \div \frac{1}{2}$

25. $\frac{3}{4} \div \frac{1}{20}$ **26.** $\frac{3}{5} \div \frac{9}{10}$

27. $\frac{7}{10} \cdot \frac{5}{21}$ **28.** $\frac{3}{35} \cdot \frac{10}{63}$

29. $2\frac{7}{9} \cdot \frac{1}{3}$ **30.** $\frac{1}{4} \cdot 5\frac{5}{6}$

The area of a plane figure is a measure of the amount of surface of the figure. Find the area of each figure below. (The area of a rectangle is the product of its length and width. The area of a triangle is $\frac{1}{2}$ the product of its base and height.)

31.
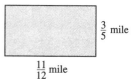

32.
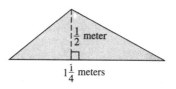

Add or subtract as indicated. Write the answer in lowest terms. See Example 5.

33. $\dfrac{4}{5} - \dfrac{1}{5}$ **34.** $\dfrac{6}{7} - \dfrac{1}{7}$

35. $\dfrac{4}{5} + \dfrac{1}{5}$ **36.** $\dfrac{6}{7} + \dfrac{1}{7}$

37. $\dfrac{17}{21} - \dfrac{10}{21}$ **38.** $\dfrac{18}{35} - \dfrac{11}{35}$

39. $\dfrac{23}{105} + \dfrac{4}{105}$ **40.** $\dfrac{13}{132} + \dfrac{35}{132}$

Write each of the following fractions as an equivalent fraction with the given denominator. See Example 6.

41. $\dfrac{7}{10}$ with a denominator of 30

42. $\dfrac{2}{3}$ with a denominator of 9

43. $\dfrac{2}{9}$ with a denominator of 18

44. $\dfrac{8}{7}$ with a denominator of 56

45. $\dfrac{4}{5}$ with a denominator of 20

46. $\dfrac{4}{5}$ with a denominator of 25

Add or subtract as indicated. Write the answer in lowest terms. See Example 7.

47. $\dfrac{2}{3} + \dfrac{3}{7}$ **48.** $\dfrac{3}{4} + \dfrac{1}{6}$

49. $2\dfrac{13}{15} - 1\dfrac{1}{5}$ **50.** $5\dfrac{2}{9} - 3\dfrac{1}{6}$

51. $\dfrac{5}{22} - \dfrac{5}{33}$ **52.** $\dfrac{7}{10} - \dfrac{8}{15}$

53. $\dfrac{12}{5} - 1$ **54.** $2 - \dfrac{3}{8}$

Each circle below represents a whole, or 1. Determine the unknown part of the circle.

55.   **56.**

57. 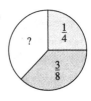   **58.**

Perform the following operations. Write answers in lowest terms.

59. $\dfrac{10}{21} + \dfrac{5}{21}$ **60.** $\dfrac{11}{35} + \dfrac{3}{35}$

61. $\dfrac{10}{3} - \dfrac{5}{21}$ **62.** $\dfrac{11}{7} - \dfrac{3}{35}$

63. $\dfrac{2}{3} \cdot \dfrac{3}{5}$ **64.** $\dfrac{2}{3} \div \dfrac{3}{4}$

65. $\dfrac{3}{4} \div \dfrac{7}{12}$ **66.** $\dfrac{3}{5} + \dfrac{2}{3}$

67. $\dfrac{5}{12} + \dfrac{4}{12}$ **68.** $\dfrac{2}{7} + \dfrac{4}{7}$

69. $5 + \dfrac{2}{3}$ **70.** $7 + \dfrac{1}{10}$

71. $\dfrac{7}{8} \div 3\dfrac{1}{4}$ **72.** $3 \div \dfrac{3}{4}$

73. $\dfrac{7}{18} \div \dfrac{14}{36}$ **74.** $4\dfrac{3}{7} \div \dfrac{31}{7}$

75. $\dfrac{23}{105} - \dfrac{2}{105}$ **76.** $\dfrac{57}{132} - \dfrac{13}{132}$

77. $1\dfrac{1}{2} + 3\dfrac{2}{3}$ **78.** $2\dfrac{3}{5} - 4\dfrac{7}{10}$

79. $\dfrac{2}{3} - \dfrac{5}{9} + \dfrac{5}{6}$ **80.** $\dfrac{8}{11} - \dfrac{1}{4} + \dfrac{1}{2}$

The perimeter of a plane figure is the total distance around the figure. Find the perimeter of each figure below.

81.

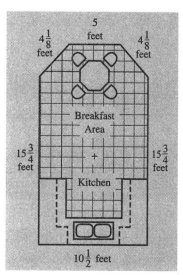

82.

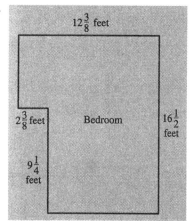

83. One of the New Orleans Saints guards was told to lose 20 pounds during the summer. So far, the guard has lost $11\frac{1}{2}$ pounds. How many more pounds must he lose?

84. In an English class at Cartez College, Stuart was told to read a 340-page book. So far, he has read $\frac{2}{5}$ of it. How many pages does he *have left to read*?

85. In your own words, explain how to add two fractions with different denominators.

86. In your own words, explain how to multiply two fractions.

The following trail chart is given to visitors at the Lakeview Forest Preserve.

Trail Name	Distance (miles)
Robin Path	$3\frac{1}{2}$
Red Falls	$5\frac{1}{2}$
Green Way	$2\frac{1}{8}$
Autumn Walk	$1\frac{3}{4}$

87. How much longer is Red Falls Trail than Green Way Trail?

88. Find the total distance traveled by someone who hiked along all four trails.

Most of the water on Earth is in the form of oceans. Only a small part is fresh water. The graph below is called a circle graph or pie chart. This particular circle graph shows the distribution of fresh water.

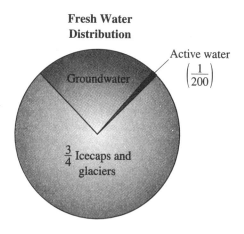

89. What fractional part of fresh water is icecaps and glaciers?

90. What fractional part of fresh water is active water?

91. What fractional part of fresh water is groundwater?

92. What fractional part of fresh water is groundwater or icecaps and glaciers?

1.3 Exponents and Order of Operations

OBJECTIVES

1. Use and simplify exponential expressions.
2. Define and use the order of operations.
3. Translate word statements to symbols.

Volume is $(2 \cdot 2 \cdot 2)$ cubic centimeters.

Frequently in algebra, products occur that contain repeated multiplication of the same factor. For example, the volume of a cube whose sides each measure 2 centimeters is $(2 \cdot 2 \cdot 2)$ cubic centimeters. We may use **exponential notation** to write such products in a more compact form. For example,

$$2 \cdot 2 \cdot 2 \quad \text{may be written as} \quad 2^3.$$

The 2 in 2^3 is called the **base**; it is the repeated factor. The 3 in 2^3 is called the **exponent** and is the number of times the base is used as a factor. The expression 2^3 is called an **exponential expression.**

$$2^3 = 2 \cdot 2 \cdot 2 = 8$$

exponent ↓ ; base ↑ 2 is a factor 3 times

EXAMPLE 1 Evaluate the following:

a. 3^2 [read as "3 squared" or as "3 to the second power"]
b. 5^3 [read as "5 cubed" or as "5 to the third power"]
c. 2^4 [read as "2 to the fourth power"]
d. 7^1 **e.** $\left(\dfrac{3}{7}\right)^2$

Solution:
a. $3^2 = 3 \cdot 3 = 9$
b. $5^3 = 5 \cdot 5 \cdot 5 = 125$
c. $2^4 = 2 \cdot 2 \cdot 2 \cdot 2 = 16$
d. $7^1 = 7$
e. $\left(\dfrac{3}{7}\right)^2 = \left(\dfrac{3}{7}\right)\left(\dfrac{3}{7}\right) = \dfrac{9}{49}$

> **REMINDER** $2^3 \neq 2 \cdot 3$ since 2^3 indicates repeated **multiplication** of the same factor.
> $2^3 = 2 \cdot 2 \cdot 2 = 8$, whereas $2 \cdot 3 = 6$.

Using symbols for mathematical operations is a great convenience. The more operation symbols presented in an expression, the more careful we must be when performing the indicated operation. For example, in the expression $2 + 3 \cdot 7$, do we add first or multiply first? To eliminate confusion, **grouping symbols** are used.

Examples of grouping symbols are parentheses ( ), brackets [ ], braces { }, and the fraction bar. If we wish $2 + 3 \cdot 7$ to be simplified by adding first, enclose $2 + 3$ in parentheses.

$$(2 + 3) \cdot 7 = 5 \cdot 7 = 35$$

If we wish to multiply first, $3 \cdot 7$ may be enclosed in parentheses.

$$2 + (3 \cdot 7) = 2 + 21 = 23$$

To eliminate confusion when no grouping symbols are present, use the following agreed upon order of operations.

ORDER OF OPERATIONS

Simplify expressions using the following order. If grouping symbols such as parentheses are present, simplify expressions within those first, starting with the innermost set. If fraction bars are present, simplify the numerator and the denominator separately.

1. Evaluate exponential expressions.
2. Perform multiplications or divisions in order from left to right.
3. Perform additions or subtractions in order from left to right.

Now simplify $2 + 3 \cdot 7$. There are no grouping symbols and no exponents, so we multiply and then add.

$$2 + 3 \cdot 7 = 2 + 21 \quad \text{Multiply.}$$
$$= 23 \quad \text{Add.}$$

EXAMPLE 2 Simplify each expression.

a. $6 \div 3 + 5^2$ **b.** $\dfrac{2(12 + 3)}{|-15|}$ **c.** $3 \cdot 10 - 7 \div 7$ **d.** $3 \cdot 4^2$ **e.** $\dfrac{3}{2} \cdot \dfrac{1}{2} - \dfrac{1}{2}$

Solution: **a.** Evaluate 5^2 first.

$$6 \div 3 + 5^2 = 6 \div 3 + 25$$

Next divide, then add.

$$6 \div 3 + 25 = 2 + 25 \quad \text{Divide.}$$
$$= 27 \quad \text{Add.}$$

b. First, simplify the numerator and the denominator separately.

$$\dfrac{2(12 + 3)}{|-15|} = \dfrac{2(15)}{15} \quad \text{Simplify numerator and denominator separately.}$$
$$= \dfrac{30}{15}$$
$$= 2 \quad \text{Simplify.}$$

c. Multiply and divide from left to right. Then subtract.

$$3 \cdot 10 - 7 \div 7 = 30 - 1$$
$$= 29 \quad \text{Subtract.}$$

d. In this example, only the 4 is squared. The factor of 3 is not part of the base because no grouping symbol includes it as part of the base.

$$3 \cdot 4^2 = 3 \cdot 16 \quad \text{Evaluate the exponential expression.}$$
$$= 48 \quad \text{Multiply.}$$

e. The order of operations applies to operations with fractions in exactly the same way as it applies to operations with whole numbers.

$$\frac{3}{2} \cdot \frac{1}{2} - \frac{1}{2} = \frac{3}{4} - \frac{1}{2} \quad \text{Multiply.}$$
$$= \frac{3}{4} - \frac{2}{4} \quad \text{The least common denominator is 4.}$$
$$= \frac{1}{4} \quad \text{Subtract.}$$

REMINDER Be careful when evaluating an exponential expression. In $3 \cdot 4^2$, the exponent 2 applies only to the base 4. In $(3 \cdot 4)^2$, we multiply first because of parentheses, so the exponent 2 applies to the product $3 \cdot 4$.

$$3 \cdot 4^2 = 3 \cdot 16 = 48 \qquad (3 \cdot 4)^2 = (12)^2 = 144$$

Expressions that include many grouping symbols can be confusing. When simplifying these expressions, keep in mind that grouping symbols separate the expression into distinct parts. Each is then simplified separately.

EXAMPLE 3 Simplify $\dfrac{3 + |4 - 3| + 2^2}{6 - 3}$.

Solution: The fraction bar serves as a grouping symbol and separates the numerator and denominator. Simplify each separately. Also, the absolute value bars here serve as a grouping symbol. We begin in the numerator by simplifying within the absolute value bars.

$$\frac{3 + |4 - 3| + 2^2}{6 - 3} = \frac{3 + |1| + 2^2}{6 - 3} \quad \text{Simplify the expression inside the absolute value bars.}$$
$$= \frac{3 + 1 + 2^2}{3} \quad \text{Find the absolute value and simplify the denominator.}$$
$$= \frac{3 + 1 + 4}{3} \quad \text{Evaluate the exponential expression.}$$
$$= \frac{8}{3} \quad \text{Simplify the numerator.}$$

EXAMPLE 4 Simplify $3[4(5 + 2) - 10]$.

22 CHAPTER 1 REVIEW OF REAL NUMBERS

Solution: Notice that both parentheses and brackets are used as grouping symbols. Start with the innermost set of grouping symbols.

$$3[4(5 + 2) - 10] = 3[4(7) - 10] \quad \text{Simplify the expression in parentheses.}$$
$$= 3[28 - 10] \quad \text{Multiply 4 and 7.}$$
$$= 3[18] \quad \text{Subtract inside the brackets.}$$
$$= 54 \quad \text{Multiply.}$$

EXAMPLE 5 Simplify $\dfrac{8 + 2 \cdot 3}{2^2 - 1}$.

Solution: $\dfrac{8 + 2 \cdot 3}{2^2 - 1} = \dfrac{8 + 6}{4 - 1} = \dfrac{14}{3}$

3 Oftentimes solving problems involves the ability to translate word phrases and sentences into symbols. Below is a list of key words and phrases to help us translate.

ADDITION (+)	SUBTRACTION (−)	MULTIPLICATION (·)	DIVISION (÷)	EQUALITY (=)
sum	difference of	product	quotient	equals
plus	minus	times	divide	gives
added to	subtracted from	multiply	into	is/was
more than	less than	twice	ratio	yields
increased by	decreased by	of	divided by	amounts to
total	less			represents
				is the same as

EXAMPLE 6 Translate each word statement into symbols.

a. Seventeen plus nine is twenty-six.
b. Fourteen decreased by four equals ten.
c. The product of two and three amounts to thirty-six divided by six.

Solution: **a.** In words: Seventeen   plus   nine   is   twenty-six
Translate: 17   +   9   =   26

b. In words: Fourteen   decreased by   four   equals   ten
Translate: 14   −   4   =   10

c. In words: The product of two and three   amounts to   thirty-six   divided by   six
Translate: 2 · 3   =   36   ÷   6

SCIENTIFIC CALCULATOR EXPLORATIONS

EXPONENTS

To evaluate exponential expressions on a scientific calculator, find the key marked y^x. To evaluate, for example, 3^5, press the following keys: [3] [y^x] [5] [=]. The display should read [243].

ORDER OF OPERATIONS

Some calculators follow the order of operations, and others do not. To see whether or not your calculator has the order of operations built in, use your calculator to find $2 + 3 \cdot 4$. To do this, press the following sequence of keys:

[2] [+] [3] [×] [4] [=].

The correct answer is 14 because the order of operations is to multiply before we add. If the calculator displays [14], then it has order of operations built in.

Even if the order of operations is built in, parentheses must sometimes be inserted. For example, to simplify $\frac{5}{12-7}$, press the keys

[5] [÷] [(] [1] [2] [−] [7] [)] [=].

The display should read [1].

Use a calculator to evaluate each expression.

1. 5^3
2. 7^4
3. 9^5
4. 8^6
5. $2(20 - 5)$
6. $3(14 - 7) + 21$
7. $24(862 - 455) + 89$
8. $99 + (401 + 962)$
9. $\dfrac{4623 + 129}{36 - 34}$
10. $\dfrac{956 - 452}{89 - 86}$

EXERCISE SET 1.3

Evaluate. See Example 1.

1. 3^5
2. 5^3
3. 3^3
4. 4^4
5. 1^5
6. 1^8
7. 5^1
8. 8^1
9. $\left(\dfrac{1}{5}\right)^3$
10. $\left(\dfrac{6}{11}\right)^2$
11. $\left(\dfrac{2}{3}\right)^4$
12. $\left(\dfrac{1}{2}\right)^5$
13. 7^2
14. 9^2
15. 2^5
16. 2^6
17. 0^3
18. 0^2
19. 4^2
20. 2^4
21. $(1.2)^2$
22. $(0.07)^2$

23. The area of a square whose sides each measure 5 meters is $(5 \cdot 5)$ square meters. Write this area using exponential notation.

5 meters

24. CHAPTER 1   REVIEW OF REAL NUMBERS

24. The volume of a solid is a measure of the space it encloses. The volume of a sphere whose radius is 5 meters is $\left(\frac{4}{3}\pi \cdot 5 \cdot 5 \cdot 5\right)$ cubic meters. Write this volume using exponential notation.

Simplify each expression. See Examples 2 through 5.

25. $5 + 6 \cdot 2$
26. $8 + 5 \cdot 3$
27. $4 \cdot 8 - 6 \cdot 2$
28. $12 \cdot 5 - 3 \cdot 6$
29. $2(8 - 3)$
30. $5(6 - 2)$
31. $2 + (5 - 2) + 4^2$
32. $6 - 2 \cdot 2 + 2^5$
33. $5 \cdot 3^2$
34. $2 \cdot 5^2$
35. $\dfrac{1}{4} \cdot \dfrac{2}{3} - \dfrac{1}{6}$
36. $\dfrac{3}{4} \cdot \dfrac{1}{2} + \dfrac{2}{3}$
37. $\dfrac{6 - 4}{9 - 2}$
38. $\dfrac{8 - 5}{24 - 20}$
39. $2[5 + 2(8 - 3)]$
40. $3[4 + 3(6 - 4)]$
41. $\dfrac{3 + 3(5 + 3)}{3^2 + 1}$
42. $\dfrac{3 + 6(8 - 5)}{4^2 + 2}$
43. $\dfrac{6 + |8 - 2| + 3^2}{18 - 3}$
44. $\dfrac{16 + |13 - 5| + 4^2}{17 - 5}$

45. Are parentheses necessary in the expression $2 + (3 \cdot 5)$? Explain your answer.
46. Are parentheses necessary in the expression $(2 + 3) \cdot 5$? Explain your answer.

Translate each word statement to symbols. See Example 6.

47. Ten added to twelve is equal to the product of eleven and two.
48. One hundred is the product of four and the sum of twenty and five.
49. The sum of five and six is greater than ten.
50. The difference of fifteen and seven is less than nine.
51. The product of three and five is greater than twelve.
52. The quotient of four and seven is less than thirty.

Insert <, >, or = in the space provided to make the following true.

53. 8^1 $8(1)$
54. 9^1 $9(1)$
55. 2^3 $2 \cdot 3$
56. 2^4 $2 \cdot 4$
57. $\left(\dfrac{2}{3}\right)^3$ $\dfrac{1}{9} + \dfrac{1}{27}$
58. $\left(\dfrac{3}{10}\right)^2$ $\dfrac{1}{10} \cdot \dfrac{2}{10}$
59. $(0.6)^2$ 36
60. $(0.3)^3$ 27

61. Why is 8^2 usually read as "eight squared"? (*Hint:* What is the area of the **square** below?)

62. Why is 4^3 usually read as "four cubed"? (*Hint:* What is the volume of the **cube** below?)

Simplify each expression.

63. $\dfrac{19 - 3 \cdot 5}{6 - 4}$
64. $\dfrac{4 \cdot 3 + 2}{4 + 3 \cdot 2}$
65. $\dfrac{|6 - 2| + 3}{8 + 2 \cdot 5}$
66. $\dfrac{15 - |3 - 1|}{12 - 3 \cdot 2}$
67. $\dfrac{3(2 + 5)}{6 + 2}$
68. $\dfrac{4(8 - 3)}{6 + 3}$
69. $5[3(0.2 + 0.1) + 4]$
70. $4[5(2.1 + 4.3) - 8]$

71. Insert parentheses so that the following expression simplifies to 32.
$$20 - 4 \cdot 4 \div 2$$

72. Insert parentheses so that the following expression simplifies to 28.
$$2 \cdot 5 + 3^2$$

Translate each word statement into symbols.

73. Three cubed is the same as the sum of twenty and seven.
74. Fifteen divided by the sum of two and three is three.
75. One increased by two equals the quotient of nine and three.
76. Four subtracted from eight is equal to two squared.
77. Nine is less than or equal to the product of eleven and two.
78. Eight is greater than one subtracted from seven.
79. Three is not equal to four divided by two.
80. The difference of sixteen and four is greater than ten.

1.4 INTRODUCTION TO VARIABLE EXPRESSIONS AND EQUATIONS

OBJECTIVES

1. Evaluate algebraic expressions, given replacement values for variables.
2. Translate word phrases into algebraic expressions.
3. Define equation and solution or root of an equation.
4. Translate word statements into equations.

TAPE BA 1.4

In algebra, we use symbols, usually letters such as x, y, or z, to represent unknown numbers. A symbol that is used to represent a number is called a **variable.** An **algebraic expression** is a collection of numbers, variables, operation symbols, and grouping symbols. For example,

$$2x, \quad -3, \quad 2x - 10, \quad 5(p^2 + 1), \quad \text{and} \quad \frac{3y^2 - 6y + 1}{5}$$

are algebraic expressions. The expression $2x$ means $2 \cdot x$. Also, $5(p^2 + 1)$ means $5 \cdot (p^2 + 1)$ and $3y^2$ means $3 \cdot y^2$. If we give a specific value to a variable, we can **evaluate an algebraic expression.** To evaluate an algebraic expression means to find its numerical value once we know the value of the variables. Make sure the order of operations is followed when evaluating an expression.

EXAMPLE 1 Evaluate each expression if $x = 3$ and $y = 2$.

    a. $2x - y$ **b.** $\dfrac{3x}{2y}$ **c.** $\dfrac{x}{y} + \dfrac{y}{2}$ **d.** $x^2 - y^2$

Solution: **a.** Replace x with 3 and y with 2.

$$\begin{aligned} 2x - y &= 2(3) - 2 &&\text{Let } x = 3 \text{ and } y = 2. \\ &= 6 - 2 &&\text{Multiply.} \\ &= 4 &&\text{Subtract.} \end{aligned}$$

b. $\dfrac{3x}{2y} = \dfrac{3 \cdot 3}{2 \cdot 2} = \dfrac{9}{4}$ Let $x = 3$ and $y = 2$.

c. Replace x with 3 and y with 2. Then simplify.

$$\frac{x}{y} + \frac{y}{2} = \frac{3}{2} + \frac{2}{2} = \frac{5}{2}$$

d. Replace x with 3 and y with 2.

$$x^2 - y^2 = 3^2 - 2^2 = 9 - 4 = 5$$

Now that we can represent an unknown number by a variable, let's practice translating phrases into algebraic expressions.

EXAMPLE 2 Write an algebraic expression that represents each of the following phrases. Let the variable x represent the unknown number.

a. The sum of a number and 3
b. The product of a 3 and a number
c. Twice a number
d. 10 decreased by a number
e. 5 times a number increased by 7

Solution:
a. $x + 3$ since "sum" means to add.
b. $3 \cdot x$ and $3x$ are both ways to denote the product of 3 and x.
c. $2 \cdot x$ or $2x$.
d. $10 - x$ because "decreased by" means to subtract.
e. $\underbrace{5x}_{\text{5 times a number}} + 7$

An **equation** is a mathematical statement that two expressions have equal value. The equal symbol "=" is used to equate the two expressions. For example, $3 + 2 = 5$, $7x = 35$, $\frac{2(x - 1)}{3} = 0$, and $I = PRT$ are all equations.

> REMINDER An equation contains the equal symbol "=". An algebraic expression does not.

When an equation contains a variable, deciding which values of the variable make an equation a true statement is called **solving** an equation for the variable. A **solution** or **root** of an equation is a value for the variable that makes the equation true. For example, 3 is a solution of the equation $x + 4 = 7$, because if x is replaced by 3 the statement is true.

$x + 4 = 7$
$\downarrow$
$3 + 4 = 7$ Replace x with 3.
$7 = 7$ True.

Similarly, 1 is not a solution of the equation $x + 4 = 7$, because $1 + 4 = 7$ is **not** a true statement. The **solution set** of an equation is the set of all solutions of the equation. The solution set of the equation $x + 4 = 7$ is {3}.

EXAMPLE 3 Decide whether 2 is a solution of $3x + 10 = 8x$.

Solution: Replace x with 2 and see if a true statement results.

$$3x + 10 = 8x \quad \text{Original equation.}$$
$$3(2) + 10 = 8(2) \quad \text{Replace } x \text{ with 2.}$$
$$6 + 10 = 16 \quad \text{Simplify each side.}$$
$$16 = 16 \quad \text{True.}$$

Since we arrived at a true statement after replacing x with 2 and simplifying both sides of the equation, 2 is a solution of the equation.

 We now practice translating sentences into equations.

EXAMPLE 4 Write each sentence as an equation. Let x represent the unknown number.

a. The quotient of 15 and a number is 4.
b. Three subtracted from 12 is a number.
c. Four times a number added to 17 is 21.

Solution: **a.** In words: | The quotient of 15 and a number | is | 4 |

Translate: $\quad \dfrac{15}{x} \quad = \quad 4$

b. In words: | Three subtracted **from** 12 | is | a number |

Translate: $\quad 12 - 3 \quad = \quad x$

Care must be taken when the operation is subtraction. The expression $3 - 12$ would be incorrect. Notice that $3 - 12 \neq 12 - 3$.

c. In words: | 4 times a number | added to | 17 | is | 21 |

Translate: $\quad 4x \quad + \quad 17 \quad = \quad 21$

EXERCISE SET 1.4

Evaluate each expression if $x = 1$, $y = 3$, and $z = 5$. See Example 1.

1. $3x - 2$
2. $6y - 8$
3. $|2x + 3y|$
4. $|5z - 2y|$
5. $xy + z$
6. $yz - x$
7. $5y^2$
8. $2z^2$
9. $|y^2 + z^2|$
10. $|10y - 3z|$

Neglecting air resistance, the expression $16t^2$ gives the distance in feet an object will fall in t seconds.

11. Complete the chart below. To evaluate $16t^2$, remember to first find t^2, then multiply by 16.

TIME t (IN SECONDS)	DISTANCE $16t^2$ (IN FEET)
1	
2	
3	
4	

12. Does an object fall the same distance *during* each second? Why or why not? (See Exercise 11.)

Write each phrase as an algebraic expression. Let x represent the unknown number. See Example 2.

13. Fifteen more than a number.
14. One-half times a number.
15. Five subtracted from a number.
16. The quotient of a number and 9.
17. Three times a number increased by 22.
18. The product of 8 and a number.

Decide whether the given number is a solution of the given equation. See Example 3.

19. $3x - 6 = 9$; 5
20. $2x + 7 = 3x$; 6
21. $2x + 6 = 5x - 1$; 0
22. $4x + 2 = x + 8$; 2
23. $x^2 + 2x + 1 = 0$; 1
24. $x^2 + 2x - 35 = 0$; 5
25. $2x - 5 = 5$; 8
26. $3x - 10 = 8$: 6
27. $x + 6 = x + 6$; 2
28. $x + 6 = x + 6$; 10
29. $x = 5x + 15$; 0
30. $4 = 1 - x$; 1

Write each of the following sentences as an equation. Use x to represent the unknown number. See Example 4.

31. The sum of 5 and a number is 20.
32. Twice a number is 17.
33. Thirteen minus three times a number is 13.
34. Seven subtracted from a number is 0.
35. The quotient of 12 and a number is $\frac{1}{2}$.
36. The sum of 8 and twice a number is 42.

Evaluate each expression if x = 2, y = 6, and z = 3.

37. $3x + 8y$
38. $|4z - 2y|$
39. $\dfrac{4x}{3y}$
40. $\dfrac{6z}{5x}$
41. $\dfrac{y}{x} + \dfrac{y}{x}$
42. $\dfrac{9}{z} + \dfrac{4z}{y}$
43. $3x^2 + y$
44. $y + 4z^2$
45. $x + yz$
46. $z + xy$

Evaluate each expression if x = 12, y = 8, and z = 4.

47. $\dfrac{x}{z} + 3y$
48. $\dfrac{y}{z} + 8x$
49. $3z + z^2$
50. $7y + y^2$
51. $x^2 - 3y + x$
52. $y^2 - 3x + y$
53. $\dfrac{2x + z}{3y - z}$
54. $\dfrac{3x - y}{4z + x}$
55. $\dfrac{x^2 + z}{y^2 + 2z}$
56. $\dfrac{y^2 + x}{x^2 + 3y}$

57. In your own words, explain the difference between an expression and an equation.
58. Determine whether each is an expression or an equation.
    a. $3x^2 - 26$
    b. $3x^2 - 26 = 1$
    c. $2x - 5 = 7x - 5$
    d. $9y + x - 8$

Write each of the following as an algebraic expression or an equation.

59. A number divided by 13
60. Seven times the sum of a number and 19
61. Ten subtracted from twice a number is 18.
62. A number subtracted from four times that number is 75.6.
63. Twenty less the product of 30 and a number
64. Twelve decreased by a number
65. A number times 0.02 equals 1.76.
66. The product of $\frac{3}{4}$ and the sum of a number and 1 equals 9.
67. A number subtracted from 19 is three times that number.
68. Three times a number added to $1\frac{11}{12}$ equals the number increased by 2.

Solve the following.

69. The perimeter of a figure is the distance around the figure. The expression $2l + 2w$ represents the perimeter of a rectangle when l is its length and w is its width. Find the perimeter of the following rectangle by substituting 8 for l and 6 for w.

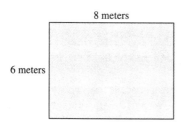

70. The expression $a + b + c$ represents the perimeter of a triangle when a, b, and c are the lengths of its sides. Find the perimeter of the following triangle.

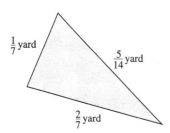

71. The area of a figure is the total enclosed surface of the figure. Area is measured in square units. The expression lw represents the area of a rectangle when l is its length and w is its width. Find the area of the following rectangular-shaped lot.

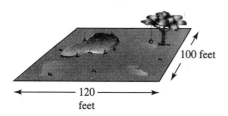

72. A trapezoid is a four-sided figure with exactly one pair of parallel sides. The expression $\frac{(B + b)h}{2}$ represents its area, when B and b are the lengths of the two parallel sides and h is the height between these sides. Find the area if $B = 15$ inches, $b = 7$ inches, and $h = 5$ inches.

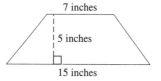

73. The expression $\frac{I}{PT}$ represents the rate of interest being charged if a loan of P dollars for T years required I dollars in interest to be paid. Find the interest rate if a $650 loan for 3 years to buy a used IBM personal computer requires $126.75 in interest to be paid.

74. The expression $\frac{d}{t}$ represents the average speed r in miles per hour if a distance of d miles is traveled in t hours. Find the rate to the nearest whole number if the distance between Dallas, Texas, and Kaw City, Oklahoma, is 432 miles, and it takes Barbara Goss 8.5 hours to drive the distance.

75. Peter Callac earns a base salary plus a commission on all sales he makes at St. Joe Brick Company. The expression $B + RS$ represents his gross income, when B is the base income, R is the commission rate, and S is the amount sold. Find Peter's income if $B = \$300$, $R = 8\%$, $S = \$500$.

Evaluate each expression if $x = 5$ and $y = 1$.

76. $2(x - 3)^2 + 5(y + 2)^2$

77. $7(x - 4)^2 - 5(y - 1)^2$

78. $\dfrac{5x}{2} + \dfrac{7y}{4}$

79. $\dfrac{3xy}{6} - \dfrac{y^3}{12}$

80. $\dfrac{4(2x - 3) - 5(y + 2)}{(x + 1)(y + 1)}$

81. $\dfrac{3[(x + 1) + (2y + 5)]}{(x + 1)^2}$

30 CHAPTER 1 REVIEW OF REAL NUMBERS

1.5 ADDING REAL NUMBERS

TAPE BA 1.5

OBJECTIVES

1. Add real numbers with the same sign.
2. Add real numbers with unlike signs.
3. Solve problems that involve addition of real numbers.
4. Find the opposite of a number.

Real numbers can be added, subtracted, multiplied, divided, and raised to powers, just as whole numbers can. We use the number line to help picture the addition of real numbers.

On a number line, a positive number can be represented anywhere by an arrow of appropriate length pointing to the right. A negative number can be represented by an arrow pointing to the left. For example, to find the sum $3 + 2$ on a number line, start at 0 and draw an arrow representing 3. This arrow should be three units long pointing to the right, since 3 is positive. From the tip of this arrow, draw another arrow representing 2. The tip of the second arrow ends at their sum, 5.

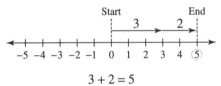

$$3 + 2 = 5$$

To find the sum $-1 + (-2)$, start at 0 and draw an arrow representing -1. This arrow is one unit long pointing to the left. At the tip of this arrow, draw an arrow representing -2. The tip of the second arrow ends at their sum, -3.

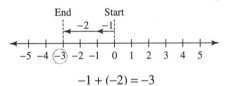

$$-1 + (-2) = -3$$

Using a number line each time we add two numbers can be time consuming. Instead, we can notice patterns in the previous examples and write rules for adding signed numbers. When adding two numbers with the same sign, notice that the sign of the sum is the same as the sign of the addends.

TO ADD TWO NUMBERS WITH THE SAME SIGN

Step 1. Find the sum of their absolute values.

Step 2. Use their common sign as the sign of the sum.

Add $(-7) + (-6)$.

Step 1. Find the sum of their absolute values.

ADDING REAL NUMBERS   SECTION 1.5   **31**

$$|-7| = 7, \quad |-6| = 6, \quad \text{and} \quad 7 + 6 = 13$$

Step 2.   Use their common sign as the sign of the sum. This means that

$$(-7) + (-6) = -13$$
$$\uparrow$$
common sign

Thinking of signed numbers as money earned or lost might help make addition more meaningful. Earnings can be thought of as positive numbers. If $1 is earned and later another $3 is earned, the total amount earned is $4. $1 + 3 = 4$.

On the other hand, losses can be thought of as negative numbers. If $1 is lost and later another $3 is lost, a total of $4 is lost. $(-1) + (-3) = -4$.

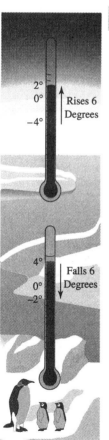

EXAMPLE 1   Find each sum.

    a. $-3 + (-7)$ **b.** $5 + (+12)$ **c.** $(-1) + (-20)$ **d.** $-2 + (-10)$

Solution: **a.** $-3 + (-7) = -10$ **b.** $5 + (+12) = 17$ **c.** $(-1) + (-20) = -21$
         d. $-2 + (-10) = -12$

Adding numbers whose signs are not the same can be pictured on the number line, also. To find the sum $-4 + 6$, begin at 0 and draw an arrow representing -4. This arrow is 4 units long and pointing to the left. At the tip of this arrow, draw an arrow representing 6. The tip of the second arrow ends at their sum, 2.

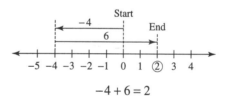

$$-4 + 6 = 2$$

Using temperature as an example, if the thermometer registers 4 degrees below 0 degrees and then rises 6 degrees, the new temperature is 2 degrees above 0 degrees. Thus, it is reasonable that $-4 + 6 = 2$. Likewise, if the temperature is 4 degrees above 0 degrees and then falls 6 degrees, the new temperature is 2 degrees below 0 degrees. On the number line, we see that $4 + (-6) = -2$.

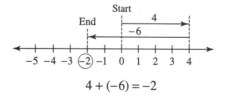

$$4 + (-6) = -2$$

Once again, we can observe a pattern: when adding two numbers with different signs, the sign of the sum is the same as the sign of the addend whose absolute value is larger.

32 CHAPTER 1 REVIEW OF REAL NUMBERS

> **TO ADD TWO NUMBERS WITH DIFFERENT SIGNS**
>
> *Step 1.* Find the difference of the larger absolute value and the smaller absolute value.
>
> *Step 2.* Use the sign of the addend whose absolute value is larger as the sign of the sum.

EXAMPLE 2 Find each sum.

    a. $3 + (-7)$ **b.** $(-2) + (10)$ **c.** $0.2 + (-0.5)$

Solution: **a.** Since $|-7|$ is larger than $|3|$, their sum has the same sign as -7. That is, the sum is negative. Then $|-7| = 7, |3| = 3$, and $7 - 3 = 4$. The sum is -4.

$$3 + (-7) = -4$$

b. Since $|10|$ is larger than $|-2|$, the sign of the sum is the same as the sign of 10, positive. Then $|10| = 10, |-2| = 2$, and $10 - 2 = 8$. The sum is $+8$ or 8.

$$(-2) + 10 = 8$$

c. $0.2 + (-0.5) = -0.3$

EXAMPLE 3 Find the sums.

    a. $3 + (-7) + (-8)$ **b.** $[7 + (-10)] + [-2 + (-4)]$

Solution: **a.** Perform the additions from left to right.

$$3 + (-7) + (-8) = -4 + (-8) \quad \text{Adding numbers with different signs.}$$
$$= -12 \quad \text{Adding numbers with like signs.}$$

b. Simplify inside brackets first.

$$[7 + (-10)] + [-2 + (-4)] = [-3] + [-6]$$
$$= -9 \quad \text{Add.}$$

Signed numbers are used in everyday life. Stock market returns show gains and losses as positive and negative numbers. Temperatures in cold climates often dip into the negative range, commonly referred to as "below zero" temperatures. Bank statements report deposits and withdrawals as positive and negative numbers.

EXAMPLE 4 During three days, a share of Lamplighter's International stock recorded the following gains and losses:

    Monday   Tuesday   Wednesday
    a gain of $2   a loss of $1   a loss of $3

Find the overall gain or loss for the stock for the three days.

Solution: The overall gain or loss is the sum of the gains and losses.

In words: gain plus loss plus loss

Translate: 2 + (−1) + (−3) = −2

The overall loss is $2.

To help us subtract real numbers in the next section, we first review the concept of opposites. The graph of 4 and −4 is shown on the number line below.

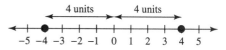

Notice that 4 and −4 lie on opposite sides of 0, and each is 4 units away from 0.

This relationship between −4 and +4 is an important one. Such numbers are known as **opposites** or **additive inverses** of each other.

> **OPPOSITES OR ADDITIVE INVERSES**
> Two numbers that are the same distance from 0 but lie on opposite sides of 0 are called opposites or additive inverses of each other. The opposite of zero is zero.

The opposite of 10 is −10.

The opposite of −3 is 3.

The opposite of $\frac{1}{2}$ is $-\frac{1}{2}$.

Notice that the sum of a number and its opposite is 0.

$$10 + (-10) = 0$$
$$-3 + 3 = 0$$
$$\frac{1}{2} + \left(-\frac{1}{2}\right) = 0$$

In general, **if a is a number, we write the opposite or additive inverse of a as $-a$.**

> **SUM OF OPPOSITES**
> The sum of a number a and its opposite $-a$ is 0.
> $$a + (-a) = 0$$

Since the opposite of a is $-a$, this means that we write the opposite of −3 as $-(-3)$. But we said above that the opposite of −3 is 3. This can be true only if $-(-3) = 3$.

> If a is a number, then $-(-a) = a$.

For example, $-(-10) = 10$, $-\left(-\frac{1}{2}\right) = \frac{1}{2}$, and $-(-2x) = 2x$.

34 CHAPTER 1 REVIEW OF REAL NUMBERS

EXAMPLE 5 Find the opposite or additive inverse of each number.

  a. 5 **b.** 0 **c.** −6

Solution: **a.** The opposite of 5 is −5. Notice that 5 and −5 are on opposite sides of 0 when plotted on a number line and are equal distances away.
  b. The opposite of 0 is 0 since $0 + 0 = 0$.
  c. The opposite of −6 is 6.

EXAMPLE 6 Simplify the following.

  a. $-(-6)$ **b.** $-|-6|$

Solution: **a.** $-(-6) = 6$
  b. $-|-6| = -6$

EXERCISE SET 1.5

Find the following sums. See Examples 1 through 3.

1. $6 + 3$
2. $9 + (-12)$
3. $-6 + (-8)$
4. $-6 + (-14)$
5. $-8 + (-7)$
6. $6 + (-4)$
7. $-52 + 36$
8. $-94 + 27$
9. $6 + (-4) + 9$
10. $-18 + |-53|$
11. $2\frac{3}{4} + \left(-\frac{1}{8}\right)$
12. $4\frac{4}{5} + \left(-\frac{3}{10}\right)$

Find the additive inverse or the opposite. See Example 5.

13. 6
14. 4
15. −2
16. −8
17. 0
18. $-\frac{1}{4}$
19. $|-6|$
20. $|-11|$

◻ 21. In your own words, explain how to find the opposite of a number.

◻ 22. In your own words, explain why 0 is the only number that is its own opposite.

Simplify the following. See Example 6.

23. $-|-2|$
24. $-(-3)$
25. $-|0|$
26. $\left|-\frac{2}{3}\right|$
27. $-\left|-\frac{2}{3}\right|$
28. $-(-7)$

Find the sums.

29. $-3 + (-5)$
30. $-7 + (-4)$
31. $-9 + (-3)$
32. $+8 + (-6)$
33. $+9 + (-3)$
34. $-9 + (+4)$
35. $-7 + (+3)$
36. $-5 + (+9)$
37. $-3 + (+10)$
38. $3 + (-8)$
39. $8 + 4$
40. $12 + (+2)$
41. $-15 + 9 + (-2)$
42. $-9 + 15 + (-5)$
43. $-21 + (-16) + (-22)$
44. $-14 + |-16|$
45. $|-8| + (-16)$
46. $|-6| + (-61)$
47. $-\frac{7}{16} + \frac{1}{4}$
48. $-\frac{5}{9} + \frac{1}{3}$
49. $-\frac{7}{10} + \left(-\frac{3}{5}\right)$
50. $-33 + (-14)$
51. $27 + (-46)$
52. $53 + (-37)$
53. $-18 + 49$
54. $-26 + 14$
55. $126 + (-67)$
56. $-\frac{5}{6} + \left(-\frac{2}{3}\right)$
57. $6.3 + (-8.4)$
58. $9.2 + (-11.4)$
59. $117 + (-79)$
60. $-114 + (-88)$
61. $-214 + (-86)$
62. $18 + (-6) + 4$
63. $-23 + 16 + (-2)$
64. $-14 + (-3) + 11$

65. $-9.6 + (-3.5)$
66. $-6.7 + (-7.6)$
67. $|5 + (-10)|$
68. $[-3 + 5] + (-11)$
69. $[-2 + (-7)] + [-11 + (-4)]$
70. $8 + [-5 + 5] + (-12)$
71. $|7 + (-10)| + |-16|$
72. $\left|-\frac{3}{10} + \frac{1}{10}\right| + \left|\frac{2}{5} - \frac{1}{10}\right|$

The following bar graph shows the daily low temperatures for a week in Sioux Falls, South Dakota.

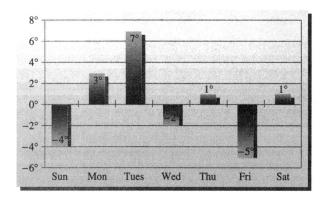

73. On what day of the week was the graphed temperature the highest?
74. On what day of the week was the graphed temperature the lowest?
75. What is the highest temperature shown on the graph?
76. What is the lowest temperature shown on the graph?
77. Find the average daily low temperature for Sunday through Thursday. (*Hint*: To find the average of the five temperatures, find their sum and divide by 5.)

Solve the following. See Example 4.

78. The low temperature in Anoka, Minnesota, was -15 degrees last night. During the day it rose only 9 degrees. Find the high temperature for the day.
79. On January 2, 1943, the temperature was -4 degrees at 7:30 A.M. in Spearfish, South Dakota. Incredibly, it got 49 degrees warmer in the next 2 minutes. To what temperature did it rise by 7:32?
80. The lowest elevation on Earth is -1312 feet (that is, 1312 feet below sea level) at the Dead Sea. If you are standing 658 feet above the Dead Sea, what is your elevation?
81. The lowest point in Africa is -512 feet at Lake Assal in Djibouti. If you are standing at a point 658 feet above Lake Assal, what is your elevation?
82. In checking the stock market results, Alexis discovers our stock posted changes of $-1\frac{5}{8}$ points and $-2\frac{1}{2}$ points over the last two days. What is the combined change?
83. Yesterday our stock posted a change of $-1\frac{1}{4}$ points, but today it showed a gain of $+\frac{7}{8}$ point. Find the overall change for the two days.
84. In golf, scores that are under par for the entire round are shown as negative scores; positive scores are shown for scores that are over par. In two rounds in the United States Open, Arnold Palmer had overall scores of -6 and $+4$. What was his total overall score?

Decide whether the given number is a solution of the given equation.

85. $x + 9 = 5; -4$
86. $7 = -x + 3; 10$
87. $y + (-3) = -7; -1$
88. $1 = y + 7; -6$
89. Explain why adding a negative number to another negative number produces a negative sum.
90. When a positive and a negative number are added, sometimes the sum is positive, sometimes it is zero, and sometimes it is negative. Explain why this happens.

1.6 SUBTRACTING REAL NUMBERS

Tape BA 1.6

OBJECTIVES

1. Subtract real numbers.
2. Add and subtract real numbers.
3. Evaluate algebraic expressions using real numbers.
4. Solve problems that involve subtraction of real numbers.

Now that addition of signed numbers has been discussed, we can explore subtraction. We know that $9 - 7 = 2$. Notice that $9 + (-7) = 2$, also. This means that

$$9 - 7 = 9 + (-7)$$

In general, the difference of two numbers a and b is the same as the sum of a and the opposite of b.

> **SUBTRACTING TWO REAL NUMBERS**
> If a and b are real numbers, then $a - b = a + (-b)$.

In other words, to find the difference of two numbers, add the first number to the opposite of the second number.

EXAMPLE 1 Find each difference.

    a. $-13 - (+4)$ **b.** $5 - (-6)$ **c.** $3 - 6$ **d.** $-1 - (-7)$

Solution:

a. $-13 - (+4) = -13 + (-4)$ Add the opposite of $+4$, which is -4.
$= -17$

b. $5 - (-6) = 5 + (6)$ Add the opposite of -6, which is 6.
$= 11$

c. $3 - 6 = 3 + (-6)$ Add 3 to the opposite of 6, which is -6.
$= -3$

d. $-1 - (-7) = -1 + (7) = 6$

EXAMPLE 2   Subtract 8 from -4.

Solution: Be careful when interpreting. The order of numbers in subtraction is important. 8 is to be subtracted **from** -4.

$$-4 - 8 = -4 + (-8) = -12$$

Expressions containing both sums and differences make good use of grouping symbols. In simplifying these expressions, remember to rewrite differences as sums and follow the standard order of operations.

EXAMPLE 3   Simplify each expression.

a. $-3 + [(-2 - 5) - 2]$ **b.** $2^3 - |10| + [-6 - (-5)]$

Solution: **a.** Start with the innermost sets of parentheses. Rewrite $-2 - 5$ as a sum.

$$\begin{aligned}
-3 + [(-2 - 5) - 2] &= -3 + [(-2 + (-5)) - 2] \\
&= -3 + [(-7) - 2] && \text{Add: } -2+(-5). \\
&= -3 + [-7 + (-2)] && \text{Write } -7-2 \text{ as a sum.} \\
&= -3 + [-9] && \text{Add.} \\
&= -12 && \text{Add.}
\end{aligned}$$

b. Start simplifying the expression inside the brackets by writing $-6 - (-5)$ as a sum.

$$\begin{aligned}
2^3 - |10| + [-6 - (-5)] &= 2^3 - |10| + [-6 + 5] \\
&= 2^3 - 10 + [-1] && \text{Add. Write } |10| \text{ as 10.} \\
&= 8 - 10 + (-1) && \text{Evaluate } 2^3. \\
&= 8 + (-10) + (-1) && \text{Write } 8-10 \text{ as a sum.} \\
&= -2 + (-1) && \text{Add.} \\
&= -3 && \text{Add.}
\end{aligned}$$

Knowing how to evaluate expressions for given replacement values is helpful when checking solutions of equations and when solving problems whose unknowns satisfy given expressions. The next example illustrates this.

EXAMPLE 4   If $x = 2$ and $y = -5$, evaluate the following expressions.

a. $\dfrac{x - y}{12 + x}$ **b.** $x^2 - y$

Solution: **a.** Replace x with 2 and y with -5. Be sure to put parentheses around -5 to separate signs. Then simplify the resulting expression.

$$\frac{x - y}{12 + x} = \frac{2 - (-5)}{12 + 2} = \frac{2 + 5}{14} = \frac{7}{14} = \frac{1}{2}$$

b. Replace the x with 2 and y with -5 and simplify.

$$x^2 - y = 2^2 - (-5) = 4 - (-5) = 4 + 5 = 9$$

4 Another use of signed numbers is in recording altitudes above and below sea level, as shown in the next example.

EXAMPLE 5 The lowest point in North America is in Death Valley, at an elevation of 282 feet below sea level. Nearby, Mount Whitney reaches 14,494 feet, the highest point in the United States outside Alaska. How much of a variation in elevation is there between these two extremes?

Solution: To find the variation in elevation between the two heights, find the difference of the high point and the low point.

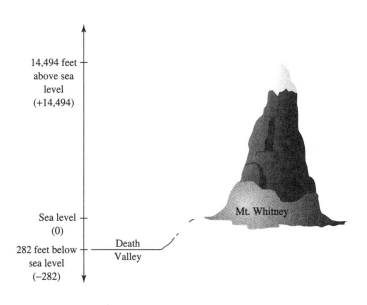

In words: high point minus low point

Translate: $14{,}494 \quad - \quad (-282) = 14{,}494 + 282$
$= 14{,}776$ feet

Thus, the variation in elevation is 14,776 feet.

EXAMPLE 6 At 6:00 P.M., the temperature at the Winter Olympics was 14 degrees; by morning the temperature dropped to -23 degrees. Find the overall change in temperature.

Solution: To find the overall change, find the difference of the morning temperature and the 6:00 P.M. temperature.

In words: morning temp. minus 6:00 P.M. temp.

Translate: $-23°$ $-$ $14°$ $= -23° + (-14°) = -37°$

Thus, the overall change is $-37°$. That is, the temperature dropped by 37 degrees.

EXERCISE SET 1.6

Find each difference. See Example 1.
1. $-6 - 4$
2. $-12 - 8$
3. $4 - 9$
4. $8 - 11$
5. $16 - (-3)$
6. $12 - (-5)$
7. $\frac{1}{2} - \frac{1}{3}$
8. $\frac{3}{4} - \frac{7}{8}$
9. $-16 - (-18)$
10. $-20 - (-48)$

Perform the operation. See Example 2.
11. Subtract -5 from 8.
12. Subtract 3 from -2.
13. Subtract -1 from -6.
14. Subtract 17 from 1.

Simplify each expression. (Remember the order of operations.) See Example 3.
15. $-6 - (2 - 11)$
16. $-9 - (3 - 8)$
17. $3^3 - 8 \cdot 9$
18. $2^3 - 6 \cdot 3$
19. $2 - 3(8 - 6)$
20. $4 - 6(7 - 3)$
21. $|-3| + 2^2 + [-4 - (-6)]$
22. $|-2| + 6^2 + (-3 - 8)$

◻ 23. If a and b are positive numbers, then is $a - b$ always positive, always negative, or sometimes positive and sometimes negative?

◻ 24. If a and b are negative numbers, then is $a - b$ always positive, always negative, or sometimes positive and sometimes negative?

If $x = -6$, $y = -3$, and $t = 2$, evaluate each expression. See Example 4.
25. $x - y$
26. $3t - x$
27. $x + t - 12$
28. $y - 3t - 8$
29. $\frac{x - (-12)}{y + 6}$
30. $\frac{x - (-9)}{t + 1}$
31. $x - t^2$
32. $y - t^3$
33. $\frac{4t - y}{3t}$
34. $\frac{5t - x}{5t}$

Simplify each expression.
35. $-6 - 5$
36. $-8 - 4$
37. $7 - (-4)$
38. $3 - (-6)$
39. $-6 - (-11)$
40. $-4 - (-16)$
41. $6\frac{2}{5} - \frac{7}{10}$
42. $-\frac{3}{5} - \frac{5}{6}$
43. $-\frac{3}{4} - \frac{1}{9}$
44. $-6.1 - (-5.3)$
45. $-2.6 - (-6.7)$
46. $6 - 11 - (-8)$
47. $4 - (-6) - 9$
48. $-6 - 14 - (-3)$
49. $-11 - (-8) - 4$
50. $-13 - 27$
51. $16 - (-21)$
52. $15 - (-33)$
53. $-44 - (-27)$
54. $-36 - (-51)$
55. $4\frac{2}{3} - \frac{5}{12}$
56. $-\frac{1}{6} - \left(-\frac{3}{4}\right)$
57. $-\frac{1}{10} - \left(-\frac{7}{8}\right)$
58. $8.3 - 11.2$
59. $9.7 - 16.1$
60. $8.3 - (-0.62)$
61. $4.3 - (-0.87)$
62. $3 - (6[3 - (-2)] - 8)$
63. $4 - (3[9 - (-8)] - 11)$
64. $3 - 6 \cdot 4^2$
65. $2 - 3 \cdot 5^2$
66. $(3 - 6) + 4^2$
67. $(2 - 3) + 5^2$
68. $3 - (6 \cdot 4)^2$
69. $2 - (3 \cdot 5)^2$
70. $-2 + [(8 - 11) - (-2 - 9)]$
71. $-5 + [(4 - 15) - (-6) - 8]$

72. Subtract 8 from 7.
73. Subtract 9 from −4.
74. Decrease −8 by 15.
75. Decrease 11 by −14.

Solve the following. See Examples 5 and 6.

76. Within 24 hours in 1916, the temperature in Browning, Montana, fell from 44 degrees to −56 degrees. How large a drop in temperature was this?
77. Much of New Orleans is just barely above sea level. If George descends 12 feet from an elevation of 5 feet above sea level, what is his new elevation?
78. In a series of plays, the San Francisco 49ers gain 2 yards, lose 5 yards, and then lose another 20 yards. What is their total gain or loss of yardage?
79. In some card games, it is possible to have a negative score. Lavonne Schultz currently has a score of 15 points. She then loses 24 points. What is her new score?

80. Aristotle died in the year −322 (or 322 B.C.). When was he born, if he was 62 years old when he died?
81. Augustus Caesar died in A.D. 14 in his 77th year. When was he born?
82. Tyson Industries stock posted a loss of $1\frac{5}{8}$ points yesterday. If it drops another $\frac{3}{4}$ points today, find its overall change for the two days.
83. A commercial jet liner hits an air pocket and drops 250 feet. After climbing 120 feet, it drops another 178 feet. What is its overall vertical change?

The following bar graph is from an earlier section and shows the daily low temperatures for a week in Sioux Falls, South Dakota.

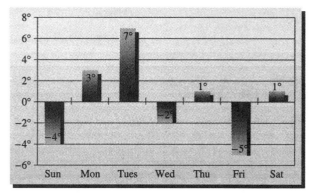

84. Use the bar graph to record the daily increases and decreases in the low temperatures from the previous day.

Day	Daily Increase or Decrease
Monday	
Tuesday	
Wednesday	
Thursday	
Friday	
Saturday	

85. Which day of the week had the greatest increase in temperature?
86. Which day of the week had the greatest decrease in temperature?

If $x = -5$, $y = 4$, and $t = 10$, simplify each expression.

87. $x - y$
88. $y - x$
89. $|x| + 2t - 8y$
90. $|x + t - 7y|$
91. $\dfrac{9 - x}{y + 6}$
92. $\dfrac{15 - x}{y + 2}$
93. $y^2 - x$
94. $t^2 - x$
95. $\dfrac{|x - (-10)|}{2t}$
96. $\dfrac{|5y - x|}{6t}$

97. In your own words, explain why $5 - 8$ simplifies to a negative number.
98. Explain why $6 - 11$ is the same as $6 + (-11)$.

Decide whether the given number is a solution of the given equation.

99. $x - 9 = 5$; -4
100. $x - 10 = -7$; 3
101. $-x + 6 = -x - 1$; -2
102. $-x - 6 = -x - 1$; -10
103. $-x - 13 = -15$; 2
104. $4 = 1 - x$; 5

1.7 MULTIPLYING AND DIVIDING REAL NUMBERS

OBJECTIVES

1. Multiply and divide real numbers.
2. Find the values of algebraic expressions.

TAPE BA 1.7

In this section, we discover patterns for multiplying and dividing real numbers. To discover sign rules for multiplication, recall that multiplication is repeated addition. Thus $3 \cdot 2$ means that 2 is an addend 3 times. That is,

$$2 + 2 + 2 = 3 \cdot 2$$

which equals 6. Similarly, $3 \cdot (-2)$ means -2 is an addend 3 times. That is,

$$(-2) + (-2) + (-2) = 3 \cdot (-2)$$

Since $(-2) + (-2) + (-2) = -6$, then $3 \cdot (-2) = -6$. This suggests that the product of a positive number and a negative number is a negative number.

What about the product of two negative numbers? To find out, consider the following pattern.

Factor decreases by 1 each time
$$\left.\begin{array}{l} -3 \cdot 2 = -6 \\ -3 \cdot 1 = -3 \\ -3 \cdot 0 = 0 \end{array}\right\} \text{Product increases by 3 each time.}$$

This pattern continues as

Factor decreases by 1 each time
$$\left.\begin{array}{l} -3 \cdot -1 = 3 \\ -3 \cdot -2 = 6 \end{array}\right\} \text{Product increases by 3 each time.}$$

This suggests that the product of two negative numbers is a positive number.

MULTIPLYING REAL NUMBERS
1. The product of two numbers with the same sign is a positive number.
2. The product of two numbers with different signs is a negative number.

EXAMPLE 1 Find the product.

    a. $(-6)(4)$ **b.** $2(-1)$ **c.** $(-5)(-10)$

Solution: **a.** $(-6)(4) = -24$ **b.** $2(-1) = -2$ **c.** $(-5)(-10) = 50$

We know that every whole number multiplied by zero equals zero. This remains true for signed numbers.

> **ZERO AS A FACTOR**
> If b is a real number, then $b \cdot 0 = 0$. Also, $0 \cdot b = 0$.

EXAMPLE 2 Perform the indicated operations.

    a. $(7)(0)(-6)$ **b.** $(-2)(-3)(-4)$ **c.** $(-1)(5)(-9)$ **d.** $(-2)^3$

    e. $(-4)(-11)-(5)(-2)$

Solution: **a.** By the order of operations, we multiply from left to right. Notice that, because one of the factors is 0, the product is 0.

$$(7)(0)(-6) = 0(-6) = 0$$

b. Multiply two factors at a time, from left to right.

$$(-2)(-3)(-4) = (6)(-4) \quad \text{Multiply } (-2)(-3).$$
$$= -24$$

c. Multiply from left to right.

$$(-1)(5)(-9) = (-5)(-9) \quad \text{Multiply } (-1)(5).$$
$$= 45$$

d. The exponent 3 means 3 factors of the base -2.

$$(-2)^3 = (-2)(-2)(-2)$$
$$= -8 \quad \text{Multiply.}$$

e. Follow the rules for order of operation.

$$(-4)(-11)-(5)(-2) = 44 - (-10) \quad \text{Find the products.}$$
$$= 44 + 10 \quad \text{Add 44 to the opposite of } -10.$$
$$= 54 \quad \text{Add.}$$

Multiplying signed decimals or fractions is carried out exactly the same way as multiplying by integers.

EXAMPLE 3 Find each product.

a. $(-1.2)(0.05)$ b. $\dfrac{2}{3} \cdot -\dfrac{7}{10}$

Solution: a. The product of two numbers with different signs is negative.

$(-1.2)(0.05) = -[(1.2)(0.05)]$
$= -0.06$

b. $\dfrac{2}{3} \cdot -\dfrac{7}{10} = -\dfrac{2 \cdot 7}{3 \cdot 10} = -\dfrac{2 \cdot 7}{3 \cdot 2 \cdot 5} = -\dfrac{7}{15}$

Just as every difference of two numbers $a - b$ can be written as the sum $a + (-b)$, so too every quotient of two numbers can be written as a product. For example, the quotient $6 \div 3$ can be written as $6 \cdot \dfrac{1}{3}$. Recall that the pair of numbers 3 and $\dfrac{1}{3}$ has a special relationship. Their product is 1 and they are called reciprocals or **multiplicative inverses** of each other.

> **RECIPROCALS OR MULTIPLICATIVE INVERSES**
> Two numbers whose product is 1 are called reciprocals or multiplicative inverses of each other.

Notice that **0 has no multiplicative inverse** since 0 multiplied by any number is never 1 but always 0.

EXAMPLE 4 Find the multiplicative inverse of each number.

a. 22 b. $\dfrac{3}{16}$ c. -10 d. $-\dfrac{9}{13}$

Solution: a. The multiplicative inverse of 22 is $\dfrac{1}{22}$ since $22 \cdot \dfrac{1}{22} = 1$.
b. The multiplicative inverse of $\dfrac{3}{16}$ is $\dfrac{16}{3}$ since $\dfrac{3}{16} \cdot \dfrac{16}{3} = 1$.
c. The multiplicative inverse of -10 is $-\dfrac{1}{10}$.
d. The multiplicative inverse of $-\dfrac{9}{13}$ is $-\dfrac{13}{9}$.

We may now write a quotient as an equivalent product.

> **QUOTIENT OF TWO REAL NUMBERS**
> If a and b are real numbers and b is not 0, then
> $$\dfrac{a}{b} = a \cdot \dfrac{1}{b}$$

In other words, the quotient of two real numbers is the product of the first number and the multiplicative inverse or reciprocal of the second number.

EXAMPLE 5 Use the definition of the quotient of two numbers to find each quotient.

a. $-18 \div 3$ b. $\dfrac{-14}{-2}$ c. $\dfrac{20}{-4}$

Solution: a. $-18 \div 3 = -18 \cdot \dfrac{1}{3} = -6$ b. $\dfrac{-14}{-2} = -14 \cdot -\dfrac{1}{2} = 7$

c. $\dfrac{20}{-4} = 20 \cdot -\dfrac{1}{4} = -5$

Since the quotient $a \div b$ can be written as the product $a \cdot \dfrac{1}{b}$, it follows that sign patterns for dividing two real numbers are the same as sign patterns for multiplying two real numbers.

> **MULTIPLYING AND DIVIDING REAL NUMBERS**
> 1. The product or quotient of two numbers with the same sign is a positive number.
> 2. The product or quotient of two numbers with different signs is a negative number.

EXAMPLE 6 Find each quotient.

a. $\dfrac{-24}{-4}$ b. $\dfrac{-36}{3}$ c. $\dfrac{2}{3} \div \left(-\dfrac{5}{4}\right)$

Solution: a. $\dfrac{-24}{-4} = 6$

b. $\dfrac{-36}{3} = -12$

c. $\dfrac{2}{3} \div \left(-\dfrac{5}{4}\right) = \dfrac{2}{3} \cdot \left(-\dfrac{4}{5}\right) = -\dfrac{8}{15}$

The definition of the quotient of two real numbers does not allow for division by 0 because 0 does not have a multiplicative inverse. There is no number we can multiply 0 by to get 1. How then do we interpret $\dfrac{3}{0}$? We say that division by 0 is not allowed or not defined and that $\dfrac{3}{0}$ does not represent a real number. The denominator of a fraction can never be 0.

Can the numerator of a fraction be 0? Can we divide 0 by a number? Yes. For example,

$$\dfrac{0}{3} = 0 \cdot \dfrac{1}{3} = 0$$

In general, the quotient of 0 and any nonzero number is 0.

MULTIPLYING AND DIVIDING REAL NUMBERS   SECTION 1.7   45

> **ZERO AS A DIVISOR OR DIVIDEND**
> 1. The quotient of any nonzero real number and 0 is undefined. In symbols, if $a \neq 0$, $\dfrac{a}{0}$ is **undefined.**
> 2. The quotient of 0 and any real number except 0 is 0. In symbols, if $a \neq 0$, $\dfrac{0}{a} = 0$.

EXAMPLE 7 Perform the indicated operations.

  a. $\dfrac{1}{0}$ **b.** $\dfrac{0}{-3}$ **c.** $\dfrac{0(-8)}{2}$

Solution: **a.** $\dfrac{1}{0}$ is undefined   **b.** $\dfrac{0}{-3} = 0$ **c.** $\dfrac{0(-8)}{2} = \dfrac{0}{2} = 0$

Notice that $\dfrac{12}{-2} = -6$, $\dfrac{12}{2} = -6$ and $\dfrac{-12}{2} = -6$. This means that

$$\dfrac{12}{-2} = -\dfrac{12}{2} = \dfrac{-12}{2}$$

In words, a single negative sign in a fraction can be written in the denominator, in the numerator, or in front of the fraction without changing the value of the fraction. Thus,

$$\dfrac{1}{-7} = \dfrac{-1}{7} = -\dfrac{1}{7}$$

In general, if a and b are real numbers, $b \neq 0$, $\dfrac{a}{-b} = \dfrac{-a}{b} = -\dfrac{a}{b}$.

Examples combining basic arithmetic operations along with the principles of order of operations help us to review these concepts.

EXAMPLE 8 Simplify each expression.

  a. $\dfrac{(-12)(-3) + 4}{-7 - (-2)}$

  b. $\dfrac{2(-3)^2 - 20}{-5 + 4}$

Solution: **a.** First, simplify the numerator and denominator separately, then divide.

$$\dfrac{(-12)(-3) + 4}{-7 - (-2)} = \dfrac{36 + 4}{-7 + 2}$$
$$= \dfrac{40}{-5}$$
$$= -8 \qquad \text{Divide.}$$

46 CHAPTER 1 REVIEW OF REAL NUMBERS

b. Simplify the numerator and denominator separately, then divide.

$$\frac{2(-3)^2 - 20}{-5 + 4} = \frac{2 \cdot 9 - 20}{-5 + 4} = \frac{18 - 20}{-5 + 4} = \frac{-2}{-1} = 2$$

EXAMPLE 9 If $x = -2$ and $y = -4$, evaluate each expression.

a. $5x - y$ **b.** $x^3 - y^2$ **c.** $\dfrac{3x}{2y}$

Solution: **a.** Replace x with -2 and y with -4 and simplify.

$$5x - y = 5(-2) - (-4) = -10 - (-4) = -10 + 4 = -6$$

b. Replace x with -2 and y with -4.

$$\begin{aligned} x^3 - y^2 &= (-2)^3 - (-4)^2 & \text{Substitute the given values for the variables.} \\ &= -8 - (16) & \text{Evaluate exponential expressions.} \\ &= -8 + (-16) & \text{Write as a sum.} \\ &= -24 & \text{Add.} \end{aligned}$$

c. Replace x with -2 and y with -4 and simplify.

$$\frac{3x}{2y} = \frac{3(-2)}{2(-4)} = \frac{-6}{-8} = \frac{3}{4}$$

SCIENTIFIC CALCULATOR EXPLORATIONS

ENTERING NEGATIVE NUMBERS

To enter a negative number on a calculator, find a key marked $\boxed{+/-}$. (On some calculators, this key is marked $\boxed{\text{CHS}}$ for "change sign.") To enter -8, for example, press the keys $\boxed{8}$ $\boxed{+/-}$. The display will read $\boxed{-8}$.

OPERATIONS WITH REAL NUMBERS

To evaluate $-2(7 - 9) - 20$ on a calculator, press the keys

$\boxed{2}$ $\boxed{+/-}$ $\boxed{\times}$ $\boxed{(}$ $\boxed{7}$ $\boxed{-}$ $\boxed{9}$ $\boxed{)}$ $\boxed{-}$ $\boxed{2}$ $\boxed{0}$ $\boxed{=}$.

The display will read $\boxed{-16}$.

Use a calculator to simplify each expression.

1. $-38(26 - 27)$
2. $-59(-8) + 1726$
3. $134 + 25(68 - 91)$
4. $45(32) - 8(218)$
5. $\dfrac{-50(294)}{175 - 265}$
6. $\dfrac{-444 - 444.8}{-181 - 324}$
7. $9^5 - 4550$
8. $5^8 - 6259$
9. $(-125)^2$ Be careful.
10. -125^2 Be careful.

EXERCISE SET 1.7

Find the following products. See Examples 1 through 3.

1. $(-3)(+4)$
2. $(+8)(-2)$
3. $-6(-7)$
4. $(-3)(-8)$
5. $(-2)(-5)(0)$
6. $(7)(0)(-3)$
7. $2(-9)$
8. $(-5)(3)$
9. $\left(-\dfrac{3}{4}\right)\left(\dfrac{8}{9}\right)$
10. $\left(\dfrac{5}{6}\right)\left(-\dfrac{3}{10}\right)$
11. $\left(-1\dfrac{1}{5}\right)\left(-1\dfrac{2}{3}\right)$
12. $\left(-\dfrac{5}{6}\right)\left(-\dfrac{3}{10}\right)$
13. $(-1)(2)(-3)(-5)$
14. $(-2)(-3)(-4)(-2)$
15. $(2)(-1)(-3)(5)(3)$
16. $(3)(-5)(-2)(-1)(-2)$
17. $(-4)^2$
18. $(-3)^3$

Decide whether each statement is true or false.

19. The product of three negative integers is negative.
20. The product of three positive integers is positive.
21. The product of four negative integers is negative.
22. The product of four positive integers is positive.

Find the multiplicative inverse or reciprocal of each number. See Example 4.

23. 9
24. 100
25. $\dfrac{2}{3}$
26. $\dfrac{1}{7}$
27. -14
28. $-\dfrac{3}{11}$

29. Find any real numbers that are their own reciprocal.
30. Explain why 0 has no reciprocal.

Find the following quotients. If the quotient is undefined, state so. See Examples 5–7.

31. $\dfrac{18}{-2}$
32. $-\dfrac{14}{7}$
33. $\dfrac{-12}{-4}$
34. $-\dfrac{20}{5}$
35. $\dfrac{-45}{-9}$
36. $\dfrac{30}{-2}$
37. $\dfrac{0}{-3}$
38. $-\dfrac{4}{0}$
39. $-\dfrac{3}{0}$
40. $\dfrac{0}{-4}$

Simplify the following. See Example 8.

41. $\dfrac{-6^2 + 4}{-2}$
42. $\dfrac{3^2 + 4}{5}$
43. $\dfrac{8 + (-4)^2}{4 - 12}$
44. $\dfrac{6 + (-2)^2}{4 - 9}$
45. $\dfrac{22 + (3)(-2)}{-5 - 2}$
46. $\dfrac{-20 + (-4)(3)}{1 - 5}$

If $x = -5$ and $y = -3$, evaluate each expression. See Example 9.

47. $3x + 2y$
48. $4x + 5y$
49. $2x^2 - y^2$
50. $x^2 - 2y^2$
51. $x^3 + 3y$
52. $y^3 + 3x$
53. $\dfrac{2x - 5}{y - 2}$
54. $\dfrac{2y - 12}{x - 4}$
55. $\dfrac{6 - y}{x - 4}$
56. $\dfrac{4 - 2x}{y + 3}$

Perform indicated operations. If the expression is undefined, state so.

57. $(-6)(-2)$
58. $5(-3)$
59. $(-7)(2)$
60. $(-3)(-9)$
61. $\dfrac{18}{-3}$
62. $\dfrac{-16}{-4}$
63. $-\dfrac{6}{0}$
64. $-\dfrac{16}{2}$
65. $-\dfrac{15}{-3}$
66. $\dfrac{48}{-12}$
67. $\dfrac{0}{-7}$
68. $-\dfrac{48}{-8}$
69. $(-6)(3)(-2)(-1)$
70. $(-3)(-2)(-1)(-2)$
71. $(-5)^3$
72. $(-2)^5$
73. $(-4)^2$
74. $(-6)^2$
75. -4^2
76. -6^2
77. $\dfrac{-3 - 5^2}{2(-7)}$
78. $\dfrac{-2 - 4^2}{3(-6)}$
79. $\dfrac{6 - 2(-3)}{4 - 3(-2)}$
80. $\dfrac{8 - 3(-2)}{2 - 5(-4)}$
81. $\dfrac{-3 - 2(-9)}{-15 - 3(-4)}$
82. $\dfrac{-4 - 8(-2)}{-9 - 2(-3)}$
83. $-3(2 - 8)$
84. $-4(3 - 9)$

48 CHAPTER 1   REVIEW OF REAL NUMBERS

85. $6(3 - 8)$ 86. $4(8 - 11)$
87. $-3[(2 - 8) - (-6 - 8)]$
88. $-2[(3 - 5) - (2 - 9)]$
89. $\left(\dfrac{2}{5}\right)\left(-1\dfrac{1}{4}\right)$ 90. $\left(-4\dfrac{2}{3}\right)\left(-\dfrac{8}{21}\right)$
91. $(1.82)(-4.6)$ 92. $(-3.6)(-0.61)$
93. $-22.4 \div (-1.6)$ 94. $12.24 \div (-2.4)$

If q is a negative number, r is a negative number, and t is a positive number, determine whether each expression simplifies to a positive or negative number. If it is not possible to determine, state so.

95. $\dfrac{q}{r \cdot t}$ 96. $q^2 \cdot r \cdot t$
97. $q + t$ 98. $t + r$
99. $t(q + r)$ 100. $r(q - t)$

Write each of the following as an expression and evaluate.

101. The sum of -2 and the quotient of -15 and 3
102. The sum of 1 and the product of -8 and -5
103. Twice the sum of -5 and -3
104. 7 subtracted from the quotient of 0 and 5

Decide whether the given number is a solution of the given equation.

105. $-5x = -35;\ 7$ 106. $2x = x - 1;\ -4$
107. $\dfrac{x}{-10} = 2;\ -20$ 108. $\dfrac{45}{x} = -15;\ -3$
109. $-3x - 5 = -20;\ 5$ 110. $2x + 4 = x + 8;\ -4$

111. The following graph shows Trader's stock consistently decreasing in value by $1.50 per share per day. If this trend continues, when will the stock be worth $20 per share?

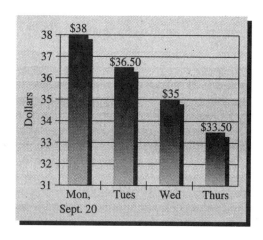

112. Explain why the product of an even number of negative numbers is a positive number.
113. If a and b are any real numbers, is the statement $a \cdot b = b \cdot a$ always true? Why or why not?
114. If a and b are any real numbers, is the statement $a - b = b - a$ always true? Why or why not?

1.8   PROPERTIES OF REAL NUMBERS

OBJECTIVES

1. Identify the commutative property.
2. Identify the associative property.
3. Identify the distributive property.
4. Identify the additive and multiplicative identities.
5. Identify the additive and multiplicative inverse properties.

TAPE BA 1.8

Specialized terms occur in every area of study. A biologist or ecologist will often be concerned with "symbiotic relationships." "Torque" is a major concern to the physicist. An economist or business analyst might be interested in "marginal revenue." A nurse needs to be familiar with "hemostasis." Mathematics, too, has its own specialized terms and principles.

This section introduces the basic properties of the real number system. Throughout this section, a, b, and c represent real numbers.

When we add, subtract, multiply, or divide two real numbers (except for division by zero), the result is a real number. This is guaranteed by the closure properties.

> **CLOSURE PROPERTIES**
> If a and b are real numbers then $a + b$, $a - b$, and ab are real numbers. Also, $\dfrac{a}{b}$, $b \neq 0$, is a real number.

Next, we look at the commutative properties. These properties state that the order in which any two real numbers are added or multiplied does not change their sum or product.

> **COMMUTATIVE PROPERTIES**
> **Addition:** $\qquad a + b = b + a$
> **Multiplication:** $\qquad a \cdot b = b \cdot a$

For example, if we let $a = 3$ and $b = 5$, then the commutative properties guarantee that

$$3 + 5 = 5 + 3 \quad \text{and} \quad 3 \cdot 5 = 5 \cdot 3$$

> REMINDER Is subtraction also commutative? Try an example. Is $3 - 2 = 2 - 3$? **No!** The left side of this statement equals 1; the right side equals -1. There is no commutative property of subtraction. Similarly, there is no commutative property for division. For example, $\frac{10}{2}$ does not equal $\frac{2}{10}$.

EXAMPLE 1   If $a = -2$, and $b = 5$, show that:

a. $a + b = b + a$ 　　　　　　　　　 **b.** $a \cdot b = b \cdot a$

Solution: **a.** Replace a with -2 and b with 5. Then

$$a + b = b + a$$

becomes

$$-2 + 5 = 5 + (-2)$$

or

$$3 = 3$$

Since both sides represent the same number, $-2 + 5 = 5 + (-2)$ is a true statement.

b. Replace a with -2 and b with 5. Then

$$a \cdot b = b \cdot a$$

becomes

$$-2 \cdot 5 = 5 \cdot (-2)$$

or

$$-10 = -10$$

Since both sides represent the same number, the statement is true.

2 When adding or multiplying three numbers, does it matter how we group the numbers? This question is answered by the associative properties. These properties state that when adding or multiplying three numbers, any two adjacent numbers may be grouped together without changing their sum or product.

> **ASSOCIATIVE PROPERTIES**
> **Addition:** $(a + b) + c = a + (b + c)$
> **Multiplication:** $(a \cdot b) \cdot c = a \cdot (b \cdot c)$

Illustrate these properties by working the following example.

EXAMPLE 2 If $a = -3$, $b = 2$, and $c = 4$, show that:

a. $(a + b) + c = a + (b + c)$ **b.** $(a \cdot b) \cdot c = a \cdot (b \cdot c)$

Solution: Replace a with -3, b with 2, and c with 4.

a.
$(a + b) + c = a + (b + c)$	
$(-3 + 2) + 4 = -3 + (2 + 4)$	Replace a with -3, b with 2, and c with 4.
$-1 + 4 = -3 + 6$	Simplify inside parentheses.
$3 = 3$	Add.

b.
$(a \cdot b) \cdot c = a \cdot (b \cdot c)$	
$(-3 \cdot 2) \cdot 4 = -3 \cdot (2 \cdot 4)$	Replace a with -3, b with 2, and c with 4.
$-6 \cdot 4 = -3 \cdot 8$	Simplify inside parentheses.
$-24 = -24$	Multiply.

3 The **distributive property of multiplication over addition** is used repeatedly throughout algebra. It is useful because it allows us to write a product as a sum or a sum as a product.

DISTRIBUTIVE PROPERTY OF MULTIPLICATION OVER ADDITION
$a(b + c) = ab + ac$

Since multiplication is commutative, the distributive property can also be written

$$(b + c)a = ba + ca$$

The truth of this property can be illustrated by letting $a = 3, b = 2$, and $c = 5$.

$$a(b + c) = ab + ac$$

becomes

$$3(2 + 5) = 3 \cdot 2 + 3 \cdot 5$$
$$3 \cdot 7 = 6 + 15$$
$$21 = 21$$

Notice in this example that 3 is "being distributed to" each addend inside the parentheses. That is, 3 is multiplied by each addend.

The distributive property can be extended so that factors can be distributed to more than two addends in parentheses. For example,

$$3(x + y + 2) = 3(x) + 3(y) + 3(2)$$
$$= 3x + 3y + 6$$

EXAMPLE 3 Use the distributive property to write each expression without parentheses.

a. $2(x + y)$ **b.** $-5(-3 + z)$ **c.** $5(x + y - z)$ **d.** $-1(2 - y)$
e. $-(3 + x - w)$

Solution:

a. $\quad 2(x + y) = 2 \cdot x + 2 \cdot y$
$\qquad\qquad\quad = 2x + 2y$

b. $\quad -5(-3 + z) = -5(-3) + (-5) \cdot z$
$\qquad\qquad\qquad\quad = 15 - 5z$

c. $\quad 5(x + y - z) = 5 \cdot x + 5 \cdot y + 5(-z)$
$\qquad\qquad\qquad\quad = 5x + 5y - 5z$

d. $\quad -1(2 - y) = (-1)(2) + (-1)(-y)$
$\qquad\qquad\qquad = -2 + y$

e. $\quad -(3 + x - w) = -1(3 + x - w)$
$\qquad\qquad\qquad\qquad = (-1)(3) + (-1)(x) + (-1)(-w)$
$\qquad\qquad\qquad\qquad = -3 - x + w$

Notice in the last example that $-(3 + x - w)$ is rewritten as $-\mathbf{1}(3 + x - w)$.

52 CHAPTER 1 REVIEW OF REAL NUMBERS

EXAMPLE 4 Use the distributive property to write each sum as a product.

a. $8 \cdot 2 + 8 \cdot x$ **b.** $7s + 7t$

Solution: **a.** $8 \cdot 2 + 8 \cdot x = 8(2 + x)$
b. $7s + 7t = 7(s + t)$

Next, we look at the **identity properties.** These properties guarantee that two special numbers exist. These numbers are called the **identity element for addition** and the **identity element for multiplication.**

IDENTITIES FOR ADDITION AND MULTIPLICATION

0 is the identity element for addition.

$$a + 0 = a \quad \text{and} \quad 0 + a = a$$

1 is the identity element for multiplication.

$$a \cdot 1 = a \quad \text{and} \quad 1 \cdot a = a$$

Notice that 0 is the only number that can be added to any real number with the result that the sum is the same real number. Also, 1 is the only number that can be multiplied by any other number with the result that the product is the same real number.

Additive inverses or **opposites** were introduced in Section 1.5. Two numbers are called additive inverses or opposites if their sum is 0. The additive inverse or opposite of 6 is -6 because $6 + (-6) = 0$. The additive inverse or opposite of -5 is 5 because $-5 + 5 = 0$.

Reciprocals or **multiplicative inverses** were introduced in Section 1.2. Two nonzero numbers are called reciprocals or multiplicative inverses if their product is 1. The reciprocal or multiplicative inverse of $\frac{2}{3}$ is $\frac{3}{2}$ because $\frac{2}{3} \cdot \frac{3}{2} = 1$. Likewise, the reciprocal of -5 is $-\frac{1}{5}$ because $-5\left(-\frac{1}{5}\right) = 1$.

ADDITIVE OR MULTIPLICATIVE INVERSES

The numbers a and $-a$ are additive inverses or opposites of each other because their sum is 0; that is,

$$a + (-a) = 0$$

The numbers b and $\frac{1}{b}$ (for $b \neq 0$) are called reciprocals or multiplicative inverses of each other because their product is 1; that is,

$$b \cdot \frac{1}{b} = 1$$

EXAMPLE 5 Find the additive inverse or opposite of each number.

 a. -3 b. 5 c. 0 d. $|-2|$

Solution: a. The additive inverse of -3 is 3 because $-3 + 3 = 0$.
 b. The additive inverse of 5 is -5 because $5 + (-5) = 0$.
 c. The additive inverse of 0 is 0 because $0 + 0 = 0$.
 d. $|-2| = 2$. The additive inverse of 2 (and $|-2|$) is -2.

EXAMPLE 6 Find the multiplicative inverse or reciprocal of each number.

 a. 7 b. $\dfrac{-1}{9}$

Solution: a. The multiplicative inverse of 7 is $\dfrac{1}{7}$ because $7 \cdot \dfrac{1}{7} = 1$.

 b. The multiplicative inverse of $\dfrac{-1}{9}$ is $\dfrac{9}{-1}$, or -9, because $\left(\dfrac{-1}{9}\right)(-9) = 1$.

EXAMPLE 7 Name the property illustrated by each true statement.

 a. $2 \cdot 3 = 3 \cdot 2$
 b. $3(x + 5) = 3x + 15$
 c. $2 + (4 + 8) = (2 + 4) + 8$

Solution: a. The commutative property of multiplication
 b. The distributive property
 c. The associative property of addition

EXAMPLE 8 Use the indicated property and the given expression to write a true statement.

 a. $2 + 9$; the commutative property of addition
 b. $(5 \cdot 8) \cdot 9$; the associative property of multiplication
 c. $x + 0$; the additive identity property

Solution: a. $2 + 9 = 9 + 2$
 b. $(5 \cdot 8) \cdot 9 = 5 \cdot (8 \cdot 9)$
 c. $x + 0 = x$

EXERCISE SET 1.8

Name the properties illustrated by each true statement. See Examples 1, 2, and 7.

1. $3 \cdot 5 = 5 \cdot 3$
2. $4(3 + 8) = 4 \cdot 3 + 4 \cdot 8$
3. $2 + (8 + 5) = (2 + 8) + 5$
4. $4 + 9 = 9 + 4$
5. $9(3 + 7) = 9 \cdot 3 + 9 \cdot 7$
6. $1 \cdot 9 = 9$
7. $(4 \cdot 8) \cdot 9 = 4 \cdot (8 \cdot 9)$
8. $6 \cdot \dfrac{1}{6} = 1$
9. $0 + 6 = 6$
10. $(4 + 9) + 6 = 4 + (9 + 6)$
11. $-4(3 + 7) = -4 \cdot 3 + (-4) \cdot 7$
12. $11 + 6 = 6 + 11$
13. $-4 \cdot (8 \cdot 3) = (-4 \cdot 8) \cdot 3$
14. $10 + 0 = 10$
15. Write an example that shows that division is not commutative.
16. Write an example that shows that subtraction is not commutative.

Use the distributive property to write each expression without parentheses. See Example 3.

17. $3(6 + x)$
18. $2(x - 5)$
19. $-2(y - z)$
20. $-3(z - y)$
21. $-7(3y - 5)$
22. $-5(2r + 11)$
23. $5(x + 4m + 2)$
24. $8(3y + z - 6)$
25. $-4(1 - 2m + n)$
26. $-4(4 + 2p + 5)$
27. $-(5x + 2)$
28. $-(9r + 5)$
29. $-(r - 3 - 7p)$
30. $-(-q - 2 + 6r)$

Use the distributive property to write each sum as a product. See Example 4.

31. $4 \cdot 1 + 4 \cdot y$
32. $14 \cdot z + 14 \cdot 5$
33. $11x + 11y$
34. $9a + 9b$
35. $(-1) \cdot 5 + (-1) \cdot x$
36. $(-3)a + (-3)y$

Find the additive inverse or opposite of each of the following numbers. See Example 5.

37. 16
38. 14
39. -8
40. -3
41. $|-9|$
42. $|11|$
43. $\dfrac{2}{3}$
44. $-\dfrac{7}{8}$
45. $-(-1.2)$
46. $-(7.9)$
47. $-|-2|$
48. $-|-9|$

Find the multiplicative inverse or reciprocal of each of the following numbers. See Example 6.

49. $\dfrac{2}{3}$
50. $\dfrac{3}{4}$
51. $-\dfrac{5}{6}$
52. $-\dfrac{7}{8}$
53. 6
54. 3
55. $-2\dfrac{1}{2}$
56. -5
57. $-\left|-\dfrac{3}{5}\right|$
58. $-\left|-\dfrac{2}{5}\right|$
59. $3\dfrac{5}{6}$
60. $2\dfrac{3}{5}$

Write the additive inverse and the multiplicative inverse of each expression. Assume that the value of each expression is not 0.

61. x
62. y
63. $-3z$
64. $5a$
65. $a + b$
66. $x - y$

Use the indicated property and the given expression to write a true statement. See Example 8.

67. $\dfrac{2}{3} \cdot \dfrac{3}{2}$; multiplicative inverse property
68. $8 + 16$; commutative property of addition
69. $(-4)(-3)$; commutative property of multiplication
70. $-4 + 4$; additive inverse property
71. $3 + (8 + 9)$; associative property of addition
72. $(4 \cdot 3) \cdot 9$; associative property of multiplication
73. $y + 0$; additive identity property
74. $1 \cdot x$; multiplicative identity property
75. $x(a + b)$; distributive property
76. $(m + n)y$; distributive property

77. $a(b+c)$; commutative property of multiplication

78. $x + 2$; commutative property of addition

Name the property illustrated by each step.

79. a. $7(2 + x) + 5 = 14 + 7x + 5$
   b. $ = 7x + 14 + 5$
       $ = 7x + 19$

80. a. $-10 + 3(y + 4) = -10 + 3y + 12$
   b. $ = -10 + 12 + 3y$
       $ = 2 + 3y$

81. a. $\triangle + (\square + \bigcirc) = (\square + \bigcirc) + \triangle$
   b. $ = (\bigcirc + \square) + \triangle$
   c. $ = \bigcirc + (\square + \triangle)$

82. a. $(x + y) + z = x + (y + z)$
   b. $ = (y + z) + x$
   c. $ = (z + y) + x$

83. Explain why 0 is called the identity element for addition.

84. Explain why 1 is called the identity element for multiplication.

1.9 READING GRAPHS

OBJECTIVES

 Read bar graphs.
 Read line graphs.

TAPE BA 1.9

 In today's world, where the exchange of information is required to be fast and entertaining, graphs are becoming increasingly popular. They provide a quick way of making comparisons, drawing conclusions, and approximating quantities. Thus far, we have practiced reading bar graphs and circle graphs or pie charts. In this section we continue our study of bar graphs and we introduce line graphs.

A bar graph consists of a series of bars arranged vertically or horizontally. The bar graph on the next page shows a comparison of the rates charged by selected electricity companies. The names of the companies are listed horizontally and a bar is shown for each company. Corresponding to the height of the bar for each company is a number along a vertical axis. These vertical numbers are cents charged for each kilowatt-hour of electricity used.

EXAMPLE 1 The bar graph on the next page shows the cents charged per kilowatt-hour for selected electricity companies.

    a. Which company charges the highest rate?

    b. Which company charges the lowest rate?

    c. Approximate the electricity rate charged by the first four companies listed.

    d. Approximate the difference in the rates charged by the companies in parts (a) and (b).

56 CHAPTER 1 REVIEW OF REAL NUMBERS

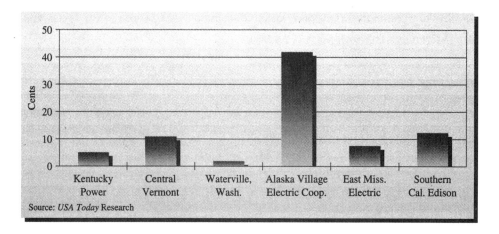

Source: *USA Today* Research

Solution: **a.** The tallest bar corresponds to the company that charges the highest rate. Alaska Village Electric Cooperative charges the highest rate.

b. The shortest bar corresponds to the company that charges the lowest rate. Waterville, Washington charges the lowest rate.

c. To approximate the rate charged by Kentucky Power, go to the top of the bar that corresponds to this company. From the top of the bar, move horizontally to the left until the vertical axis is reached.

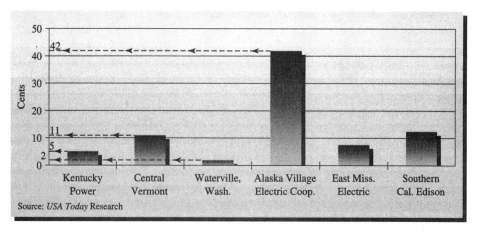

Source: *USA Today* Research

The height of the bar is approximately halfway between the 0 and 10 marks. We therefore conclude that

Kentucky Power charges approximately 5¢ per kilowatt-hour.

Central Vermont charges approximately 11¢ per kilowatt-hour.

Waterville, Washington charges approximately 2¢ per kilowatt-hour.

Alaska Village Electric charges approximately 42¢ per kilowatt-hour.

d. The difference in rates for Alaska Village Electric Cooperative and Waterville, Washington is approximately 42¢ − 2¢ or 40¢.

As mentioned earlier, a bar graph can consist of vertically arranged bars as in the graph above or horizontally arranged bars as in the next graph.

EXAMPLE 2 The following bar graph shows Disney's top six animated films before 1995 and the amount of money they generated at theaters.

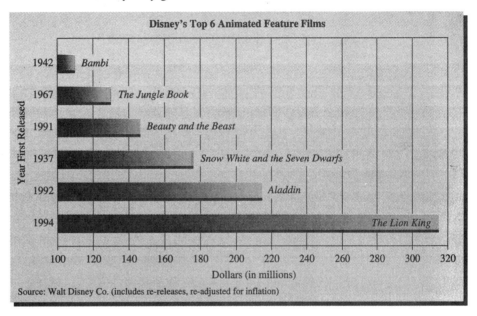

a. Find the film shown that generated the most income for Disney and approximate the income.

b. How much more money did the film *Aladdin* make than the film *Beauty and the Beast*?

Solution: a. Since these bars are arranged horizontally, look for the longest bar, which is the bar representing the film *The Lion King*. To approximate the income from this film, from the right edge of this bar move vertically downward to the dollars axis. This film generated approximately 315 million dollars, or $315,000,000, the most income for Disney.

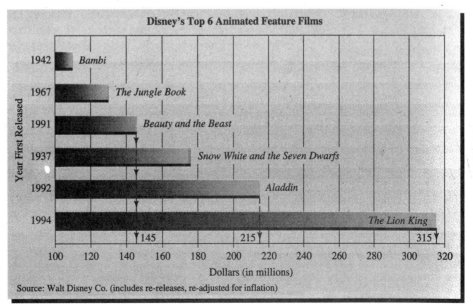

b. *Aladdin* generated approximately 215 million dollars. *Beauty and the Beast* generated approximately 145 million dollars. To find how much more money *Aladdin* generated than *Beauty and the Beast,* subtract: 215 − 145 = 70 million dollars, or $70,000,000.

 The next graph is called a **line graph.**

EXAMPLE 3 The line graph below shows the relationship between two sets of measurements: the distance driven in a 14-foot U-Haul truck in one day and the total cost of renting this truck for that day. Notice that the horizontal axis is labeled Distance and the vertical axis is labeled Total Cost.

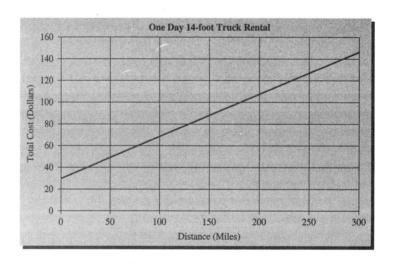

a. Find the total cost of renting the truck if 100 miles are driven.
b. Find the number of miles driven if the total cost of renting is $140.

Solution: **a.** Find the number 100 on the horizontal scale and move vertically upward until the line is reached. From this point on the line, move horizontally to the left until the vertical scale is reached. The total cost of renting the truck if 100 miles are driven is approximately $70.

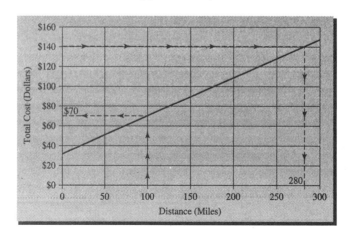

b. Find the number 140 on the vertical scale and move horizontally to the right until the line is reached. From this point on the line, move vertically downward until the horizontal scale is reached. The truck is driven approximately 280 miles.

From the previous example, we can see that graphing provides a quick way to approximate quantities. In Chapter 3 we show how we can use equations to find exact answers to the questions posed in Example 3. The next graph is another example of a line graph. It is also sometimes called a **broken line graph.**

EXAMPLE 4 This line graph shows the relationship between time spent smoking a cigarette and pulse rate. Time is recorded along the horizontal axis in minutes, with 0 minutes being the moment a smoker lights a cigarette. Pulse is recorded along the vertical axis in heartbeats per minute.

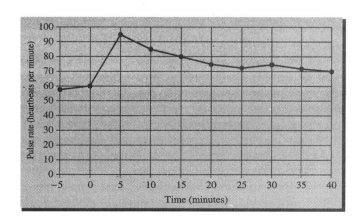

a. What is the pulse rate 15 minutes after lighting a cigarette?
b. When is the pulse rate the lowest?
c. When does the pulse rate show the greatest change?

Solution: **a.** Locate the number 15 along the time axis and move vertically upward until the line is reached. From this point on the line, move horizontally to the left until the pulse rate axis is reached. Read the number of beats per minute. The pulse rate is 80 beats per minute 15 minutes after lighting a cigarette.

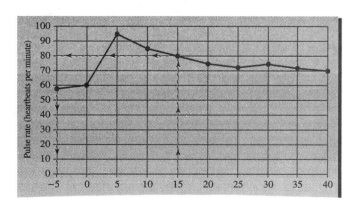

b. Find the lowest point of the line graph, which represents the lowest pulse rate. From this point, move vertically downward to the time axis. The pulse rate is the lowest at -5 minutes, which means 5 minutes *before* lighting a cigarette.

c. The pulse rate shows the greatest change during the 5 minutes between 0 and 5. Notice that the line graph is *steepest* between 0 and 5 minutes.

EXERCISE SET 1.9

Use the bar graph in Example 1 to answer the following.

1. Approximate the electricity rate charged by East Miss. Electric.
2. Approximate the electricity rate charged by Southern Cal. Edison.
3. Which companies shown charge more than 12¢ per kilowatt-hour?
4. Which companies shown charge less than 10¢ per kilowatt hour?
5. Find the rate charged by the company shown that is the nearest to your home.

Use the bar graph in Example 2 to answer the following.

6. Approximate the income generated by the film *Bambi*.
7. Approximate the income generated by the film *The Jungle Book*.
8. Approximate the income generated by the film *Snow White and the Seven Dwarfs*.
9. How much more money did the film *Snow White and the Seven Dwarfs* generate than *Bambi*?
10. Before 1990, which Disney film generated the most income?
11. After 1990, which Disney Film generated the most income?

Use the line graph in Example 3 to answer the following.

12. Find the total cost of renting the truck if 50 miles are driven.
13. Find the total cost of renting the truck if 230 miles are driven.
14. Find the number of miles driven if the total cost of renting is $80.
15. Find the number of miles driven if the total cost of renting is $50.

Use the line graph in Example 4 to answer the following.

16. Approximate the pulse rate 5 minutes before lighting a cigarette.
17. Approximate the pulse rate 10 minutes after lighting a cigarette.
18. Find the difference in pulse rate between 5 minutes before and 10 minutes after lighting a cigarette.
19. What is the highest pulse rate shown on the graph?

The following bar graph shows the team in each sport that has gone the longest time without being in a playoff.

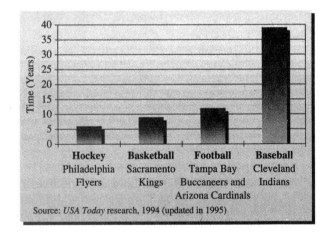

20. Which team for the sports shown has gone the longest time without being in a playoff?
21. Which hockey team has gone the longest without being in a playoff?
22. Why are 2 football teams listed?
23. Approximate the greatest number of years that a basketball team has gone without being in the playoffs.

24. How many more years has a football team gone without being in the playoffs than a basketball team?

25. How many more years has a football team gone without being in the playoffs than a hockey team?

The line graph below shows the average cost of newsprint per metric ton since 1984.

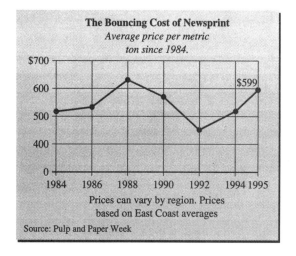

26. During what year was the cost of newsprint less than $500 per metric ton?

27. What year shown had the highest cost of newsprint? Estimate this cost.

28. What year shown had the lowest cost of newsprint? Estimate this cost.

29. Estimate the cost of newsprint in 1986.

30. In April 1995, the *Houston Post* newspaper closed. One reason given for this close is shown by a trend on this line graph. What is that trend?

Geographic locations can be described by a gridwork of lines called latitudes and longitudes as shown below. For example, the location of Houston, Texas, can be described by latitude 30° north and longitude 95° west.

31. Using latitude and longitude, describe the location of New Orleans, Louisiana.

32. Using latitude and longitude, describe the location of Denver, Colorado.

33. Use an atlas and describe the location of your hometown.

34. Give another name for 0° latitude.

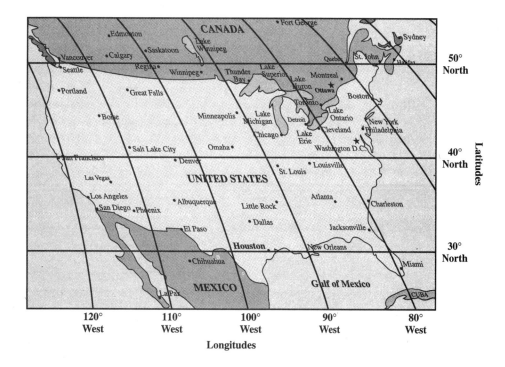

GROUP ACTIVITY

CREATING AND INTERPRETING GRAPHS

MATERIALS:
- Compass or circle template
- Grapher with bar graph and circle graph capabilities (optional)

Conduct the following survey during class and record the results.

a. Into which of the following categories does your age fall?
   - age ≤ 20
   - 20 < age ≤ 25
   - 25 < age ≤ 30
   - 30 < age ≤ 35
   - 35 < age ≤ 40
   - age > 40

b. What is your gender?
   - female
   - male

c. Which of the following brands of breakfast cereal did you eat today? (Choose only one.)
   - Kellogg's
   - Post
   - General Mills
   - Quaker
   - Ralston
   - Other
   - Did not eat breakfast cereal today

1. For each survey question, tally the results for each category. Present the results in a table.

2. For each survey question, find the fraction of the total number of responses that fall in each answer category.

3. Using a compass (or circle template), create a circle graph for each set of responses to questions (a)–(c). Label the graph.

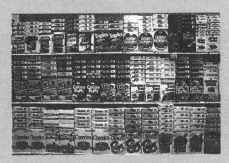

4. Create a bar graph for each set of responses to questions (a)–(c). Label the graph.

5. Study your graphs and discuss what you may conclude from them. What do the graphs tell you about your survey respondents? Write a paragraph summarizing your group's conclusions. For any of the survey question results, do you prefer one type of graph over the other? Why?

6. (Optional) Use a grapher to create the circle graphs and bar graphs in Questions 3 and 4. Compare them to your own graphs. Are there any differences? Explain.

Chapter 1 Highlights

Definitions and Concepts	Examples
Section 1.1 Symbols and Sets of Numbers	
A **set** is called a collection of objects, called **elements**, enclosed in braces.	$\{a, c, e\}$
Natural Numbers: $\{1, 2, 3, 4, \ldots\}$ **Whole Numbers:** $\{0, 1, 2, 3, 4, \ldots\}$ **Integers:** $\{\ldots, -3, -2, -1, 0, 1, 2, 3, \ldots\}$ **Rational Numbers:** {real numbers that can be expressed as a quotient of integers} **Irrational Numbers:** {real numbers that cannot be expressed as a quotient of integers} **Real Numbers:** {all numbers that correspond to a point on the number line}	Given the set $\{-3.4, \sqrt{3}, 0, \frac{2}{3}, 5, -4\}$ list the numbers that belong to the set of Natural numbers 5 Whole numbers 0, 5 Integers $-4, 0, 5$ Rational numbers $-3.4, 0, \frac{2}{3}, 5, -4$ Irrational numbers $\sqrt{3}$ Real numbers $-3.4, \sqrt{3}, 0, \frac{2}{3}, 5, -4$
A line used to picture numbers is called a **number line.**	$\xleftarrow{\quad\mid\quad\mid\quad\mid\quad\mid\quad\mid\quad\mid\quad\mid\quad}\rightarrow$ $-3\;-2\;-1\;\;0\;\;1\;\;2\;\;3$
The **absolute value** of a real number a denoted by $|a|$ is the distance between a and 0 on the number line.	$\|5\| = 5 \quad \|0\| = 0 \quad \|-2\| = 2$
Symbols: $=$ is equal to $\neq$ is not equal to $>$ is greater than $<$ is less than $\leq$ is less than or equal to $\geq$ is greater than or equal to	$-7 = -7$ $3 \neq -3$ $4 > 1$ $1 < 4$ $6 \leq 6$ $18 \geq -\dfrac{1}{3}$
Order Property for Real Numbers For any two real numbers a and b, a is less than b if a is to the left of b on the number line.	$\xleftarrow{\;\bullet\;\mid\;\mid\;\bullet\;\mid\;\mid\;\bullet\;\mid}\rightarrow$ $-3\;-2\;-1\;\;0\;\;1\;\;2\;\;3$ $-3 < 0 \quad 0 > -3 \quad 0 < 2.5 \quad 2.5 > 0$
Section 1.2 Fractions	
A quotient of two integers is called a **fraction.** The **numerator** of a fraction is the top number. The **denominator** of a fraction is the bottom number.	$\dfrac{13}{17}$ ← numerator ← denominator
If $a \cdot b = c$, then a and b are **factors** and c is the **product.**	$\underset{\text{factor}}{7} \cdot \underset{\text{factor}}{9} = \underset{\text{product}}{63}$

(continued)

Definitions and Concepts	Examples
Section 1.2 Fractions	
A fraction is in **lowest terms** when the numerator and the denominator have no factors in common other than 1.	$\frac{13}{17}$ is in lowest terms.
To write a fraction in lowest terms, factor the numerator and the denominator; then apply the fundamental property.	Write in lowest terms. $\frac{6}{14} = \frac{2 \cdot 3}{2 \cdot 7} = \frac{3}{7}$
Two fractions are **reciprocals** if their product is 1. The reciprocal of $\frac{a}{b}$ is $\frac{b}{a}$.	The reciprocal of $\frac{6}{25}$ is $\frac{25}{6}$
To multiply fractions, numerator times numerator is the numerator of the product and denominator times denominator is the denominator of the product.	Perform the indicated operations. $\frac{2}{5} \cdot \frac{3}{7} = \frac{6}{35}$
To divide fractions, multiply the first fraction by the reciprocal of the second fraction.	$\frac{5}{9} \div \frac{2}{7} = \frac{5}{9} \cdot \frac{7}{2} = \frac{35}{18}$
To add fractions with the same denominator, add the numerators and place the sum over the common denominator.	$\frac{5}{11} + \frac{3}{11} = \frac{8}{11}$
To subtract fractions with the same denominator, subtract the numerators and place the difference over the common denominator.	$\frac{13}{15} - \frac{3}{15} = \frac{10}{15} = \frac{2}{3}$
Fractions that represent the same quantity are called **equivalent fractions.**	$\frac{1}{5} = \frac{1 \cdot 4}{5 \cdot 4} = \frac{4}{20}$ $\frac{1}{5}$ and $\frac{4}{20}$ are equivalent fractions.
Section 1.3 Exponents and Order of Operations	
The expression a^n is an **exponential expression.** The number a is called the **base;** it is the repeated factor. The number n is called the **exponent;** it is the number of times that the base is a factor.	$4^3 = 4 \cdot 4 \cdot 4 = 64$ $7^2 = 7 \cdot 7 = 49$
Order of Operations Simplify expressions in the following order. If grouping symbols are present, simplify expressions within those first, starting with the innermost set. Also, simplify the numerator and the denominator of a fraction separately. 1. Simplify exponential expressions. 2. Multiply or divide in order from left to right. 3. Add or subtract in order from left to right.	$\frac{8^2 + 5(7-3)}{3 \cdot 7} = \frac{8^2 + 5(4)}{21}$ $= \frac{64 + 5(4)}{21}$ $= \frac{64 + 20}{21}$ $= \frac{84}{21}$ $= 4$

(continued)

Definitions and Concepts	**Examples**
Section 1.4 Introduction to Variable Expressions and Equations	
A symbol used to represent a number is called a **variable**.	Examples of variables are: $$q, x, z$$
An **algebraic expression** is a collection of numbers, variables, operation symbols, and grouping symbols.	Examples of algebraic expressions are: $$5x,\ 2(y-6),\ \frac{q^2 - 3q + 1}{6}$$
To **evaluate an algebraic expression** containing a variable, substitute a given number for the variable and simplify.	Evaluate $x^2 - y^2$ if $x = 5$ and $y = 3$. $$x^2 - y^2 = (5)^2 - 3^2$$ $$= 25 - 9$$ $$= 16$$
A mathematical statement that two expressions are equal is called an **equation**.	Equations: $$3x - 9 = 20$$ $$A = \pi r^2$$
A **solution** or **root** of an equation is a value for the variable that makes the equation a true statement.	Determine whether 4 is a solution of $5x + 7 = 27$. $$5x + 7 = 27$$ $$5(4) + 7 = 27$$ $$20 + 7 = 27$$ $$27 = 27 \quad \text{True}$$ 4 is a solution.
Section 1.5 Adding Real Numbers	
To Add Two Numbers with the Same Sign	Add.
1. Add their absolute values.	$$10 + 7 = 17$$
2. Use their common sign as the sign of the sum.	$$-3 + (-8) = -11$$
To Add Two Numbers with Different Signs	
1. Subtract their absolute values.	$$-25 + 5 = -20$$
2. Use the sign of the number whose absolute value is larger as the sign of the sum.	$$14 + (-9) = 5$$
Two numbers that are the same distance from 0 but lie on opposite sides of 0 are called **opposites** or **additive inverses**. The opposite of a number a is denoted by $-a$.	The opposite of -7 is 7. The opposite of 123 is -123.
The sum of a number a and its opposite, $-a$, is 0. $$a + (-a) = 0$$ If a is a number, then $-(-a) = a$	$$-4 + 4 = 0$$ $$12 + (-12) = 0$$ $$-(-8) = 8$$ $$-(-14) = 14$$

(continued)

66 CHAPTER 1 REVIEW OF REAL NUMBERS

DEFINITIONS AND CONCEPTS	EXAMPLES
SECTION 1.6 SUBTRACTING REAL NUMBERS	
To subtract two numbers a and b, add the first number a to the opposite of the second number b. $$a - b = a + (-b)$$	Subtract. $$3 - (-44) = 3 + 44 = 47$$ $$-5 - 22 = -5 + (-22) = -27$$ $$-30 - (-30) = -30 + 30 = 0$$
SECTION 1.7 MULTIPLYING AND DIVIDING REAL NUMBERS	
Quotient of two real numbers $$\frac{a}{b} = a \cdot \frac{1}{b}$$ **Multiplying and Dividing Real Numbers** The product or quotient of two numbers with the same sign is a positive number. The product or quotient of two numbers with different signs is a negative number. **Products and Quotients Involving Zero** The product of 0 and any number is 0. $$b \cdot 0 = 0 \text{ and } 0 \cdot b = 0$$ The quotient of a nonzero number and 0 is undefined. $\frac{b}{0}$ is undefined The quotient of 0 and any nonzero number is 0. $$\frac{0}{b} = 0$$	Multiply or divide. $$\frac{42}{2} = 42 \cdot \frac{1}{2} = 21$$ $7 \cdot 8 = 56 \qquad -7 \cdot (-8) = 56$ $-2 \cdot 4 = -8 \qquad 2 \cdot (-4) = -8$ $\frac{90}{10} = 9 \qquad \frac{-90}{-10} = 9$ $\frac{42}{-6} = -7 \qquad \frac{-42}{6} = -7$ $-4 \cdot 0 = 0 \qquad 0 \cdot \left(-\frac{3}{4}\right) = 0$ $\frac{-85}{0}$ is undefined. $\frac{0}{18} = 0 \qquad \frac{0}{-47} = 0$
SECTION 1.8 PROPERTIES OF REAL NUMBERS	
Commutative Properties Addition: $\quad a + b = b + a$ Multiplication: $\quad a \cdot b = b \cdot a$ **Associative Properties** Addition: $\quad (a + b) + c = a + (b + c)$ Multiplication: $\quad (a \cdot b) \cdot c = a \cdot (b \cdot c)$ Two numbers whose product is 1 are called **multiplicative inverses** or **reciprocals**. The reciprocal of a nonzero number a is $\frac{1}{a}$ because $a \cdot \frac{1}{a} = 1$.	$3 + (-7) = -7 + 3$ $-8 \cdot 5 = 5 \cdot (-8)$ $(5 + 10) + 20 = 5 + (10 + 20)$ $(-3 \cdot 2) \cdot 11 = -3 \cdot (2 \cdot 11)$ The reciprocal of 3 is $\frac{1}{3}$. The reciprocal of $-\frac{2}{5}$ is $-\frac{5}{2}$.

(continued)

Definitions and Concepts	Examples
Section 1.8 Properties of Real Numbers	
Distributive Property $a(+c) = a \cdot b + a \cdot c$ **Identities** $a + 0 = a \quad\quad 0 + a = a$ $a \cdot 1 = a \quad\quad 1 \cdot a = a$ **Inverses** Addition or opposite: $\quad a + (-a) = 0$ Multiplication or reciprocal: $\quad b \cdot \frac{1}{b} = 1$	$5(6 + 10) = 5 \cdot 6 + 5 \cdot 10$ $-2(3 + x) = -2 \cdot 3 + (-2)(x)$ $5 + 0 = 5 \quad\quad 0 + (-2) = -2$ $-14 \cdot 1 = -14 \quad\quad 1 \cdot 27 = 27$ $7 + (-7) = 0$ $3 \cdot \frac{1}{3} = 1$
Section 1.9 Reading Graphs	
To find the value on the vertical axis representing a location on a graph, move horizontally from the location on the graph until the vertical axis is reached. To find the value on the horizontal axis representing a location on a graph, move vertically from the location on the graph until the horizontal axis is reached. The broken line graph to the right shows the average public classroom teachers' salaries for the school year ending in the years shown. Estimate the average public teacher's salary for the school year ending in 1989. The average salary is approximately $29,500. Find the earliest year that the average salary rose above $32,000. The year was 1991.	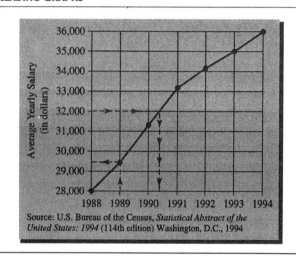 Source: U.S. Bureau of the Census, *Statistical Abstract of the United States: 1994* (114th edition) Washington, D.C., 1994

CHAPTER 1 REVIEW

(1.1) Insert $<$, $>$, or $=$ in the appropriate space to make the following statements true.

1. 8   10

2. 7   2

3. -4 -5

4. $\frac{12}{2}$ -8

5. $|-7|$ $|-8|$

6. $|-9|$ -9

7. $-|-1|$ -1

8. $|-14|$ $-(-14)$

9. 1.2   1.02

10. $-\frac{3}{2}$ $-\frac{3}{4}$

Translate each statement into symbols.

11. Four is greater than or equal to negative three.

12. Six is not equal to five.

13. 0.03 is less than 0.3.

14. Lions and hyenas were featured in the Disney film *The Lion King*. For short distances, lions can run at a rate of 50 miles per hour whereas hyenas can run at a rate of 40 miles per hour. Write an inequality statement comparing the numbers 50 and 40.

Given the following sets of numbers, list the numbers in each set that also belong to the set of:

a. Natural numbers b. Whole numbers
c. Integers d. Rational numbers
e. Irrational numbers f. Real numbers

15. $\{-6, 0, 1, 1\frac{1}{2}, 3, \pi, 9.62\}$

16. $\{-3, -1.6, 2, 5, \frac{11}{2}, 15.1, \sqrt{5}, 2\pi\}$

The following chart shows the gains and losses in dollars of Density Oil and Gas stock for a particular week.

Day	Gain or Loss in Dollars
Monday	+1
Tuesday	−2
Wednesday	+5
Thursday	+1
Friday	−4

17. Which day showed the greatest loss?

18. Which day showed the greatest gain?

(1.2) *Write the number as a product of prime factors.*

19. 36 **20.** 120

Perform the indicated operations. Write results in lowest terms.

21. $\frac{8}{15} \cdot \frac{27}{30}$ **22.** $\frac{7}{8} \div \frac{21}{32}$

23. $\frac{7}{15} + \frac{5}{6}$ **24.** $\frac{3}{4} - \frac{3}{20}$

25. $2\frac{3}{4} + 6\frac{5}{8}$ **26.** $7\frac{1}{6} - 2\frac{2}{3}$

27. $5 \div \frac{1}{3}$ **28.** $2 \cdot 8\frac{3}{4}$

29. Determine the unknown part of the given circle.

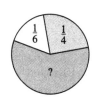

Find the area and the perimeter of each figure.

30.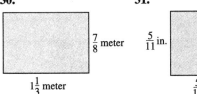

31.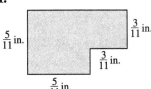

32. A trim carpenter needs a piece of quarter round molding $6\frac{1}{8}$ feet long for a bathroom. She finds a piece $7\frac{1}{2}$ feet long. How long a piece does she need to cut from the $7\frac{1}{2}$-foot-long molding in order to use it in the bathroom?

(1.3) *Simplify each expression.*

33. $6 \cdot 3^2 + 2 \cdot 8$ **34.** $68 - 5 \cdot 2^3$

35. $3(1 + 2 \cdot 5) + 4$ **36.** $8 + 3(2 \cdot 6 - 1)$

37. $\dfrac{4 + |6 - 2| + 8^2}{4 + 6 \cdot 4}$ **38.** $5[3(2 + 5) - 5]$

Translate each word statement to symbols.

39. The difference of twenty and twelve is equal to the product of two and four.

40. The quotient of nine and two is greater than negative five.

(1.4) *Evaluate each expression if $x = 6$, $y = 2$, $z = 8$.*

41. $2x + 3y$ **42.** $x(y + 2z)$

43. $\dfrac{x}{y} + \dfrac{z}{2y}$ **44.** $x^2 - 3y^2$

45. The expression $180 - a - b$ represents the measure of the unknown angle of the given triangle. Replace a with 37 and b with 80 to find the measure of the unknown angle.

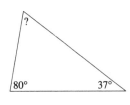

Decide whether the given number is a solution to the given equation.

46. $7x - 3 = 18;\ 3$

47. $3x^2 + 4 = x - 1;\ 1$

(1.5) *Find the additive inverse or the opposite.*

48. -9 **49.** $\dfrac{2}{3}$

50. $|-2|$ **51.** $-|-7|$

Find the following sums.

52. $-15 + 4$ **53.** $-6 + (-11)$

54. $\dfrac{1}{16} + \left(-\dfrac{1}{4}\right)$ **55.** $-8 + |-3|$

56. $-4.6 + (-9.3)$ **57.** $-2.8 + 6.7$

(1.6) *Perform the indicated operations.*

58. $6 - 20$ **59.** $-3.1 - 8.4$

60. $-6 - (-11)$ **61.** $4 - 15$

62. $-21 - 16 + 3(8 - 2)$ **63.** $\dfrac{11 - (-9) + 6(8 - 2)}{2 + 3 \cdot 4}$

If $x = 3$, $y = -6$, and $z = -9$, evaluate each expression.

64. $2x^2 - y + z$ **65.** $\dfrac{y - x + 5x}{2x}$

66. At the beginning of the week the price of Density Oil and Gas stock from Exercises 17 and 18 is $50 per share. Find the price of a share of stock at the end of the week.

Find the multiplicative inverse or reciprocal.

67. -6 **68.** $\dfrac{3}{5}$

(1.7) *Simplify each expression.*

69. $6(-8)$ **70.** $(-2)(-14)$

71. $\dfrac{-18}{-6}$ **72.** $\dfrac{42}{-3}$

73. $-3(-6)(-2)$ **74.** $(-4)(-3)(0)(-6)$

75. $\dfrac{4 \cdot (-3) + (-8)}{2 + (-2)}$ **76.** $\dfrac{3(-2)^2 - 5}{-14}$

(1.8) *Name the property illustrated.*

77. $-6 + 5 = 5 + (-6)$

78. $6 \cdot 1 = 6$

79. $3(8 - 5) = 3 \cdot 8 + 3 \cdot (-5)$

80. $4 + (-4) = 0$

81. $2 + (3 + 9) = (2 + 3) + 9$

82. $2 \cdot 8 = 8 \cdot 2$

83. $6(8 + 5) = 6 \cdot 8 + 6 \cdot 5$

84. $(3 \cdot 8) \cdot 4 = 3 \cdot (8 \cdot 4)$

85. $4 \cdot \dfrac{1}{4} = 1$

86. $8 + 0 = 8$

87. $4(8 + 3) = 4(3 + 8)$

(1.9) *Use the graph below showing Disney's consumer products revenues to answer the exercises.*

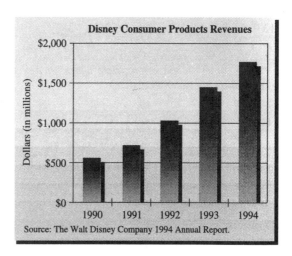

Source: The Walt Disney Company 1994 Annual Report.

88. Approximate Disney's consumer products revenue in 1994.

89. Approximate the increase in consumer products revenue in 1992.

90. What year shows the greatest revenue?

91. What trend is shown by this graph?

CHAPTER 1 TEST

Translate the statement into symbols.

1. The absolute value of negative seven is greater than five.
2. The sum of nine and five is greater than or equal to four.

Simplify the expression.

3. $-13 + 8$
4. $-13 - (-2)$
5. $6 \cdot 3 - 8 \cdot 4$
6. $(13)(-3)$
7. $(-6)(-2)$
8. $\dfrac{|-16|}{-8}$
9. $\dfrac{-8}{0}$
10. $\dfrac{|-6| + 2}{5 - 6}$
11. $\dfrac{1}{2} - \dfrac{5}{6} \cdot \dfrac{1}{3}$
12. $-1\dfrac{1}{8} + 5\dfrac{3}{4}$
13. $-\dfrac{3}{6} + \dfrac{15}{8}$
14. $3(-4)^2 - 80$
15. $6[5 + 2(3 - 8) - 3]$
16. $\dfrac{-12 + 3 \cdot 8}{4}$
17. $\dfrac{(-2)(0)(-3)}{-6}$

Insert <, >, or = in the appropriate space to make each of the following statements true.

18. -3 ___ -7
19. 4 ___ -8
20. $|-3|$ ___ 2
21. $|-2|$ ___ $-1 - (-3)$

22. Given $\{-5, -1, \frac{1}{4}, 0, 1, 7, 11.6, \sqrt{7}, 3\pi\}$, list the numbers in this set that also belong to the set of:
    a. Natural numbers
    b. Whole numbers
    c. Integers
    d. Rational numbers
    e. Irrational numbers
    f. Real numbers

If $x = 6$, $y = -2$, and $z = -3$, evaluate each expression.

23. $x^2 + y^2$
24. $x + yz$
25. $2 + 3x - y$
26. $\dfrac{y + z - 1}{x}$

Identify the property illustrated by each expression.

27. $8 + (9 + 3) = (8 + 9) + 3$
28. $6 \cdot 8 = 8 \cdot 6$
29. $-6(2 + 4) = -6 \cdot 2 + (-6) \cdot 4$
30. $\dfrac{1}{6}(6) = 1$

31. Find the opposite of -9.
32. Find the reciprocal of $-\dfrac{1}{3}$.

The New Orleans Saints were 22 yards from the goal when the following series of gains and losses occurred.

	GAINS AND LOSSES IN YARDS
First Down	5
Second Down	-10
Third Down	-2
Fourth Down	29

33. During which down did the greatest loss of yardage occur?
34. Was a touchdown scored?
35. The temperature at the Winter Olympics was a frigid 14 degrees below zero in the morning, but by noon it had risen 31 degrees. What was the temperature at noon?

36. Jean Avarez decided to sell 280 shares of stock, which decreased in value by $1.50 per share yesterday. How much money did she lose?

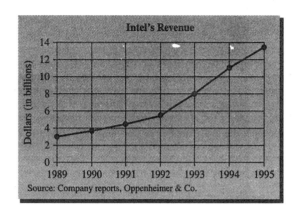

Intel is a semiconductor manufacturer that makes almost one-third of the world's computer chips. (You may have seen the slogan "Intel Inside" in commercials on television.) The line graph to the right shows Intel's net revenues in billions of dollars. Use this figure to answer the questions below.

37. Estimate Intel's revenue in 1993.

38. Estimate Intel's revenue in 1989.

39. Find the increase in Intel's revenue from 1993 to 1995.

40. What year shows the greatest increase in revenue?

CHAPTER 2

Equations, Inequalities, and Problem Solving

- **2.1** Simplifying Algebraic Expressions
- **2.2** The Addition Property of Equality
- **2.3** The Multiplication Property of Equality
- **2.4** Solving Linear Equations
- **2.5** An Introduction to Problem Solving
- **2.6** Formulas and Problem Solving
- **2.7** Percent and Problem Solving
- **2.8** Further Problem Solving
- **2.9** Solving Linear Inequalities

Calculating Price Per Unit

Business people, such as caterers and printers, provide services as well as products to their customers. They must often calculate a price per unit (item, person, hour, etc.) to charge their customers that covers the cost of all materials as well as payment for their time and labor. As a first step in calculating price per unit, it is often helpful to start by figuring the business person's actual cost per unit of the materials, which can be described by an algebraic formula.

In the Chapter Group Activity on page 149, you will have the opportunity to help a caterer figure a price per person for fruit salad to quote to a prospective customer.

74 CHAPTER 2 EQUATIONS, INEQUALITIES, AND PROBLEM SOLVING

Much of mathematics relates to deciding which statements are true and which are false. When a statement, such as an equation, contains variables, it is usually not possible to decide whether the equation is true or false until the variable has been replaced by a value. For example, the statement $x + 7 = 15$ is an equation stating that the sum $x + 7$ has the same value as 15. Is this statement true or false? It is false for some values of x and true for just one value of x, namely 8. Our purpose in this chapter is to learn ways of deciding which values make an equation or an inequality true. To begin, we spend a bit more time learning about algebraic expressions.

2.1 SIMPLIFYING ALGEBRAIC EXPRESSIONS

OBJECTIVES

1. Identify terms, like terms, and unlike terms.
2. Combine like terms.
3. Use the distributive property to remove parentheses.
4. Write word phrases as algebraic expressions.

TAPE BA 2.1

As we explore in this section, an expression such as $3x + 2x$ is not as simple as possible, because—even without replacing x by a value—we can perform the indicated addition.

Before we practice simplifying expressions, some new language is presented. A **term** is a number or the product of a number and variables raised to powers. For example,

$$-y, \quad 2x^3, \quad -5, \quad 3xz^2, \quad \frac{2}{y}, \quad 0.8z$$

are terms. The **numerical coefficient** of a term is the numerical factor. The numerical coefficient of $3x$ is 3. Recall that $3x$ means $3 \cdot x$.

TERM	NUMERICAL COEFFICIENT	
$3x$	3	
$\dfrac{y^3}{5}$	$\dfrac{1}{5}$	since $\dfrac{y^3}{5}$ means $\dfrac{1}{5} \cdot y^3$
$-0.7ab^3c^5$	-0.7	
z	1	
$-y$	-1	
-5	-5	

> REMINDER The term $-y$ means $-1y$ and thus has a numerical coefficient of -1. The term z means $1z$ and thus has a numerical coefficient of 1.

EXAMPLE 1 Identify the numerical coefficient.

   a. $-3y$ **b.** $22z^4$ **c.** y **d.** $-x$ **e.** $\dfrac{x}{7}$

Solution:
a. The numerical coefficient of $-3y$ is -3.
b. The numerical coefficient of $22z^4$ is 22.
c. The numerical coefficient of y is 1, since y is $1y$.
d. The numerical coefficient of $-x$ is -1, since $-x$ is $-1x$.
e. The numerical coefficient of $\dfrac{x}{7}$ is $\dfrac{1}{7}$, since $\dfrac{x}{7}$ is $\dfrac{1}{7} \cdot x$.

Terms with the same variables raised to exactly the same powers are called **like terms**.

LIKE TERMS	UNLIKE TERMS	
$3x, 2x$	$5x, 5x^2$	Why? Same variable x, but different powers x and x^2
$-6x^2y, 2x^2y, 4x^2y$	$7y, 3z, 8x^2$	Why? Different variables
$2ab^2c^3, ac^3b^2$	$6abc^3, 6ab^2$	Why? Different variables and different powers

Each variable and its exponent must match exactly in like terms, but like terms need not have the same numerical coefficients, nor do their factors need to be in the same order. For example, $2x^2y$ and $-yx^2$ are like terms.

EXAMPLE 2 Tell whether the terms are like or unlike.

   a. $-x^2, 3x^3$ **b.** $4x^2y, x^2y, -2x^2y$ **c.** $-2yz, -3zy$ **d.** $-x^4, x^4$

Solution:
a. Unlike terms, since the exponents on x are not the same.
b. Like terms, since each variable and its exponent match.
c. Like terms, since $zy = yz$ by the commutative property.
d. Like terms.

An algebraic expression containing the sum or difference of like terms can be simplified by applying the distributive property. For example, by the distributive property, we rewrite the sum of the like terms $3x + 2x$ as

$$3x + 2x = (3 + 2)x = 5x$$

Also,

$$-y^2 + 5y^2 = (-1 + 5)y^2 = 4y^2$$

Simplifying the sum or difference of like terms is called **combining like terms**.

EXAMPLE 3 Simplify the following by combining like terms.

    a. $7x - 3x$ **b.** $10y^2 + y^2$ **c.** $8x^2 + 2x - 3x$

Solution: **a.** $7x - 3x = (7 - 3)x = 4x$
            b. $10y^2 + y^2 = (10 + 1)y^2 = 11y^2$
            c. $8x^2 + 2x - 3x = 8x^2 + (2 - 3)x = 8x^2 - x$

EXAMPLE 4 Simplify each expression by combining like terms.

    a. $2x + 3x + 5 + 2$ **b.** $-5a - 3 + a + 2$ **c.** $4y - 3y^2$ **d.** $2.3x + 5x - 6$

Solution: Use the distributive property to combine the numerical coefficients of like terms.

    a. $2x + 3x + 5 + 2 = (2 + 3)x + (5 + 2)$
                                 $= 5x + 7$

    b. $-5a - 3 + a + 2 = -5a + 1a + (-3 + 2)$
                                  $= (-5 + 1)a + (-3 + 2)$
                                  $= -4a - 1$

    c. $4y - 3y^2$ These two terms cannot be combined because they are unlike terms.

    d. $2.3x + 5x - 6 = (2.3 + 5)x - 6$
                            $= 7.3x - 6$

The examples above suggest the following:

> To **combine like terms,** add the numerical coefficients and multiply the result by the common variable factors.

3 Simplifying expressions makes frequent use of the distributive property to remove parentheses.

EXAMPLE 5 Find each product by using the distributive property to remove parentheses.

    a. $5(x + 2)$ **b.** $-2(y + 0.3z - 1)$ **c.** $-(x + y - 2z + 6)$

Solution: **a.** $5(x + 2) = 5(x) + 5(2)$ Apply the distributive property.
                      $= 5x + 10$ Multiply.

b. $-2(y + 0.3z - 1) = -2(y) + (-2)(0.3z) + (-2)(-1)$ Apply the distributive property

$\qquad = -2y - 0.6z + 2$ Multiply.

c. $-(x + y - 2z + 6) = -1(x + y - 2z + 6)$ Distribute -1 over each term.

$\qquad = -1(x) - 1(y) - 1(-2z) - 1(6)$

$\qquad = -x - y + 2z - 6$

REMINDER If a "$-$" sign precedes parentheses, the sign of each term inside the parentheses is changed when the distributive property is applied to remove parentheses.

Examples:

$-(2x + 1) = -2x - 1$ $\qquad$ $-(x - 2y) = -x + 2y$

$-(-5x + y - z) = 5x - y + z$ $\qquad$ $-(-3x - 4y - 1) = 3x + 4y + 1$

To simplify an expression containing parentheses, we use the distributive property to remove parentheses and then the distributive property to combine any like terms.

EXAMPLE 6 Simplify the following expressions.

a. $3(2x - 5) + 1$ $\quad$ **b.** $8 - (7x + 2) + 3x$ $\quad$ **c.** $-2(4x + 7) - (3x - 1)$

Solution: **a.** $3(2x - 5) + 1 = 6x - 15 + 1$ Apply the distributive property.

$\qquad = 6x - 14$ Combine like terms.

b. $8 - (7x + 2) + 3x = 8 - 7x - 2 + 3x$ Apply the distributive property.

$\qquad = -7x + 3x + 8 - 2$

$\qquad = -4x + 6$ Combine like terms.

c. $-2(4x + 7) - (3x - 1) = -8x - 14 - 3x + 1$ Apply the distributive property.

$\qquad = -11x - 13$ Combine like terms.

EXAMPLE 7 Subtract $4x - 2$ from $2x - 3$.

Solution: "Subtract $4x - 2$ **from** $2x - 3$" translates to $(2x - 3) - (4x - 2)$. Next, simplify the algebraic expression.

$(2x - 3) - (4x - 2) = 2x - 3 - 4x + 2$ Apply the distributive property.

$\qquad = -2x - 1$ Combine like terms.

78 CHAPTER 2 EQUATIONS, INEQUALITIES, AND PROBLEM SOLVING

4 Next, we practice writing word phrases as algebraic expressions.

EXAMPLE 8 Write the following phrases as algebraic expressions and simplify if possible. Let x represent the unknown number.

a. Twice a number, added to 6.
b. The difference of a number and 4, divided by 7.
c. Five added to 3 times the sum of a number and 1.

Solution: a.

In words:	twice a number	added to	6
	↓	↓	↓
Translate:	$2x$	$+$	6

b.

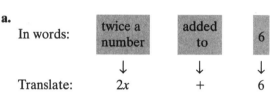

c.

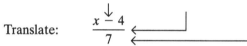

	↓	↓	↓	↓
Translate:	5	+	3 ·	$(x+1)$

Next, we simplify this expression.

$$5 + 3(x+1) = 5 + 3x + 3$$
$$= 8 + 3x$$

MENTAL MATH

Identify the numerical coefficient of each term. See Example 1.

1. $-7y$ 2. $3x$ 3. x
4. $-y$ 5. $17x^2y$ 6. $1.2xyz$

Indicate whether the following lists of terms are like or unlike. See Example 2.

7. $5y, -y$ 8. $-2x^2y, 6xy$ 9. $2z, 3z^2$

10. $ab^2, -7ab^2$ 11. $8wz, \frac{1}{7}zw$ 12. $7.4p^3q^2, 6.2p^3q^2r$

Exercise Set 2.1

Simplify each expression by combining any like terms. See Examples 3 and 4.

1. $7y + 8y$
2. $5x - 2x$
3. $8w - w + 6w$
4. $c - 7c + 2c$
5. $3b - 5 - 10b - 4$
6. $6g + 5 - 3g - 7$
7. $m - 4m + 2m - 6$
8. $a + 3a - 2 - 7a$

Simplify each expression. Use the distributive property to remove any parentheses. See Examples 5 and 6.

9. $5(y - 4)$
10. $7(r - 3)$
11. $7(d - 3) + 10$
12. $9(z + 7) - 15$
13. $-(3x - 2y + 1)$
14. $-(y + 5z - 7)$
15. $5(x + 2) - (3x - 4)$
16. $4(2x - 3) - 2(x + 1)$

17. In your own words, explain how to combine like terms.
18. Do like terms contain the same numerical coefficients? Explain your answer.

Write each of the following as an algebraic expression. Simplify if possible. See Example 7.

19. Add $6x + 7$ to $4x - 10$.
20. Add $3y - 5$ to $y + 16$.
21. Subtract $7x + 1$ from $3x - 8$.
22. Subtract $4x - 7$ from $12 + x$.

Write each of the following phrases as an algebraic expression and simplify if possible. Let x represent the unknown number. See Example 8.

23. Twice a number decreased by four.
24. The difference of a number and two, divided by five.
25. Three-fourths of a number increased by twelve.
26. Eight more than triple the number.
27. The sum of -2 and 5 times a number, added to 7 times a number.
28. The sum of 3 times a number and 10, **subtracted from** 9 times a number.

Simplify each expression.

29. $7x^2 + 8x^2 - 10x^2$
30. $8x + x - 11x$
31. $6x - 5x + x - 3 + 2x$
32. $8h + 13h - 6 + 7h - h$
33. $-5 + 8(x - 6)$
34. $-6 + 5(r - 10)$
35. $5g - 3 - 5 - 5g$
36. $8p + 4 - 8p - 15$
37. $6.2x - 4 + x - 1.2$
38. $7.9y - 0.7 - y + 0.2$
39. $2k - k - 6$
40. $7c - 8 - c$
41. $0.5(m + 2) + 0.4m$
42. $0.2(k + 8) - 0.1k$
43. $-4(3y - 4)$
44. $-3(2x + 5)$
45. $3(2x - 5) - 5(x - 4)$
46. $2(6x - 1) - (x - 7)$
47. $3.4m - 4 - 3.4m - 7$
48. $2.8w - 0.9 - 0.5 - 2.8w$
49. $6x + 0.5 - 4.3x - 0.4x + 3$
50. $0.4y - 6.7 + y - 0.3 - 2.6y$
51. $-2(3x - 4) + 7x - 6$
52. $8y - 2 - 3(y + 4)$
53. $-9x + 4x + 18 - 10x$
54. $5y - 14 + 7y - 20y$
55. $5k - (3k - 10)$
56. $-11c - (4 - 2c)$
57. $(3x + 4) - (6x - 1)$
58. $(8 - 5y) - (4 + 3y)$

59. Recall that the perimeter of a figure is the total distance around the figure. Given the following rectangle, express the perimeter as an algebraic expression containing the variable x.

```
              5x feet
        ┌──────────────┐
(4x - 1) feet          │ (4x - 1)
        │              │  feet
        └──────────────┘
              5x feet
```

60. Given the following triangle, express its perimeter as an algebraic expression containing the variable x.

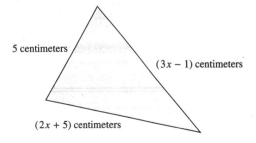

Write each of the following as an algebraic expression. Simplify if possible.

61. Subtract $5m - 6$ from $m - 9$.

62. Subtract $m - 3$ from $2m - 6$.

63. Eight times the sum of a number and six.

64. Five less than four times the number.

65. Double a number minus the sum of the number and ten.

66. Half a number minus the product of the number and eight.

67. Seven multiplied by the quotient of a number and six.

68. The product of a number and ten, less twenty.

Given the following, determine whether each scale is balanced or not.

1 cone balances 1 cube

1 cylinder balances 2 cubes

69.

70.

71.

72.

73. To convert from feet to inches, we multiply by 12. For example, the number of inches in 2 feet is $12 \cdot 2$ inches. If one board has a length of $(x + 2)$ *feet* and a second board has a length of $3x - 1$ *inches*, express their total length in inches as an algebraic expression.

74. The value of 7 nickels is $5 \cdot 7$ cents. Likewise, the value of x nickels is $5x$ cents. If the money box in a drink machine contains x *nickels*, $3x$ *dimes*, and $30x - 1$ *quarters*, express their total value in cents as an algebraic expression.

Review Exercises

Evaluate the following expressions for the given values. See Section 1.7.

75. If $x = -1$ and $y = 3$, find $y - x^2$.

76. If $g = 0$ and $h = -4$, find $gh - h^2$.

77. If $a = 2$ and $b = -5$, find $a - b^2$.

78. If $x = -3$, find $x^3 - x^2 + 4$.

79. If $y = -5$ and $z = 0$, find $yz - y^2$.

80. If $x = -2$, find $x^3 - x^2 - x$.

A Look Ahead

EXAMPLE

Simplify $-3xy + 2x^2y - (2xy - 1)$.

Solution:

$$-3xy + 2x^2y - (2xy - 1) = -3xy + 2x^2y - 2xy + 1$$
$$= -5xy + 2x^2y + 1$$

Simplify each expression.

81. $5b^2c^3 + 8b^3c^2 - 7b^3c^2$

82. $4m^4p^2 + m^4p^2 - 5m^2p^4$

83. $3x - (2x^2 - 6x) + 7x^2$

84. $9y^2 - (6xy^2 - 5y^2) - 8xy^2$

85. $-(2x^2y + 3z) + 3z - 5x^2y$

86. $-(7c^3d - 8c) - 5c - 4c^3d$

2.2 THE ADDITION PROPERTY OF EQUALITY

TAPE BA 2.2

OBJECTIVES

 Define linear equation in one variable and equivalent equations.
 Use the addition property of equality to solve linear equations.
 Write word phrases as algebraic expressions.

 Recall from Section 1.4 that an equation is a statement that two expressions have the same value. Also, a value of the variable that makes an equation a true statement is called a solution or root of the equation. The process of finding the solution of an equation is called **solving** the equation for the variable. In this section we concentrate on solving **linear equations** in one variable.

> **LINEAR EQUATION IN ONE VARIABLE**
> A linear equation in one variable can be written in the form
> $$ax + b = c$$
> where a, b, and c are real numbers and $a \neq 0$.

Evaluating a linear equation for a given value of the variable, as we did in Section 1.4, can tell us whether that value is a solution, but we can't rely on evaluating an equation as our method of solving it.

Instead, to solve a linear equation in x, we write a series of simpler equations, all *equivalent* to the original equation, so that the final equation has the form

$x =$ **number**   or   **number** $= x$

Equivalent equations are equations that have the same solution. This means that the "number" above is the solution to the original equation.

The first property of equality that helps us write simpler equivalent equations is the **addition property of equality.**

> **ADDITION PROPERTY OF EQUALITY**
> If a, b, and c are real numbers, then
> $$a = b \quad \text{and} \quad a + c = b + c$$
> are equivalent equations.

This property guarantees that adding the same number to both sides of an equation does not change the solution of the equation. Since subtraction is defined in terms of addition, we may also **subtract the same number from both sides** without changing the solution.

A good way to picture a true equation is as a balanced scale. Since it is balanced, each side of the scale weighs the same amount.

If the same weight is added to or subtracted from each side, the scale remains balanced.

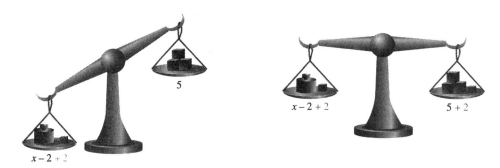

We use the addition property of equality to write equivalent equations until the variable is **isolated** (by itself on one side of the equation) and the equation looks like "$x =$ number" or "number $= x$."

EXAMPLE 1 Solve $x - 7 = 10$ for x.

Solution: To solve for x, isolate x on one side of the equation. To do this, add 7 to both sides of the equation.

$$x - 7 = 10$$
$$x - 7 + 7 = 10 + 7 \quad \text{Add 7 to both sides.}$$
$$x = 17 \quad \text{Simplify.}$$

The solution of the equation $x = 17$ is obviously 17. Since we are writing equivalent equations, the solution of the equation $x - 7 = 10$ is also 17.

To check, replace x with 17 in the original equation.

$$x - 7 = 10$$
$$17 - 7 = 10 \quad \text{Replace } x \text{ with 17 in the original equation.}$$
$$10 = 10 \quad \text{True.}$$

Since the statement is true, 17 is the solution and the solution set is {17}.

EXAMPLE 2   Solve $y + 0.6 = -1.0$.

Solution: To solve for y, subtract 0.6 from both sides of the equation.

$$y + 0.6 = -1.0$$
$$y + 0.6 - 0.6 = -1.0 - 0.6 \quad \text{Subtract 0.6 from both sides.}$$
$$y = -1.6 \quad \text{Combine like terms.}$$

To check the proposed solution, -1.6, replace y with -1.6 in the original equation.

Check:

$$y + 0.6 = -1.0$$
$$-1.6 + 0.6 = -1.0 \quad \text{Replace } y \text{ with } -1.6 \text{ in the original equation.}$$
$$-1.0 = -1.0 \quad \text{True.}$$

The solution set is $\{-1.6\}$.

EXAMPLE 3   Solve $5t - 5 = 6t + 2$ for t.

Solution: To solve for t, we first want all terms containing t on one side of the equation and all other terms on the other side of the equation. To do this, first subtract $5t$ from both sides of the equation.

$$5t - 5 = 6t + 2$$
$$5t - 5 - 5t = 6t + 2 - 5t \quad \text{Subtract } 5t \text{ from both sides.}$$
$$-5 = t + 2 \quad \text{Combine like terms.}$$

Next, subtract 2 from both sides and the variable t will be isolated.

$$-5 = t + 2$$
$$-5 - 2 = t + 2 - 2 \quad \text{Subtract 2 from both sides.}$$
$$-7 = t$$

Check the solution, -7, in the original equation. The solution set is $\{-7\}$.

> REMINDER   We may isolate the variable on either side of the equation. $-7 = t$ is equivalent to $t = -7$.

Many times, it is best to simplify one or both sides of an equation before applying the addition property of equality.

EXAMPLE 4   Solve $2x + 3x - 5 + 7 = 10x + 3 - 6x - 4$ for x.

Solution: First, simplify both sides of the equation.

$$2x + 3x - 5 + 7 = 10x + 3 - 6x - 4$$
$$5x + 2 = 4x - 1 \qquad \text{Combine like terms on each side of the equation.}$$
$$5x + 2 - 4x = 4x - 1 - 4x \qquad \text{Subtract } 4x \text{ from both sides.}$$
$$x + 2 = -1 \qquad \text{Combine like terms.}$$
$$x + 2 - 2 = -1 - 2 \qquad \text{Subtract 2 from both sides.}$$
$$x = -3 \qquad \text{Combine like terms.}$$

Check by replacing x with -3 in the original equation.

$$2x + 3x - 5 + 7 = 10x + 3 - 6x - 4$$
$$2(-3) + 3(-3) - 5 + 7 = 10(-3) + 3 - 6(-3) - 4 \qquad \text{Replace } x \text{ with } -3.$$
$$-6 - 9 - 5 + 7 = -30 + 3 + 18 - 4 \qquad \text{Multiply.}$$
$$-13 = -13 \qquad \text{True.}$$

The solution set is $\{-3\}$.

If an equation contains parentheses, use the distributive property to remove them.

EXAMPLE 5 Solve $-5(2a - 1) - (-11a + 6) = 7$ for a.

Solution:
$$-5(2a - 1) - (-11a + 6) = 7$$
$$-10a + 5 + 11a - 6 = 7 \qquad \text{Apply the distributive property.}$$
$$a - 1 = 7 \qquad \text{Combine like terms.}$$
$$a - 1 + 1 = 7 + 1 \qquad \text{Add 1 to both sides to isolate } a$$
$$a = 8 \qquad \text{Combine like terms.}$$

Check to see that 8 is the solution or the solution set is $\{8\}$.

When solving equations, we may sometimes encounter an equation such as

$$-x = 5$$

This equation is not solved for x because x is not isolated. To solve this equation for x, recall that

"$-$" can be read as "the opposite of"

We can read the equation $-x = 5$, then, as "the opposite of x is 5." If the opposite of x is 5, this means that x is the opposite of 5 or -5.

In summary,

$$-x = 5 \quad \text{and} \quad x = -5$$

are equivalent equations and $x = -5$ is solved for x.

EXAMPLE 6 Solve $3 - x = 7$ for x.

Solution: First, subtract 3 from both sides.

$$3 - x = 7$$
$$3 - x - 3 = 7 - 3 \quad \text{Subtract 3 from both sides.}$$
$$-x = 4 \quad \text{Simplify.}$$

Since the opposite of x is 4, this means that x is the opposite of 4, or -4.

$$x = -4$$

Check by replacing x with -4 in the original equation.

$$3 - x = 7$$
$$3 - (-4) = 7 \quad \text{Replace } x \text{ with } -4.$$
$$7 = 7 \quad \text{True.}$$

The solution set is $\{-4\}$.

Next, we practice writing algebraic expressions.

EXAMPLE 7
a. The sum of two numbers is 8. If one number is 3, find the other number.
b. The sum of two numbers is 8. If one number is x, write an expression representing the other number.

Solution: a. If the sum of two numbers is 8 and one number is 3, we find the other number by subtracting 3 from 8. The other number is $8 - 3$ or 5.

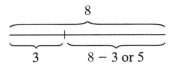

b. If the sum of two numbers is 8 and one number is x, we find the other number by subtracting x from 8. The other number is represented by $8 - x$.

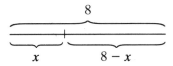

MENTAL MATH

Solve each equation mentally. See Examples 1 and 2.

1. $x + 4 = 6$
2. $x + 7 = 10$
3. $n + 18 = 30$
4. $z + 22 = 40$
5. $b - 11 = 6$
6. $d - 16 = 5$

EXERCISE SET 2.2

Solve each equation. See Examples 1 and 2.

1. $x + 11 = -2$
2. $y - 5 = -9$
3. $5y + 14 = 4y$
4. $8x - 7 = 9x$
5. $8x = 7x - 8$
6. $x = 2x + 3$
7. $x - 2 = -4$
8. $y + 7 = 5$
9. $\frac{1}{2} + f = \frac{3}{4}$
10. $c + \frac{1}{4} = \frac{3}{8}$

Solve each equation. See Examples 3 and 4.

11. $3x - 6 = 2x + 5$
12. $7y + 2 = 6y + 2$
13. $3t - t - 7 = t - 7$
14. $4c + 8 - c = 8 + 2c$
15. $7x + 2x = 8x - 3$
16. $3n + 2n = 7 + 4n$
17. $2y + 10 = y$
18. $4x - 4 = 3x$
19. $y + 0.8 = 9.7$
20. $w + 0.9 = 3.6$
21. $5b - 0.7 = 6b$
22. $8n + 1.5 = 9n$
23. $5x - 6 = 6x - 5$
24. $2x + 7 = x - 10$
25. $7t - 12 = 6t$
26. $9m + 14 = 8m$
27. $y - 5y + 0.6 = 0.8 - 5y$
28. $6z + z - 0.9 = 6z + 0.9$

29. In your own words, explain what is meant by the solution of an equation.
30. In your own words, explain how to check a solution of an equation.

Solve each equation. See Examples 5 and 6.

31. $2(x - 4) = x + 3$
32. $3(y + 7) = 2y - 5$
33. $7(6 + w) = 6(2 + w)$
34. $6(5 + c) = 5(c - 4)$
35. $10 - (2x - 4) = 7 - 3x$
36. $15 - (6 - 7k) = 2 + 6k$
37. $-5(n - 2) = 8 - 4n$
38. $-4(z - 3) = 2 - 3z$
39. $-3(x - 4) = -4x$
40. $-2(x - 1) = -3x$
41. $3(n - 5) - (6 - 2n) = 4n$
42. $5(3 + z) - (8z + 9) = -4z$
43. $-2(t - 1) - 3t = 8 - 4t$
44. $-4(r + 5) - 7r = 12 - 10r$
45. $4y - 6(y + 4) = 1 - y$
46. $-7k + 3(k - 1) = 6 - 5k$
47. $7(m - 2) - 6(m + 1) = -20$
48. $-4(x - 1) - 5(2 - x) = -6$
49. $0.8t + 0.2(t - 0.4) = 1.75$
50. $0.6v + 0.4(0.3 + v) = 2.34$

See Example 7.

51. Two numbers have a sum of 20. If one number is p, express the other number in terms of p.
52. Two numbers have a sum of 13. If one number is y, express the other number in terms of y.
53. A 10-foot board is cut into two pieces. If one piece is x feet long, express the other length in terms of x.

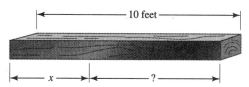

54. A 5-foot piece of string is cut into two pieces. If one piece is x feet long, express the other length in terms of x.

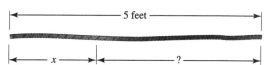

55. Two angles are *supplementary* if their sum is 180°. If one angle measures $x°$, express the measure of its supplement in terms of x.

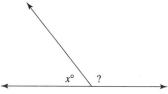

56. Two angles are *complementary* if their sum is 90°. If one angle measures $x°$, express the measure of its complement in terms of x.

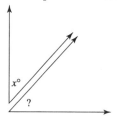

57. The sum of the angles of a triangle is 180°. If one angle of a triangle measures $x°$ and a second angle measures $(2x + 7)°$, express the measure of the third angle in terms of x. Simplify the expression.

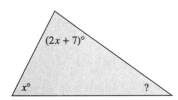

58. A quadrilateral is a four-sided figure like the one shown next whose angle sum is 360°. If one angle measures $x°$, a second angle measures $3x°$, and a third angle measures $5x°$, express the measure of the fourth angle in terms of x. Simplify the expression.

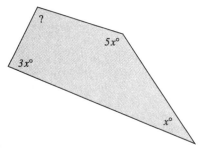

59. In a mayoral election, April Catarella received 284 more votes than Charles Pecot. If Charles received n votes, how many votes did April receive?

60. The length of the top of a computer desk is $1\frac{1}{2}$ feet longer than its width. If its width measures m feet, express its length as an algebraic expression in m.

Use a calculator to determine whether the given value is a solution of the given equation.

61. $1.23x - 0.06 = 2.6x - 0.1285; x = 0.05$

62. $8.13 + 5.85y = 20.05y - 8.91; y = 1.2$

63. $3(a + 4.6) = 5a + 2.5; a = 6.3$

64. $7(z - 1.7) + 9.5 = 5(z + 3.2) - 9.2; z = 4.8$

Review Exercises

Find the multiplicative inverse or reciprocal of each. See Section 1.7.

65. $\dfrac{5}{8}$ **66.** $\dfrac{7}{6}$ **67.** 2

68. 5   **69.** $-\dfrac{1}{9}$ **70.** $-\dfrac{3}{5}$

Bryan, a 4-year old child, was admitted to Slidell Memorial Hospital with pneumonia and a fever of 104°. His temperature was taken every hour and the results are recorded on the bar graph shown. (See Section 1.9.)

71. How many hours after arrival at the hospital was Bryan's fever the highest?

72. What was Bryan's highest recorded temperature?

73. How many hours after arrival at the hospital was Bryan's temperature 101°?

74. When did Bryan's temperature show the greatest increase?

75. When did Bryan's temperature show the greatest decrease?

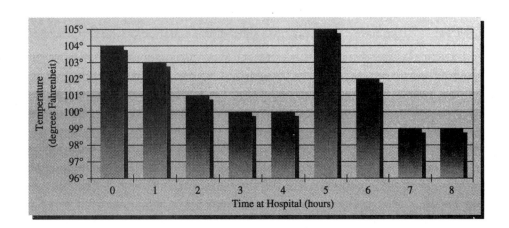

2.3 THE MULTIPLICATION PROPERTY OF EQUALITY

TAPE BA 2.3

OBJECTIVES

1. Use the multiplication property of equality to solve linear equations.
2. Use both the addition and multiplication properties of equality to solve linear equations.
3. Write word phrases as algebraic expressions.

As useful as the addition property of equality is, it cannot help us solve every type of linear equation in one variable. For example, adding or subtracting a value on both sides of the equation does not help solve

$$\frac{5}{2}x = 15.$$

Instead, we apply another important property of equality, the **multiplication property of equality.**

> **MULTIPLICATION PROPERTY OF EQUALITY**
> If a, b, and c are real numbers and $c \neq 0$, then
> $$a = b \quad \text{and} \quad ac = bc$$
> are equivalent equations.

This property guarantees that multiplying both sides of an equation by the same nonzero number does not change the solution of the equation. Since division is defined in terms of multiplication, we may also divide both sides of the equation by the same nonzero number without changing the solution.

EXAMPLE 1 Solve for x: $\frac{5}{2}x = 15$.

Solution: To isolate x, multiply both sides of the equation by the reciprocal of $\frac{5}{2}$, which is $\frac{2}{5}$.

$$\frac{5}{2}x = 15$$

$$\frac{2}{5} \cdot \frac{5}{2}x = \frac{2}{5} \cdot 15 \quad \text{Multiply both sides by } \frac{2}{5}.$$

$$\left(\frac{2}{5} \cdot \frac{5}{2}\right)x = \frac{2}{5} \cdot 15 \quad \text{Apply the associative property.}$$

$$1x = 6 \quad \text{Simplify.}$$

or

$$x = 6$$

To check, replace x with 6 in the original equation.

$$\frac{5}{2}x = 15$$

$$\frac{5}{2}(6) = 15 \quad \text{Replace } x \text{ with 6.}$$

$$15 = 15 \quad \text{True.}$$

The solution set is $\{6\}$.

Why did we multiply both sides by the reciprocal of the coefficient of x? Multiplying the coefficient $\frac{5}{2}$ by $\frac{2}{5}$ leaves a coefficient of 1 and thus isolates the variable.

In general, multiplying by the reciprocal of the variable's coefficient is a way to isolate a variable. Multiplying by the reciprocal of a number is, of course, the same as dividing by the number.

EXAMPLE 2 Solve $-3x = 33$ for x.

Solution: Recall that $-3x$ means $-3 \cdot x$. To isolate x, divide both sides by the coefficient of x, that is, -3.

$$-3x = 33$$

$$\frac{-3x}{-3} = \frac{33}{-3} \quad \text{Divide both sides by } -3.$$

$$1x = -11 \quad \text{Simplify.}$$

$$x = -11$$

To check, replace x with -11 in the original equation.

$$-3x = 33$$

$$-3(-11) = 33 \quad \text{Replace } x \text{ with } -11 \text{ in the original equation.}$$

$$33 = 33 \quad \text{True.}$$

The solution set is $\{-11\}$.

EXAMPLE 3 Solve $\frac{y}{7} = 20$ for y.

Solution: Recall that $\frac{y}{7} = \frac{1}{7}y$. To isolate y, multiply both sides of the equation by 7, the reciprocal of $\frac{1}{7}$.

$$\frac{y}{7} = 20$$

$$7 \cdot \frac{y}{7} = 7 \cdot 20 \quad \text{Multiply each side by 7.}$$

$$y = 140 \quad \text{Simplify.}$$

To check, replace y with 140 in the original equation.

$$\frac{y}{7} = 20 \quad \text{Original equation.}$$

$$\frac{140}{7} = 20 \quad \text{Replace } y \text{ with 140.}$$

$$20 = 20 \quad \text{True.}$$

The solution set is {140}.

EXAMPLE 4 Solve $-\frac{2}{3}x = -5$ for x.

Solution: To isolate x, multiply both sides of the equation by $-\frac{3}{2}$, the reciprocal of the coefficient of x.

$$-\frac{2}{3}x = -5$$

$$-\frac{3}{2} \cdot -\frac{2}{3}x = -\frac{3}{2} \cdot -5 \quad \text{Multiply both sides by the reciprocal of } -\frac{2}{3}.$$

$$x = \frac{15}{2} \quad \text{Simplify.}$$

The solution set is $\left\{\frac{15}{2}\right\}$. Check this solution in the original equation.

We are now ready to combine the skills learned in the last section with the skills learned from this section to solve equations by applying more than one property.

EXAMPLE 5 Solve $-z - 4 = 6$ for z.

Solution: First, isolate $-z$, the term containing the variable. To do so, add 4 to both sides of the equation.

$$-z - 4 + 4 = 6 + 4 \quad \text{Add 4 to both sides.}$$

$$-z = 10 \quad \text{Simplify.}$$

Next, recall that $-z$ means $-1 \cdot z$. To isolate z, either multiply or divide both sides of the equation by -1. In this example, we divide.

$$-z = 10$$

$$\frac{-z}{-1} = \frac{10}{-1} \quad \text{Divide both sides by the coefficient } -1.$$

$$z = -10 \quad \text{Simplify.}$$

Check: $-z - 4 = 6$
$-(-10) - 4 = 6$ Replace z with -10.
$10 - 4 = 6$
$6 = 6$ True.

Since this is a true statement, -10 is the solution and the solution set is $\{-10\}$.

EXAMPLE 6   Solve $5x - 2 = 18$ for x.

Solution: First, isolate $5x$, the term containing the variable. To do so, add 2 to both sides of the equation.

$$5x - 2 = 18$$
$$5x - 2 + 2 = 18 + 2 \quad \text{Add 2 to both sides.}$$
$$5x = 20 \quad \text{Simplify.}$$

Having isolated the variable term, we now use the multiplication property of equality to achieve a coefficient of 1.

$$\frac{5x}{5} = \frac{20}{5} \quad \text{Divide both sides by 5.}$$
$$x = 4 \quad \text{Simplify.}$$

The solution is 4. As usual, we can check this solution by replacing x with 4 in the original equation.

$$5x - 2 = 18$$
$$5(4) - 2 = 18 \quad \text{Replace } x \text{ with 4.}$$
$$20 - 2 = 18 \quad \text{Simplify.}$$
$$18 = 18 \quad \text{True.}$$

The solution set is $\{4\}$.

EXAMPLE 7   Solve for a: $2a + 5a - 10 + 7 = 5a - 13$.

Solution: First, simplify both sides of the equation by combining like terms.

$$2a + 5a - 10 + 7 = 5a - 13$$
$$7a - 3 = 5a - 13 \quad \text{Combine like terms.}$$
$$7a - 3 - 5a = 5a - 13 - 5a \quad \text{Subtract } 5a \text{ from both sides.}$$

$$2a - 3 = -13 \quad \text{Combine like terms.}$$
$$2a - 3 + 3 = -13 + 3 \quad \text{Add 3 to both sides.}$$
$$2a = -10 \quad \text{Simplify.}$$
$$\frac{2a}{2} = \frac{-10}{2} \quad \text{Divide both sides by 2.}$$
$$a = -5 \quad \text{Simplify.}$$

To check, replace a with -5 in the original equation. The solution set is $\{-5\}$.

3 Next, we continue to sharpen our problem-solving skills by writing algebraic expressions.

EXAMPLE 8 If x is the first of three consecutive integers, express the sum of the three integers in terms of x. Simplify if possible.

Solution: An example of three consecutive integers is 7, 8, 9. The second consecutive integer is always 1 more than the first, and the third consecutive integer is 2 more than the first. If x is the first of three consecutive integers, the three consecutive integers are

$$x, \quad x + 1, \quad x + 2$$

Their sum is

In words: first integer + second integer + third integer

Translate: $\quad x \quad + \quad (x + 1) \quad + \quad (x + 2)$

which simplifies to $3x + 3$.

The exercise set mentions consecutive even and odd integers.

An example of three consecutive **even** integers is 14, 16, and 18. Notice that each integer is *two more* than the previous integer.

An example of three consecutive **odd** integers is 5, 7, and 9. Notice that each integer is *again two more* than the previous integer.

In general, if x is the first integer, we have the following:

Three consecutive integers: $\quad x, x + 1, x + 2$
Three consecutive odd integers: $\quad x, x + 2, x + 4$, if x is odd
Three consecutive even integers: $\quad x, x + 2, x + 4$, if x is even

MENTAL MATH

Solve the following equations mentally. See Example 1.

1. $3a = 27$
2. $9c = 54$
3. $5b = 10$
4. $7t = 14$
5. $6x = -30$
6. $8r = -64$

EXERCISE SET 2.3

Solve the following equations. See Examples 1 through 4.

1. $-5x = 20$
2. $-7x = -49$
3. $3x = 0$
4. $-2x = 0$
5. $-x = -12$
6. $-y = 8$
7. $\frac{2}{3}x = -8$
8. $\frac{3}{4}n = -15$
9. $\frac{1}{6}d = \frac{1}{2}$
10. $\frac{1}{8}v = \frac{1}{4}$
11. $\frac{a}{-2} = 1$
12. $\frac{d}{15} = 2$
13. $\frac{k}{7} = 0$
14. $\frac{f}{-5} = 0$

Solve the following equations. See Examples 5 and 6.

15. $2x - 4 = 16$
16. $3x - 1 = 26$
17. $-5x + 2 = 22$
18. $7x + 4 = -24$
19. $6x + 10 = -20$
20. $-10y + 15 = 5$

Solve the following equations. See Example 7.

21. $-4y + 10 = -6y - 2$
22. $-3z + 1 = -2z + 4$
23. $9x - 8 = 10 + 15x$
24. $15t - 5 = 7 + 12t$
25. $2x - 7 = 6x - 27$
26. $3 + 8y = 3y - 2$
27. $6 - 2x + 8 = 10$
28. $-5 - 6y + 6 = 19$
29. $-3a + 6 + 5a = 7a - 8a$
30. $4b - 8 - b = 10b - 3b$

31. The equation $3x + 6 = 2x + 10 + x - 4$ is true for all real numbers. Substitute a few real numbers for x to see that this is so and then try solving the equation.

32. The equation $6x + 2 - 2x = 4x + 1$ has no solution. Try solving this equation for x and see what happens.

33. From the results of Exercises 31 and 32, when do you think an equation has all real numbers as its solution set?

34. From the results of Exercises 31 and 32, when do you think an equation has no solution?

Solve the following equations.

35. $-3w = 18$
36. $5j = -45$
37. $-0.2z = -0.8$
38. $-0.1m = 3.6$
39. $-h = -\frac{3}{4}$
40. $-b = \frac{4}{7}$
41. $6a + 3 = 3$
42. $8t + 5 = 5$
43. $5 - 0.3k = 5$
44. $2 + 0.4p = 2$
45. $2x + \frac{1}{2} = \frac{7}{2}$
46. $3n - \frac{1}{3} = \frac{8}{3}$
47. $\frac{x}{3} - 2 = 5$
48. $\frac{b}{4} + 1 = 7$
49. $10 = 2x - 1$
50. $12 = 3j - 4$
51. $4 - 12x = 7$
52. $24 + 20b = 10$
53. $-\frac{2}{3}x = \frac{5}{9}$
54. $-\frac{3}{8}y = -\frac{1}{16}$
55. $10 = -6n + 16$
56. $-5 = -2m + 7$
57. $z - 5z = 7z - 9 - z$
58. $t - 6t = -13 + t - 3t$
59. $5x + 20 = 8 - x$
60. $6t - 18 = t - 3$

61. $5y - y = 2y - 14$ **62.** $k - 4k = 20 - 5k$
63. $6z - 8 - z + 3 = 0$ **64.** $4a + 1 + a - 11 = 0$
65. $10 - n - 2 = 2n + 2$
66. $4 - 10 - 5c = 3c - 12$
67. $0.4x - 0.6x - 5 = 1$ **68.** $0.4x - 0.9x - 6 = 19$

Write each algebraic expression described. Simplify if possible. See Example 8.

69. If x represents the first odd integer, express the next odd integer in terms of x.

70. If x represents the first even integer, express the next even integer in terms of x.

71. If x represents the first of two consecutive even integers, express the sum of the two integers in terms of x.

72. If x represents the first of two consecutive odd integers, express the sum of the two integers in terms of x.

73. If x is the first of three consecutive integers, express the sum of the first integer and the third integer as an algebraic expression containing the variable x.

74. If x is the first of two consecutive integers, express the sum of 20 and the second consecutive integer as an algebraic expression containing the variable x.

75. Express the sum of three odd consecutive integers as an algebraic expression in x. Let x be the first odd integer.

76. If x is the first of four consecutive even integers, write their sum as an algebraic expression in x.

Solve each equation.

77. $-3.6x = 10.62$ **78.** $4.95y = -31.185$
79. $7x - 5.06 = -4.92$
80. $0.06y + 2.63 = 2.5562$

Review Exercises

Simplify each expression. See Section 2.1.

81. $5x + 2(x - 6)$ **82.** $-7y + 2y - 3(y + 1)$
83. $6(2z + 4) + 20$ **84.** $-(3a - 3) + 2a - 6$
85. $-(x - 1) + x$ **86.** $8(z - 6) + 7z - 1$

Insert $<$, $>$, or $=$ in the appropriate space to make each statement true. See Sections 1.5 and 1.7.

87. $(-3)^2$ -3^2 **88.** $(-2)^4$ -2^4
89. $(-2)^3$ -2^3 **90.** $(-4)^3$ -4^3
91. $-|-6|$ 6 **92.** $-|-0.7|$ -0.7

2.4 SOLVING LINEAR EQUATIONS

TAPE BA 2.4

OBJECTIVES

1. Apply the general strategy for solving a linear equation.
2. Solve equations containing fractions.
3. Solve equations containing decimals.
4. Recognize identities and equations with no solution.
5. Write sentences as equations and solve.

We now present a general strategy for solving linear equations. One new piece of strategy is a suggestion to "clear an equation of fractions" as a first step. Doing so makes the equation more manageable, since operating on integers is more convenient than operating on fractions.

To Solve Linear Equations in One Variable

Step 1. Clear the equation of fractions by multiplying both sides of the equation by the lowest common denominator (LCD) of all denominators in the equation.

Step 2. Remove any grouping symbols, such as parentheses, by using the distributive property.

Step 3. Simplify each side of the equation by combining like terms.

Step 4. Write the equation with variable terms on one side and numbers on the other side by using the addition property of equality.

Step 5. Isolate the variable by using the multiplication property of equality.

Step 6. Check the solution by substituting it in the original equation.

EXAMPLE 1 Solve $4(2x - 3) + 7 = 3x + 5$.

Solution: There are no fractions to clear, so begin with step 2.

$$4(2x - 3) + 7 = 3x + 5$$

Step 2. $\quad 8x - 12 + 7 = 3x + 5 \quad$ Apply the distributive property.

Step 3. $\quad 8x - 5 = 3x + 5 \quad$ Combine like terms.

Step 4. Get variable terms on the same side of the equation by subtracting $3x$ from both sides; then add 5 to both sides.

$$8x - 5 - 3x = 3x + 5 - 3x \quad \text{Subtract } 3x \text{ from both sides.}$$
$$5x - 5 = 5 \quad \text{Simplify.}$$
$$5x - 5 + 5 = 5 + 5 \quad \text{Add 5 to both sides.}$$
$$5x = 10 \quad \text{Simplify.}$$

Step 5. Use the multiplication property of equality to isolate x.

$$\frac{5x}{5} = \frac{10}{5} \quad \text{Divide both sides by 5.}$$
$$x = 2 \quad \text{Simplify.}$$

Step 6. To check, replace x with 2 in the original equation.

Check: $\quad 4(2x - 3) + 7 = 3x + 5$
$\quad\quad\quad 4[2(2) - 3] + 7 = 3(2) + 5 \quad$ Replace x with 2.
$\quad\quad\quad 4(4 - 3) + 7 = 6 + 5$
$\quad\quad\quad 4(1) + 7 = 11$
$\quad\quad\quad 4 + 7 = 11$
$\quad\quad\quad 11 = 11 \quad$ True.

The solution set is $\{2\}$.

EXAMPLE 2 Solve $8(2 - t) = -5t$.

Solution: First, apply the distributive property.

$$8(2 - t) = -5t$$

Step 2.	$16 - 8t = -5t$	Use the distributive property.
Step 4.	$16 - 8t + 8t = -5t + 8t$	Add $8t$ to both sides.
	$16 = 3t$	Combine like terms.
Step 5.	$\dfrac{16}{3} = \dfrac{3t}{3}$	Divide both sides by 3.
	$\dfrac{16}{3} = t$	Simplify.

Step 6. Check to see that the solution is $\dfrac{16}{3}$.

$$8(2 - t) = -5t$$

$8\left(2 - \dfrac{16}{3}\right) = -5\left(\dfrac{16}{3}\right)$	Replace t with $\dfrac{16}{3}$.
$8\left(\dfrac{6}{3} - \dfrac{16}{3}\right) = -\dfrac{80}{3}$	The LCD is 3.
$8\left(-\dfrac{10}{3}\right) = -\dfrac{80}{3}$	Subtract fractions.
$-\dfrac{80}{3} = -\dfrac{80}{3}$	True.

The solution set is $\left\{\dfrac{16}{3}\right\}$.

2 Next, we solve an equation containing fractions.

EXAMPLE 3 Solve for x: $\dfrac{x}{2} - 1 = \dfrac{2}{3}x - 3$.

Solution: This equation contains fractions, so we begin by clearing fractions. To do this, multiply both sides of the equation by the LCD of 2 and 3, which is 6.

$$\dfrac{x}{2} - 1 = \dfrac{2}{3}x - 3$$

Step 1.	$6\left(\dfrac{x}{2} - 1\right) = 6\left(\dfrac{2}{3}x - 3\right)$	Multiply both sides by 6.
Step 2.	$6\left(\dfrac{x}{2}\right) - 6(1) = 6\left(\dfrac{2}{3}x\right) - 6(3)$	Apply the distributive property.
	$3x - 6 = 4x - 18$	Simplify.

There are no longer grouping symbols and no like terms on either side of the equation, so we continue with step 4.

$$3x - 6 = 4x - 18$$

Step 4. $\quad 3x - 6 - 3x = 4x - 18 - 3x \quad$ Subtract $3x$ from both sides.

$$-6 = x - 18 \quad \text{Simplify.}$$
$$-6 + 18 = x - 18 + 18 \quad \text{Add 18 to both sides.}$$
$$12 = x \quad \text{Simplify.}$$

Step 5. The variable is isolated so there is no need to apply the multiplication property of equality.

Step 6. The equation is solved for x and the solution is 12. To check, replace x with 12 in the original equation.

$$\frac{x}{2} - 1 = \frac{2}{3}x - 3 \quad \text{Original equation.}$$
$$\frac{12}{2} - 1 = \frac{2}{3} \cdot 12 - 3 \quad \text{Replace } x \text{ with 12.}$$
$$6 - 1 = 8 - 3 \quad \text{Simplify.}$$
$$5 = 5 \quad \text{True.}$$

The solution set is {12}.

EXAMPLE 4 Solve $\dfrac{2(a + 3)}{3} = 6a + 2$.

Solution: Clear the equation of fractions first.

$$\frac{2(a + 3)}{3} = 6a + 2$$

Step 1. $\quad 3 \cdot \dfrac{2(a + 3)}{3} = 3(6a + 2) \quad$ Clear fraction by multiplying both sides by the LCD 3.

Step 2. Next, use the distributive property and remove parentheses.

$$2a + 6 = 18a + 6 \quad \text{Apply the distributive property.}$$

Step 4. $\quad 2a + 6 - 6 = 18a + 6 - 6 \quad$ Subtract 6 from both sides.

$$2a = 18a$$
$$2a - 18a = 18a - 18a \quad \text{Subtract } 18a \text{ from both sides.}$$
$$-16a = 0$$

Step 5. $\quad \dfrac{-16a}{-16} = \dfrac{0}{-16} \quad$ Divide both sides by -16.

$$a = 0 \quad \text{Write the fraction in simplest form.}$$

Step 6. To check, replace a with 0 in the original equation. The solution set is {0}.

3 Oftentimes, especially when solving a problem having to do with money, we encounter an equation containing decimals. An equation such as this may be cleared of decimals by multiplying both sides by an appropriate power of 10 as shown in the next example.

EXAMPLE 5 Solve $.25x + .10(x - 3) = .05(x + 18)$

Solution: First, clear this equation of decimals by multiplying both sides of the equation by 100. Recall that multiplying a decimal number by 100 has the effect of moving the decimal point 2 places to the right.

$$.25x + .10(x - 3) = .05(x + 18)$$

Step 1. $.25x + .10(x - 3) = .05(x + 18)$ Multiply both sides by 100.
$25x + 10(x - 3) = 5(x + 18)$

Step 2. $25x + 10x - 30 = 5x + 90$ Apply the distributive property.

Step 3. $35x - 30 = 5x + 90$ Combine like terms.

Step 4. $35x - 30 - 5x = 5x + 90 - 5x$ Subtract $5x$.
$30x - 30 = 90$ Combine like terms.
$30x - 30 + 30 = 90 + 30$ Add 30.
$30x = 120$ Combine like terms.

Step 5. $\dfrac{30x}{30} = \dfrac{120}{30}$ Divide by 30.
$x = 4$

Step 6. To check, replace x with 4 in the original equation. The solution set is $\{4\}$.

4 So far, each equation that we have solved has had a single solution.
Not every equation in one variable has a single solution. Some equations have no solution, while others have an infinite number of solutions. For example,

$$x + 5 = x + 7$$

has no solution since, no matter which **real number** we replace x with, the equation is false.

real number $+ 5 =$ same **real number** $+ 7$ **FALSE**

On the other hand,

$$x + 6 = x + 6$$

has infinitely many solutions since x can be replaced by any real number and the equation is always true.

real number $+ 6 =$ same **real number** $+ 6$ **TRUE**

The equation $x + 6 = x + 6$ is called an **identity**. The next few examples illustrate equations like these.

EXAMPLE 6 Solve $-2(x - 5) + 10 = -3(x + 2) + x$.

Solution:
$$-2(x - 5) + 10 = -3(x + 2) + x$$
$$-2x + 10 + 10 = -3x - 6 + x \quad \text{Distribute on both sides.}$$
$$-2x + 20 = -2x - 6 \quad \text{Combine like terms.}$$
$$-2x + 20 + 2x = -2x - 6 + 2x \quad \text{Add } 2x \text{ to both sides.}$$
$$20 = -6 \quad \text{Combine like terms.}$$

Notice that no value for x makes $20 = -6$ a true equation. We conclude that there is **no solution** to this equation. Its solution set is written as either $\{\ \}$ or $\varnothing$.

EXAMPLE 7 Solve $3(x - 4) = 3x - 12$.

Solution:
$$3(x - 4) = 3x - 12$$
$$3x - 12 = 3x - 12 \quad \text{Apply the distributive property.}$$

The left side of the equation is now identical to the right side. Every real number may be substituted for x and a true statement will result. We arrive at the same conclusion if we continue.

$$3x - 12 = 3x - 12$$
$$3x - 12 + 12 = 3x - 12 + 12 \quad \text{Add 12 to both sides.}$$
$$3x = 3x \quad \text{Combine like terms.}$$
$$0 = 0 \quad \text{Subtract } 3x \text{ from both sides.}$$

Again, one side of the equation is identical to the other side. Thus, $3(x - 4) = 3x - 12$ is an **identity** and every real number is a solution. The solution set may be written as

$$\{x \mid x \text{ is a real number}\}. \text{ This is read as}$$

↑ ↑
"the   such
set of   that x is a real number"
all x

5 We can apply our equation-solving skills to solving problems written in words.
Many times, writing an equation that describes or models a problem involves a direct translation from a word sentence to an equation.

EXAMPLE 8 Twice a number added to seven is the same as three subtracted from the number. Find the number.

Solution: Translate the sentence into an equation and solve.

In words:	twice a number	added to	seven	is the same as	three subtracted from the number
Translate:	$2x$	$+$	7	$=$	$x - 3$

To solve, begin by subtracting x on both sides to isolate the variable term.

$2x + 7 = x - 3$
$2x + 7 - x = x - 3 - x$ Subtract x from both sides.
$x + 7 = -3$ Combine like terms.
$x + 7 - 7 = -3 - 7$ Subtract 7 from both sides.
$x = -10$ Combine like terms.

Check the solution in the problem as it was originally stated. To do so, replace "number" in the sentence with -10. Twice "-10" added to 7 is the same as 3 subtracted from "-10."

$2(-10) + 7 = -10 - 3$
$-13 = -13$

The unknown number is -10.

> **REMINDER** When checking solutions, go back to the original stated problem, rather than to your equation in case errors have been made in translating to an equation.

SCIENTIFIC CALCULATOR EXPLORATIONS

Checking Equations
We can use a calculator to check possible solutions of equations. To do this, replace the variable by the possible solution and evaluate both sides of the equation separately.

Equation: $3x - 4 = 2(x + 6)$ Solution: $x = 16$
$3x - 4 = 2(x + 6)$ Original equation.
$3(16) - 4 \stackrel{?}{=} 2(16 + 6)$ Replace x with 16.

(continued)

Now evaluate each side with your calculator.

Evaluate left side: [3] [×] [16] [−] [4] [=]   Display: 44

Evaluate right side: [2] [(] [16] [+] [6] [)] [=]   Display: 44

Since the left side equals the right side, the equation checks.

Use a calculator to check the possible solutions to each of the following equations.
1. $2x = 48 + 6x$; $x = -12$
2. $-3x - 7 = 3x - 1$; $x = -1$
3. $5x - 2.6 = 2(x + 0.8)$; $x = 4.4$
4. $-1.6x - 3.9 = -6.9x - 25.6$; $x = 5$
5. $\dfrac{564x}{4} = 200x - 11(649)$; $x = 121$
6. $20(x - 39) = 5x - 432$; $x = 23.2$

EXERCISE SET 2.4

Solve each equation. See Examples 1 and 2.
1. $-2(3x - 4) = 2x$
2. $-(5x - 1) = 9$
3. $4(2n - 1) = (6n + 4) + 1$
4. $3(4y + 2) = 2(1 + 6y) + 8$
5. $5(2x - 1) - 2(3x) = 4$
6. $3(2 - 5x) + 4(6x) = 12$
7. $6(x - 3) + 10 = -8$
8. $-4(2 + n) + 9 = 1$

Solve each equation. See Examples 3 through 5.
9. $\dfrac{3}{4}x - \dfrac{1}{2} = 1$
10. $\dfrac{2}{3}x + \dfrac{5}{3} = \dfrac{5}{3}$
11. $x + \dfrac{5}{4} = \dfrac{3}{4}x$
12. $\dfrac{7}{8}x + \dfrac{1}{4} = \dfrac{3}{4}x$
13. $\dfrac{x}{2} - 1 = \dfrac{x}{5} + 2$
14. $\dfrac{x}{5} - 2 = \dfrac{x}{3}$
15. $\dfrac{6(3 - z)}{5} = -z$
16. $\dfrac{4(5 - w)}{3} = -w$
17. $\dfrac{2(x + 1)}{4} = 3x - 2$
18. $\dfrac{3(y + 3)}{5} = 2y + 6$

19. $.50x + .15(70) = .25(142)$
20. $.40x + .06(30) = .20(49)$
21. $.12(y - 6) + .06y = .08y - .07(10)$
22. $.60(z - 300) + .05z = .70z - .41(500)$

Solve each equation. See Examples 6 and 7.
23. $5x - 5 = 2(x + 1) + 3x - 7$
24. $3(2x - 1) + 5 = 6x + 2$
25. $\dfrac{x}{4} + 1 = \dfrac{x}{4}$
26. $\dfrac{x}{3} - 2 = \dfrac{x}{3}$
27. $3x - 7 = 3(x + 1)$
28. $2(x - 5) = 2x + 10$

29. Explain the difference between simplifying an expression and solving an equation.
30. When solving an equation, if the final equivalent equation is $0 = 5$, what can we conclude? If the final equivalent equation is $-2 = -2$, what can we conclude?
31. On your own, construct an equation for which every real number is a solution.
32. On your own, construct an equation that has no solution.

Solve each equation.

33. $4x + 3 = 2x + 11$
34. $6y - 8 = 3y + 7$
35. $-2y - 10 = 5y + 18$
36. $7n + 5 = 10n - 10$
37. $.6x - .1 = .5x + .2$
38. $.2x - .1 = .6x - 2.1$
39. $2y + 2 = y$
40. $7y + 4 = -3$
41. $3(5c - 1) - 2 = 13c + 3$
42. $4(3t + 4) - 20 = 3 + 5t$
43. $x + \frac{7}{6} = 2x - \frac{7}{6}$
44. $\frac{5}{2}x - 1 = x + \frac{1}{4}$
45. $2(x - 5) = 7 + 2x$
46. $-3(1 - 3x) = 9x - 3$
47. $\frac{2(z + 3)}{3} = 5 - z$
48. $\frac{3(w + 2)}{4} = 2w + 3$
49. $\frac{4(y - 1)}{5} = -3y$
50. $\frac{5(1 - x)}{6} = -4x$
51. $8 - 2(a - 1) = 7 + a$
52. $5 - 6(2 + b) = b - 14$
53. $2(x + 3) - 5 = 5x - 3(1 + x)$
54. $4(2 + x) + 1 = 7x - 3(x - 2)$
55. $\frac{5x - 7}{3} = x$
56. $\frac{7n + 3}{5} = -n$
57. $\frac{9 + 5v}{2} = 2v - 4$
58. $\frac{6 - c}{2} = 5c - 8$
59. $-3(t - 5) + 2t = 5t - 4$
60. $-(4a - 7) - 5a = 10 + a$
61. $.02(6t - 3) = .05(t - 2) + .02$
62. $.03(m + 7) = .02(5 - m) + .03$
63. $.06 - .01(x + 1) = -.02(2 - x)$
64. $-.01(5x + 4) = .04 - .01(x + 4)$
65. $\frac{3(x - 5)}{2} = \frac{2(x + 5)}{3}$
66. $\frac{5(x - 1)}{4} = \frac{3(x + 1)}{2}$
67. $1000(7x - 10) = 50(412 + 100x)$
68. $10{,}000(x + 4) = 100(16 + 7x)$
69. $.035x + 5.112 = .010x + 5.107$
70. $.127x - 2.685 = .027x - 2.38$

Write each of the following as equations. Then solve. See Example 8.

71. The sum of twice a number and $\frac{1}{5}$ is equal to the difference between three times a number and $\frac{4}{5}$. Find the number.

72. The sum of four times a number and $\frac{2}{3}$ is equal to the difference of five times the number and $\frac{5}{6}$. Find the number.

73. The sum of twice a number and 7 is equal to the sum of a number and 6. Find the number.

74. The difference of three times a number and 1 is the same as twice a number. Find the number.

75. Three times a number minus 6 is equal to two times a number plus 8. Find the number.

76. The sum of 4 times a number and -2 is equal to the sum of 5 times a number and -2. Find the number.

77. One-third of a number is five-sixths. Find the number.

78. Seven-eighths of a number is one-half. Find the number.

79. The difference of a number and four is twice the number. Find the number.

80. The sum of double a number and six is four times the number. Find the number.

81. If the quotient of a number and 4 is added to $\frac{1}{2}$, the result is $\frac{3}{4}$. Find the number.

82. If $\frac{3}{4}$ is added to three times a number, the result is $\frac{1}{2}$ subtracted from twice a number. Find the number.

83. The perimeter of a geometric figure is the sum of the lengths of its sides. If the perimeter of the following pentagon (five-sided figure) is 28 centimeters, find the length of each side.

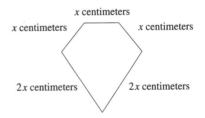

84. The perimeter of the following triangle is 35 meters. Find the length of each side.

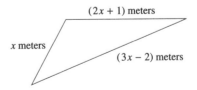

85. Five times a number subtracted from ten is triple the number. Find the number.

86. Nine is equal to ten subtracted from double a number. Find the number.

The graph below is called a 3-dimensional bar graph. It shows the five most common names of cities, towns, or villages in the United States.

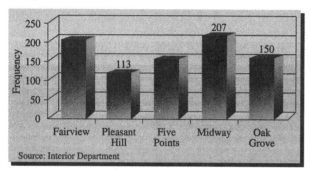

87. What is the most popular name of a city, town, or village in the United States?

88. How many more cities, towns, or villages are named Oak Grove than named Pleasant Hill?

89. Let x represent "the number of towns, cities, or villages named Five Points" and use the information given to determine the unknown number. "The number of towns, cities, or villages named Five Points" added to 55 is equal to twice "the number of towns, cities, or villages named Five Points" minus 90. Check your answer by noticing the height of the bar representing Five Points. Is your answer reasonable?

90. Let x represent "the number of towns, cities, or villages named Fairview" and use the information given to determine the unknown number. Three times "the number of towns, cities, or villages named Fairview" added to 24 is equal to 168 subtracted from 4 times "the number of towns, cities, or villages named Fairview." Check your answer by noticing the height of the bar representing Fairview. Is your answer reasonable?

Review Exercises

Evaluate. See Section 1.7.

91. $|2^3 - 3^2| - |5 - 7|$

92. $|5^2 - 2^2| + |9 \div (-3)|$

93. $\dfrac{5}{4 + 3 \cdot 7}$

94. $\dfrac{8}{24 - 8 \cdot 2}$

See Section 2.1.

95. A plot of land is in the shape of a triangle. If one side is x meters, a second side is $2x - 3$ meters and a third side is $3x - 5$ meters, express the perimeter of the lot as a simplified expression in x.

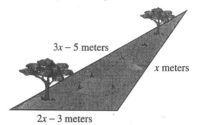

96. A portion of a board has length x feet. The other part has length $7x - 9$ feet. Express the total length of the board as a simplified expression in x.

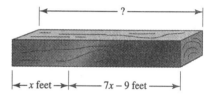

A Look Ahead

EXAMPLE

Solve $t(t + 4) = t^2 - 2t + 10$.

Solution:
$$t(t + 4) = t^2 - 2t + 10$$
$$t^2 + 4t = t^2 - 2t + 10$$
$$t^2 + 4t - t^2 = t^2 - 2t + 10 - t^2$$
$$4t = -2t + 10$$
$$4t + 2t = -2t + 10 + 2t$$
$$6t = 10$$
$$\dfrac{6t}{6} = \dfrac{10}{6}$$
$$t = \dfrac{5}{3}$$

Solve each equation.

97. $x(x - 3) = x^2 + 5x + 7$

98. $t^2 - 6t = t(8 + t)$

99. $2z(z + 6) = 2z^2 + 12z - 8$

100. $y^2 - 4y + 10 = y(y - 5)$

101. $n(3 + n) = n^2 + 4n$

102. $3c^2 - 8c + 2 = c(3c - 8)$

2.5 AN INTRODUCTION TO PROBLEM SOLVING

OBJECTIVE

1. Apply the steps for problem solving

TAPE BA 2.5

In previous sections, you practiced writing word phrases and sentences as algebraic expressions and equations to help prepare for problem solving. We now use these translations to help write equations that model a problem. The problem-solving steps given next may be helpful.

> **PROBLEM-SOLVING STEPS**
>
> 1. UNDERSTAND the problem. During this step don't work with variables, but simply become comfortable with the problem. Some ways of accomplishing this are listed below.
>   - Read and reread the problem.
>   - Construct a drawing.
>   - Propose a solution and check. Pay careful attention to how you check your proposed solution. This will help later when writing an equation to model the problem.
> 2. ASSIGN a variable to an unknown in the problem. Use this variable to represent any other unknown quantities.
> 3. ILLUSTRATE the problem. A diagram or chart using the assigned variables can often help visualize the known facts.
> 4. TRANSLATE the problem into a mathematical model. This is often an equation.
> 5. COMPLETE the work. This often means to solve the equation.
> 6. INTERPRET the results: *Check* the proposed solution in the stated problem and *state* your conclusion.

EXAMPLE 1 A 10-foot board is to be cut into two pieces so that the longer piece is 4 times the shorter. Find the length of each piece.

Solution: 1. UNDERSTAND the problem. To do so, read and reread the problem, construct a drawing, and propose a solution. For example, if 3 feet represents the length of the shorter piece, then $4(3) = 12$ feet is the length of the longer piece, since it is 4 times the length of the shorter piece.

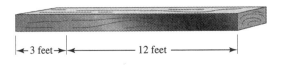

This guess gives a total board length of 3 feet + 12 feet = 15 feet, too long. The purpose of guessing a solution is not to guess correctly, but to help better understand the problem and how to model it. At this point, we have a better understanding of the problem. Although it may be possible now to guess the solution, for purposes of practicing problem-solving skills, we continue with the problem-solving steps.

2. ASSIGN a variable. Use this variable to represent any other known quantities. If we let

x = length of shorter piece,

then $4x$ = length of longer piece.

3. ILLUSTRATE the problem. Draw a picture of the board and label the picture with the assigned variables.

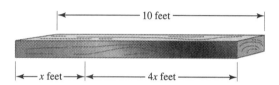

4. TRANSLATE the problem. First, write the equation in words.

In words:	length of shorter piece	added to	length of longer	equals	total length of board
Translate:	x	$+$	$4x$	$=$	10

5. COMPLETE the work. Here, we solve the equation.

$x + 4x = 10$

$5x = 10$ Combine like terms.

$\dfrac{5x}{5} = \dfrac{10}{5}$ Divide both sides by 5.

$x = 2$

6. INTERPRET the results. First, *check* the solution in the stated problem. If the shorter piece of board is 2 feet, the longer piece is $4 \cdot (2 \text{ feet}) = 8$ feet and the sum of the two pieces is 2 feet + 8 feet = 10 feet. Next, *state* the conclusions. The shorter piece of board is 2 feet and the longer piece of board is 8 feet.

EXAMPLE 2 In 1996, Congress had 8 more Republican senators than Democratic. If the total number of senators is 100, how many senators of each party were there?

Solution: 1. UNDERSTAND the problem. Read and reread the problem. Let's guess that there are 40 Democratic senators. Since there are 8 more Republicans than Democrats, there must be 40 + 8 = 48 Republicans. The total number of Democrats and Republicans is then 40 + 48 = 88. This is an incorrect guess, since the total should be 100, but we now have a better understanding of the problem. This was the purpose for guessing.

2. ASSIGN a variable. Let

x = number of Democrats,

then

$x + 8$ = number of Republicans.

3. ILLUSTRATE the problem. No diagram or chart is needed.
4. TRANSLATE the problem. First write the equation in words.

In words:	number of Democrats	added to	number of Republicans	equals	100
Translate:	x	+	$(x + 8)$	=	100

5. COMPLETE the work. We solve the equation.

$x + (x + 8) = 100$
$2x + 8 = 100$ Combine like terms.
$2x + 8 - 8 = 100 - 8$ Subtract 8 from both sides.
$2x = 92$
$\dfrac{2x}{2} = \dfrac{92}{2}$ Divide both sides by 2.
$x = 46$

6. INTERPRET the results.
Check: If there are 46 Democratic senators, then there are 46 + 8 = 54 Republican senators. The total number of senators is then 46 + 54 = 100. The results check.
State: In 1996, there were 46 Democratic and 54 Republican senators.

Much of problem solving involves a direct translation from a sentence to an equation.

EXAMPLE 3 Twice the sum of a number and 4 is the same as four times the number decreased by 12. Find the number.

Solution: 1. UNDERSTAND. Read and reread the problem. Propose a solution and check.
2. ASSIGN. Let

x = the unknown number.

3. ILLUSTRATE. No illustration is needed.
4. TRANSLATE.

In words:	twice	sum of a number and 4	is the same as	four times the number	decreased by	12
Translate:	2	$(x + 4)$	$=$	$4x$	$-$	12

5. COMPLETE. Now solve the equation.

$$2(x + 4) = 4x - 12$$
$$2x + 8 = 4x - 12 \qquad \text{Apply the distributive property.}$$
$$2x + 8 - 4x = 4x - 12 - 4x \qquad \text{Subtract } 4x \text{ from both sides.}$$
$$-2x + 8 = -12$$
$$-2x + 8 - 8 = -12 - 8 \qquad \text{Subtract 8 from both sides.}$$
$$-2x = -20$$
$$\frac{-2x}{-2} = \frac{-20}{-2} \qquad \text{Divide both sides by } -2.$$
$$x = 10$$

6. INTERPRET.
   Check: Check this solution in the problem as it was originally stated. To do so, replace "number" with 10. Twice the sum of "10" and 4 is 28, which is the same as 4 times "10" decreased by 12.
   State: The number is 10.

EXAMPLE 4 A local cellular phone company charges Elaine Chapoton $50 per month and $0.36 per minute of phone use in her usage category. If Elaine was charged $99.68 for a month's cellular phone use, determine the number of whole minutes of phone use.

Solution:
1. UNDERSTAND. Read and reread the problem. Next, guess an answer. Let's guess 70 minutes, and pay careful attention as to how we check this guess. For 70 minutes of use, Elaine's phone bill will be $50 plus $0.36 per minute of use. This is $50 + 0.36(70) = $75.20, less than $99.68. We now understand the problem and know that the number of minutes is greater than 70.

2. ASSIGN. Let

   x represent the unknown in this problem,
   or let x = number of minutes.

3. ILLUSTRATE. No diagram is needed.
4. TRANSLATE.

In words:	$50	added to	minute charge	is equal to	$99.68
Translate:	50	$+$	$.36x$	$=$	99.68

5. COMPLETE.

$$50 + 0.36x = 99.68$$
$$50 + 0.36x - 50 = 99.68 - 50$$
$$0.36x = 49.68$$
$$\frac{0.36x}{0.36} = \frac{49.68}{0.36}$$
$$x = 138$$

6. INTERPRET.

Check: If Elaine spends 138 minutes on her cellular phone, her bill is $50 + $0.36(138) = $99.68.

State: Elaine spent 138 minutes on her cellular phone this month.

EXERCISE SET 2.5

Solve the following. See Examples 1 and 2.

1. The governor of New York makes twice as much money as the governor of Nebraska. If the total of their salaries is $195,000, find the salary of each.

2. In the 1992 Summer Olympics, the Unified Team, consisting of athletes from 12 former Soviet republics, won 8 more gold medals than the United States Team. If the total number of gold medals for both is 82, find the number of gold medals that each team won.

3. A 40-inch board is to be cut into three pieces so that the second piece is twice as long as the first piece and the third piece is 5 times as long as the first piece. If x represents the length of the first piece, find the lengths of all three pieces.

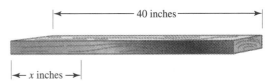

4. A 21-foot beam is to be divided so that the longer piece is 1 foot more than 3 times the shorter piece. If x represents the length of the shorter piece, find the lengths of both pieces.

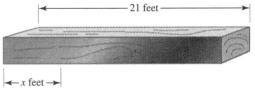

Solve. See Example 3.

5. The product of twice a number and three is the same as the difference of five times the number and $\frac{3}{4}$. Find the number.

In the '92 Olympics, Michael Marsh won the 200-m run in 20.01 seconds

6. If the difference of a number and four is doubled, the result is $\frac{1}{4}$ less than the number. Find the number.

7. If the sum of a number and five is tripled, the result is one less than twice the number. Find the number.

8. Twice the sum of a number and six equals three times the sum of the number and four. Find the number.

Solve. See Example 4.

9. A car rental agency advertised renting a Buick Century for $24.95 per day and $0.29 per mile. If you rent this car for 2 days, how many whole miles can be driven on a $100 budget?

10. A plumber gave an estimate for the renovation of a kitchen. Her hourly pay is $27 per hour and the plumber's parts will cost $80. If her total estimate is $404, how many hours does she expect this job to take?

Solve.

11. A 17-foot piece of string is cut into two pieces so that one piece is 2 feet longer than twice the shorter piece. If the shorter piece is x feet long, find the lengths of both pieces.

12. In a recent election in Florida for a seat in the United States House of Representatives, Corrine Brown received 16,950 more votes than Marc Little. If the total number of votes was 110,740 find the number of votes for each candidate.

13. Two angles are supplementary if their sum is 180°. One angle measures three times the measure of a smaller angle. If x represents the measure of the smaller angle and these two angles are supplementary, find the measure of each angle.

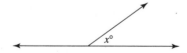

14. Two angles are complementary if their sum is 90°. Given the measures of the complementary angles shown, find the measure of each angle.

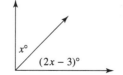

15. On June 20, 1994, John Paxson sank a 3-point shot with 3.9 seconds left to give the Chicago Bulls their third straight National Basketball Association championship. The opposing team was the Phoenix Suns. If the final score of the game was 2 consecutive integers whose sum is 197, find each final score.

16. To make an international telephone call, you need the code for the country you are calling. The codes for Mali Republic, Côte d'Ivoire, and Niger are three consecutive odd integers whose sum is 675. Find the code for each country.

17. Determine whether there are two consecutive odd integers such that 7 times the first exceeds 5 times the second by 54.

18. The sum of three consecutive integers is 13 more than twice the smallest integer. Find the integers.

19. Twice the difference of a number and 8 is equal to three times the sum of a number and 3. Find the number.

20. Five times the sum of a number and −1 is the same as 6 times a number. Find the number.

21. Find two consecutive odd integers such that twice the larger is 15 more than three times the smaller.

22. Find three consecutive even integers whose sum is negative 114.

23. On December 7, 1995, a probe launched from the robot explorer called Galileo entered the atmosphere of Jupiter at 100,000 miles per hour. The diameter of the probe is 19 inches less than twice its height. If the sum of the height and the diameter is 83 inches, find each dimension.

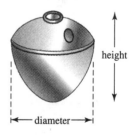

24. Over the past few years the satellite Voyager II has passed by the planets Saturn, Uranus, and Neptune, continually updating information about these planets, including the number of moons for each.

Uranus is now believed to have 7 more moons than Neptune. Also, Saturn is now believed to have three times the number of moons of Neptune. If the total number of moons for these planets is 47, find the number of moons for each planet.

25. A woman's $15,000 estate is to be divided so that her husband receives twice as much as her son. If x represents the amount of money that her son receives, find the amount of money that her husband receives and the amount of money that her son receives.

26. The sum of two consecutive integers is 31. What are the numbers?

27. The flag of Brazil contains a parallelogram. One angle of the parallelogram is 15° less than twice the measure of the angle next to it. Find the measure of each angle of the parallelogram. (*Hint:* Recall that opposite angles of a parallelogram have the same measure and that the sum of the angles is 360°.)

28. The flag of Equatorial Guinea contains an isosceles triangle. (Recall that an isosceles triangle contains two angles with the same measure.) If the measure of the third angle of the triangle is 30° more than twice the measure of either of the other two angles, find the measure of each angle of the triangle. (*Hint:* Recall that the sum of the measure of the angles of a triangle is 180°.)

29. The measures of the angles of a triangle are 3 consecutive even integers. Find the measure of each angle.

30. The golden rectangle is a rectangle whose length is approximately 1.6 times its width. The early Greeks thought that a rectangle with these dimensions was the most pleasing to the eye and examples of the golden rectangle are found in many early works of art. For example, the Parthenon in Athens contains many examples of golden rectangles. Mike Hallahan would like to plant a rectangular garden in the shape of a golden rectangle. If he has 78 feet of fencing available, find the dimensions of the garden.

The length-width rectangle approximates the golden rectangle as well as the width-height rectangle.

31. Dr. Dorothy Smith gave the students in her geometry class at the University of New Orleans the fol-

lowing question. Is it possible to construct a triangle such that the second angle of the triangle has a measure that is twice the measure of the first angle and the measure of the third angle is 3 times the measure of the first? If so, find the measure of each angle. (*Hint:* Recall that the sum of the measures of the angles of a triangle is 180°.)

32. Give an example of how you recently solved a problem using mathematics.

33. In your own words, explain why a solution of a word problem should be checked using the original wording of the problem and not the equation written from the wording.

Recall from Exercise 30 that the golden rectangle is a rectangle whose length is approximately 1.6 times its width.

34. It is thought that for about 75% of adults, a rectangle in the shape of the golden rectangle is the most pleasing to the eye. Draw 3 rectangles, one in the shape of the golden rectangle, and poll your class. Do the results agree with the percentage given above?

35. Examples of golden rectangles can be found today in architecture and manufacturing packaging. Find an example of a golden rectangle in your home. A few suggestions: the front face of a book, the floor of a room, the front of a box of food.

36. Measure the dimensions of each rectangle and decide which one best approximates the shape of the golden rectangle.

a.

b.

c.

Review Exercises

Perform indicated operations. See Sections 1.5 and 1.6.

37. $3 + (-7)$ **38.** $-2 + (-8)$
39. $4 - 10$ **40.** $-11 + 2$
41. $-5 - (-1)$ **42.** $-12 - 3$

Translate each sentence into an equation. See Sections 1.4 and 2.5.

43. Half of the difference of a number and one is thirty-seven.

44. Five times the opposite of a number is the number plus sixty.

45. If three times the sum of a number and 2 is divided by 5, the quotient is 0.

46. If the sum of a number and 9 is subtracted from 50, the result is 0.

2.6 FORMULAS AND PROBLEM SOLVING

OBJECTIVES

 Use formulas to solve problems.
2. Solve a formula or equation for one of its variables.

TAPE
BA 2.6

An equation that describes a known relationship among quantities, such as distance, time, volume, weight, and money is called a **formula.** These quantities are represented by letters and are thus variables of the formula. Here are some common formulas and their meanings.

$A = lw$
Area of a rectangle = length · width

$I = PRT$
Simple Interest = Principal · Rate · Time

$P = a + b + c$
Perimeter of a triangle = side a + side b + side c

$d = rt$
distance = rate · time

$V = lwh$
Volume of a rectangular solid = length · width · height

$F = \left(\frac{9}{5}\right)C + 32$

degrees Fahrenheit = $\left(\frac{9}{5}\right)$ · degrees Celsius + 32

Formulas are valuable tools because they allow us to calculate measurements as long as we know certain other measurements. For example, if we know we traveled a distance of 100 miles at a rate of 40 miles per hour, we can replace the variables d and r in the formula $d = rt$ and find our time, t.

$d = rt$ Formula.
$100 = 40t$ Replace d with 100 and r with 40.

This is a linear equation in one variable, t. To solve for t, divide both sides of the equation by 40.

$\dfrac{100}{40} = \dfrac{40t}{40}$ Divide both sides by 40.

$\dfrac{5}{2} = t$ Simplify.

The time traveled is $\frac{5}{2}$ hours or $2\frac{1}{2}$ hours.

In this section we solve problems that can be modeled by known formulas. We use the same problem-solving steps that were introduced in the previous section. These steps have been slightly revised to include formulas.

PROBLEM-SOLVING STEPS

1. **UNDERSTAND** the problem. During this step don't work with variables (except for known formulas), but simply become comfortable with the problem. Some ways of accomplishing this are listed below.
   - Read and reread the problem.
   - Construct a drawing.
   - Look up an unknown formula.
   - Propose a solution and check. Pay careful attention to how to check your proposed solution. This will help later when writing an equation to model the problem.
2. **ASSIGN** a variable to an unknown in the problem. Use this variable to represent any other unknown quantities.
3. **ILLUSTRATE** the problem. A diagram or chart using the assigned variables can often help to visualize the known facts.
4. **TRANSLATE** the problem into a mathematical model. This is often an equation.
5. **COMPLETE** the work. This often means to solve the equation.
6. **INTERPRET** the results. *Check* the proposed solution in the stated problem and *state* your conclusion.

EXAMPLE 1 A glacier is a giant mass of rocks and ice that flows downhill like a river. Portage Glacier in Alaska is about 6 miles, or 31,680 feet, long and moves 400 feet per year. Icebergs are created when the front end of the glacier flows into Portage Lake. How long does it take for ice at the head (beginning) of the glacier to reach the lake?

Solution: 1. **UNDERSTAND.** Read and reread the problem. The appropriate formula needed to solve this problem is the distance formula, $d = rt$. To become familiar with this formula, let's find the distance that ice traveling at a rate of 400 feet per year travels in 100 years. To do so, we let t time be 100 years, r rate be the given 400 feet per year, and substitute these values into the formula $d = rt$. We then have that distance $d = 400(100) = 40{,}000$ feet. Since we are interested in finding how long it takes ice to travel 31,680 feet, we now know that it is less than 100 years.

2. **ASSIGN.** Since we are using the formula $d = rt$, we let

   t = the time in years for ice to reach the lake,
   r = rate or speed of ice, and
   d = distance from beginning of glacier to lake.

3. **ILLUSTRATE.** No other illustration needed.

4. **TRANSLATE.** To translate to an equation, we use the formula $d = rt$ and let distance $d = 31{,}680$ feet and rate $r = 400$ feet per year.

   Formula: $d = r \cdot t$
   Substitute: $31{,}680 = 400 \cdot t$ Let $d = 31{,}680$ and $r = 400$.

5. **COMPLETE.** Solve the equation for t. To solve for t, divide both sides by 400.

   $$\frac{31{,}680}{400} = \frac{400 \cdot t}{400} \quad \text{Divide both sides by 400.}$$
   $$79.2 = t \quad \text{Simplify.}$$

6. **INTERPRET.**
   Check: To check, substitute 79.2 for t, and 400 for r in the distance formula and check to see that the distance is 31,680 feet.
   State: It takes 79.2 years for the ice at the head of Portage Glacier to reach the lake.

EXAMPLE 2 Charles Pecot can afford enough fencing to enclose a rectangular garden with a perimeter of 140 feet. If the width of his garden is to be 30 feet, find the length.

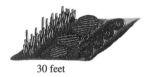

30 feet

Solution: 1. UNDERSTAND. Read and reread the problem. The formula needed to solve this problem is the formula for the perimeter of a rectangle, $P = 2l + 2w$. Before continuing, become familiar with this formula.

2. ASSIGN. Let

   l = the length of the rectangular garden

   w = the width of the rectangular garden, and

   P = perimeter of the garden.

3. ILLUSTRATE.

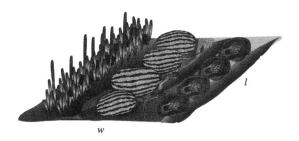

4. TRANSLATE. To translate to an equation, we use the formula $P = 2l + 2w$ and let perimeter $P = 140$ feet and width $w = 30$ feet.

   Formula: $P = 2l + 2w$
   Substitute: $140 = 2l + 2(30)$ Let $P = 140$ and $w = 30$.

5. COMPLETE.

   $140 = 2l + 2(30)$

   $140 = 2l + 60$

   $140 - 60 = 2l + 60 - 60$ Subtract 60 from both sides.

   $80 = 2l$ Combine like terms.

   $40 = l$ Divide both sides by 2.

6. INTERPRET. To *check*, substitute 40 for l and 30 for w in the perimeter formula and check to see that the perimeter is 140 feet.
   State: The length of the rectangular garden is 40 feet.

EXAMPLE 3   The average maximum temperature for January in Algerias, Algeria, is 59° Fahrenheit. Find the equivalent temperature in degrees Celsius.

Solution: 1. UNDERSTAND. Read and reread the problem. A formula that can be used to solve this problem is the formula for converting degrees Celsius to degrees Fahrenheit, $F = \frac{9}{5}C + 32$. Before continuing, become familiar with this formula.

2. ASSIGN. Let

   C = temperature in degrees Celsius, and
   F = temperature in degrees Fahrenheit.

3. ILLUSTRATE. No illustration is needed.

4. TRANSLATE. To translate to an equation, we use the formula $F = \frac{9}{5}C + 32$ and let degrees Fahrenheit $F = 59$.

   Formula: $\quad F = \frac{9}{5}C + 32$

   Substitute: $\quad 59 = \frac{9}{5}C + 32 \quad$ Let $F = 59$

5. COMPLETE.

$$59 = \frac{9}{5}C + 32$$
$$59 - 32 = \frac{9}{5}C + 32 - 32 \quad \text{Subtract 32 from both sides.}$$
$$27 = \frac{9}{5}C \quad \text{Combine like terms.}$$
$$\frac{5}{9} \cdot 27 = \frac{5}{9} \cdot \frac{9}{5}C \quad \text{Multiply both sides by } \tfrac{5}{9}.$$
$$15 = C \quad \text{Combine like terms.}$$

6. INTERPRET

   Check: To check, replace C with 15 and F with 59 in the formula and see that a true statement results.

   State: Thus, 59° Fahrenheit is equivalent to 15° Celsius.

We say that the formula $F = \frac{9}{5}C + 32$ is solved for F because F is isolated on one side of the equation and the other side of the equation contains no F's. Suppose that we need to convert many Fahrenheit temperatures to equivalent degrees Celsius. In this case, it is easier to perform this task by solving the formula $F = \frac{9}{5}C + 32$ for C. (See Example 7.) For this reason, it is important to be able to solve an equation for any one of its specified variables. For example, the formula $d = rt$ is solved for d in terms of r and t. We can also solve $d = rt$ for t in terms of d and r. To solve for t, divide both sides of the equation by r.

$$d = rt$$
$$\frac{d}{r} = \frac{rt}{r} \quad \text{Divide both sides by } r.$$
$$\frac{d}{r} = t \quad \text{Simplify.}$$

To solve a formula or an equation for a specified variable, we use the same steps as for solving a linear equation. These steps are listed next.

To Solve Equations for a Specified Variable

Step 1. Clear the equation of fractions by multiplying each side of the equation by the lowest common denominator.

Step 2. Remove grouping symbols such as parentheses by using the distributive property.

Step 3. Simplify each side of the equation by combining like terms.

Step 4. Write the equation with terms containing the specified variable on one side and all other terms on the other side by using the addition property of equality.

Step 5. Isolate the specified variable by using the distributive property and the multiplication property of equality.

EXAMPLE 4 Solve $V = lwh$ for l.

Solution: This formula is used to find the volume of a box. To solve for l, divide both sides by wh.

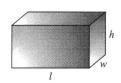

$$V = lwh$$
$$\frac{V}{wh} = \frac{lwh}{wh} \quad \text{Divide both sides by } wh.$$
$$\frac{V}{wh} = l \quad \text{Simplify.}$$

Since we have isolated l on one side of the equation, we have solved for l in terms of V, w, and h. Remember that it does not matter on which side of the equation we isolate the variable.

EXAMPLE 5 Solve $y = mx + b$ for x.

Solution: First, isolate mx by subtracting b from both sides.

$$y = mx + b$$
$$y - b = mx + b - b \quad \text{Subtract } b \text{ from both sides.}$$
$$y - b = mx \quad \text{Combine like terms.}$$

Next, solve for x by dividing both sides by m.

$$\frac{y - b}{m} = \frac{mx}{m}$$
$$\frac{y - b}{m} = x \quad \text{Simplify.}$$

EXAMPLE 6 Solve $P = 2l + 2w$ for w.

Solution: This formula relates the perimeter of a rectangle to its length and width. To solve for w, begin by subtracting $2l$ from both sides.

$$P = 2l + 2w$$
$$P - 2l = 2l + 2w - 2l \quad \text{Subtract } 2l \text{ from both sides.}$$
$$P - 2l = 2w \quad \text{Combine like terms.}$$
$$\frac{P - 2l}{2} = \frac{2w}{2} \quad \text{Divide both sides by 2.}$$
$$\frac{P - 2l}{2} = w \quad \text{Simplify.}$$

The next example has an equation containing a fraction. We will first clear the equation of fractions and then solve for the specified variable.

EXAMPLE 7 Solve $F = \frac{9}{5}C + 32$ for C.

Solution:
$$F = \frac{9}{5}C + 32$$
$$5(F) = 5\left(\frac{9}{5}C + 32\right) \quad \text{Clear the fraction by multiplying both sides by the LCD.}$$
$$5F = 9C + 160 \quad \text{Distribute the 5.}$$
$$5F - 160 = 9C + 160 - 160 \quad \text{Subtract 160 from both sides.}$$
$$5F - 160 = 9C \quad \text{Combine like terms.}$$
$$\frac{5F - 160}{9} = \frac{9C}{9} \quad \text{Divide both sides by 9.}$$
$$\frac{5F - 160}{9} = C \quad \text{Simplify.}$$

EXERCISE SET 2.6

Substitute the given values into the given formulas and solve for the unknown variable. See Examples 1 through 7.

1. $A = bh$; $A = 45$, $b = 15$
2. $D = rt$; $D = 195$, $t = 3$
3. $S = 4lw + 2wh$; $S = 102$, $l = 7$, $w = 3$
4. $V = lwh$; $l = 14$, $w = 8$, $h = 3$
5. $C = 2\pi r$; $C = 15.7$ (use the approximation 3.14 for π)
6. $A = \pi r^2$; $r = 4.5$ (use the approximation 3.14 for π)
7. $I = PRT$; $I = 3750$, $P = 25{,}000$, $R = .05$
8. $I = PRT$; $I = 1{,}056{,}000$, $R = .055$, $T = 6$

9. $A = \frac{1}{2}(B + b)h$; $A = 180, B = 11, b = 7$

10. $A = \frac{1}{2}(B + b)h$; $A = 60, B = 7, b = 3$

11. $P = a + b + c$; $P = 30, a = 8, b = 10$

12. $V = \frac{1}{3}Ah$; $V = 45, h = 5$

13. $V = \frac{1}{3}\pi r^2 h$; $V = 565.2, r = 6$ (use the approximation 3.14 for π)

14. $V = \frac{4}{3}\pi r^3$; $r = 3$ (use the approximation 3.14 for π)

Solve each formula for the specified variable. See Examples 4 through 7.

15. $f = 5gh$ for h
16. $C = 2\pi r$ for r
17. $V = LWH$ for W
18. $T = mnr$ for n
19. $3x + y = 7$ for y
20. $-x + y = 13$ for y
21. $A = p + PRT$ for R
22. $A = p + PRT$ for T
23. $V = \frac{1}{3}Ah$ for A
24. $D = \frac{1}{4}fk$ for k
25. $P = a + b + c$ for a
26. $PR = s_1 + s_2 + s_3 + s_4$ for s_3
27. $S = 2\pi rh + 2\pi r^2$ for h
28. $S = 4lw + 2wh$ for h

29. The formula $V = LWH$ is used to find the volume of a box. If the length of a box is doubled, the width is doubled, and the height is doubled, how does this affect the volume?

30. The formula $A = bh$ is used to find the area of a parallelogram. If the base of a parallelogram is doubled and its height is doubled, how does this affect the area?

*Delta Airlines awards "Frequency Flyer" miles equal to the number of miles traveled rounded **up** to the nearest thousand. Use this for Exercises 31 and 32.*

31. A 5.5-hour nonstop flight from Orlando, Florida, to San Francisco, California, averages 470 mph. Find the "Frequency Flyer" miles earned on this flight.

32. A 45-minute ($\frac{3}{4}$-hour) nonstop flight from New Orleans, Louisiana, to Houston, Texas, averages 500 mph. Find the "Frequent Flyer" miles earned on this flight.

Solve by using a known formula.

33. The distance from the sun to the Earth is approximately 93,000,000 miles. If light travels at a rate of 186,000 miles per second, how long does it take light from the sun to reach us?

34. Light travels at a rate of 186,000 miles per second. If our moon is 238,860 miles from the Earth, how long does it take light from the moon to reach us? (Round to the nearest tenth of a second.)

35. Find how much rope is needed to wrap around the Earth at the equator, if the radius of the Earth is 4000 miles. (*Hint:* Use 3.14 for π and the formula for circumference.)

36. If the length of a rectangularly shaped garden is 6 meters and its width is 4.5 meters, find the amount of fencing required.

37. Dry Ice is a name given to solidified carbon dioxide. At $-78.5°$ Celsius it changes directly from a solid to a gas. Convert this temperature to Fahrenheit.

38. Lightning bolts can reach a temperature of 50,000° Fahrenheit. Convert this to degrees Celsius.

39. Convert Nome, Alaska's 14°F high temperature to Celsius.

40. Convert Paris, France's low temperature of −5°C to Fahrenheit.

41. The SR-71 is a top secret spy plane. It is capable of traveling from Rochester, New York, to San Francisco, California, a distance of approximately 3000 miles, in $1\frac{1}{2}$ hours. Find the rate of the SR-71.

42. A limousine built in 1968 for the president cost $500,000 and weighed 5.5 tons. This Lincoln Continental Executive could travel at 50 miles per hour with all of its tires shot away. At this rate, how long would it take to travel from Charleston, Virginia, to Washington, D.C., a distance of 375 miles?

The Dante II is a spider-like robot that is used to map the depths of an active Alaskan volcano.

43. The dimensions of the Dante II are 10 feet long by 8 feet wide by 10 feet high. Find the volume of the smallest box needed to store this robot.

44. The Dante II traveled 600 feet into an active Alaskan volcano in $3\frac{1}{3}$ hours. Find the traveling rate of Dante II in feet per minute. (*Hint:* First convert $3\frac{1}{3}$ hours to minutes.)

45. Piranha fish require 1.5 cubic feet of water per fish to maintain a healthy environment. Find the maximum number of piranhas you could put in a tank measuring 8 feet by 3 feet by 6 feet.

46. Maria's Pizza sells one 16-inch cheese pizza or two 10-inch cheese pizzas for $9.99. Determine which size gives more pizza.

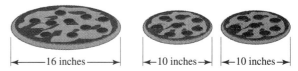

47. Find how long it takes Tran Nguyen to drive 135 miles on I-10 if he merges onto I-10 at 10 A.M. and drives nonstop with his cruise control set on 60 mph.

48. Beaumont, Texas, is about 150 miles from Toledo Bend. If Leo Miller leaves Beaumont at 4 A.M. and averages 45 mph, when should he arrive at Toledo Bend?

49. Find the temperature at which the Celsius measurement and Fahrenheit measurement are the same number.

50. Find how many goldfish you can put in a cylindrical tank whose diameter is 8 meters and whose height is 3 meters, if each goldfish needs 2 cubic meters of water.

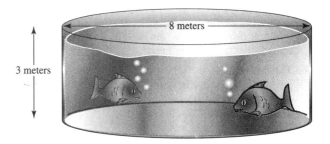

51. Bolts of lightning can travel at 270,000 miles per second. How many times can a lightning bolt travel around the world in one second? (See Exercise 35. Round to the nearest tenth.)

52. A glacier is a giant mass of rocks and ice that flows downhill like a river. Exit Glacier, near Seward, Alaska moves at a rate of 20 inches a day. Find the distance in feet the glacier moves in a year. (Assume 365 days a year. Round to 2 decimal places.)

53. On July 16, 1994, the Shoemaker-Levy 9 comet collided with Jupiter. The impact of the largest fragment of the comet, a massive chunk of rock and ice, created a fireball with a radius of 2000 miles. Find the volume of this spherical fireball. (Use 3.14 for π. Round to the nearest whole cubic mile.)

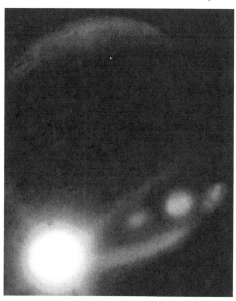

54. The fireball from the largest fragment of the comet (see Exercise 53) immediately collapsed, as it was pulled down by gravity. As it fell, it cooled to approximately $-350°$ Fahrenheit. Convert this to degrees Celsius.

55. Stalactites join stalagmites to form columns. A column found at Natural Bridge Caverns near San Antonio, Texas, rises 15 feet and has a *diameter* of only 2 inches. Find the volume of this column in cubic inches. (*Hint:* Use the formula for volume of a cylinder and use 3.14 for π.)

56. A lawn is in the shape of a trapezoid with a height of 60 feet and bases of 70 feet and 130 feet. How many bags of fertilizer must be purchased to cover the lawn if each bag covers 4000 square feet?

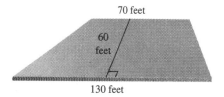

57. Normal room temperature is about 78°F. Write this temperature as degrees Celsius.

58. Flying fish do not fly, but actually glide. (See photo at the top of next page.) They have been known to travel a distance of 1300 feet at a rate of 20 miles per hour. How many seconds did it take to travel

Color image made by the Hubble Space Telescope of the impact site of Fragment G of comet.

this distance? (*Hint:* First convert miles per hour to feet per second. Recall that 1 mile = 5280 feet. Round to the nearest tenth of a second.)

59. The X-30 is a new "space plane" being developed that will skim the edge of space at 4000 miles per hour. Neglecting altitude, if the circumference of the earth is approximately 25,000 miles, how long will it take for the X-30 to travel around the Earth?

60. In the United States, the longest hang glider flight was a 303-mile, $8\frac{1}{2}$ hour flight from New Mexico to Kansas. What was the average rate during this flight?

61. If the area of a right-triangularly shaped sail is 20 square feet and its base is 5 feet, find the height of the sail.

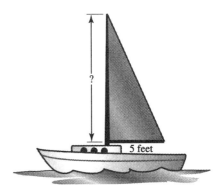

Review Exercises

Write the following phrases as algebraic expressions. See Section 2.1.

62. Nine divided by the sum of a number and 5.

63. Half the product of a number and five.

64. Three times the sum of a number and four.

65. One-third of the quotient of a number and six.

66. Double the sum of ten and four times the number.

67. Twice a number divided by three times the number.

68. Triple the difference of a number and twelve.

69. A number minus the sum of the number and six.

2.7 PERCENT AND PROBLEM SOLVING

TAPE
BA 2.7

OBJECTIVES

1. Write percents as decimals and decimals as percents.
2. Find percents of numbers.
3. Read and interpret graphs containing percents.
4. Solve percent equations.
5. Solve problems containing percents.

 Much of today's statistics is given in terms of percent: a basketball player's free throw percent, current interest rates, stock market trends, and nutrition labeling, just to name a few. In this section, we explore percent and applications involving percents.

If 37 percent of all households in the United States have a home computer,

Percent and Problem Solving Section 2.7

what does this percent mean? It means that 37 households out of every 100 households have a home computer. In other words,

the word **percent** means **per hundred** so that

37 **percent** means 37 **per hundred** or $\frac{37}{100}$.

We use the symbol % to denote percent, and we can write a percent as a fraction or a decimal.

$$37\% = \frac{37}{100} = 0.37$$

$$8\% = \frac{8}{100} = 0.08$$

$$100\% = \frac{100}{100} = 1.00$$

This suggests the following:

TO WRITE A PERCENT AS A DECIMAL

Drop the percent symbol and move the decimal point two places to the left.

EXAMPLE 1 Write each percent as a decimal.

  a. 35%   **b.** 89.5%   **c.** 150%

Solution: **a.** 35% = 0.35   **b.** 89.5% = 0.895   **c.** 150% = 1.5

To write a decimal as a percent, we reverse the procedure above.

TO WRITE A DECIMAL AS A PERCENT

Move the decimal point two places to the right and attach the percent symbol, %.

EXAMPLE 2 Write each number as a percent.

  a. 0.73   **b.** 1.39   **c.** $\frac{1}{4}$

Solution: **a.** 0.73 = 73%   **b.** 1.39 = 139%

124 CHAPTER 2 EQUATIONS, INEQUALITIES, AND PROBLEM SOLVING

c. First, write $\frac{1}{4}$ as a decimal.

$$\frac{1}{4} = 0.25 = 25\%.$$

To find a percent of a number, recall that the word "of" means multiply.

EXAMPLE 3 Find 72% of 200.

Solution: To find 72% of 200, we multiply.

$$72\% \text{ of } 200 = 72\%(200)$$
$$= 0.72(200)$$
$$= 144.$$

Thus, 72% of 200 is 144.

As mentioned earlier, percents are often used in statistics. Recall that the graph below is called a circle graph or a pie chart. The circle or pie represents a whole, or 100%. Each circle is divided into sectors (shaped like pieces of a pie) which represent various parts of the whole 100%.

EXAMPLE 4 The circle graph below shows how much money homeowners in the United States spend annually on maintaining their homes. Use this graph to answer the questions below.

Yearly Home Maintenance in the U.S.

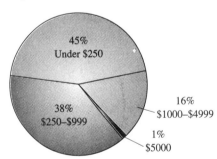

a. What percent of homeowners spend under $250 on yearly home maintenance?
b. What percent of homeowners spend less than $1000 per year on maintenance?
c. How many of the 22,000 homeowners in a town called Fairview might we expect to spend under $250 a year on home maintenance?
d. Find the number of degrees in the 16% sector.

Solution: **a.** From the circle graph, we see that 45% of homeowners spend under $250 per year on home maintenance.

b. From the circle graph, we know that 45% of homeowners spend under $250 per year and 38% of homeowners spend $250–$999 per year, so that the sum 45% + 38% or 83% of homeowners spend less than $1000 per year.

c. Since 45% of homeowners spend under $250 per year on maintenance, we find 45% of 22,000.

$$45\% \text{ of } 22{,}000 = 0.45(22{,}000)$$
$$= 9900$$

We might then expect that 9900 homeowners in Fairview spend under $250 per year on home maintenance.

d. To find the number of degrees in the 16% sector, recall that the number of degrees around a circle is 360°. Thus, to find the number of degrees in the 16% sector, we find 16% of 360°.

$$16\% \text{ of } 360° = 0.16(360°)$$
$$= 57.6°$$

The 16% sector contains 57.6°.

4 Next, we practice writing sentences as percent equations. Since these problems involve direct translations, all of our previous problem-solving steps are not needed.

EXAMPLE 5 The number 63 is what percent of 72?

Solution: 1. **UNDERSTAND.** Read and reread the problem. Next, let's guess a solution. Suppose we guess that 63 is 80% of 72. We may check our guess by finding 80% of 72. 80% of 72 = 0.80(72) = 57.6. Close, but not 63. At this point, though, we have a better understanding of the problem, we know the correct answer is close to and greater than 80%, and we know how to check our proposed solution later.

2. **ASSIGN.** Let x = the unknown percent.

4. **TRANSLATE.** Recall that "is" means "equals" and "of" signifies multiplying. Translate the sentence directly.

In words:	The number 63	is	what percent	of	72	?
Translate:	63	=	x	·	72	

5. **COMPLETE.**

$$63 = 72x$$
$$0.875 = x \quad \text{Divide both sides by 72.}$$
$$87.5\% = x \quad \text{Write as a percent.}$$

6. **INTERPRET.**
Check by verifying that 87.5% of 72 is 63.
State the results. The number 63 is 87.5% of 72.

EXAMPLE 6 The number 120 is 15% of what number?

Solution: 1. **UNDERSTAND.** Read and reread the problem. Guess a solution and check your guess.

2. **ASSIGN.** Let x = the unknown number.

4. TRANSLATE.

In words: The number 120 is 15% of what number?

Translate: 120 = 15% · x

5. COMPLETE.

$$120 = 0.15x$$
$$800 = x \quad \text{Divide both sides by 0.15.}$$

6. INTERPRET.
Check the proposed solution of 800 by finding 15% of 800 and verifying that the result is 120.
State: Thus, 120 is 15% of 800.

5 Percent increase or percent decrease is a common way to describe how some measurement has increased or decreased. For example, crime increased by 8%, teachers received a 5.5% increase in salary, or a company decreased its employees by 10%. The next example is a review of percent increase.

EXAMPLE 7 The cost of a hand-tossed pepperoni pizza at Domino's recently increased from $5.80 to $7.03. Find the percent increase.

Solution:

1. UNDERSTAND. Read and reread the problem. Let's guess that the percent increase is 10%. To see if this is the case, we find 10% of $5.80 to find the *increase* in price. Then we add this increase to $5.80 to find the *new price*. In other words, 10%($5.80) = 0.10($5.80) = $0.58, the *increase* in price. The new price then would be $5.80 + $0.58 = $6.38, less than the actual new price of $7.03. We now know that the increase is greater than 10% and we know how to check our proposed solution.

2. ASSIGN. Let x = the percent increase.

4. TRANSLATE. First, find the **increase,** and then the **percent increase.** The increase in price is found by:

In words: increase = new price − old price or

Translate: increase = $7.03 − $5.80
= $1.23

Next, find the percent increase. The percent increase or percent decrease is always a percent of the original number or in this case, the old price.

In words: increase is what percent increase of old price

Translate: $1.23 = x $5.80

5. COMPLETE.

$$1.23 = 5.80x$$
$$0.212 \approx x \quad \text{Divide both sides by 5.80 and round to 3 decimal places.}$$
$$21.2\% \approx x \quad \text{Write as a percent.}$$

6. INTERPRET.
   Check the proposed solution.
   State: The percent increase in price is approximately 21.2%.

EXERCISE SET 2.7

Write each percent as a decimal. See Example 1.
1. 120%
2. 73%
3. 22.5%
4. 4.2%
5. 0.12%
6. 0.86%

Write each number as a percent. See Example 2.
7. 0.75
8. 0.3
9. 2
10. 5.1
11. $\frac{1}{8}$
12. $\frac{3}{5}$

Use the home maintenance graph to answer each question. See Example 4.

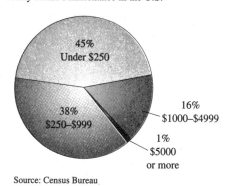

Yearly Home Maintenance in the U.S.

45% Under $250
38% $250–$999
16% $1000–$4999
1% $5000 or more

Source: Census Bureau

13. What percent of homeowners spend $250–$999 on yearly home maintenance?
14. What percent of homeowners spend $5000 or more on yearly home maintenance?
15. What percent of homeowners spend $250–$4999 on yearly home maintenance?
16. What percent of homeowners spend $250 or more on yearly home maintenance?
17. Find the number of degrees in the 38% sector.
18. Find the number of degrees in the 1% sector.
19. How many homeowners in your town might you expect to spend $250–$999 on yearly home maintenance?
20. How many homeowners in your town might you expect to spend $5000 or more on yearly home maintenance?

Solve the following. See Examples 3, 5, and 6.
21. What number is 16% of 70?
22. What number is 88% of 1000?
23. The number 28.6 is what percent of 52?
24. The number 87.2 is what percent of 436?
25. The number 45 is 25% of what number?
26. The number 126 is 35% of what number?
27. Find 23% of 20.
28. Find 140% of 86.
29. The number 40 is 80% of what number?
30. The number 56.25 is 45% of what number?
31. The number 144 is what percent of 480?
32. The number 42 is what percent of 35?

Solve. See Example 7. Many applications in this exercise set may be solved more efficiently with the use of a calculator.

33. Dillard's advertised a 25% off sale. If a London Fog coat originally sold for $156, find the decrease and the sale price.
34. Time Saver increased the price of a $0.75 cola by 15%. Find the increase and the new price.
35. Hallahan's Construction Company increased their estimate for building a new house from $95,500 to $110,000. Find the percent increase.
36. By buying in quantity, the Cannon family was able to decrease their weekly food bill from $150 a week to $130 a week. Find the percent decrease.

37. At this writing, the women's world record for throwing a disc (like a Frisbee) is held by Anni Kreml of the USA. Her throw was 447.2 feet. The men's record is held by Niclas Bergehamn of Sweden. His throw was 44.8% further than Anni's. Find the length of his throw. (Round to the nearest tenth of a foot.) (*Source:* World Flying Disc Federation.)

38. Scoville units are used to measure the hotness of a pepper. An alkaloid, capsaicin, is the ingredient that makes a pepper hot and liquid chromatography measures the amount of capsaicin in parts per million. The jalapeno measures around 5000 Scoville units, while the hottest pepper, the habanero, measures around 3000% of the measure of the jalapeno. Find the measure of the habanero pepper.

43. Do the percents shown in the graph below have a sum of 100%? Why or why not?

44. Survey your algebra class and find what percent of the class has used over-the-counter drugs for each of the categories listed. Draw a bar graph of the results.

45. Iceberg lettuce is grown and shipped to stores for about 40 cents a head, and consumers purchase it for about 70 cents a head. Find the percent increase.

46. The lettuce consumption per capita in 1968 was about 21.5 pounds, and in 1992 the consumption rose to 26.1 pounds. Find the percent increase. (Round to the nearest tenth of a percent.)

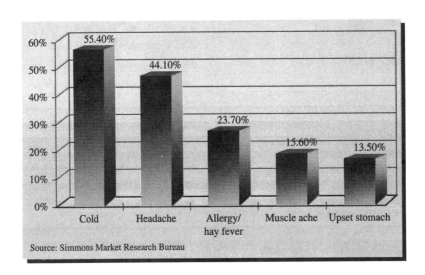

The above graph shows the percent of people in a survey who have used various over-the-counter drugs in a twelve-month period.

39. What percent of those surveyed used over-the-counter drugs to combat the common cold?

40. What percent of those surveyed used over-the-counter drugs to combat an upset stomach?

41. If 230 people were surveyed, how many of these used over-the-counter drugs for allergies?

42. The city of Chattanooga has a population of approximately 152,000. How many of these people would you expect to have used over-the-counter drugs for relief of a headache?

47. A recent study showed that 26% of men have dozed off at their place of work. If you currently employ 121 men, how many of these men might you expect to have dozed off at work? (*Source:* Better Sleep Council.)

48. A recent study showed that women and girls spend 41% of the household clothes budget. If a family spent $2000 last year on clothing, how much might have been spent on clothing for women and girls? (*Source:* The Interep Radio Store.)

49. The table on the next page shows where lightning strikes. Use this table to draw a circle graph or pie chart of this information.

Fields, Ballparks	Under Trees	Bodies of Water	Golf Courses	Near Heavy Equipment	Telephone Poles	Other
28%	17%	13%	4%	6%	1%	31%

50. During a recent 5-year period, bank fees for bounced checks have risen from $12.62 to $15.65. Find the percent increase. (Round to the nearest whole percent.) If inflation over the same period has been 16%, do you think the increase in bank fees is fair? (*Source:* Federal Reserve.)

51. During a recent 5-year period, the bank fee for depositing a bad check has risen from $5.38 to $6.08. Find the percent increase. (Round to the nearest whole percent.) Given the inflation rate of 16%, do you think that this increase in bank fees is fair?

52. The first Barbie doll was introduced in March 1959 and cost $3. This same 1959 Barbie doll now costs up to $5000. Find the percent increase rounded to the nearest whole percent.

53. The ACT Assessment is a college entrance exam taken by about 60% of college-bound students. The national average score was 20.7 in 1993 and rose to 20.8 in 1994. Find the percent increase. (Round to the nearest hundredth of a percent.)

The double bar graph below shows selected services and products and the percent of supermarkets offering each.

54. What percent of supermarkets offered plastic grocery bags in 1990?

55. What percent of supermarkets accepted credit cards in 1992?

56. Suppose that there are 20 supermarkets in your city. How many of these 20 supermarkets might have offered customers plastic grocery bags in 1992?

57. How many more supermarkets (in percent) offered reusable grocery bags in 1992 than in 1990?

58. Do you notice any trends shown in this graph?

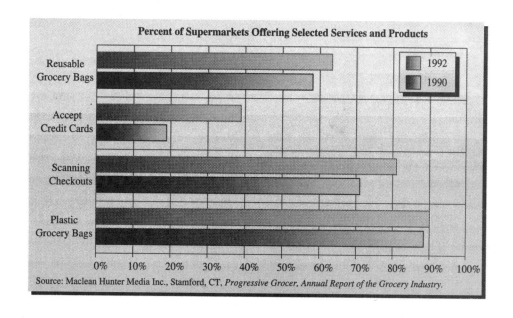

Standardized nutrition labels like the ones below have been displayed on food items since 1994. The percent column on the right shows the percent of daily values based on a 2000-calorie diet shown at the bottom of the label. For example, a serving of this food contains 4 grams of total fat, where the recommended daily fat based on a 2000 calorie diet is 65 grams of fat. This means that $\frac{4}{65}$ or approximately 6% (as shown) of your daily recommended fat is taken in by eating a serving of this food.

Nutrition Facts
Serving Size 18 crackers (31g)
Servings Per Container About 9

Amount Per Serving

Calories 130 Calories from Fat 35

	% Daily Value*
Total Fat 4g	6%
Saturated Fat 0.5g	3%
Polyunsaturated Fat 0g	
Monounsaturated Fat 1.5g	
Cholesterol 0mg	0%
Sodium 230mg	x
Total Carbohydrate 23g	y
Dietary Fiber 2g	8%
Sugars 3g	
Protein 2g	

Vitamin A 0%	•	Vitamin C 0%
Calcium 2%	•	Iron 6%

*Percent Daily Values are based on a 2,000 calorie diet. Your daily values may be higher or lower depending on your calorie needs.

	Calories:	2,000	2,500
Total Fat	Less than	65g	80g
Sat Fat	Less than	20g	25g
Cholesterol	Less than	300mg	300mg
Sodium	Less than	2400mg	2400mg
Total Carbohydrate		300g	375g
Dietary Fiber		25g	30g

59. Based on a 2000-calorie diet, what percent of daily values of sodium is contained in a serving of this food? In other words, find x. (Round to the nearest tenth of a percent.)

60. Based on a 2000-calorie diet, what percent of daily values of total carbohydrate is contained in a serving of this food? In other words, find y. (Round to the nearest tenth of a percent.)

61. Notice on the nutrition label that one serving of this food contains 130 calories and 35 of these calories are from fat. Find the percent of calories from fat. (Round to the nearest tenth of a percent.) It is recommended that no more than 30% of calorie intake come from fat. Does this food satisfy this recommendation?

Below is a nutrition label for a particular food.

NUTRITIONAL INFORMATION PER SERVING
Serving Size: 9.8 oz.   Servings Per Container: 1

Calories	280	Polyunsaturated Fat	1g
Protein	12g	Saturated Fat	.3g
Carbohydrate	45g	Cholesterol	20mg
Fat	.6g	Sodium	520mg
Percent of Calories from Fat	?	Potassium	220mg

62. If fat contains approximately 9 calories per gram, find the percent of calories from fat in one serving of this food. (Round to the nearest tenth of a percent.)

63. If protein contains approximately 4 calories per gram, find the percent of calories from protein from one serving of this food. (Round to the nearest tenth of a percent.)

64. Find a food that contains more than 30% of its calories per serving from fat. Analyze the nutrition label and verify that the percents shown are correct.

Review Exercises

Evaluate the following expressions for the given values. See Section 1.7.

65. $2a + b - c$; $a = 5, b = -1$, and $c = 3$

66. $-3a + 2c - b$; $a = -2, b = 6$, and $c = -7$

67. $4ab - 3bc$; $a = -5, b = -8$, and $c = 2$

68. $ab + 6bc$; $a = 0, b = -1$, and $c = 9$

69. $n^2 - m^2$; $n = -3$ and $m = -8$

70. $2n^2 + 3m^2$; $n = -2$ and $m = 7$

Solve. See Sections 2.6 and 2.7.

71. Find how much interest $2000 earns in 3 years in a savings account paying 4% simple interest annually.

72. Find the amount of principal that must be invested in a CD paying 5% simple interest annually to earn $125 in $2\frac{1}{2}$ years.

73. Find the distance traveled if driving at a speed of 57 miles per hour for 3.5 hours.

74. The distance from Kansas City, Missouri, to Duluth, Minnesota, is approximately 590 miles. How long does it take to travel from Kansas City to Duluth at an average speed of 55 miles per hour? (Round to the nearest tenth of an hour.)

2.8 FURTHER PROBLEM SOLVING

TAPE
BA 2.8

OBJECTIVES

1. Solve problems involving geometry concepts.
2. Solve problems involving distance.
3. Solve problems involving mixtures.
4. Solve problems involving interest.

This section is devoted to solving problems in the categories listed. The same problem-solving steps used in previous sections are also followed in this section. They are listed below for review.

PROBLEM-SOLVING STEPS

1. **UNDERSTAND** the problem. During this step don't work with variables (except for known formulas), but simply become comfortable with the problem. Some ways of accomplishing this are listed below.
   - Read and reread the problem.
   - Construct a drawing.
   - Look up an unknown formula.
   - Propose a solution and check. Pay careful attention to how to check your proposed solution. This will help later when writing an equation to model the problem.
2. **ASSIGN** a variable to an unknown in the problem. Use this variable to represent any other unknown quantities.
3. **ILLUSTRATE** the problem. A diagram or chart using the assigned variables can often help visualize the known facts.
4. **TRANSLATE** the problem into a mathematical model. This is often an equation.
5. **COMPLETE** the work. This often means to solve the equation.
6. **INTERPRET** the results. *Check* the proposed solution in the stated problem and *state* your conclusion.

EXAMPLE 1 The length of a rectangular road sign is 2 feet less than three times its width. Find the dimensions if the perimeter is 28 feet.

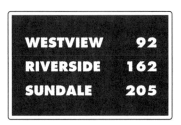

Solution: **1. UNDERSTAND.** Read and reread the problem. Recall that the formula for the perimeter of a rectangle is $P = 2l + 2w$. Draw a rectangle and guess the solution. If the width of the rectangular sign is 5 feet, its length is 2 feet less than 3 times the width or $3(5 \text{ feet}) - 2 \text{ feet} = 13 \text{ feet}$. The perimeter P of the rectangle is then $2(13 \text{ feet}) + 2(5 \text{ feet}) = 36 \text{ feet}$, too much. We now know that the width is less than 5 feet.

2. ASSIGN. Let

w = the width of the rectangular sign; then

$3w - 2$ = the length of the sign.

3. ILLUSTRATE. Draw a rectangle and label it with the assigned variables.

4. TRANSLATE.

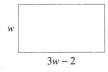

Formula: $\quad P = 2l \quad + \quad 2w$ or

Substitute: $\quad 28 = 2(3w - 2) + 2w$.

5. COMPLETE. $28 = 2(3w - 2) + 2w$

$\qquad 28 = 6w - 4 + 2w \qquad$ Apply the distributive property.

$\qquad 28 = 8w - 4$

$\qquad 28 + 4 = 8w - 4 + 4 \qquad$ Add 4 to both sides.

$\qquad 32 = 8w$

$\qquad \dfrac{32}{8} = \dfrac{8w}{8} \qquad$ Divide both sides by 8.

$\qquad 4 = w$

6. INTERPRET.

Check: If the width of the sign is 4 feet, the length of the sign is $3(4 \text{ feet}) - 2 \text{ feet} = 10 \text{ feet}$. This gives a perimeter of $P = 2(4 \text{ feet}) + 2(10 \text{ feet}) = 28 \text{ feet}$, the correct perimeter.

State: The width of the sign is 4 feet and the length of the sign is 10 feet.

EXAMPLE 2 Marie Antonio, a bicycling enthusiast, rode her 10-speed at an average speed of 18 miles per hour on level roads and then slowed down to an average of 10 miles per hour on the hilly roads of the trip. If she covered a distance of 98 miles, how long did the entire trip take if traveling the level roads took the same time as traveling the hilly roads?

Solution:
1. UNDERSTAND the problem. To do so, read and reread the problem. The formula $d = r \cdot t$ is needed. At this time, let's guess a solution. Suppose that she spent 2 hours traveling on the level roads. This means that she also spent 2 hours traveling on the hilly roads, since the times spent were the same. What is her total distance? Her distance on the level road is rate $\cdot$ time $= 18(2) = 36$ miles. Her distance on the hilly roads is rate $\cdot$ time $= 10(2) = 20$ miles. This gives a total distance of 36 miles + 20 miles = 56 miles, not the correct distance of 98 miles. Remember that the purpose of guessing a solution is not to guess correctly (although this may happen) but to help better understand the problem and how to model it with an equation.

2. ASSIGN a variable to an unknown in the problem. We are looking for the length of the entire trip so begin by letting

   x = the time spent on level roads.

   Because the same amount of time is spent on hilly roads, then also

   x = the time spent on hilly roads.

3. ILLUSTRATE the problem. We summarize the information from the problem on the following chart. Fill in the rates given, the variables used to represent the times, and use the formula $d = r \cdot t$ to fill in the distance column.

	RATE	· TIME	= DISTANCE
LEVEL	18	x	$18x$
HILLY	10	x	$10x$

4. TRANSLATE the problem into a mathematical model. Since the entire trip covered 98 miles, we have that

   In words: total distance = level distance + hilly distance
   Translate: 98   = $18x$ + $10x$

5. COMPLETE the work by solving the equation.

   $98 = 28x$

   $\dfrac{98}{28} = \dfrac{28x}{28}$

   $3.5 = x$

6. INTERPRET the results.

Check: Recall that x represents the time spent on the level portion of the trip and also the time spent on the hilly portion. If Marie rides for 3.5 hours at 18 mph, her distance is $18(3.5) = 63$ miles. If Marie rides for 3.5 hours at 10 mph, her distance is $10(3.5) = 35$ miles. The total distance is 63 miles + 35 miles = 98 miles, the required distance.

State: The time of the entire trip is then 3.5 hours + 3.5 hours or 7 hours.

Mixture problems involve two or more different quantities being combined to form a new mixture. These applications range from Dow Chemical's need to form a chemical mixture of a required strength to Planter's Peanut Company's need to find the correct mixture of peanuts and cashews, given taste and price constraints.

EXAMPLE 3   A chemist working on his doctoral degree at Massachusetts Institute of Technology needs 12 liters of a 50% acid solution for a lab experiment. The stockroom has only 40% and 70% solutions. How much of each solution should be mixed together to form 12 liters of a 50% solution?

Solution: **1. UNDERSTAND.** First, read and reread the problem a few times. Next, guess a solution. Suppose that we need 7 liters of the 40% solution. Then we need $12 - 7 = 5$ liters of the 70% solution. To see if this is indeed the solution, find the amount of pure acid in 7 liters of the 40% solution, in 5 liters of the 70% solution, and in 12 liters of a 50% solution, the required amount and strength.

number of liters	×	acid strength	=	amount of pure acid
7 liters	×	40%	=	7(0.40) or 2.8 liters
5 liters	×	70%	=	5(0.70) or 3.5 liters
12 liters	×	50%	=	12(0.50) or 6 liters

Since 2.8 liters + 3.5 liters = 6.3 liters and not 6, our guess is incorrect, but we have gained some invaluable insight into how to model and check this problem.

2. ASSIGN. Let

x = number of liters of 40% solution; then
$12 - x$ = number of liters of 70% solution.

3. ILLUSTRATE. The following table summarizes the information given. Recall that the amount of acid in each solution is found by multiplying the acid strength of each solution by the number of liters.

FURTHER PROBLEM SOLVING SECTION 2.8 **135**

	No. of Liters	·	Acid Strength	=	Amount of Acid
40% SOLUTION	x		40%		$0.40x$
70% SOLUTION	$12 - x$		70%		$0.70(12 - x)$
50% SOLUTION NEEDED	12		50%		$0.50(12)$

TRANSLATE. The amount of acid in the final solution is the sum of the amounts of acid in the two beginning solutions.

In words: acid in 40% solution + acid in 70% solution = acid in 50% mixture

Translate: $0.40x$ $+$ $0.70(12 - x)$ $=$ $0.50(12)$

5. COMPLETE.

$$0.40x + 0.70(12 - x) = 0.50(12)$$
$0.4x + 8.4 - 0.7x = 6$ Apply the distributive property.
$-0.3x + 8.4 = 6$ Combine like terms.
$-0.3x = -2.4$ Subtract 8.4 from both sides.
$x = 8$ Divide both sides by -0.3.

6. INTERPRET.
  Check: To check, recall how we checked our guess.
  State: If 8 liters of the 40% solution are mixed with $12 - 8$ or 4 liters of the 70% solution, the result is 12 liters of a 50% solution. ■

The next example is an investment problem.

EXAMPLE 4 Rajiv Puri invested part of his $20,000 inheritance in a mutual funds account that pays 7% simple interest yearly and the rest in a certificate of deposit that pays 9% simple interest yearly. At the end of one year, Rajiv's investments earned $1550. Find the amount he invested at each rate.

Solution: **1. UNDERSTAND:** Read and reread the problem. Next, guess a solution. Suppose that Rajiv invested $8000 in the 7% fund and the rest, $12,000, in the fund paying 9%. To check, find his interest after one year. Recall the formula, $I = PRT$, so the interest from the 7% fund = $8000(0.07)(1) = 560. The interest from the 9% fund = $12,000(0.09)(1) = 1080. The sum of the interests is $560 + $1080 = 1640. Our guess is incorrect, since the sum of the interests is not $1550, but we now have a better understanding of the problem.

2. ASSIGN. Let

x = amount of money in the account paying 7%. The rest of the money is $20,000 less x or

$20,000 - x$ = amount of money in the account paying 9%.

3. ILLUSTRATE. We apply the simple interest formula $I = PRT$ and organize our information in the following chart. Since there are two different rates of interest and two different amounts invested, we apply the formula twice.

	PRINCIPAL ·	RATE ·	TIME	=	INTEREST
7% FUND	x	0.07	1		$x(0.07)(1)$ or $0.07x$
9% FUND	$20,000 - x$	0.09	1		$(20,000 - x)(0.09)(1)$ or $0.09(20,000 - x)$
TOTAL	20,000				1550

4. TRANSLATE. The total interest earned, $1550, is the sum of the interest earned at 7% and the interest earned at 9%.

In words: interest at 7%   +   interest at 9%   =   total interest

Translate: $0.07x$ + $0.09(20,000 - x)$ =   1550

5. COMPLETE.

$$0.07x + 0.09(20,000 - x) = 1550$$
$$0.07x + 1800 - 0.09x = 1550 \quad \text{Apply the distributive property.}$$
$$1800 - 0.02x = 1550 \quad \text{Combine like terms.}$$
$$-0.02x = -250 \quad \text{Subtract 1800 from both sides.}$$
$$x = 12,500 \quad \text{Divide both sides by } -0.02.$$

6. INTERPRET.

Check: If $x = 12,500$, then $20,000 - x = 20,000 - 12,500$ or 7500. These solutions are reasonable, since their sum is $20,000 as required. The annual interest on $12,500 at 7% is $875; the annual interest on $7500 at 9% is $675, and $875 + $675 = $1550.

State: The amount invested at 7% is $12,500. The amount invested at 9% is $7500.

EXERCISE SET 2.8

Solve each word problem. See Example 1.

1. An architect designs a rectangular flower garden such that the width is exactly two-thirds of the length. If 260 feet of antique picket fencing are to be used, find the dimensions of the garden.

2. If the length of a rectangular parking lot is 10 me-

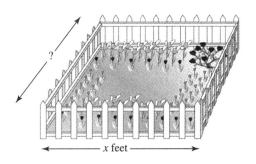

ters less than twice its width, and the perimeter is 400 meters, find the length of the parking lot.

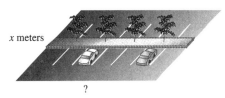

3. A flower bed is in the shape of a triangle with one side twice the length of the shortest side, and the third side is 30 feet more than the length of the shortest side. Find the dimensions if the perimeter is 102 feet.

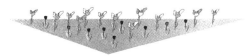

4. The perimeter of a yield sign in the shape of an isosceles triangle is 22 feet. If the shortest side is 2 feet less than the other two sides, find the length of the shortest side. (*Hint:* An isosceles triangle has two sides the same length.)

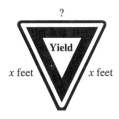

Solve. See Example 2.

5. A jet plane traveling at 500 mph overtakes a propeller plane traveling at 200 mph that had a 2-hour head start. How far from the starting point are the planes?

6. How long will it take a bus traveling at 60 miles per hour to overtake a car traveling at 40 mph if the car had a 1.5-hour head start?

7. The Jones family drove to Disneyland at 50 miles per hour and returned on the same route at 40 mph. Find the distance to Disneyland if the total driving time was 7.2 hours.

8. A bus traveled on a level road for 3 hours at an average speed 20 miles per hour faster than it traveled on a winding road. The time spent on the winding road was 4 hours. Find the average speed on the level road if the entire trip was 305 miles.

Solve. See Example 3.

9. How much pure acid should be mixed with 2 gallons of a 40% acid solution in order to get a 70% acid solution?

10. How many cubic centimeters of a 25% antibiotic solution should be added to 10 cubic centimeters of a 60% antibiotic solution in order to get a 30% antibiotic solution?

11. Planter's Peanut Company wants to mix 20 pounds of peanuts worth $3 a pound with cashews worth $5 a pound in order to make an experimental mix worth $3.50 a pound. How many pounds of cashews should be added to the peanuts?

12. Community Coffee Company wants a new flavor of Cajun coffee. How many pounds of coffee worth $7 a pound should be added to 14 pounds of coffee worth $4 a pound to get a mixture worth $5 a pound?

13. Is it possible to mix a 30% antifreeze solution with a 50% antifreeze solution and obtain a 70% antifreeze solution? Why or why not?

14. A trail mix is made by combining peanuts worth $3 a pound, raisins worth $2 a pound, and M & M's worth $4 a pound. Would it make good business sense to sell the trail mix for $1.98 a pound? Why or why not?

15. Zoya invested part of her $25,000 advance at 8% annual simple interest and the rest at 9% annual simple interest. If her total yearly interest from both accounts was $2135, find the amount invested at each rate.

16. Shirley invested some money at 9% annual simple interest and $250 more than that amount at 10% annual simple interest. If her total yearly interest was $101, how much was invested at each rate?

17. Michael invested part of his $10,000 bonus in a fund that paid an 11% profit and invested the rest in stock that suffered a 4% loss. Find the amount of each investment if his overall net profit was $650.

18. Bruce invested a sum of money at 10% annual simple interest and invested twice that amount at 12% annual simple interest. If his total yearly income from both investments was $2890, how much was invested at each rate?

Solve.

19. The perimeter of an equilateral triangle is 7 inches more than the perimeter of a square, and the side of the triangle is 5 inches longer than the side of the square. Find the side of the triangle.

20. A square animal pen and a pen shaped like an equilateral triangle have equal perimeters. Find the length of the sides of each pen if the sides of the triangular pen are fifteen less than twice a side of the square pen. (*Hint:* An equilateral triangle has three sides the same length.)

21. How can $54,000 be invested, part at 8% annual simple interest and the remainder at 10% annual simple interest, so that the interest earned by the two accounts will be equal?

22. Kathleen and Cade Williams leave simultaneously from the same point hiking in opposite directions, Kathleen walking at 4 miles per hour and Cade at 5 mph. How long can they talk on their walkie-talkies if the walkie-talkies have a 20-mile radius?

23. Ms. Mills invested her $20,000 bonus in two accounts. She took a 4% loss on one investment and made a 12% profit on another investment, but ended up breaking even. How much was invested in each account?

24. Alan and Dave Schaferkötter leave from the same point driving in opposite directions, Alan driving at 55 miles per hour and Dave at 65 mph. Alan has a one-hour head start. How long will they be able to talk on their car phones if the phones have a 250-mile range?

25. If $3000 is invested at 6% annual simple interest, how much should be invested at 9% annual simple interest so that the total yearly income from both investments is $585?

26. How much of an alloy that is 20% copper should be mixed with 200 ounces of an alloy that is 50% copper in order to get an alloy that is 30% copper?

27. Trudy Waterbury, a financial planner, invested a certain amount of money at 9% annual simple interest, twice that amount at 10% annual simple interest, and three times that amount at 11% annual simple interest. Find the amount invested at each rate if her total yearly income from the investments was $2790.

28. How much water should be added to 30 gallons of a solution that is 70% antifreeze in order to get a mixture that is 60% antifreeze?

29. April Thrower spent $32.25 to take her daughter's birthday party guests to the movies. Adult tickets cost $5.75 and children tickets cost $3.00. If 8 persons were at the party, how many adult tickets were bought?

30. Nedra and Latonya Dominguez are 12 miles apart hiking toward each other. How long will it take them to meet if Nedra walks at 3 miles per hour and Latonya walks 1 mph faster?

31. Two hikers are 11 miles apart and walking toward each other. They meet in 2 hours. Find the rate of each hiker if one hiker walks 1.1 miles per hour faster than the other.

32. On a 255-mile trip, Gary Alessandrini traveled at an average speed of 70 miles per hour, got a speeding ticket, and then traveled at 60 mph for the remainder of the trip. If the entire trip took 4.5 hours and the speeding ticket stop took 30 minutes, how long did Gary speed before getting stopped?

33. Mark Martin can row upstream at 5 miles per hour and downstream at 11 mph. If Mark starts rowing upstream until he gets tired and then rows downstream to his starting point, how far did Mark row if the entire trip took 4 hours?

 *To "break even" in a manufacturing business, revenue R (income) **must equal** the cost C of production, or R = C.*

34. The cost C to produce x number of skateboards is given by $C = 100 + 20x$. The skateboards are sold wholesale for $24 each, so revenue R is given by $R = 24x$. Find how many skateboards the manufacturer needs to produce and sell to break even. (*Hint:* Set the expression for R equal to the expression for C, then solve for x.)

35. The revenue R from selling x number of computer boards is given by $R = 60x$, and the cost C of producing them is given by $C = 50x + 5000$. Find how many boards must be sold to break even. Find how much money is needed to produce the break-even number of boards.

36. The cost C of producing x number of paperback books is given by $C = 4.50x + 2400$. Income R from these books is given by $R = 7.50x$. Find how many books should be produced and sold to break even.

37. Find the break-even quantity for a company that makes x number of computer monitors at a cost C given by $C = 870 + 70x$ and receives revenue R given by $R = 105x$.

38. Problems 34 through 37 involve finding the break-even point for manufacturing. Discuss what happens if a company makes and sells fewer products than the break-even point. Discuss what happens if more products than the break-even point are made and sold.

Review Exercises

Perform the indicated operations. See Section 1.6.

39. $3 + (-7)$ **40.** $(-2) + (-8)$

41. $\dfrac{3}{4} - \dfrac{3}{16} \cdot \dfrac{9}{16}$ **42.** $-11 + 2.9$

43. $-5 - (-1)$ **44.** $-12 - 3$

Place $<$, $>$, or $=$ in the appropriate space to make each a true statement. See Sections 1.1 and 1.7.

45. $-5 \quad -7$ **46.** $\dfrac{12}{3} \quad 2^2$

47. $|-5| \quad -(-5)$ **48.** $-3^3 \quad (-3)^3$

2.9   Solving Linear Inequalities

TAPE
BA 2.9

OBJECTIVES

1. Define linear inequality in one variable.
2. Graph solution sets on a number line.
3. Solve linear inequalities.
4. Solve compound inequalities.
5. Solve inequality applications.

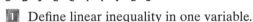

In Chapter 1, we reviewed these inequality symbols and their meanings:

$<$ means "is less than" $\leq$ means "is less than or equal to"
$>$ means "is greater than" $\geq$ means "is greater than or equal to"

A linear inequality is similar to a linear equation except that the equality symbol is replaced with an inequality symbol.

LINEAR EQUATIONS	LINEAR INEQUALITIES
$x = 3$	$x < 3$
$5n - 6 = 14$	$5n - 6 \geq 14$
$12 = 7 - 3y$	$12 \leq 7 - 3y$
$\dfrac{x}{4} - 6 = 1$	$\dfrac{x}{4} - 6 > 1$

140 CHAPTER 2 EQUATIONS, INEQUALITIES, AND PROBLEM SOLVING

> **LINEAR INEQUALITY IN ONE VARIABLE**
>
> A linear inequality in one variable is an inequality that can be written in the form
>
> $ax + b < c$
>
> where a, b, and c are real numbers and a is not 0.

This definition and all other definitions, properties, and steps in this section also hold true for the inequality symbols, $>$, $\geq$, and $\leq$.

A **solution of an inequality** is a value of the variable that makes the inequality a true statement. The solution set is the set of all solutions. In the inequality $x < 3$, replacing x with any number less than 3, that is, to the left of 3 on the number line, makes the resulting inequality true. This means that any number less than 3 is a solution of the inequality $x < 3$. Since there are infinitely many such numbers, we cannot list all the solutions of the inequality. We *can* use set notation and write

$\{\ x\ \ |\ \ x < 3\ \}$. Recall that this is read
↑ ↑   ⌣
the such
set of that x is less than 3.
all x

We can also picture the solution set on a number line. To do so, shade the portion of the number line corresponding to numbers less than 3.

Recall that all the numbers less than 3 lie to the left of 3 on the number line. An open circle about the point representing 3 indicates that 3 **is not** a solution of the inequality: 3 **is not** less than 3. The shaded arrow indicates that the solutions of $x < 3$ continue indefinitely to the left of 3.

Picturing the solutions of an inequality on a number line is called **graphing** the solutions or graphing the inequality, and the picture is called the **graph** of the inequality.

To graph $x \leq 3$, shade the numbers to the left of 3 and place a closed circle on the point representing 3. The closed circle indicates that 3 **is** a solution: 3 **is** less than or equal to 3.

EXAMPLE 1 Graph $x \geq -1$.

Solution: We place a closed circle at -1 since the inequality symbol is $\geq$ and -1 is greater than or equal to -1. Then shade to the right of -1.

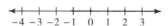

Inequalities containing one inequality symbol are called **simple inequalities,** while inequalities containing two inequality symbols are called **compound inequalities.** A compound inequality is really two simple inequalities in one. The compound inequality

$$3 < x < 5 \quad \text{means} \quad 3 < x \text{ and } x < 5$$

This can be read "x is greater than 3 and less than 5."

A solution of a compound inequality is a value that is a solution of both of the simple inequalities that make up the compound inequality. For example,

$4\frac{1}{2}$ is a solution of $3 < x < 5$ since $3 < 4\frac{1}{2}$ **and** $4\frac{1}{2} < 5$.

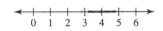

To graph $3 < x < 5$, place open circles at both 3 and 5 and shade between.

EXAMPLE 2 Graph $2 < x \leq 4$.

Solution: Graph all numbers greater than 2 and less than or equal to 4. Place an open circle at 2, a closed circle at 4, and shade between.

When solutions of a linear inequality are not immediately obvious, they are found through a process similar to the one used to solve a linear equation. Our goal is to isolate the variable, and we use properties of inequality similar to properties of equality.

> **ADDITION PROPERTY OF INEQUALITY**
> If a, b, and c are real numbers, then
> $$a < b \quad \text{and} \quad a + c < b + c$$
> are equivalent inequalities.

This property also holds true for subtracting values, since subtraction is defined in terms of addition. In other words, adding or subtracting the same quantity from both sides of an inequality does not change the solutions of the inequality.

EXAMPLE 3 Solve $x + 4 \leq -6$ for x. Graph the solution set.

Solution: To solve for x, subtract 4 from both sides of the inequality.

$$x + 4 \leq -6 \quad \text{Original inequality.}$$
$$x + 4 - 4 \leq -6 - 4 \quad \text{Subtract 4 from both sides.}$$

$x \leq -10$ Simplify.

The solution set is $\{x \mid x \leq -10\}$.

An important difference between linear equations and linear inequalities is shown when we multiply or divide both sides of an inequality by a nonzero real number. For example, start with the true statement $6 < 8$ and multiply both sides by 2. As we see below, the resulting inequality is also true.

$6 < 8$ True.
$2(6) < 2(8)$ Multiply both sides by 2.
$12 < 16$ True.

But if we start with the same true statement $6 < 8$ and multiply both sides by -2, the resulting inequality is not a true statement.

$6 < 8$ True.
$-2(6) < -2(8)$ Multiply both sides by -2.
$-12 < -16$ False.

Notice, however, that if we reverse the direction of the inequality symbol, the resulting inequality is true.

$-12 < -16$ False.
$-12 > -16$ True.

This demonstrates the multiplication property of inequality.

> **MULTIPLICATION PROPERTY OF INEQUALITY**
> 1. If a, b, and c are real numbers, and c is **positive,** then
> $a < b$ and $ac < bc$
>   are equivalent inequalities.
> 2. If a, b, and c are real numbers, and c is **negative,** then
> $a < b$ and $ac > bc$
>   are equivalent inequalities.

Because division is defined in terms of multiplication, this property also holds true when dividing both sides of an inequality by a nonzero number: If we multiply or divide both sides of an inequality by a negative number, **the direction of the inequality sign must be reversed for the inequalities to remain equivalent.**

> REMINDER Whenever both sides of an inequality are multiplied or divided by a negative number, the direction of the inequality symbol **must be** reversed to form an equivalent inequality.

EXAMPLE 4 Solve $-2x \leq -4$, and graph the solution set.

Solution: Remember to reverse the direction of the inequality symbol when dividing by a negative number.

$$-2x \leq -4$$

$$\frac{-2x}{-2} \geq \frac{-4}{-2} \qquad \text{Divide both sides by } -2 \text{ and reverse the inequality sign.}$$

$$x \geq 2 \qquad \text{Simplify.}$$

The solution set $\{x \mid x \geq 2\}$ is graphed as shown.

EXAMPLE 5 Solve $2x < -4$, and graph the solution set.

Solution:
$$2x < -4$$

$$\frac{2x}{2} < \frac{-4}{2} \qquad \text{Divide both sides by 2.} \\ \text{Do not reverse the inequality sign.}$$

$$x < -2 \qquad \text{Simplify.}$$

The graph of $\{x \mid x < -2\}$ is shown.

Follow these steps to solve linear inequalities.

TO SOLVE LINEAR INEQUALITIES IN ONE VARIABLE

Step 1. Clear the inequality of fractions by multiplying both sides of the inequality by the lowest common denominator (LCD) of all fractions in the inequality.

Step 2. Remove grouping symbols such as parentheses by using the distributive property.

Step 3. Simplify each side of the inequality by combining like terms.

Step 4. Write the inequality with variable terms on one side and numbers on the other side by using the addition property of inequality.

Step 5. Isolate the variable by using the multiplication property of inequality.

Don't forget that if both sides of an inequality are multiplied or divided by a negative number, the direction of the inequality sign must be reversed.

EXAMPLE 6 Solve $-4x + 7 \geq -9$, and graph the solution set.

Solution:
$$-4x + 7 \geq -9$$
$$-4x + 7 - 7 \geq -9 - 7 \quad \text{Subtract 7 from both sides.}$$
$$-4x \geq -16 \quad \text{Simplify.}$$
$$\frac{-4x}{-4} \leq \frac{-16}{-4} \quad \text{Divide both sides by } -4 \text{ and reverse the direction of the inequality sign.}$$
$$x \leq 4 \quad \text{Simplify.}$$

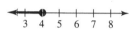

The graph of $\{x \mid x \leq 4\}$ is shown.

EXAMPLE 7 Solve $2x + 7 \leq x - 11$, and graph the solution set.

Solution:
$$2x + 7 \leq x - 11$$
$$2x + 7 - x \leq x - 11 - x \quad \text{Subtract } x \text{ from both sides.}$$
$$x + 7 \leq -11 \quad \text{Combine like terms.}$$
$$x + 7 - 7 \leq -11 - 7 \quad \text{Subtract 7 from both sides.}$$
$$x \leq -18 \quad \text{Combine like terms.}$$

The graph of the solution set $\{x \mid x \leq -18\}$ is shown.

EXAMPLE 8 Solve $-5x + 7 < 2(x - 3)$, and graph the solution set.

Solution:
$$-5x + 7 < 2(x - 3)$$
$$-5x + 7 < 2x - 6 \quad \text{Apply the distributive property.}$$
$$-5x + 7 - 2x < 2x - 6 - 2x \quad \text{Subtract } 2x \text{ from both sides.}$$
$$-7x + 7 < -6 \quad \text{Combine like terms.}$$
$$-7x + 7 - 7 < -6 - 7 \quad \text{Subtract 7 from both sides.}$$
$$-7x < -13 \quad \text{Combine like terms.}$$
$$\frac{-7x}{-7} > \frac{-13}{-7} \quad \text{Divide both sides by } -7 \text{ and reverse the direction of the inequality sign.}$$
$$x > \frac{13}{7} \quad \text{Simplify.}$$

The graph of the solution set $\left\{x \mid x > \frac{13}{7}\right\}$ is shown.

EXAMPLE 9 Solve $2(x - 3) - 5 \leq 3(x + 2) - 18$, and graph the solution set.

Solution:
$$2(x - 3) - 5 \leq 3(x + 2) - 18$$
$$2x - 6 - 5 \leq 3x + 6 - 18 \quad \text{Apply the distributive property.}$$
$$2x - 11 \leq 3x - 12 \quad \text{Combine like terms.}$$

$$-x - 11 \leq -12 \qquad \text{Subtract } 3x \text{ from both sides.}$$
$$-x \leq -1 \qquad \text{Add 11 to both sides.}$$
$$\frac{-x}{-1} \geq \frac{-1}{-1} \qquad \text{Divide both sides by } -1 \text{ and reverse the direction of the inequality sign.}$$
$$x \geq 1 \qquad \text{Simplify.}$$

The graph of the solution set $\{x \mid x \geq 1\}$ is shown.

When we solve a simple inequality, we isolate the variable on one side of the inequality. When we solve a compound inequality, we isolate the variable in the middle part of the inequality. Also, when solving a compound inequality, we must perform the same operation to all **three** parts of the inequality: left, middle, and right.

EXAMPLE 10 Solve $-1 \leq 2x - 3 < 5$, and graph the solution set.

Solution:
$$-1 \leq 2x - 3 < 5$$
$$-1 + 3 \leq 2x - 3 + 3 < 5 + 3 \qquad \text{Add 3 to all three parts.}$$
$$2 \leq 2x < 8 \qquad \text{Combine like terms.}$$
$$\frac{2}{2} \leq \frac{2x}{2} < \frac{8}{2} \qquad \text{Divide all three parts by 2.}$$
$$1 \leq x < 4 \qquad \text{Simplify.}$$

The solution set $\{x \mid 1 \leq x < 4\}$ is graphed.

EXAMPLE 11 Solve $3 \leq \frac{3x}{2} + 4 \leq 5$, and graph the solution set.

Solution:
$$3 \leq \frac{3x}{2} + 4 \leq 5$$
$$2(3) \leq 2\left(\frac{3x}{2} + 4\right) \leq 2(5) \qquad \text{Multiply all three parts by 2 to clear the fraction.}$$
$$6 \leq 3x + 8 \leq 10 \qquad \text{Distribute.}$$
$$-2 \leq 3x \leq 2 \qquad \text{Subtract 8 from all three parts.}$$
$$\frac{-2}{3} \leq \frac{3x}{3} \leq \frac{2}{3} \qquad \text{Divide all three parts by 3.}$$
$$\frac{-2}{3} \leq x \leq \frac{2}{3} \qquad \text{Simplify.}$$

The graph of the solution set $\{x \mid -\frac{2}{3} \leq x \leq \frac{2}{3}\}$ is shown.

Problems containing words such as "at least," "at most," "between," "no more than," and "no less than" usually indicate that an inequality should be solved instead of an equation. In solving applications involving linear inequalities, use the same procedure you use to solve applications involving linear equations.

EXAMPLE 12 Marie Chase and Jonathan Edwards are having their wedding reception at the Gallery reception hall. They may spend at most $1000 for the reception. If the reception hall charges a $100 cleanup fee plus $14 per person, find the greatest number of people that they can invite and still stay within their budget.

Solution:
1. UNDERSTAND. Read and reread the problem. Next, guess a solution. If 50 people attend the reception, the cost is $100 + $14(50) = $100 + $700 = $800.
2. ASSIGN. Let x = the number of people who attend the reception.
4. TRANSLATE.

In words:	cleanup fee	+	cost per person	must be less than or equal to	$1000
Translate:	100	+	14x	≤	1000

5. COMPLETE.

$$100 + 14x \leq 1000$$
$$14x \leq 900 \quad \text{Subtract 100 from both sides.}$$
$$x \leq 64\frac{2}{7} \quad \text{Divide both sides by 14.}$$

6. INTERPRET.
   Check: Since x represents the number of people, we round down to the nearest whole, or 64. Notice that if 64 people attend, the cost is $100 + $14(64) = $996. If 65 people attend, the cost is $100 + $14(65) = $1010, which is more than the given $1000.
   State: Marie Chase and Jonathan Edwards can invite at most 64 people to the reception.

MENTAL MATH

Solve each of the following inequalities.

1. $5x > 10$ **2.** $4x < 20$ **3.** $2x \geq 16$ **4.** $9x \leq 63$

EXERCISE SET 2.9

Graph each on a number line. See Examples 1 and 2.
1. $x \leq -1$
2. $y < 0$
3. $x > \dfrac{1}{2}$
4. $z \geq -\dfrac{2}{3}$
5. $-1 < x < 3$
6. $2 \leq y \leq 3$
7. $0 \leq y < 2$
8. $-1 \leq x \leq 4$

Solve each inequality and graph the solution set. See Examples 3 through 5.
9. $2x < -6$
10. $3x > -9$
11. $x - 2 \geq -7$
12. $x + 4 \leq 1$
13. $-8x \leq 16$
14. $-5x < 20$

Solve each inequality and graph the solution set. See Examples 6 and 7.
15. $3x - 5 > 2x - 8$
16. $3 - 7x \geq 10 - 8x$
17. $4x - 1 \leq 5x - 2x$
18. $7x + 3 < 9x - 3x$
19. $x - 7 < 3(x + 1)$
20. $3x + 9 \geq 5(x - 1)$

Solve each inequality and graph the solution set. See Examples 8 and 9.
21. $-6x + 2 \geq 2(5 - x)$
22. $-7x + 4 > 3(4 - x)$
23. $4(3x - 1) \leq 5(2x - 4)$
24. $3(5x - 4) \leq 4(3x - 2)$
25. $3(x + 2) - 6 > -2(x - 3) + 14$
26. $7(x - 2) + x \leq -4(5 - x) - 12$

Solve each inequality, and then graph the solution set. See Examples 10 and 11.
27. $-3 < 3x < 6$
28. $-5 < 2x < -2$
29. $2 \leq 3x - 10 \leq 5$
30. $4 \leq 5x - 6 \leq 19$
31. $-4 < 2(x - 3) < 4$
32. $0 < 4(x + 5) < 8$

33. Explain how solving a linear inequality is similar to solving a linear equation.
34. Explain how solving a linear inequality is different from solving a linear equation.

Solve the following inequalities. Graph each solution set.
35. $-2x \leq -40$
36. $-7x > 21$
37. $-9 + x > 7$
38. $y - 4 \leq 1$
39. $3x - 7 < 6x + 2$
40. $2x - 1 \geq 4x - 5$
41. $5x - 7x \leq x + 2$
42. $4 - x < 8x + 2x$
43. $\dfrac{3}{4}x > 2$
44. $\dfrac{5}{6}x \geq -8$
45. $3(x - 5) < 2(2x - 1)$
46. $5(x + 4) < 4(2x + 3)$
47. $4(2x + 1) > 4$
48. $6(2 - x) \geq 12$
49. $-5x + 4 \leq -4(x - 1)$
50. $-6x + 2 < -3(x + 4)$
51. $-2 < 3x - 5 < 7$
52. $1 < 4 + 2x \leq 7$
53. $-2(x - 4) - 3x < -(4x + 1) + 2x$
54. $-5(1 - x) + x \leq -(6 - 2x) + 6$
55. $-3x + 6 \geq 2x + 6$
56. $-(x - 4) < 4$
57. $-6 < 3(x - 2) < 8$
58. $-5 \leq 2(x + 4) < 8$

Solve the following. See Example 12.

59. Six more than twice a number is greater than negative fourteen. Find all numbers that make this statement true.
60. Five times a number increased by one is less than or equal to ten. Find all such numbers.
61. The perimeter of a rectangle is to be no greater than 100 centimeters and the width must be 15 centimeters. Find the maximum length of the rectangle.

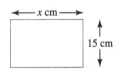

62. One side of a triangle is four times as long as another side, and the third side is 12 inches long. If the perimeter can be no longer than 87 inches, find the maximum lengths of the other two sides.

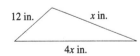

63. The temperatures in Ohio range from $-39°C$ to $45°C$. Use a compound inequality to convert these temperatures to Fahrenheit temperatures. (*Hint:* Use $C = \dfrac{5}{9}(F - 32)$.)

64. Mario Lipco has scores of 85, 95, and 92 on his algebra tests. Use a compound inequality to find the range of scores he can make on his final exam in order to receive an A in the course. The final exam counts as three tests, and an A is received if the final course average is from 90 to 100. (*Hint:* The average of a list of numbers is their sum divided by the number of numbers in the list.)

65. The formula $C = 3.14d$ can be used to approximate the circumference of a circle given its diameter. Waldo Manufacturing manufactures and sells a certain washer with an outside circumference of 3 centimeters. The company has decided that a washer whose actual circumference is in the interval $2.9 \leq C \leq 3.1$ centimeters is acceptable. Use a compound inequality and find the corresponding interval for diameters of these washers. (Round to 3 decimal places.)

66. Bunnie Supplies manufactures plastic Easter eggs that open. The company has determined that if the circumference of the opening of each part of the egg is in the interval $118 \leq C \leq 122$ millimeters, the eggs will open and close comfortably. Use a compound inequality and find the corresponding interval for diameters of these openings. (Round to 2 decimal places.)

67. Twice a number increased by one is between negative five and seven. Find all such numbers.

68. Half a number decreased by four is between two and three. Find all such numbers.

69. A financial planner has a client with $15,000 to invest. If he invests $10,000 in a certificate of deposit paying 11% annual simple interest, at what rate does the remainder of the money need to be invested so that the two investments together yield at least $1600 in yearly interest?

70. Alex earns $600 per month plus 4% of all his sales over $1000. Find the minimum sales that will allow Alex to earn at least $3000 per month.

71. Ben Holladay bowled 146 and 201 in his first two games. What must he bowl in his third game to have an average of at least 180?

72. On an NBA team the two forwards measure 6'8" and 6'6" and the two guards measure 6'0" and 5'9" tall. How tall a center should they hire if they wish to have a starting team average height of at least 6'5"?

Review Exercises

Evaluate the following. See Section 1.3.

73. $(2)^3$ **74.** $(3)^3$ **75.** $(1)^{12}$

76. 0^5 **77.** $\left(\dfrac{4}{7}\right)^2$ **78.** $\left(\dfrac{2}{3}\right)^3$

This broken line graph shows the enrollment of people (members) in a Health Maintenance Organization (HMO). The height of each dot corresponds to the number of members (in millions).

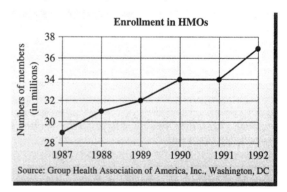

Source: Group Health Association of America, Inc., Washington, DC

79. How many people were enrolled in Health Maintenance Organizations in 1989?

80. How many people were enrolled in Health Maintenance Organizations in 1992?

81. Which year shows the greatest increase in number of members?

82. In what year were there 31,000,000 members of HMOs?

A Look Ahead

EXAMPLE

Solve $x(x - 6) > x^2 - 5x + 6$, and then graph the solution.

Solution:
$$x(x - 6) > x^2 - 5x + 6$$
$$x^2 - 6x > x^2 - 5x + 6$$
$$x^2 - 6x - x^2 > x^2 - 5x + 6 - x^2$$
$$-6x > -5x + 6$$
$$-x > 6$$
$$\frac{-x}{-1} < \frac{6}{-1}$$
$$x < -6$$

The solution set $\{x \mid x < -6\}$ is graphed as shown.

$\leftarrow\!\!+\!\!\!+\!\!\!+\!\!\!+\!\!\!+\!\!\!+\!\!\!+\!\!\rightarrow$
$-7\ -6\ -5\ -4\ -3\ -2\ -1$

Solve each inequality and then graph the solution set.

83. $x(x + 4) > x^2 - 2x + 6$
84. $x(x - 3) \geq x^2 - 5x - 8$
85. $x^2 + 6x - 10 < x(x - 10)$
86. $x^2 - 4x + 8 < x(x + 8)$
87. $x(2x - 3) \leq 2x^2 - 5x$
88. $x(4x + 1) < 4x^2 - 3x$

GROUP ACTIVITY

CALCULATING PRICE PER UNIT

MATERIALS:
- Grocery store circulars
- Supermarket newspaper advertisements

Sabrina owns a catering business, and she has been asked to quote a price per person for catering fruit salad at an afternoon reception for 20 people. (*Price per person* is the amount per person Sabrina actually charges her customer.) The recipe she plans to use is given in the table.

1. Study the recipe in the figure and give a description of the general process necessary to figure the fruit salad cost per person. (*Cost per person* is the amount Sabrina pays, per person, for her raw materials.)

FRUIT SALAD RECIPE	
2	pounds green grapes, halved
3	pounds watermelon, cubed
1	pint fresh blueberries
1.25	pounds nectarines or mangos, sliced
1	6 ounce can frozen pineapple juice concentrate, thawed
	Combine ingredients, and serve chilled. Serves 10 guests.

(continued)

2. Assign variables to the unknowns and translate your verbal description from Question 1 into an algebraic formula for the fruit salad cost per person.
3. Collect actual prices for the fruit salad ingredients from grocery store circulars, newspaper advertisements, or trips to a supermarket. Organize the ingredient prices you have found in a table. Use this real data and the algebraic formula from Question 2 to compute an estimate of the fruit salad cost per person. (*Note:* If you are unable to find prices for the exact ingredients listed in the recipe, you may substitute a new ingredient and use the new ingredient's price. In your calculations, make a note of any substitutions you are making.)
4. What *additional* amount should Sabrina charge per person if she hopes to receive a total of $90 (including both cost of raw materials and profit) for the catering job? What percent increase does the price Sabrina ultimately charges per person represent over her actual cost per person?

CHAPTER 2   HIGHLIGHTS

DEFINITIONS AND CONCEPTS	EXAMPLES	
SECTION 2.1 SIMPLIFYING ALGEBRAIC EXPRESSIONS		
The **numerical coefficient** of a **term** is its numerical factor.	TERM	NUMERICAL COEFFICIENT
	$-7y$	-7
	x	1
	$\dfrac{1}{5}a^2b$	$\dfrac{1}{5}$
Terms with the same variables raised to exactly the same powers are **like terms.**	LIKE TERMS	UNLIKE TERMS
	$12x, -x$	$3y, 3y^2$
	$-2xy, 5yx$	$7a^2b, -2ab^2$
To combine like terms, add the numerical coefficients and multiply the result by the common variable factor.	$9y + 3y = 12y$	
	$-4z^2 + 5z^2 - 6z^2 = -5z^2$	
To remove parentheses, apply the distributive property.	$-4(x + 7) + 10(3x - 1)$	
	$\quad = -4x - 28 + 30x - 10$	
	$\quad = 26x - 38$	

(continued)

Chapter 2 Highlights

Definitions and Concepts	Examples

Section 2.2 The Addition Property of Equality

A **linear equation in one variable** can be written in the form $ax + b = c$ where a, b, and c are real numbers and $a \neq 0$.

Linear Equations

$$-3x + 7 = 2$$
$$3(x - 1) = -8(x + 5) + 4$$

Equivalent equations are equations that have the same solution.

$x - 7 = 10$ and $x = 17$ are equivalent equations.

Addition Property of Equality

Adding the same number to or subtracting the same number from both sides of an equation does not change its solution.

$$y + 9 = 3$$
$$y + 9 - 9 = 3 - 9$$
$$y = -6$$

Section 2.3 The Multiplication Property of Equality

Multiplication Property of Equality

Multiplying both sides or dividing both sides of an equation by the same nonzero number does not change its solution.

$$\frac{2}{3}a = 18$$
$$\frac{3}{2}\left(\frac{2}{3}a\right) = \frac{3}{2}(18)$$
$$a = 27$$

Section 2.4 Solving Linear Equations

To Solve Linear Equations

Solve: $\dfrac{5(-2x + 9)}{6} + 3 = \dfrac{1}{2}$

1. Clear the equation of fractions.

   1. $6 \cdot \dfrac{5(-2x + 9)}{6} + 6 \cdot 3 = 6 \cdot \dfrac{1}{2}$

      $5(-2x + 9) + 18 = 3$

2. Remove any grouping symbols such as parentheses.

   2. $-10x + 45 + 18 = 3$ Distributive property.

3. Simplify each side by combining like terms.

   3. $-10x + 63 = 3$ Combine like terms.

4. Write variable terms on one side and numbers on the other side using the addition property of equality.

   4. $-10x + 63 - 63 = 3 - 63$ Subtract 63.
      $-10x = -60$

5. Isolate the variable using the multiplication property of equality.

   5. $\dfrac{-10x}{-10} = \dfrac{-60}{-10}$ Divide by -10.
      $x = 6$

(continued)

Definitions and Concepts	Examples

Section 2.4 Solving Linear Equations

6. Check by substituting in the original equation.

6. $\dfrac{5(-2x+9)}{6} + 3 = \dfrac{1}{2}$

$\dfrac{5(-2 \cdot 6 + 9)}{6} + 3 = \dfrac{1}{2}$

$\dfrac{5(-3)}{6} + 3 = \dfrac{1}{2}$

$-\dfrac{5}{2} + \dfrac{6}{2} = \dfrac{1}{2}$

$\dfrac{1}{2} = \dfrac{1}{2}$ True.

Section 2.5 An Introduction to Problem Solving

PROBLEM-SOLVING STEPS

The height of the Hudson volcano in Chili is twice the height of the Kiska volcano in the Aleutian Islands. If the sum of their heights is 12,870 feet, find the height of each.

1. UNDERSTAND the problem.

1. Read and reread the problem. Guess a solution and check your guess.

2. ASSIGN a variable.

2. Let x be the height of the Kiska volcano. Then $2x$ is the height of the Hudson volcano.

3. ILLUSTRATE the problem.

3.

x — Kiska $2x$ — Hudson

4. TRANSLATE the problem.

4. In words: | height of Kiska | added to | height of Hudson | is | 12,870 |

   Translate: x + $2x$ =   12,870

5. COMPLETE by solving.

5. $x + 2x = 12{,}870$

$3x = 12{,}870$

$x = 4290$

6. INTERPRET the results.

6. *Check:* If x is 4290 then $2x$ is 2(4290) or 8580. Their sum is 4290 + 8580 or 12,870, the required amount.

State: Kiska volcano is 4290 feet high and Hudson volcano is 8580 feet high.

Section 2.6 Formulas and Problem Solving

FORMULAS

An equation that describes a known relationship among quantities is called a **formula.**

$A = lw$ (area of a rectangle)

$I = PRT$ (simple interest)

(continued)

Definitions and Concepts	Examples

Section 2.6 Formulas and Problem Solving

To solve a formula for a specified variable, use the same steps as for solving a linear equation. Treat the specified variable as the only variable of the equation.

Solve $P = 2l + 2w$ for l.

$$P = 2l + 2w$$
$$P = 2l + 2w \qquad \text{Subtract } 2w.$$
$$P - 2w = 2l$$
$$\frac{P - 2w}{2} = \frac{2l}{2} \qquad \text{Divide by 2.}$$
$$\frac{P - 2w}{2} = l \qquad \text{Simplify.}$$

If all values for the variables in a formula are known except for one, this unknown value may be found by substituting in the known values and solving.

If $d = 182$ miles and $r = 52$ miles per hour in the formula $d = r \cdot t$, find t.

$$d = \cdot t$$
$$182 = \cdot t \qquad \text{Let } d = 182 \text{ and } r = 52.$$
$$3.5 = t$$

The time is 3.5 hours.

Section 2.7 Percent and Problem Solving

The word **percent** means **per hundred.** The symbol % is used to denote percent.

To write a percent as a decimal, drop the percent symbol and move the decimal point two places to the left.

$49\% = \dfrac{49}{100}, \quad 1\% = \dfrac{1}{100}$

$85.\% = .85, \quad 3.5\% = .035$

To write a decimal as a percent, move the decimal point two places to the right and attach the percent symbol, %.

$.35 = 35\%, \quad 10.1 = 1010\%$

$\dfrac{1}{8} = 0.125 = 12.5\%$

Use the same problem-solving steps to solve a problem containing percents.

1. UNDERSTAND.
2. ASSIGN.

4. TRANSLATE.

5. COMPLETE.

32% of what number is 36.8?

1. Read and reread. Guess a solution and check.
2. Let x = the unknown number.

4. In words: | 32% | of | what number | is | 36.8 |

   Translate: $\quad 32\% \quad \cdot \quad x \quad = \quad 36.8$

5. Solve $\quad 32\% \cdot x = 36.8$

$$.32x = 36.8$$
$$\frac{.32x}{.32} = \frac{36.8}{.32} \qquad \text{Divide by .32.}$$
$$x = 115 \qquad \text{Simplify.}$$

(continued)

Definitions and Concepts	Examples
Section 2.7 Percent and Problem Solving	
6. INTERPRET.	6. *Check:* 32% of 115 is .32(115) = 36.8. *State:* The unknown number is 115.
Section 2.8 Further Problem Solving	
	How many liters of a 20% acid solution must be mixed with a 50% acid solution in order to obtain 12 liters of a 30% solution?
1. UNDERSTAND.	1. Read and reread. Guess a solution and check.
2. ASSIGN.	2. Let x = number of liters of 20% solution. Then $12 - x$ = number of liters of 50% solution.
3. ILLUSTRATE.	3.

	No. of Liters · Acid Strength		= Amount of Acid
20% solution	x	20%	$0.20x$
50% solution	$12 - x$	50%	$0.50(12 - x)$
30% solution needed	12	30%	$0.30(12)$

4. TRANSLATE.	4. In words: (acid in 20% solution) + (acid in 50% solution) = (acid in 30% solution) Translate: $\quad 0.20x \;+\; 0.50(12 - x) \;=\; 0.30(12)$
5. COMPLETE.	5. Solve $0.20x + 0.50(12 - x) = 0.30(12)$ $\quad 0.20x + 6 - 0.50x = 3.6 \quad$ Apply the distributive property. $\quad -0.30x + 6 = 3.6$ $\quad -0.30x = -2.4 \quad$ Subtract 6. $\quad x = 8 \quad$ Divide by -0.30.
6. INTERPRET.	6. *Check,* then *state.* If 8 liters of a 20% acid solution are mixed with $12 - 8$ or 4 liters of a 50% acid solution, the result is 12 liters of a 30% solution.
Section 2.9 Solving Linear Inequalities	
A **linear inequality in one variable** is an inequality that can be written in one of the forms: $ax + b < c \qquad ax + b \leq c$ $ax + b > c \qquad ax + b \geq c$ where a, b, and c are real numbers and a is not 0.	Linear Inequalities $2x + 3 < 6 \qquad\qquad 5(x - 6) \geq 10$ $\dfrac{x - 2}{5} > \dfrac{5x + 7}{2} \qquad \dfrac{-(x + 8)}{9} \leq \dfrac{-2x}{11}$

Chapter 2 Highlights

Definitions and Concepts	Examples
Section 2.9 Solving Linear Inequalities	
Addition Property of Inequality Adding the same number to or subtracting the same number from both sides of an inequality does not change the solutions.	$y + 4 \leq -1$ $y + 4 - 4 \leq -1 - 4$ Subtract 4. $y \leq -5$ 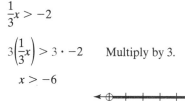
Multiplication Property of Inequality Multiplying or dividing both sides of an inequality by the same *positive number* does not change its solutions.	$\frac{1}{3}x > -2$ $3\left(\frac{1}{3}x\right) > 3 \cdot -2$ Multiply by 3. $x > -6$ 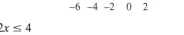
Multiplying or dividing both sides of an inequality by the same **negative number and reversing the direction of the inequality symbol** does not change its solutions.	$-2x \leq 4$ $\dfrac{-2x}{-2} \geq \dfrac{4}{-2}$ Divide by -2, reverse inequality symbol. $x \geq -2$ 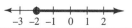
To Solve Linear Inequalities 1. Clear the equation of fractions. 2. Remove grouping symbols. 3. Simplify each side by combining like terms. 4. Write variable terms on one side and numbers on the other side using the addition property of inequality. 5. Isolate the variable using the multiplication property of inequality.	Solve: $3(x + 2) \leq -2 + 8$ 1. No fractions to clear. $3(x + 2) \leq -2 + 8$ 2. $3x + 6 \leq -2 + 8$ Distributive property 3. $3x + 6 \leq 6$ Combine like terms. 4. $3x + 6 - 6 \leq 6 - 6$ Subtract 6. $3x \leq 0$ 5. $\dfrac{3x}{3} \leq \dfrac{0}{3}$ Divide by 3. $x \leq 0$ 
	Compound Inequalities
Inequalities containing two inequality symbols are called **compound inequalities.**	$-2 < x < 6$ $5 \leq 3(x - 6) < \dfrac{20}{3}$

(continued)

Definitions and Concepts	Examples
Section 2.9 Solving Linear Inequalities	
To solve a compound inequality, isolate the variable in the middle part of the inequality. Perform the same operation to all three parts of the inequality: left, middle, right.	Solve: $-2 < 3x + 1 < 7$ $-2 < 3x + 1 < 7$ Subtract 1. $ -3 < 3x < 6$ $\dfrac{-3}{3} < \dfrac{3x}{3} < \dfrac{6}{3}$ Divide by 3. $-1 < x < 2$ ⟵—+—+—+—+—+—⟶ $-2-10123$

Chapter 2 Review

Simplify the following expressions.

1. $5x - x + 2x$
2. $0.2z - 4.6x - 7.4z$
3. $\dfrac{1}{2}x + 3 + \dfrac{7}{2}x - 5$
4. $\dfrac{4}{5}y + 1 + \dfrac{6}{5}y + 2$
5. $2(n - 4) + n - 10$
6. $3(w + 2) - (12 - w)$
7. Subtract $7x - 2$ from $x + 5$
8. Subtract $1.4y - 3$ from $y - 0.7$

Write each of the following as algebraic expressions.

9. Three times a number decreased by 7.
10. Twice the sum of a number and 2.8 added to 3 times a number.

Solve the following.

11. $8x + 4 = 9x$
12. $5y - 3 = 6y$
13. $3x - 5 = 4x + 1$
14. $2x - 6 = x - 6$
15. $4(x + 3) = 3(1 + x)$
16. $6(3 + n) = 5(n - 1)$

Write each as an algebraic expression.

17. The sum of two numbers is 10. If one number is x, express the other number in terms of x.
18. Mandy is 5 inches taller than Melissa. If x inches represents the height of Mandy, express Melissa's height in terms of x.
19. If one angle measures $(x + 5)°$, express the measure of its supplement in terms of x.

$(x + 5)°?$

Solve each equation.

20. $\dfrac{3}{4}x = -9$
21. $\dfrac{x}{6} = \dfrac{2}{3}$
22. $-3x + 1 = 19$
23. $5x + 25 = 20$
24. $5x - 6 + x = 9 + 4x - 1$
25. $8 - y + 4y = 7 - y - 3$
26. Express the sum of three even consecutive integers as an expression in x. Let x be the first even integer.

(2.4) *Solve the following.*

27. $\frac{2}{7}x - \frac{5}{7} = 1$
28. $\frac{5}{3}x + 4 = \frac{2}{3}x$
29. $-(5x + 1) = -7x + 3$
30. $-4(2x + 1) = -5x + 5$
31. $-6(2x - 5) = -3(9 + 4x)$
32. $3(8y - 1) = 6(5 + 4y)$
33. $\frac{3(2 - z)}{5} = z$
34. $\frac{4(n + 2)}{5} = -n$
35. $5(2n - 3) - 1 = 4(6 + 2n)$
36. $-2(4y - 3) + 4 = 3(5 - y)$
37. $9z - z + 1 = 6(z - 1) + 7$
38. $5t - 3 - t = 3(t + 4) - 15$
39. $-n + 10 = 2(3n - 5)$
40. $-9 - 5a = 3(6a - 1)$
41. $\frac{5(c + 1)}{6} = 2c - 3$
42. $\frac{2(8 - a)}{3} = 4 - 4a$
43. $200(70x - 3560) = -179(150x - 19{,}300)$
44. $1.72y - .04y = 0.42$
45. The quotient of a number and 3 is the same as the difference of the number and two. Find the number.
46. Double the sum of a number and six is the opposite of the number. Find the number.

(2.5) *Solve each of the following.*

47. The height of the Eiffel Tower is 68 feet more than three times a side of its square base. If the sum of these two dimensions is 1380 feet, find the height of the Eiffel Tower.

48. A 12-foot board is to be divided into two pieces so that one piece is twice as long as the other. If x represents the length of the shorter piece, find the length of each piece.

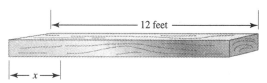

49. One area code in Ohio is 34 more than three times another area code used in Ohio. If the sum of these area codes is 1262, find the two area codes.
50. Find three consecutive even integers whose sum is negative 114.

(2.6) *Substitute the given values into the given formulas and solve for the unknown variable.*

51. $P = 2l + 2w$; $P = 46$, $l = 14$
52. $V = lwh$; $V = 192$, $l = 8$, $w = 6$

Solve each of the following for the indicated variable.

53. $y = mx + b$ for m
54. $r = vst - 5$ for s
55. $2y - 5x = 7$ for x
56. $3x - 6y = -2$ for y
57. $C = \pi D$ for π
58. $C = 2\pi r$ for π
59. A swimming pool holds 900 cubic meters of water. If its length is 20 meters and its height is 3 meters, find its width.

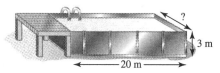

60. The highest temperature on record in Rome, Italy, is 104° Fahrenheit. Convert this temperature to degrees Celsius.
61. A charity 10K race is given annually to benefit a local hospice organization. How long will it take to run/walk a 10K race (10 kilometers or 10,000 meters) if your average pace is 125 **meters** per minute?

(2.7) *Solve.*

62. Find 12% of 250.

63. Find 110% of 85.
64. The number 9 is what percent of 45?
65. The number 59.5 is what percent of 85?
66. The number 137.5 is 125% of what number?
67. The number 768 is 60% of what number?
68. The state of Mississippi has the highest phoneless rate in the United States, 12.6% of households. If a city in Mississippi has 50,000 households, how many of these would you expect to be phoneless?

The graph below shows how business travelers relax when in their hotel rooms.

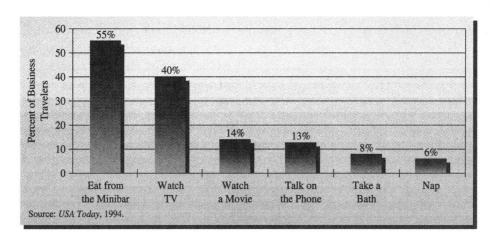

69. What percent of business travelers surveyed relax by taking a nap?
70. What is the most popular way to relax according to the survey?
71. If a hotel in New York currently has 300 business travelers, how many might you expect to relax by watching TV?
72. Do the percents in the graph above have a sum of 100%? Why or why not?
73. The number of employees at Arnold's Box Manufacturers just decreased from 210 to 180. Find the percent decrease. Round to the nearest tenth of a percent.

(2.8) *Solve each of the following.*

74. A $50,000 retirement pension is to be invested into two accounts: a money market fund that pays 8.5% and a certificate of deposit that pays 10.5%. How much should be invested at each rate in order to provide a yearly interest income of $4550?

75. A pay phone is holding its maximum number of 500 coins consisting of nickels, dimes, and quarters. The number of quarters is twice the number of dimes. If the value of all the coins is $88.00, how many nickels were in the pay phone?
76. How long will it take an Amtrak passenger train to catch up to a freight train if their speeds are 60 and 45 miles per hour and the freight train had an hour and a half head start?
77. Fabio Casartelli, from Italy, won a gold medal in cycling during the 1992 Summer Olympics. Suppose he rides a bicycle up a mountain trail at 8 miles per hour and down the same trail at 12 mph. Find the round-trip distance traveled if the total travel time was 5 hours.

(2.9) *Solve and graph the solution of each of the following inequalities.*

78. $x \leq -2$
79. $x > 0$
80. $-1 < x < 1$
81. $0.5 \leq y < 1.5$
82. $-2x \geq -20$
83. $-3x > 12$
84. $5x - 7 > 8x + 5$
85. $x + 4 \geq 6x - 16$
86. $2 \leq 3x - 4 < 6$
87. $-3 < 4x - 1 < 2$
88. $-2(x - 5) > 2(3x - 2)$
89. $4(2x - 5) \leq 5x - 1$

90. Tina earns $175 per week plus a 5% commission on all her sales. Find the minimum amount of sales to ensure that she earns at least $300 per week.
91. Ellen shot rounds of 76, 82, and 79 golfing. What must she shoot on her next round so that her average will be below 80?

CHAPTER 2 TEST

Simplify each of the following expressions.

1. $2y - 6 - y - 4$
2. $2.7x + 6.1 + 3.2x - 4.9$
3. $4(x - 2) - 3(2x - 6)$
4. $-5(y + 1) + 2(3 - 5y)$

Solve each of the following equations.

5. $-\frac{4}{5}x = 4$
6. $4(n - 5) = -(4 - 2n)$
7. $5y - 7 + y = -(y + 3y)$
8. $4z + 1 - z = 1 + z$
9. $\frac{2(x + 6)}{3} = x - 5$
10. $\frac{4(y - 1)}{5} = 2y + 3$
11. $\frac{1}{2} - x + \frac{3}{2} = x - 4$
12. $\frac{1}{3}(y + 3) = 4y$
13. $-.3(x - 4) + x = .5(3 - x)$
14. $-4(a + 1) - 3a = -7(2a - 3)$

Solve each of the following applications.

15. A number increased by two-thirds of the number is 35. Find the number.

16. A gallon of water seal covers 200 square feet. How many gallons are needed to paint two coats of water seal on a deck that measures 20 feet by 35 feet?

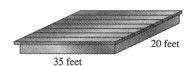

17. Sedric Angell invested an amount of money in Amoxil stock that earned an annual 10% return, and then he invested twice the original amount in IBM stock that earned an annual 12% return. If his total return from both investments was $2890, find how much he invested in each stock.

18. Two trains leave Los Angeles simultaneously traveling on the same track in opposite directions at speeds of 50 and 64 miles per hour. How long will it take before they are 285 miles apart?

19. Find the value of x if $y = -14$, $m = -2$, and $b = -2$ in the formula $y = mx + b$.

Solve each of the following equations for the indicated variable.

20. $V = \pi r^2 h$ for h
21. $3x - 4y = 10$ for y

Solve and graph each of the following inequalities.

22. $3x - 5 > 7x + 3$
23. $x + 6 > 4x - 6$
24. $-2 < 3x + 1 < 8$
25. $0 < 4x - 7 < 9$
26. $\frac{2(5x + 1)}{3} > 2$

The following graph shows the source of income for charities.

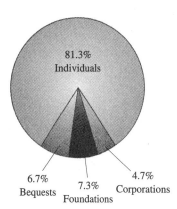

27. What percent of charity income comes from individuals?

28. If the total annual income for charities is $126.2 billion, find the amount that comes from corporations.

29. Find the number of degrees in the Bequests sector.

CHAPTER 2 CUMULATIVE REVIEW

1. Tell whether each statement is true or false.
   a. $8 \geq 8$ b. $8 \leq 8$
   c. $23 \leq 0$ d. $23 \geq 0$

2. Insert $<$, $>$, or $=$ in the appropriate space to make the statement true.
   a. $|0|$ 2
   b. $|-5|$ 5
   c. $|-3|$ $|-2|$
   d. $|5|$ $|6|$
   e. $|-7|$ $|6|$

3. Find the product of $\frac{2}{15}$ and $\frac{5}{13}$. Write the product in lowest terms.

4. Simplify $\dfrac{3 + |4 - 3| + 2^2}{6 - 3}$.

5. Find each sum.
   a. $3 + (-7)$ b. $(-2) + (10)$
   c. $0.2 + (-0.5)$

6. Simplify each expression.
   a. $-3 + [(-2 - 5) - 2]$
   b. $2^3 - |10| + [-6 \div (-5)]$

7. Find each product.
   a. $(-1.2)(0.05)$ b. $\dfrac{2}{3} \cdot -\dfrac{7}{10}$

8. Find the additive inverse or opposite of each number.
   a. -3 b. 5 c. 0 d. $|-2|$

9. The line graph below shows the relationship between two sets of measurements: the distance driven in a 14-foot U-Haul truck in one day and the total costs of renting this truck for that day. Notice that the horizontal axis is labeled Distance and the vertical axis is labeled Total Cost.

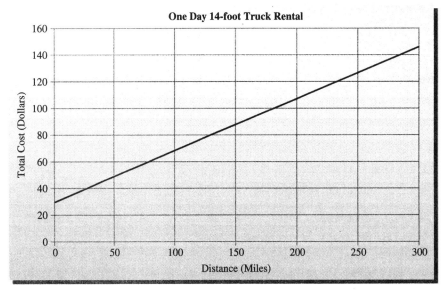

   a. Find the total cost of renting the truck if 100 miles are driven.
   b. Find the number of miles driven if the total cost of renting is $140.

10. Tell whether the terms are like or unlike.
    a. $-x^2, 3x^3$ b. $4x^2y, x^2y, -2x^2y$
    c. $-2yz, -3zy$ d. $-x^4, x^4$

11. Subtract $4x - 2$ from $2x - 3$.
12. Solve $x - 7 = 10$ for x.
13. Solve $5x - 2 = 18$ for x.
14. Solve $\dfrac{2(a + 3)}{3} = 6a + 2$.
15. In 1996, Congress had 8 more Republican senators than Democratic. If the number of senators is 100, how many senators of each party were there?
16. A glacier is a giant mass of rocks and ice that flows downhill like a river. Portage Glacier in Alaska is about 6 miles, or 31,680 feet, long and moves 400 feet per year. Icebergs are created when the front end of the glacier flows into Portage Lake. How long does it take for ice at the head (beginning) of the glacier to reach the lake?
17. Write each percent as a decimal.
    a. 35%   b. 89.5%   c. 150%
18. The number 63 is what percent of 72?
19. Marie Antonio, a bicycling enthusiast, rode her 10-speed at an average speed of 18 miles per hour on level roads and then slowed down to an average of 10 miles per hour on the hilly roads of the trip. If she covered a distance of 98 miles, how long did the entire trip take if traveling the level roads took the same time as traveling the hilly roads?
20. Graph $2 < x \leq 4$.
21. Solve $2(x - 3) - 5 \leq 3(x + 2) - 18$, and graph the solution set.

CHAPTER

3

GRAPHING

3.1 THE RECTANGULAR COORDINATE SYSTEM
3.2 GRAPHING LINEAR EQUATIONS
3.3 INTERCEPTS
3.4 SLOPE
3.5 GRAPHING LINEAR INEQUALITIES

FINANCIAL ANALYSIS

Investment analysts must investigate a company's financial data, such as sales, profit margin, debt, and assets, to evaluate whether investing in it is a wise choice. One way to analyze such data is to graph the data and identify trends in it visually over time. Another way to analyze such data is to find algebraically the rate at which it is changing over time.

IN THE CHAPTER GROUP ACTIVITY ON PAGE 217, YOU WILL HAVE THE OPPORTUNITY TO ANALYZE THE SALES OF SEVERAL COMPANIES IN THE AEROSPACE INDUSTRY.

3.1 THE RECTANGULAR COORDINATE SYSTEM

OBJECTIVES

 Define the rectangular coordinate system.
2. Plot ordered pairs of numbers.
3. Determine whether an ordered pair is a solution of an equation in two variables.
4. Find the missing coordinate of an ordered pair solution, given one coordinate of the pair.

In Section 1.9, we learned how to read graphs. Example 4 in Section 1.9 presented the graph below showing the relationship between time spent smoking a cigarette and pulse rate. Notice in this graph that there are two numbers associated with each point of the graph. For example, we discussed earlier that 15 minutes after "lighting up," the pulse rate is 80 beats per minute. If we agree to write the time first and the pulse rate second, we can say there is a point on the graph corresponding to the **ordered pair** of numbers (15, 80). A few more ordered pairs are listed alongside their corresponding points.

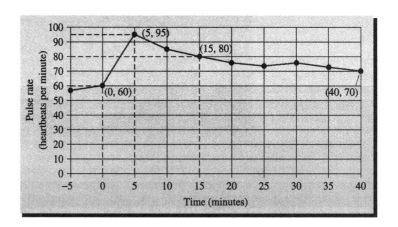

In general, we use this same ordered pair idea to describe the location of a point in a plane (such as a piece of paper). We start with a horizontal and a vertical axis. Each axis is a number line, and for the sake of consistency we construct our axes to intersect at the 0 coordinate of both. This point of intersection is called the

origin. Notice that these two number lines or axes divide the plane into four regions called **quadrants.** The quadrants are usually numbered with Roman numerals as shown. The axes are not considered to be in any quadrant.

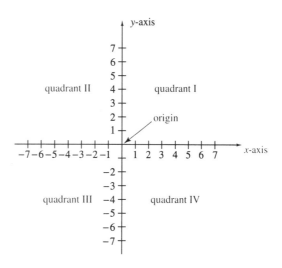

It is helpful to label axes, so we label the horizontal axis the **x-axis** and the vertical axis the **y-axis.** We call the system described above the **rectangular coordinate system.**

Just as with the pulse rate graph, we can then describe the locations of points by ordered pairs of numbers. We list the horizontal **x-axis** measurement first and the vertical **y-axis** measurement second.

The location of the point shown below can be described by the ordered pair of numbers (3, 2). Here, the x-value or **x-coordinate** is 3 and the y-value or **y-coordinate** is 2.

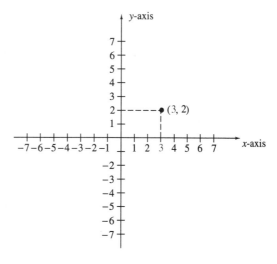

Does the order in which the coordinates are listed matter? Yes! Notice that the point corresponding to the ordered pair (2, 3) is in a different location than the point corresponding to (3, 2). These two ordered pairs of numbers describe two different points of the plane.

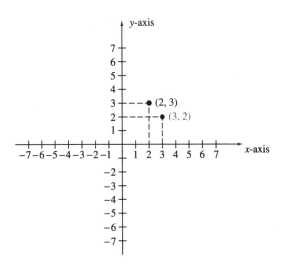

Given an ordered pair, how do we **graph** or **plot** it? That is, how do we find its corresponding point? To see, let's graph the ordered pair $(-2, 5)$. Start at the origin and move 2 units in the negative *x*-direction; from there, move 5 units in the positive *y*-direction. The ending location is the location of the point corresponding to the ordered pair $(-2, 5)$.

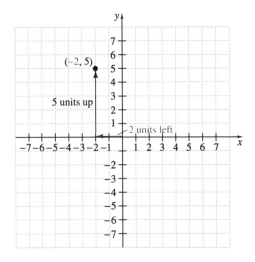

Here are some more ordered pairs that have been plotted.

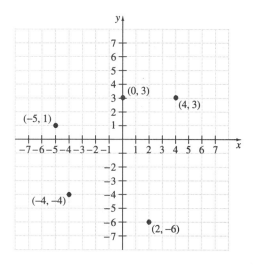

Keep in mind that **each ordered pair corresponds to exactly one point in the real plane and that each point in the plane corresponds to exactly one ordered pair.** Because of this correspondence, we may refer to the point corresponding to the ordered pair (2, 5) as simply the point (2, 5), for example.

EXAMPLE 1 On a single coordinate system, plot the ordered pairs. State in which quadrant, if any, each point lies.

a. (3, 2) b. (−2, −4) c. (1, −2) d. (−5, 3)

e. (0, 0) f. (0, 2) g. (−5, 0) h. $\left(0, -1\frac{1}{2}\right)$

Solution:

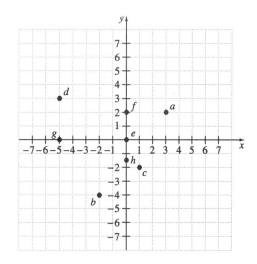

Point **a** lies in quadrant I.
Point **d** lies in quadrant II.
Point **b** lies in quadrant III.
Point **c** lies in quadrant IV.
Points **e, f, g,** and **h** lie on an axis, so they are not in any quadrant.

Notice that the *y*-coordinate of any point on the *x*-axis is 0. For example, the coordinates of point *g* are $(-5, 0)$. Also, the *x*-coordinate of any point on the *y*-axis is 0. For example, the coordinates of point *f* are $(0, 2)$.

 Let's see how we can use ordered pairs to record solutions of equations containing two variables. An equation in one variable such as $x + 1 = 5$ has one solution, which is 4: the number 4 is the value of the variable *x* that makes the equation true.

An equation in two variables, such as $2x + y = 8$, has solutions consisting of two values, one for *x* and one for *y*. For example, $x = 3$ and $y = 2$ is a solution of $2x + y = 8$ because, if *x* is replaced with 3 and *y* with 2, we get a true statement.

$$2x + y = 8$$
$$2(3) + 2 = 8$$
$$8 = 8 \quad \text{True.}$$

The solution $x = 3$ and $y = 2$ can be written as $(3, 2)$, an **ordered pair** of numbers. The first number, 3, is the *x*-value and the second number, 2, is the *y*-value.

In general, an ordered pair is a **solution** of an equation in two variables if replacing the variables by the values of the ordered pair results in a true statement.

EXAMPLE 2 Determine whether each ordered pair is a solution of the equation $x - 2y = 6$.

    a. $(6, 0)$ **b.** $(0, 3)$ **c.** $(2, -2)$

Solution: **a.** Let $x = 6$ and $y = 0$ in the equation $x - 2y = 6$.

$$x - 2y = 6$$
$$6 - 2(0) = 6 \quad \text{Replace } x \text{ with 6 and } y \text{ with 0.}$$
$$6 - 0 = 6 \quad \text{Simplify.}$$
$$6 = 6 \quad \text{True.}$$

$(6, 0)$ is a solution, since $6 = 6$ is a true statement.

b. Let $x = 0$ and $y = 3$.

$$x - 2y = 6$$
$$0 - 2(3) = 6 \quad \text{Replace } x \text{ with 0 and } y \text{ with 3.}$$
$$0 - 6 = 6$$
$$-6 = 6 \quad \text{False.}$$

(0, 3) is *not* a solution, since $-6 = 6$ is a false statement.

c. Let $x = 2$ and $y = -2$ in the equation.

$$x - 2y = 6$$
$$2 - 2(-2) = 6 \quad \text{Replace } x \text{ with 2 and } y \text{ with } -2.$$
$$2 + 4 = 6$$
$$6 = 6 \quad \text{True.}$$

$(2, -2)$ is a solution, since $6 = 6$ is a true statement.

 If one value of an ordered pair solution of an equation is known, the other value can be determined. To find the unknown value, replace one variable in the equation by its known value. Doing so results in an equation with just one variable that can be solved for the variable using the methods of Chapter 2.

EXAMPLE 3 Complete the following ordered pair solutions for the equation $3x + y = 12$.

    a. (0,)   **b.** (, 6)   **c.** $(-1, \ )$

Solution: **a.** In the ordered pair (0,), the x-value is 0. Let $x = 0$ in the equation and solve for y.

$$3x + y = 12$$
$$3(0) + y = 12 \quad \text{Replace } x \text{ with 0.}$$
$$0 + y = 12$$
$$y = 12$$

The completed ordered pair is (0, 12).

b. In the ordered pair (, 6), the y-value is 6. Let $y = 6$ in the equation and solve for x.

$$3x + y = 12$$
$$3x + 6 = 12 \quad \text{Replace } y \text{ with 6.}$$
$$3x = 6 \quad \text{Subtract 6 from both sides.}$$
$$x = 2 \quad \text{Divide both sides by 3.}$$

The ordered pair is (2, 6).

c. In the ordered pair $(-1, \ )$, the x-value is -1. Let $x = -1$ in the equation and solve for y.

$$3x + y = 12$$
$$3(-1) + y = 12 \quad \text{Replace } x \text{ with } -1.$$
$$-3 + y = 12$$
$$y = 15 \quad \text{Add 3 to both sides.}$$

The ordered pair is $(-1, 15)$.

Solutions of equations in two variables can also be recorded in a **table of values,** as shown in the next example.

EXAMPLE 4 Complete the table for the equation $y = 3x$.

	x	y
a.	-1	
b.		0
c.		-9

Solution: **a.** Replace x with -1 in the equation and solve for y.

$y = 3x$
$y = 3(-1)$ Let $x = -1$.
$y = -3$

The ordered pair is $(-1, -3)$.

b. Replace y with 0 in the equation and solve for x.

$y = 3x$
$0 = 3x$ Let $y = 0$.
$0 = x$ Divide both sides by 3.

The completed ordered pair is $(0, 0)$.

c. Replace y with -9 in the equation and solve for x.

$y = 3x$
$-9 = 3x$ Let $y = -9$.
$-3 = x$ Divide both sides by 3.

x	y
-1	-3
0	0
-3	-9

The completed ordered pair is $(-3, -9)$. The completed table is shown to the right.

EXAMPLE 5 Complete the table for the equation $y = 3$.

x	y
-2	
0	
-5	

Solution: The equation $y = 3$ is the same as $0x + y = 3$. No matter what value we replace x by, y always equals 3. The completed table is:

x	y
-2	3
0	3
-5	3

EXAMPLE 6

A small business purchased a computer for $2000. The business predicts that the computer will be used for 5 years and the value in dollars y of the computer in x years is $y = -300x + 2000$. Complete the table.

x	0	1	2	3	4	5
y						

Solution: To find the value of y when x is 0, replace x with 0 in the equation. We use this same procedure to find y when x is 1 and when x is 2.

WHEN $x = 0$,
$y = -300x + 2000$
$y = -300 \cdot 0 + 2000$
$y = 0 + 2000$
$y = 2000$

WHEN $x = 1$,
$y = -300x + 2000$
$y = -300 (1) + 2000$
$y = -300 + 2000$
$y = 1700$

WHEN $x = 2$,
$y = -300x + 2000$
$y = -300 \cdot 2 + 2000$
$y = -600 + 2000$
$y = 1400$

We have the ordered pairs (0, 2000), (1, 1700), and (2, 1400). This means that in 0 years the value of the computer is $2000, in 1 year the value of the computer is $1700, and in 2 years the value is $1400. Complete the table of values.

WHEN $x = 3$,
$y = -300x + 2000$
$y = -300 \cdot 3 + 2000$
$y = -900 + 2000$
$y = 1100$

WHEN $x = 4$,
$y = -300x + 2000$
$y = -300 \cdot 4 + 2000$
$y = -1200 + 2000$
$y = 800$

WHEN $x = 5$,
$y = -300x + 2000$
$y = -300 \cdot 5 + 2000$
$y = -1500 + 2000$
$y = 500$

The completed table is

x	0	1	2	3	4	5
y	2000	1700	1400	1100	800	500

The ordered pair solutions recorded in the completed table for the example above are graphed on the following page. Notice that the graph gives a visual picture of the decrease in value of the computer.

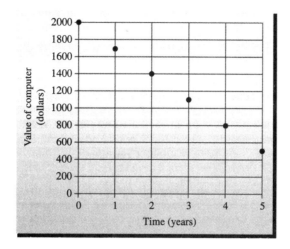

x	y
0	2000
1	1700
2	1400
3	1100
4	800
5	500

MENTAL MATH

Give two ordered pair solutions for each of the following linear equations.

1. $x + y = 10$
2. $x + y = 6$
3. $x = 3$
4. $y = -2$

EXERCISE SET 3.1

Plot the ordered pairs. State in which quadrant, if any, each point lies. See Example 1.

1. $(1, 5)$
2. $(-5, -2)$
3. $(-6, 0)$
4. $(0, -1)$
5. $(2, -4)$
6. $(-1, 4)$
7. $\left(4\frac{3}{4}, 0\right)$
8. $\left(0, \frac{7}{8}\right)$
9. $(0, 0)$
10. $(5, 0)$
11. $(0, 4)$
12. $(-3, -3)$

13. When is the graph of the ordered pair (a, b) the same as the graph of the ordered pair (b, a)?

14. In your own words, describe how to plot an ordered pair.

15. Find the perimeter of the rectangle whose vertices are the points with coordinates $(-1, 5)$, $(3, 5)$, $(3, -4)$, and $(-1, -4)$.

16. Find the area of the rectangle whose vertices are the points with coordinates $(5, 2)$, $(5, -6)$, $(0, -6)$, and $(0, 2)$.

Determine whether each ordered pair is a solution of the given linear equation. See Example 2.

17. $2x + y = 7$; $(3, 1)$, $(7, 0)$, $(0, 7)$
18. $x - y = 6$; $(5, -1)$, $(7, 1)$, $(0, -6)$
19. $y = -5x$; $(-1, -5)$, $(0, 0)$, $(2, -10)$
20. $x = 2y$; $(0, 0)$, $(2, 1)$, $(-2, -1)$

21. $x = 5$; $(4, 5)$, $(5, 4)$, $(5, 0)$
22. $y = 2$; $(-2, 2)$, $(2, 2)$, $(0, 2)$
23. $x + 2y = 9$; $(5, 2)$, $(0, 9)$
24. $3x + y = 8$; $(2, 3)$, $(0, 8)$
25. $2x - y = 11$; $(3, -4)$, $(9, 8)$
26. $x - 4y = 14$; $(2, -3)$, $(14, 6)$
27. $x = \frac{1}{3}y$; $(0, 0)$, $(3, 9)$
28. $y = -\frac{1}{2}x$; $(0, 0)$, $(4, 2)$
29. $y = -2$; $(-2, -2)$, $(5, -2)$
30. $x = 4$; $(4, 0)$, $(4, 4)$

Find the x- and y-coordinates of the following labeled points.

31.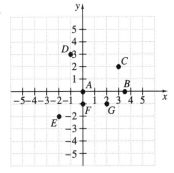

32.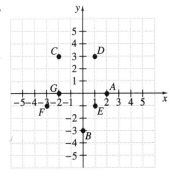

Determine the quadrant or quadrants in which the points described below lie.

33. The first coordinate is positive and the second coordinate is negative.

34. Both coordinates are negative.
35. The first coordinate is negative.
36. The second coordinate is positive.

Complete each ordered pair so that it is a solution of the given linear equation. See Examples 3 through 5.

37. $x - 4y = 4$; $(\ , -2)$, $(4, \ )$
38. $x - 5y = -1$; $(\ , -2)$, $(4, \ )$
39. $3x + y = 9$; $(0, \ )$, $(\ , 0)$
40. $x + 5y = 15$; $(0, \ )$, $(\ , 0)$
41. $y = -7$; $(11, \ )$, $(\ , -7)$
42. $x = \frac{1}{2}$; $(\ , 0)$, $\left(\frac{1}{2}, \ \right)$

Complete the table of values for each given linear equation; then plot each solution. Use a single coordinate system for each equation. See Examples 4 through 6.

43. $x + 3y = 6$

x	y
0	
	0
	1

44. $2x + y = 4$

x	y
0	
	0
	2

45. $2x - y = 12$

x	y
0	
	-2
-3	

46. $-5x + y = 10$

x	y
	0
	5
2	

47. $2x + 7y = 5$

x	y
0	
	0
	1

48. $x - 6y = 3$

x	y
0	
1	
	-1

49. $x = 3$

x	y
	0
	-0.5
	$\frac{1}{4}$

50. $y = -1$

x	y
-2	
0	
-1	

51. $x = -5y$

x	y
	0
	1
10	

52. $y = -3x$

x	y
0	
-2	
	9

❑ **53.** Discuss any similarities in the graphs of the ordered pair solutions for Exercises 43–52.

❑ **54.** Explain why equations in two variables have more than one solution.

55. The cost in dollars y of producing x computer desks is given by $y = 80x + 5000$.

   a. Complete the following table and graph the results.

x	100	200	300
y			

   b. Find the number of computer desks that can be produced for $8600. (*Hint:* Find x when $y = 8600$.)

56. The hourly wage y of an employee at a certain production company is given by $y = 0.25x + 9$ where x is the number of units produced in an hour.

   a. Complete the table and graph the results.

x	0	1	5	10
y				

   b. Find the number of units that must be produced each hour to earn an hourly wage of $12.25. (*Hint:* Find x when $y = 12.25$.)

The graph below shows Walt Disney Company's annual revenues.

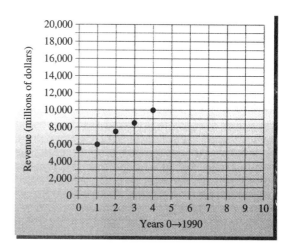

❑ **57.** Estimate the increase in revenues for years 1, 2, 3, and 4.

❑ **58.** Use a straight edge or ruler and this graph to predict Disney's revenue in the year 2000.

Review Exercises

Solve each equation for y. See Section 2.4.

59. $x + y = 5$ **60.** $x - y = 3$

61. $2x + 4y = 5$ **62.** $5x + 2y = 7$

63. $10x = -5y$ **64.** $4y = -8x$

65. $x - 3y = 6$ **66.** $2x - 9y = -20$

3.2 GRAPHING LINEAR EQUATIONS

OBJECTIVES

1. Identify linear equations.
2. Graph a linear equation by finding and plotting ordered pair solutions.

TAPE
BA 3.2

In the previous section, we found that equations in two variables may have more than one solution. For example, both $(6, 0)$ and $(2, -2)$ are solutions of the equation $x - 2y = 6$. In fact, this equation has an infinite number of solutions. Other solutions include $(0, -3)$, $(4, -1)$, $(-2, -4)$, and $(8, 1)$. If we graph these solutions, notice that a pattern appears.

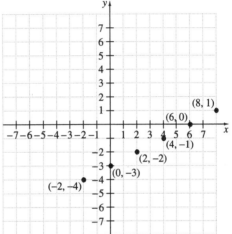

These solutions all appear to lie on the same line, which has been filled in below. It can be shown that every ordered pair solution of the equation corresponds to a point on this line, and every point on this line corresponds to an ordered pair solution. Thus, we say that this line is the graph of the equation $x - 2y = 6$.

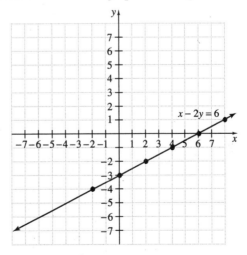

The equation $x - 2y = 6$ is called a **linear equation in two variables** and **the graph of every linear equation in two variables is a line.**

> **LINEAR EQUATION IN TWO VARIABLES**
> A linear equation in two variables is an equation that can be written in the form
> $$Ax + By = C$$
> where A, B, and C are real numbers and A and B are not both 0.

The form $Ax + By = C$ is called **standard form.**

EXAMPLES OF LINEAR EQUATIONS IN TWO VARIABLES

$$2x + y = 8 \quad -2x = 7y \quad y = \frac{1}{3}x + 2 \quad y = 7$$

Before we graph linear equations in two variables, let's practice identifying these equations.

EXAMPLE 1 Identify the linear equations in two variables.

a. $x - 1.5y = -1.6$ **b.** $y = -2x$ **c.** $x + y^2 = 9$ **d.** $x = 5$

Solution: **a.** This is a linear equation in two variables because it is written in the form $Ax + By = C$ with $A = 1$, $B = -1.5$, and $C = -1.6$.
b. This is a linear equation in two variables because it can be written in the form $Ax + By = C$.

$$y = -2x$$
$$2x + y = 0 \qquad \text{Add } 2x \text{ to both sides.}$$

c. This is *not* a linear equation in two variables because y is squared.
d. This is a linear equation in two variables because it can be written in the form $Ax + By = C$.

$$x = 5$$
$$x + 0y = 5 \qquad \text{Add } 0 \cdot y.$$

2 From geometry, we know that a straight line is determined by just two points. Graphing a linear equation in two variables, then, requires that we find just two of its infinitely many solutions. Once we do so, we plot the solution points and draw the line connecting the points. Usually, we find a third solution as well, as a check.

EXAMPLE 2 Graph the linear equation $2x + y = 5$.

Solution: Find three ordered pair solutions of $2x + y = 5$. To do this, choose a value for one variable, x or y, and solve for the other variable. For example, let $x = 1$. Then $2x + y = 5$ becomes

$$2x + y = 5$$
$$2(\mathbf{1}) + y = 5 \quad \text{Replace } x \text{ with } 1.$$
$$2 + y = 5 \quad \text{Multiply.}$$
$$y = \mathbf{3} \quad \text{Subtract 2 from both sides.}$$

Since $y = 3$ when $x = 1$, the ordered pair $(1, 3)$ is a solution of $2x + y = 5$. Next, let $x = 0$.

$$2x + y = 5$$
$$2(\mathbf{0}) + y = 5 \quad \text{Replace } x \text{ with } 0.$$
$$0 + y = 5$$
$$y = \mathbf{5}$$

The ordered pair $(0, 5)$ is a second solution.

The two solutions found so far allow us to draw the straight line that is the graph of all solutions of $2x + y = 5$. However, we find a third ordered pair as a check. Let $y = -1$.

$$2x + y = 5$$
$$2x + (\mathbf{-1}) = 5 \quad \text{Replace } y \text{ with } -1.$$
$$2x - 1 = 5$$
$$2x = 6 \quad \text{Add 1 to both sides.}$$
$$x = \mathbf{3} \quad \text{Divide both sides by 2.}$$

The third solution is $(3, -1)$. These three ordered pair solutions are listed in table form as shown. The graph of $2x + y = 5$ is the line through the three points.

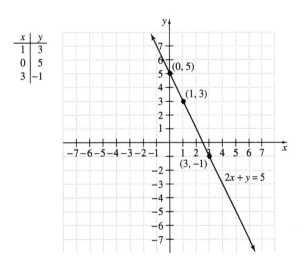

EXAMPLE 3 Graph the linear equation $-5x + 3y = 15$.

Solution: Find three ordered pair solutions of $-5x + 3y = 15$.

Let $x = 0$.
$-5x + 3y = 15$
$-5 \cdot 0 + 3y = 15$
$0 + 3y = 15$
$3y = 15$
$y = 5$

Let $y = 0$.
$-5x + 3y = 15$
$-5x + 3 \cdot 0 = 15$
$-5x + 0 = 15$
$-5x = 15$
$x = -3$

Let $x = -2$.
$-5x + 3y = 15$
$-5(-2) + 3y = 15$
$10 + 3y = 15$
$3y = 5$
$y = \dfrac{5}{3}$

The ordered pairs are $(0, 5)$, $(-3, 0)$, and $\left(-2, \dfrac{5}{3}\right)$. The graph of $-5x + 3y = 15$ is the line through the three points.

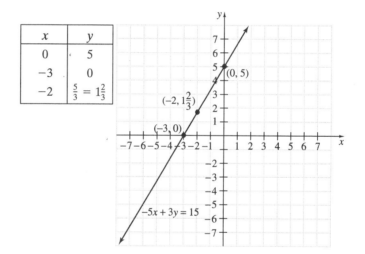

x	y
0	5
-3	0
-2	$\frac{5}{3} = 1\frac{2}{3}$

EXAMPLE 4 Graph the linear equation $y = 3x$.

Solution: To graph this linear equation, we find three ordered pair solutions. Since this equation is solved for y, choose three x values.

If $x = 2$, $y = 3 \cdot 2 = 6$.
If $x = 0$, $y = 3 \cdot 0 = 0$.
If $x = -3$, $y = 3 \cdot -3 = -9$.

x	y
2	6
0	0
-3	-9

Next, graph the ordered pair solutions listed in the table above and draw a line through the plotted points. The line is the graph of $y = 3x$. Every point on the graph represents an ordered pair solution of the equation and every ordered pair solution is a point on this line.

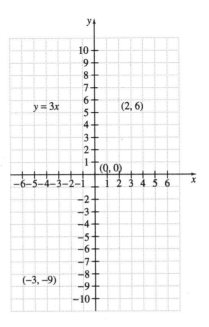

EXAMPLE 5   Graph the linear equation $y = -\frac{1}{3}x$.

Solution: Find three ordered pair solutions, graph the solutions, and draw a line through the plotted solutions. To avoid fractions, choose x values that are multiples of 3 to substitute in the equation.

If $x = 6$, then $y = -\frac{1}{3} \cdot 6 = -2$.

If $x = 0$, then $y = -\frac{1}{3} \cdot 0 = 0$.

If $x = -3$, then $y = -\frac{1}{3} \cdot -3 = 1$.

x	y
6	-2
0	0
-3	1

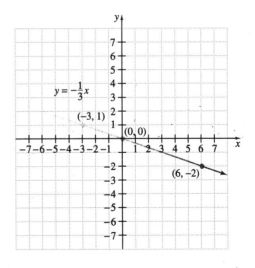

Let's compare the graphs in Examples 4 and 5. The graph of $y = 3x$ tilts upward (as we follow the line from left to right) and the graph of $y = -\frac{1}{3}x$ tilts downward (as we follow the line from left to right). Also notice that both lines go through the origin or that (0, 0) is an ordered pair solution of both equations. This

m is a constant. The graph of an equation in this form goes through the origin $(0, 0)$ because when x is 0, $y = mx$ becomes $y = m \cdot 0 = 0$.

EXAMPLE 6 Graph the linear equation $y = 3x + 6$ and compare this graph with the graph of $y = 3x$ in Example 4.

Solution: Find ordered pair solutions, graph the solutions, and draw a line through the plotted solutions. We choose x values and substitute in the equation $y = 3x + 6$.

If $x = -3$, then $y = 3(-3) + 6 = -3$.
If $x = 0$, then $y = 3(0) + 6 = 6$.
If $x = 1$, then $y = 3(1) + 6 = 9$.

x	y
-3	-3
0	6
1	9

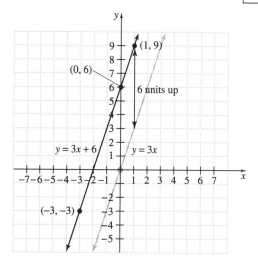

The most startling similarity is that both graphs appear to have the same upward tilt as we move from left to right. Also, the graph of $y = 3x$ crosses the y-axis at the origin, while the graph of $y = 3x + 6$ crosses the y-axis at 6. In fact, the graph of $y = 3x + 6$ is the same as the graph of $y = 3x$ moved vertically upward 6 units.

Notice above that the graph of $y = 3x + 6$ crosses the y-axis at 6. This happens because when $x = 0$, $y = 3x + 6$ becomes $y = 3 \cdot 0 + 6 = 6$. The graph contains the point $(0, 6)$, which is on the y-axis.

In general, if a linear equation in two variables is solved for y, we say that it is written in the form $y = mx + b$. The graph of this equation contains the point $(0, b)$ because when $x = 0$, $y = mx + b$ is $y = m \cdot 0 + b = b$.

> The graph of $y = mx + b$ crosses the y-axis at b.

GRAPHING CALCULATOR EXPLORATIONS

In this section, we begin a study of graphing calculators and graphing software packages for computers. These graphers use the same point plotting technique that was introduced in this section. The advantage of this graphing technology is, of course, that graphing calculators and computers can find and plot ordered pair solutions much faster than we can. Note, however, that the features described in these boxes may not be available on all graphing calculators.

The rectangular screen where a portion of the rectangular coordinate system is displayed is called a **window**. We call it a **standard window** for graphing when both the x- and y-axes show coordinates between -10 and 10. This information is often displayed in the window menu on a graphing calculator as

Xmin = −10
Xmax = 10
Xscl = 1   The scale on the x-axis is one unit per tick mark.
Ymin = −10
Ymax = 10
Yscl = 1   The scale on the y-axis is one unit per tick mark.

To use a graphing calculator to graph the equation $y = 2x + 3$, press the $\boxed{Y=}$ key and enter the keystrokes $\boxed{2}$ $\boxed{x}$ $\boxed{+}$ $\boxed{3}$. The top row should now read $Y_1 = 2x + 3$. Next press the $\boxed{\text{GRAPH}}$ key, and the display should look like this:

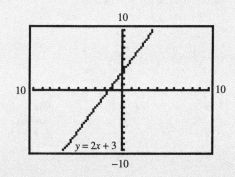

Use a standard window and graph the following linear equations. (Unless otherwise stated, use a standard window when graphing.)

1. $y = -3x + 7$
2. $y = -x + 5$
3. $y = \frac{1}{4}x - 2$
4. $y = \frac{2}{3}x - 1$
5. $y = 2.5x - 7.9$
6. $y = -1.3x + 5.2$
7. $y = -\frac{3}{10}x + \frac{32}{5}$
8. $y = \frac{2}{9}x - \frac{22}{3}$

Exercise Set 3.2

Determine whether each equation is a linear equation in two variables. See Example 1.

1. $-x = 3y + 10$
2. $y = x - 15$
3. $x = y$
4. $x = y^3$
5. $x^2 + 2y = 0$
6. $0.01x - 0.2y = 8.8$
7. $y = -1$
8. $x = 25$

Graph each linear equation. See Examples 2 through 5.

9. $x + y = 4$
10. $x + y = 7$
11. $x - y = -2$
12. $-x + y = 6$
13. $x - 2y = 4$
14. $-x + 5y = 5$
15. $y = 6x + 3$
16. $y = -2x + 7$

Write each statement as an equation in two variables. Then graph the equation.

17. The y-value is 5 more than the x-value.
18. The y-value is twice the x-value.
19. Two times the x-value added to three times the y-value is 6.
20. Five times the x-value added to twice the y-value is −10.

Graph each pair of linear equations on the same set of axes. Discuss how the graphs are similar and how they are different. See Example 6.

21. $y = 5x$; $y = 5x + 4$
22. $y = 2x$; $y = 2x + 5$
23. $y = -2x$; $y = -2x - 3$
24. $y = x$; $y = x - 7$
25. $y = \frac{1}{2}x$; $y = \frac{1}{2}x + 2$
26. $y = -\frac{1}{4}x$; $y = -\frac{1}{4}x + 3$

Graph each linear equation.

27. $x - 2y = -6$
28. $-x + 2y = 5$
29. $y = 6x$
30. $x = -2y$
31. $3y - 10 = 5x$
32. $-2x + 7 = 2y$
33. $x + 3y = 9$
34. $2x + y = -2$
35. $y - x = -1$
36. $x - y = 5$
37. $x = -3y$
38. $y = -x$
39. $5x - y = 10$
40. $7x - y = 2$
41. $y = \frac{1}{2}x + 2$
42. $y = -\frac{1}{5}x - 1$

43. Graph the nonlinear equation $y = x^2$ by completing the table shown. Plot the ordered pairs and connect them with a smooth curve.

x	y
0	
1	
−1	
2	
−2	

44. Graph the nonlinear equation $y = |x|$ by completing the table shown. Plot the ordered pairs and connect them. This curve is "V" shaped.

$$y = |x|$$

x	y
0	
1	
−1	
2	
−2	

The graph of $y = 5x$ is below as well as Figures A–D. For Exercises 45 through 48, match each equation with its graph.

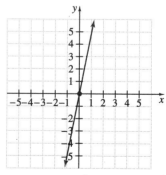

$y = 5x$

45. $y = 5x + 5$ **46.** $y = 5x - 4$
47. $y = 5x - 1$ **48.** $y = 5x + 2$

Recall that if an equation is written in the form $y = mx + b$, its graph crosses the y-axis at b. Use this to match each graph with its corresponding equation in Exercises 49 through 54.

A.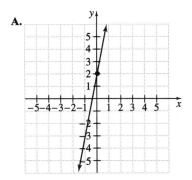

B.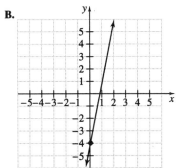

C.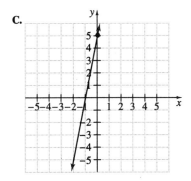

D.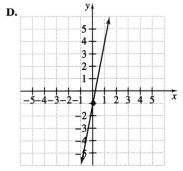

A.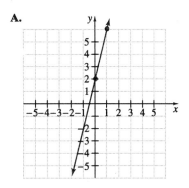

B.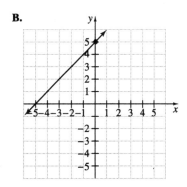

C.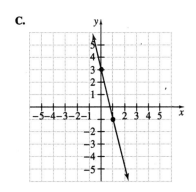

D.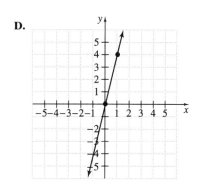

E.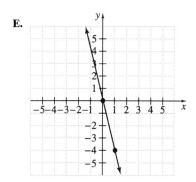

F.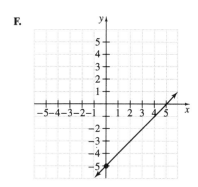

49. $y = 4x$
50. $y = -4x$
51. $y = 4x + 2$
52. $y = -4x + 3$
53. $y = x + 5$
54. $y = x - 5$

55. Explain how to find ordered pair solutions of linear equations in two variables.

56. The perimeter of the trapezoid below is 22 centimeters. Write a linear equation in two variables for the perimeter. Find y if x is 3 cm.

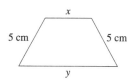

57. If (a, b) is an ordered pair solution of $x + y = 5$, is (b, a) also a solution? Explain why or why not.

58. The perimeter of the rectangle below is 50 miles. Write a linear equation in two variables for this perimeter. Use this equation to find x when y is 20.

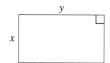

Review Exercises

59. The coordinates of three vertices of a rectangle are $(-2, 5)$, $(4, 5)$, and $(-2, -1)$. Find the coordinates of the fourth vertex. See Section 3.1.

60. The coordinates of two vertices of square are $(-3, -1)$ and $(2, -1)$. Find the coordinates of two pairs of points possible for the third and fourth vertices. See Section 3.1.

Solve the following equations. See Section 2.4.

61. $3(x - 2) + 5x = 6x - 16$

62. $5 + 7(x + 1) = 12 + 10x$

63. $3x + \dfrac{2}{5} = \dfrac{1}{10}$

64. $\dfrac{1}{6} + 2x = \dfrac{2}{3}$

Complete each table. See Section 3.1.

65. $x - y = -3$

x	y
0	
	0

66. $y - x = 5$

x	y
0	
	0

67. $y = 2x$

x	y
0	
	0

68. $x = -3y$

x	y
0	
	0

3.3 INTERCEPTS

TAPE
BA 3.3

OBJECTIVES

 Identify intercepts of a graph.
 Graph a line given its intercepts.
 Graph a linear equation by finding and plotting intercepts.
 Identify and graph vertical and horizontal lines.

 In this section, we graph linear equations in two variables by identifying intercepts. For example, the graph of $y = 4x - 8$ is shown below. Notice that this graph crosses the y-axis at the point $(0, -8)$. The y-coordinate of this point, -8, is called the **y-intercept.** Likewise, the graph crosses the x-axis at $(2, 0)$ and the x-coordinate, 2, is called the **x-intercept.**

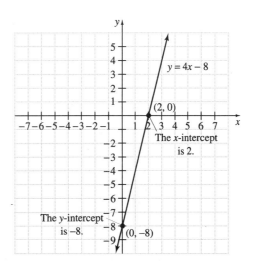

In general, an **intercept point** of a graph is a point where the graph intersects an axis. When an intercept point is on the x-axis, the x-coordinate of the point is called an x-intercept. When an intercept point is on the y-axis, the y-coordinate of the point is called a y-intercept.

EXAMPLE 1 Identify the x- and y-intercepts and the intercept points.

a. 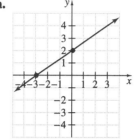   b. 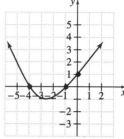   c.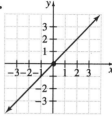

d. 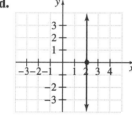   e.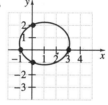

Solution:
a. The graph crosses the x-axis at -3, so the x-intercept is -3. The graph crosses the y-axis at 2, so the y-intercept is 2. The intercept points are $(-3, 0)$ and $(0, 2)$.
b. The graph crosses the x-axis at -4 and -1, so the x-intercepts are -4 and -1. The graph crosses the y-axis at 1, so the y-intercept is 1. The intercept points are $(-4, 0)$, $(-1, 0)$, and $(0, 1)$.
c. The x-intercept is 0 and the y-intercept is 0. The intercept point is $(0, 0)$.
d. The x-intercept is 2. There are no y-intercepts. The intercept point is $(2, 0)$.
e. The x-intercepts are -1 and 3. The y-intercepts are -1 and 2. The intercept points are $(-1, 0)$, $(3, 0)$, $(0, -1)$, and $(0, 2)$.

2 Since a line is determined by two points, we can usually graph a line when we know its x-intercept and its y-intercept. For example, if the x-intercept is 3 and the y-intercept is -5, we graph the line containing the points $(3, 0)$ and $(0, -5)$.

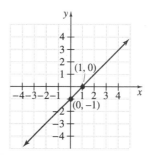

EXAMPLE 2   Graph the line with x-intercept 1 and y-intercept -1.

Solution: To graph the line, we identify and graph the intercept points. The intercept points are $(1, 0)$ and $(0, -1)$. Plot these points and draw a line through them.

Given an equation of a line, intercept points are usually easy to find since one coordinate is 0.

One way to find the y-intercept of a line, given its equation, is to let $x = 0$, since a point on the y-axis has an x-coordinate of 0. To find the x-intercept of a line, let $y = 0$, since a point on the x-axis has a y-coordinate of 0.

> **FINDING X- AND Y-INTERCEPTS**
> To find the x-intercept, let $y = 0$ and solve for x.
> To find the y-intercept, let $x = 0$ and solve for y.

EXAMPLE 3   Graph $x - 3y = 6$ by finding and plotting intercept points.

Solution: Let $y = 0$ to find the x-intercept and let $x = 0$ to find the y-intercept.

$$\begin{array}{ll} \text{Let } y = 0 & \text{Let } x = 0 \\ x - 3y = 6 & x - 3y = 6 \\ x - 3(0) = 6 & 0 - 3y = 6 \\ x - 0 = 6 & -3y = 6 \\ x = 6 & y = -2 \end{array}$$

The x-intercept is 6 and the y-intercept is -2. We find a third ordered pair solution to check our work. If we let $y = -1$, then $x = 3$. Plot the points $(6, 0)$, $(0, -2)$, and $(3, -1)$. The graph of $x - 3y = 6$ is the line drawn through these points, as shown.

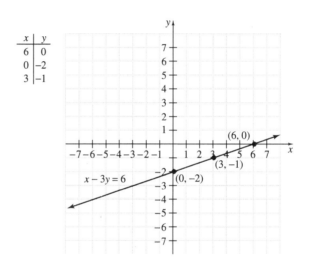

EXAMPLE 4   Graph $x = -2y$ by plotting intercept points.

Solution: Let $y = 0$ to find the x-intercept and $x = 0$ to find the y-intercept.

$$\begin{array}{lll} \text{If } y = 0 \quad \text{then} & & \text{If } x = 0 \quad \text{then} \\ x = -2y & & \\ x = -2(0) \quad \text{or} & & 0 = -2y \quad \text{or} \\ x = 0 & & 0 = y \end{array}$$

Both the x-intercept and y-intercept are 0. In other words, when $x = 0$, then $y = 0$, which gives the ordered pair $(0, 0)$. Also, when $y = 0$, then $x = 0$, which gives the same ordered pair $(0, 0)$. This happens when the graph passes through the origin. Since two points are needed to determine a line, we must find at least one more ordered pair that satisfies $x = -2y$. Let $y = -1$ to find a second ordered pair solution and let $y = 1$ as a checkpoint.

If $y = -1$ then $\quad$ If $y = 1$ then
$x = -2(-1)$ or $\quad$ $x = -2(1)$ or
$x = 2$ $\quad\quad\quad\quad\quad\quad$ $x = -2$

The ordered pairs are $(0, 0)$, $(2, -1)$, and $(-2, 1)$. Plot these points to graph $x = -2y$.

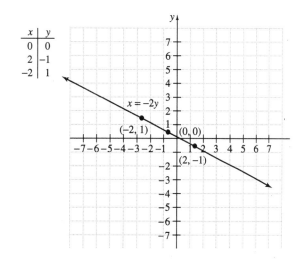

EXAMPLE 5 $\quad$ Graph $4x = 3y - 9$.

Solution: $\quad$ Find the x- and y-intercepts, and then choose $x = 2$ to find a third checkpoint.

If $y = 0$ then $\quad\quad$ If $x = 0$ then $\quad\quad$ If $x = 2$ then
$4x = 3(0) - 9$ or $\quad$ $4 \cdot 0 = 3y - 9$ or $\quad$ $4(2) = 3y - 9$ or
$4x = -9$ $\quad\quad\quad\quad\quad$ $9 = 3y$ $\quad\quad\quad\quad\quad\quad$ $8 = 3y - 9$
Solve for x. $\quad\quad\quad\quad$ Solve for y. $\quad\quad\quad\quad$ Solve for y.
$x = -\dfrac{9}{4}$ or $-2\dfrac{1}{4}$ $\quad$ $3 = y$ $\quad\quad\quad\quad\quad\quad$ $17 = 3y$

$\quad\quad\quad\quad\quad\quad\quad\quad\quad\quad\quad\quad\quad\quad\quad\quad\quad\quad\quad\quad\quad\quad$ $\dfrac{17}{3} = y$ or $y = 5\dfrac{2}{3}$.

The ordered pairs are $\left(-2\dfrac{1}{4}, 0\right)$, $(0, 3)$, and $\left(2, 5\dfrac{2}{3}\right)$. The equation $4x = 3y - 9$ is graphed as follows.

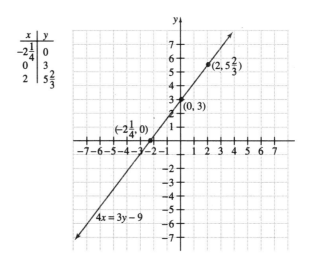

4 The equation $x = c$, where c is a real number constant, is a linear equation in two variables because it can be written in the form $x + 0y = c$. The graph of this equation is a vertical line as shown in the next example.

EXAMPLE 6 Graph $x = 2$.

Solution: The equation $x = 2$ can be written as $x + 0y = 2$. For any y-value chosen, notice that x is 2. No other value for x satisfies $x + 0y = 2$. Any ordered pair whose x-coordinate is 2 is a solution of $x + 0y = 2$. We will use the ordered pair solutions $(2, 3)$, $(2, 0)$, and $(2, -3)$ to graph $x = 2$.

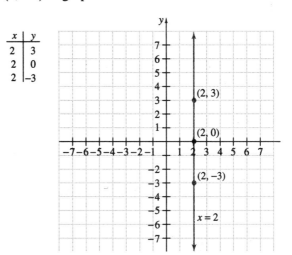

The graph is a vertical line with x-intercept 2. Note that this graph has no y-intercept because x is never 0.

VERTICAL LINES

The graph of $x = c$, where c is a real number, is a vertical line with x-intercept c.

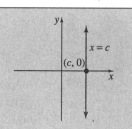

EXAMPLE 7 Graph $y = -3$.

Solution: The equation $y = -3$ can be written as $0x + y = -3$. For any x-value chosen, y is -3. If we choose 4, 1, and -2 as x-values, the ordered pair solutions are $(4, -3)$, $(1, -3)$, and $(-2, -3)$. Use these ordered pairs to graph $y = -3$. The graph is a horizontal line with y-intercept -3 and no x-intercept.

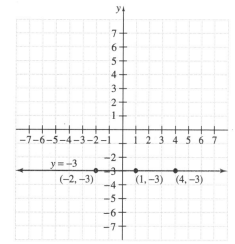

HORIZONTAL LINES

The graph of $y = c$, where c is a real number, is a horizontal line with y-intercept c.

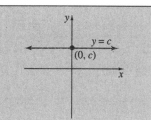

GRAPHING CALCULATOR EXPLORATIONS

You may have noticed that to use the $\boxed{Y=}$ key on a grapher to graph an equation, the equation must be solved for y. For example, to graph $2x + 3y = 7$, we solve this equation for y.

$$2x + 3y = 7$$
$$3y = -2x + 7 \quad \text{Subtract } 2x \text{ from both sides.}$$
$$\frac{3y}{3} = -\frac{2x}{3} + \frac{7}{3} \quad \text{Divide both sides by 3.}$$
$$y = -\frac{2}{3}x + \frac{7}{3} \quad \text{Simplify.}$$

To graph $2x + 3y = 7$ or $y = -\frac{2}{3}x + \frac{7}{3}$, press the $\boxed{Y=}$ key and enter

$$Y_1 = -\frac{2}{3}x + \frac{7}{3}$$

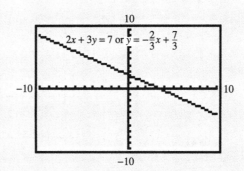

Graph each linear equation.

1. $x = 3.78y$ **2.** $-2.61y = x$
3. $3x + 7y = 21$ **4.** $-4x + 6y = 12$
5. $-2.2x + 6.8y = 15.5$ **6.** $5.9x - 0.8y = -10.4$

EXERCISE SET 3.3

Identify the intercepts and intercept points. See Example 1.

1.
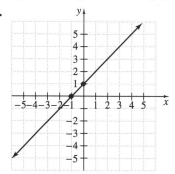

2.
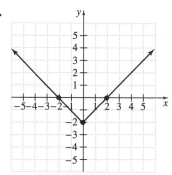

3.
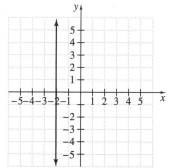

4.
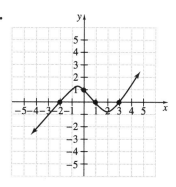

5.
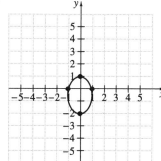

6.
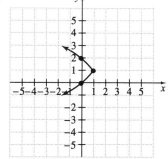

❑ **7.** What is the greatest number of intercepts for a line?
❑ **8.** What is the least number of intercepts for a line?
❑ **9.** What is the least number of intercepts for a circle?
❑ **10.** What is the greatest number of intercepts for a circle?

194 CHAPTER 3 GRAPHING

Graph each line with given x- and y-intercept. See Example 2.

11. x-intercept: 5
y-intercept: −3

12. x-intercept: 0
y-intercept: 4

13. x-intercept: 7/2
y-intercept: 0

14. x-intercept: −1/2
y-intercept: −9

Graph each linear equation by finding x- and y-intercepts. See Examples 3 through 5.

15. $x - y = 3$
16. $x - y = -4$
17. $x = 5y$
18. $2x = y$
19. $-x + 2y = 6$
20. $x - 2y = -8$
21. $2x - 4y = 8$
22. $2x + 3y = 6$

Graph each linear equation. See Examples 6 and 7.

23. $x = -1$
24. $y = 5$
25. $y = 0$
26. $x = 0$
27. $y + 7 = 0$
28. $x - 2 = 0$

*Two lines in the same plane that do not intersect are called **parallel lines**.*

 29. Draw a line parallel to the line $x = 5$ that intersects the x-axis at 1. What is the equation of this line?

 30. Draw a line parallel to the line $y = -1$ that intersects the y-axis at −4. What is the equation of this line?

Graph each linear equation.

31. $x + 2y = 8$
32. $x - 3y = 3$
33. $x - 7 = 3y$
34. $y - 3x = 2$
35. $x = -3$
36. $y = 3$
37. $3x + 5y = 7$
38. $3x - 2y = 5$
39. $x = y$
40. $x = -y$
41. $x + 8y = 8$
42. $x - 3y = 9$
43. $5 = 6x - y$
44. $4 = x - 3y$
45. $-x + 10y = 11$
46. $-x + 9 = -y$
47. $y = 1$
48. $x = 1$
49. $x = 2y$
50. $y = -2x$
51. $x + 3 = 0$
52. $y - 6 = 0$
53. $x = 4y - \dfrac{1}{3}$
54. $y = -3x + \dfrac{3}{4}$
55. $2x + 3y = 6$
56. $4x + y = 5$

For Exercises 57 through 62, match each equation with its graph.

A.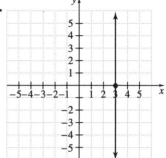

B.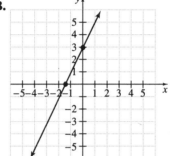

C.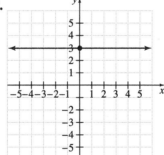

D.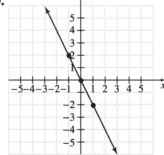

E.
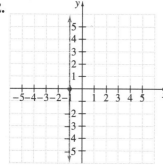

F.
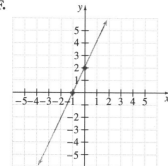

57. $y = 3$
58. $y = 2x + 2$
59. $x = -1$
60. $x = 3$
61. $y = 2x + 3$
62. $y = -2x$

63. A computer purchased for $4000 loses value or depreciates in the amount of $800 per year. The equation $y = 4000 - 800x$ models the depreciated value of the computer, where y is the depreciated value after x years. Graph this equation for x values from 0 to 5.

64. The production supervisor at Alexandra's Office Products finds that it takes 3 hours to manufacture a particular office chair and 6 hours to manufacture an office desk. A total of 1200 hours is available to produce office chairs and desks of this style. The linear equation that models this situation is $3x + 6y = 1200$, where x represents the number of chairs produced and y the number of desks manufactured.

   a. Complete the ordered pair solution (0,) of this equation. Describe the manufacturing situation that corresponds to this solution.
   b. Complete the ordered pair solution (, 0) of this equation. Describe the manufacturing situation that corresponds to this solution.
   c. Use the ordered pairs found above and graph the equation $3x + 6y = 1200$.
   d. If 50 desks are manufactured, find the greatest number of chairs that they can make.

65. Discuss whether a vertical line ever has a y-intercept.
66. Explain why it is a good idea to use three points to graph a linear equation.
67. Discuss whether a horizontal line ever has an x-intercept.
68. Explain how to find intercepts.

Review Exercises

Simplify.

69. $\dfrac{-6 - 3}{2 - 8}$
70. $\dfrac{4 - 5}{-1 - 0}$
71. $\dfrac{-8 - (-2)}{-3 - (-2)}$
72. $\dfrac{12 - 3}{10 - 9}$
73. $\dfrac{0 - 6}{5 - 0}$
74. $\dfrac{2 - 2}{3 - 5}$

3.4 SLOPE

TAPE
BA 3.4

OBJECTIVES

1. Find the slope of a line given two points of the line.
2. Graph a line given its slope and a point of the line.
3. Find the slope of a line given its equation.
4. Find the slopes of horizontal and vertical lines.
5. Compare the shapes of parallel and perpendicular lines.

Thus far, much of this chapter has been devoted to graphing lines. You have probably noticed by now that a key feature of a line is its slant or steepness. In mathematics, the slant or steepness of a line is formally known as its **slope.** We measure the slope of a line by the ratio of vertical change to the corresponding horizontal change as we move along the line.

On the line below, for example, suppose that we begin at the point (1, 2) and move to the point (4, 6). The vertical change is the change in y-coordinates: $6 - 2$ or 4 units. The corresponding horizontal change is the change in x-coordinates: $4 - 1 = 3$ units. The ratio of these changes is

$$\text{slope} = \frac{\text{change in } y \text{ (vertical change)}}{\text{change in } x \text{ (horizontal change)}} = \frac{4}{3}$$

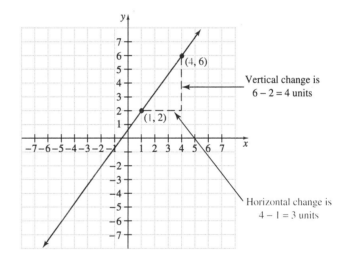

The slope of this line, then, is $\frac{4}{3}$: for every 4 units of change in y-coordinates, there is a corresponding change of 3 units in x-coordinates. You may wonder if the slope depends on what two points of a line are chosen. The answer is no, it makes no difference at all which two points we choose: the ratio of change is constant, and this is why we call this constant ratio slope. Slope, or steepness of a line, after all, is the same everywhere on a line.

To find the slope of a line, then, choose two points of the line. Label the two x-coordinates of two points, x_1 and x_2 (read "x sub one" and "x sub two"), and label the corresponding y-coordinates y_1 and y_2.

The vertical change or **rise** between these points is the difference in the y-coordinates: $y_2 - y_1$. The horizontal change or **run** between the points is the difference of the x-coordinates: $x_2 - x_1$. The slope of the line is the ratio of $y_2 - y_1$ to $x_2 - x_1$ and we traditionally use the letter m to denote slope.

$$m = \frac{y_2 - y_1}{x_2 - x_1}$$

SLOPE SECTION 3.4 **197**

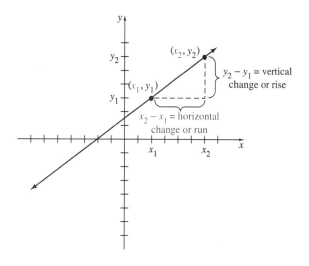

SLOPE OF A LINE

The slope m of the line through points (x_1, y_1) and (x_2, y_2), is given by

$$m = \frac{\text{rise}}{\text{run}} = \frac{y_2 - y_1}{x_2 - x_1}, \qquad \text{as long as } x_2 \neq x_1.$$

EXAMPLE 1 Find the slope of the line through $(-1, 5)$ and $(2, -3)$. Graph the line.

Solution: If we let (x_1, y_1) be $(-1, 5)$, then $x_1 = -1$ and $y_1 = 5$. Also, let (x_2, y_2) be point $(2, -3)$ so that $x_2 = 2$ and $y_2 = -3$. Then, by the definition of slope,

$$m = \frac{y_2 - y_1}{x_2 - x_1}$$

$$= \frac{-3 - 5}{2 - (-1)}$$

$$= \frac{-8}{3} = -\frac{8}{3}$$

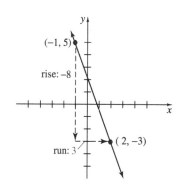

The slope of the line is $-\frac{8}{3}$.

In Example 1, we could just as well have identified (x_1, y_1) with $(2, -3)$ and (x_2, y_2) with $(-1, 5)$. It makes no difference which point is called (x_1, y_1) or (x_2, y_2).

REMINDER When finding the slope of a line through two given points, it makes no difference which given point is called (x_1, y_1) and which is called (x_2, y_2). However, once an x-coordinate is called x_1, make sure its corresponding y-coordinate is called y_1.

EXAMPLE 2 Find the slope of the line through $(-1, -2)$ and $(2, 4)$. Graph the line.

Solution: Let (x_1, y_1) be $(2, 4)$ and let (x_2, y_2) be $(-1, -2)$.

$$m = \frac{y_2 - y_1}{x_2 - x_1}$$

$$= \frac{-2 - 4}{-1 - 2}$$

$$= \frac{-6}{-3} = 2$$

The slope is 2.

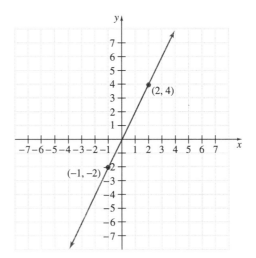

Notice that the slope of the line in Example 1 is negative, whereas the slope of the line in Example 2 is positive. Let your eye follow the line with negative slope from left to right and notice that the line "goes down." Following the line with positive slope from left to right, notice that the line "goes up." This is true in general.

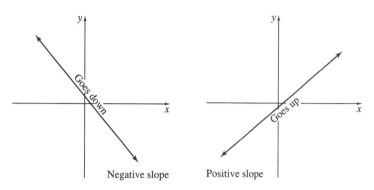

Negative slope   Positive slope

From geometry, we know that a line is uniquely determined by 2 points. A line is also uniquely determined by 1 point and a slope.

EXAMPLE 3 Graph the line through $(-1, 5)$ with slope -2.

Solution: To graph the line, we need two points. One point is $(-1, 5)$, and we use the slope -2, which can be written as $\frac{-2}{1}$, to find another point.

$$m = \frac{\text{rise}}{\text{run}} = \frac{-2}{1}$$

To find another point, start at $(-1, 5)$ and move vertically two units down, since the numerator of the slope is -2; then move horizontally 1 unit to the right since the denominator of the slope is 1. We stop at the point $(0, 3)$. The line through $(-1, 5)$ and $(0, 3)$ has the required slope of -2.

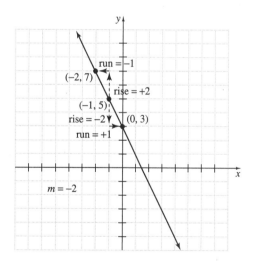

The slope -2 can also be written as $\frac{2}{-1}$, so to find another point we could start at $(-1, 5)$ and move two units up and then one unit left. We would stop at the point $(-2, 7)$. The line through $(-1, 5)$ and $(-2, 7)$ has the required slope and is the same line as shown previously through $(-1, 5)$ and $(0, 3)$.

As we have seen, the slope of a line is defined by two points on the line. Thus, if we know the equation of a line, we can find its slope by finding two of its points.

EXAMPLE 4 Find the slope of the line whose equation is $-2x + 3y = 12$.

Solution: To find the slope of the line defined by $-2x + 3y = 12$, find any two points on the line. Let's find and use intercept points as our two points.

IF $x = 0$	IF $y = 0$
$-2 \cdot 0 + 3y = 12$	$-2x + 3 \cdot 0 = 12$
$3y = 12$	$-2x = 12$
$y = 4$	$x = -6$

If $x = 0$, the corresponding y-value is 4. The y-intercept point is $(0, 4)$.

If $y = 0$, the corresponding x-value is -6. The x-intercept point is $(-6, 0)$.

Use the points $(0, 4)$ and $(-6, 0)$ to find the slope. Let (x_1, y_1) be $(0, 4)$ and (x_2, y_2) be $(-6, 0)$. Then

$$m = \frac{y_2 - y_1}{x_2 - x_1} = \frac{0 - 4}{-6 - 0} = \frac{-4}{-6} = \frac{2}{3}$$

The slope is $\frac{2}{3}$.

Next, we find the slopes of horizontal and vertical lines.

EXAMPLE 5 Find the slope of the line $y = -1$.

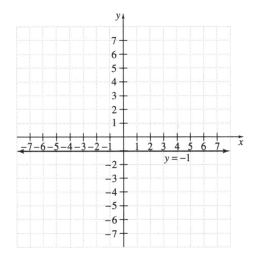

Solution: Recall that $y = -1$ is a horizontal line with y-intercept -1. To find the slope, find two ordered pair solutions of $y = -1$. Solutions of $y = -1$ must have a y-value of -1.

Let $(x_1, y_1) = (2, -1)$ and $(x_2, y_2) = (-3, -1)$. Then
$$m = \frac{y_2 - y_1}{x_2 - x_1} = \frac{-1 - (-1)}{-3 - 2} = \frac{0}{-5} = 0$$

The slope of the line $y = -1$ is 0. Since the y-values will have a difference of 0 for all horizontal lines, we can say that all **horizontal lines have a slope of 0.**

EXAMPLE 6 Find the slope of the line $x = 5$.

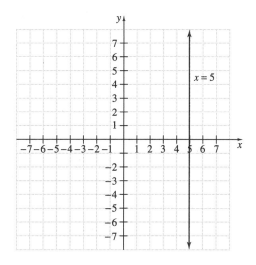

Solution: Recall that the graph of $x = 5$ is a vertical line with x-intercept 5.

To find the slope, find two ordered pair solutions of $x = 5$. Solutions of $x = 5$ must have an x-value of 5.

Let $(x_1, y_1) = (5, 0)$ and $(x_2, y_2) = (5, 4)$. Then
$$m = \frac{y_2 - y_1}{x_x - x_1} = \frac{4 - 0}{5 - 5} = \frac{4}{0}$$

Since $\frac{4}{0}$ is undefined, we say the slope of the vertical line $x = 5$ is undefined. Since the x-values will have a difference of 0 for all vertical lines, we can say that all **vertical lines have undefined slope.**

> **REMINDER** Slope of 0 and undefined slope are not the same. Vertical lines have undefined slope or no slope, while horizontal lines have a slope of 0.

Here is a general review of slope.

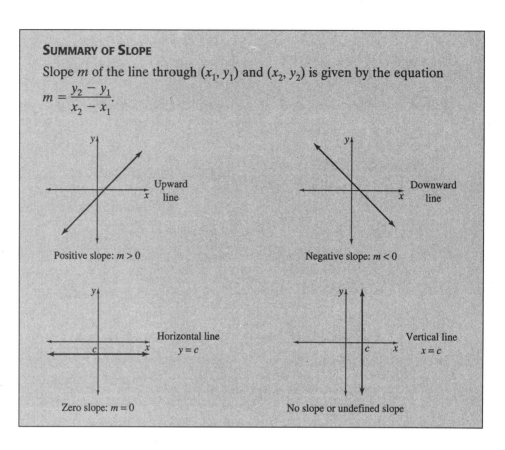

SUMMARY OF SLOPE
Slope m of the line through (x_1, y_1) and (x_2, y_2) is given by the equation $m = \dfrac{y_2 - y_1}{x_2 - x_1}$.

5 Two lines in the same plane are parallel if they do not intersect. Slopes of lines can help us determine whether lines are parallel. Parallel lines have the same steepness, so it follows that they have the same slope.

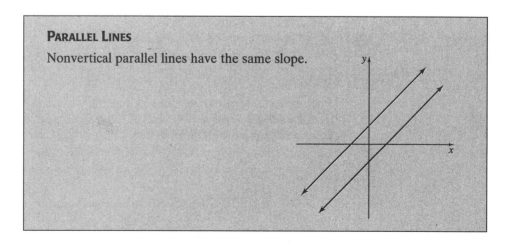

PARALLEL LINES
Nonvertical parallel lines have the same slope.

How do the slopes of perpendicular lines compare? Two lines that intersect at right angles are said to be **perpendicular.** The product of the slopes of two perpendicular lines is -1.

PERPENDICULAR LINES

If the product of the slopes of two lines is -1, then the lines are perpendicular.

(Two nonvertical lines are perpendicular if the slope of one is the negative reciprocal of the slope of the other.)

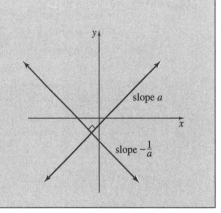

EXAMPLE 7 Is the line passing through the points $(-6, 0)$ and $(-2, 3)$ parallel to the line passing through the points $(5, 4)$ and $(7, 5)$?

Solution: To see if these lines are parallel, we find and compare slopes. The line passing through the points $(-6, 0)$ and $(-2, 3)$ has slope

$$m = \frac{3 - 0}{-2 - (-6)} = \frac{3}{4}$$

The line passing through the points $(5, 4)$ and $(7, 5)$ has slope

$$m = \frac{5 - 4}{7 - 5} = \frac{1}{2}$$

Since the slopes are not the same, these lines are not parallel.

EXAMPLE 8 Find the slope of a line perpendicular to the line passing through the points $(-1, 7)$ and $(2, 2)$.

Solution: First, let's find the slope of the line through $(-1, 7)$ and $(2, 2)$. This line has slope

$$m = \frac{2 - 7}{2 - (-1)} = \frac{-5}{3}$$

The slope of every line perpendicular to the given line has a slope equal to the negative reciprocal of

$$-\frac{5}{3} \quad \text{or} \quad -\left(-\frac{3}{5}\right) = \frac{3}{5}$$
$$\phantom{-\frac{5}{3} \quad \text{or} \quad }\uparrow \quad \uparrow$$
$$\phantom{-\frac{5}{3} \quad \text{or} \quad }\text{negative reciprocal}$$

The slope of a line perpendicular to the given line has slope $\frac{3}{5}$.

GRAPHING CALCULATOR EXPLORATIONS

It is possible to use a grapher and sketch the graph of more than one equation on the same set of axes. This feature can be used to confirm our findings from Section 3.2 when we learned that the graph of an equation written in the form $y = mx + b$ has a y-intercept of b. For example, graph the equations $y = \frac{2}{5}x$, $y = \frac{2}{5}x + 7$, and $y = \frac{2}{5}x - 4$ on the same set of axes. To do so, press the $\boxed{Y=}$ key and enter the equations on the first three lines.

$$Y_1 = \left(\frac{2}{5}\right)x$$

$$Y_2 = \left(\frac{2}{5}\right)x + 7$$

$$Y_3 = \left(\frac{2}{5}\right)x - 4$$

The screen should look like:

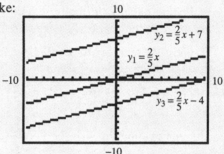

Notice that all three graphs appear to have the same positive slope. The graph of $y = \frac{2}{5}x + 7$ is the graph of $y = \frac{2}{5}x$ moved 7 units upward with a y-intercept of 7. Also, the graph of $y = \frac{2}{5}x - 4$ is the graph of $y = \frac{2}{5}x$ moved 4 units downward with a y-intercept of -4.

Graph the equations on the same set of axes. Describe the similarities and differences in their graphs.

1. $y = 3.8x$, $y = 3.8x - 3$, $y = 3.8x + 9$
2. $y = -4.9x$, $y = -4.9x + 1$, $y = -4.9x + 8$
3. $y = \frac{1}{4}x$; $y = \frac{1}{4}x + 5$, $y = \frac{1}{4}x - 8$
4. $y = -\frac{3}{4}x$, $y = -\frac{3}{4}x - 5$, $y = -\frac{3}{4}x + 6$

MENTAL MATH

Decide whether a line with the given slope is upward, downward, horizontal or vertical.

1. $m = \dfrac{7}{6}$
2. $m = -3$
3. $m = 0$
4. m is undefined.

EXERCISE SET 3.4

Find the slope of each line if it exists. See Example 1.

1.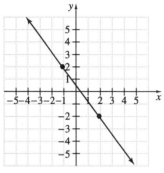

2.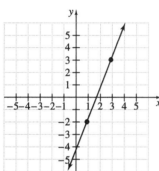

3.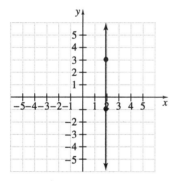

4.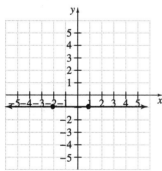

5.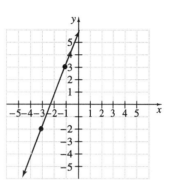

6.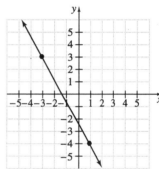

Find the slope of the line that goes through the given points. See Examples 1 and 2.

7. $(0, 0)$ and $(7, 8)$
8. $(-1, 5)$ and $(0, 0)$
9. $(-1, 5)$ and $(6, -2)$
10. $(-1, 9)$ and $(-3, 4)$
11. $(1, 4)$ and $(5, 3)$
12. $(3, 1)$ and $(2, 6)$
13. $(-4, 3)$ and $(-4, 5)$
14. $(6, -6)$ and $(6, 2)$
15. $(-2, 8)$ and $(1, 6)$
16. $(4, -3)$ and $(2, 2)$
17. $(1, 0)$ and $(1, 1)$
18. $(0, 13)$ and $(-4, 13)$
19. $(5, -11)$ and $(1, -11)$
20. $(5, 4)$ and $(0, 5)$

For each graph, determine which line has the greater slope.

21.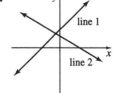

22.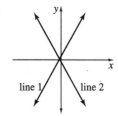

23. 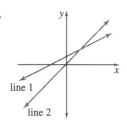 **24.**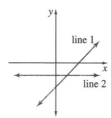

35. Through $(-2, -7)$, with slope -1.

36. Through $(-4, -3)$, with slope -2.

37. Through $(-3, 4)$, with slope $-\dfrac{3}{5}$.

38. Through $(6, -2)$, with slope $-\dfrac{1}{3}$.

Match each line with its slope.

A. $m = 0$ **B.** no slope **C.** $m = 3$
D. $m = 1$ **E.** $m = -\dfrac{1}{2}$ **F.** $m = -\dfrac{3}{4}$

Find the slope of each line. See Examples 4–6.

39. $y = 5x - 2$ **40.** $y = -2x + 6$
41. $2x + y = 7$ **42.** $-5x + y = 10$
43. $x = 1$ **44.** $y = -2$
45. $y = -3$ **46.** $x = 5$
47. $2x - 3y = 10$ **48.** $-3x - 4y = 6$
49. $x = 2y$ **50.** $x = -4y$

25. 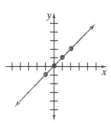 **26.**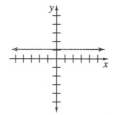

The pitch of a roof is its slope. Find the pitch of each roof shown.

51.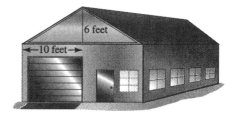

27. 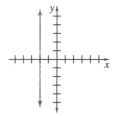 **28.**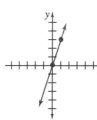

52.

29. 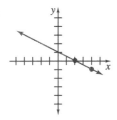 **30.**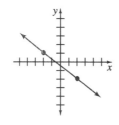

53. Find x so that the pitch of the roof is $\dfrac{1}{3}$.

Graph each line passing through the given point with the given slope. See Example 3.

31. Through $(2, 3)$, with slope $\dfrac{1}{4}$.

32. Through $(-1, 5)$, with slope $\dfrac{2}{3}$.

33. Through $(0, -4)$, with slope 3.

34. Through $(3, 1)$, with slope 4.

54. Find x so that the pitch of the roof is $\frac{2}{5}$.

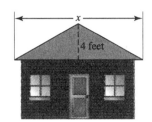

The grade of a road is its slope written as a percent. Find the grade of the road shown.

55.

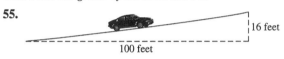

56.

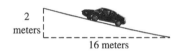

Find the slope of the line a. parallel and b. perpendicular to the line through each pair of points. See Examples 7 and 8.

57. $(-3, -3)$ and $(0, 0)$

58. $(6, -2)$ and $(1, 4)$

59. $(-8, -4)$ and $(3, 5)$

60. $(6, -1)$ and $(-4, -10)$

Determine whether the lines through each pair of points are parallel, perpendicular, or neither. See Examples 7 and 8.

61. $(0, 6)$ and $(-2, 0)$
$(0, 5)$ and $(1, 8)$

62. $(1, -1)$ and $(-1, -11)$
$(-2, -8)$ and $(0, 2)$

63. $(2, 6)$ and $(-2, 8)$
$(0, 3)$ and $(1, 5)$

64. $(-1, 7)$ and $(1, 10)$
$(0, 3)$ and $(1, 5)$

65. $(3, 6)$ and $(7, 8)$
$(0, 6)$ and $(2, 7)$

66. $(4, 2)$ and $(6, 6)$
$(0, -2)$ and $(1, 0)$

67. $(2, -3)$ and $(6, -5)$
$(5, -2)$ and $(-3, -4)$

68. $(-1, -5)$ and $(4, 4)$
$(10, 8)$ and $(-7, 4)$

69. $(-4, -3)$ and $(-1, 0)$
$(4, -4)$ and $(0, 0)$

70. $(2, -2)$ and $(4, -8)$
$(0, 7)$ and $(3, 8)$

71. Show that a triangle with vertices at the points $(2, 5)$, $(-4, 4)$, and $(-3, 0)$ is a right triangle.

72. Show that the quadrilateral with vertices $(1, 3)$, $(2, 1)$, $(-4, 0)$, and $(-3, -2)$ is a parallelogram.

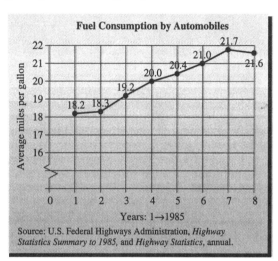

The following bar graph shows average miles per gallon for cars during the years shown.

73. In 1988, cars averaged how many miles per gallon of fuel?

74. In 1991, cars averaged how many miles per gallon of fuel?

75. Find the increase of average miles per gallon for automobiles for the years 1985 to 1992.

76. Do you notice any trends from this graph?

77. What line segment has the greatest slope?

78. What line segment has the least slope?

79. What year showed the greatest increase in miles per gallon?

80. What year showed the least increase in miles per gallon?

81. Find the slope of a line parallel to the line passing through $(-7, -5)$ and $(-2, -6)$.

82. Find the slope of a line parallel to the line passing through the origin and $(-2, 10)$.

83. Find the slope of a line perpendicular to the line passing through the origin and $(1, -3)$.

84. Find the slope of the line perpendicular to the line passing through $(-1, 3)$ and $(2, -8)$.

85. Find the slope of a line parallel to the line $y = x$.

86. Find the slope of a line parallel to the line $x + 2y = 6$.

Find the slope of the line through the given points.

87. $(2.1, 6.7)$ and $(-8.3, 9.3)$

88. $(-3.8, 1.2)$ and $(-2.2, 4.5)$

89. $(2.3, 0.2)$ and $(7.9, 5.1)$

90. $(14.3, -10.1)$ and $(9.8, -2.9)$

91. Find the slope and the y-intercept of the line defined by $x + 3y = 6$. Next, solve the equation $x + 3y = 6$ for y. Compare the slope and the y-intercept with the equation solved for y. Write down any observations.

92. Find the slope and the y-intercept of the line defined by $5x + 2y = 7$. Next, solve the equation $5x + 2y = 7$ for y. Compare the slope and the y-intercept with the equation solved for y. Write down any observations.

93. The graph of $y = \frac{1}{2}x$ has a slope of $\frac{1}{2}$. The graph of $y = 3x$ has a slope of 3. The graph of $y = 5x$ has a slope of 5. Graph all three equations on a single coordinate system. As slope becomes larger, how does the steepness of the line change?

94. The graph of $y = -\frac{1}{3}x + 2$ has a slope of $-\frac{1}{3}$. The graph of $y = -2x + 2$ has a slope of -2. The graph of $y = -4x + 2$ has a slope of -4. Graph all three equations on a single coordinate system. As the absolute value of the slope becomes larger, how does the steepness of the line change?

Review Exercises

Solve each linear inequality. See Section 2.9.

95. $-3x \leq -9$ **96.** $-x > -16$

97. $\dfrac{x - 6}{2} < 3$ **98.** $\dfrac{2x + 1}{3} \geq -1$

Graph each linear equation in two variables. See Sections 3.2 and 3.3.

99. $x - y = 6$ **100.** $x = -2y$

101. $y = 3x$ **102.** $5x + 3y = 15$

103. $x = -2$ **104.** $y = 5$

3.5 GRAPHING LINEAR INEQUALITIES

OBJECTIVES

1. Determine whether an ordered pair is a solution of a linear inequality in two variables.
2. Graph a linear inequality in two variables.

TAPE
BA 3.5

1 Recall that a linear equation in two variables is an equation that can be written in the form $Ax + By = C$ where A, B, and C are real numbers and A and B are not both 0. The definition of a linear inequality is the same except that the equal sign is replaced with an inequality sign.

A **linear inequality in two variables** is an inequality that can be written in one of the forms:

$$Ax + By < C \qquad Ax + By \leq C$$
$$Ax + By > C \qquad Ax + By \geq C$$

where A, B, and C are real numbers and A and B are not both 0. Just as for linear equations in x and y, an ordered pair is a **solution** of an inequality in x and y if replacing the variables by coordinates of the ordered pair results in a true statement.

EXAMPLE 1 Determine whether each ordered pair is a solution of the equation $2x - y < 6$.

    a. $(5, -1)$ **b.** $(2, 7)$ **c.** $(0, -6)$

Solution: **a.** Replace x with 5 and y with -1 and see if a true statement results.

$$2x - y < 6$$
$$2(5) - (-1) < 6 \quad \text{Replace } x \text{ with 5 and } y \text{ with } -1.$$
$$10 + 1 < 6$$
$$11 < 6 \quad \text{False.}$$

The ordered pair $(5, -1)$ is not a solution since $11 < 6$ is a false statement.

b. Replace x with 2 and y with 7 and see if a true statement results.

$$2x - y < 6$$
$$2(2) - 7 < 6 \quad \text{Replace } x \text{ with 2 and } y \text{ with 7.}$$
$$4 - 7 < 6$$
$$-3 < 6 \quad \text{True.}$$

The ordered pair $(2, 7)$ is a solution since $-3 < 6$ is a true statement.

c. Replace x with 0 and y with -6 and see if a true statement results.

$$2x - y < 6$$
$$2(0) - (-6) < 6 \quad \text{Replace } x \text{ with 0 and } y \text{ with } -6$$
$$0 + 6 < 6$$
$$6 < 6 \quad \text{False.}$$

The ordered pair $(0, -6)$ is not a solution since $6 < 6$ is a false statement.

The linear equation $x + y = 5$ is graphed next. Recall that all points on the line correspond to ordered pairs that satisfy the equation $x + y = 5$. It can be shown that all the points above the line $x + y = 5$ have coordinates that satisfy the inequality $x + y > 5$. Similarly, all points below the line have coordinates that satisfy the inequality $x + y < 5$.

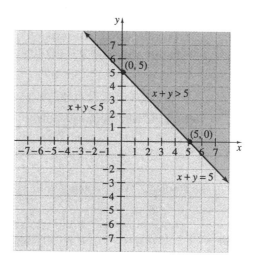

The region above the line and the region below the line are called **half-planes.** Every line divides the plane (similar to a sheet of paper extending indefinitely in all directions) into two half-planes; the line is called the **boundary.**

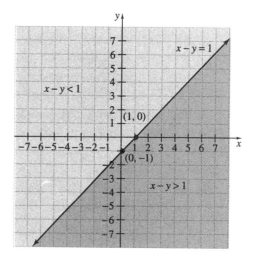

Similarly, the half-plane above the boundary $x - y = 1$ is the graph of the inequality $x - y < 1$. The half-plane below the boundary $x - y = 1$ is the graph of the inequality $x - y > 1$. Since the inequality $x - y \leq 1$ means

$$x - y = 1 \text{ or } x - y < 1$$

the graph of $x - y \leq 1$ is the half-plane $x - y < 1$ along with the boundary line $x - y = 1$.

The steps to graph a linear inequality are given next.

To Graph a Linear Inequality in Two Variables

Step 1. Graph the boundary line found by replacing the inequality sign with an equal sign. If the inequality sign is > or <, graph a dashed boundary line indicating that the points on the line are not solutions of the inequality. If the inequality sign is ≥ or ≤, graph a solid boundary line indicating that the points on the line are solutions of the inequality.

Step 2. Choose a point, not on the boundary line, as a test point. Substitute the coordinates of this test point into the original inequality.

Step 3. If a true statement is obtained in step 2, shade the half-plane that contains the test point. If a false statement is obtained, shade the half-plane that does not contain the test point.

EXAMPLE 2 Graph $x + y < 7$.

Solution: First, graph the boundary line by graphing the equation $x + y = 7$. Graph this boundary as a dashed line because the inequality sign is <, and thus the points on the line are not solutions of the inequality $x + y < 7$.

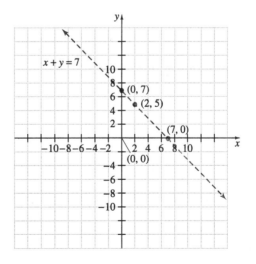

Next, choose a test point, being careful not to choose a point on the boundary line. We choose (0, 0). Substitute the coordinates of (0, 0) into $x + y < 7$.

$x + y < 7$ Original inequality.
$0 + 0 < 7$ Replace x with 0 and y with 0.
$0 < 7$ True.

Since the result is a true statement, (0, 0) is a solution of $x + y < 7$, and every point in the same half-plane as (0, 0) is also a solution. To indicate this, shade the entire half-plane containing (0, 0), as shown.

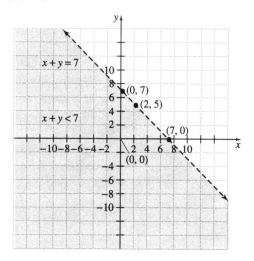

EXAMPLE 3 Graph $2x - y \geq 3$.

Solution: Graph the boundary line by graphing $2x - y = 3$. Draw this line as a solid line since the inequality sign is $\geq$, and thus the points on the line are solutions of $2x - y \geq 3$. Once again, (0, 0) is a convenient test point since it is not on the boundary line.

Substitute 0 for x and 0 for y into the **original inequality.**

$$2x - y \geq 3$$
$$2(0) - 0 \geq 3 \quad \text{Let } x = 0 \text{ and } y = 0.$$
$$0 \geq 3 \quad \text{False.}$$

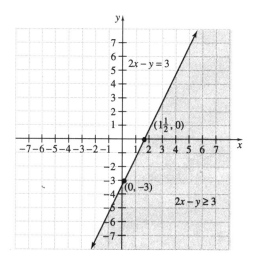

Since the statement is false, no point in the half-plane containing (0, 0) is a solution. Shade the half-plane that does not contain (0, 0). Every point in the shaded half-plane and every point on the boundary line satisfies $2x - y \geq 3$.

REMINDER When graphing an inequality, make sure the test point is substituted in the **original inequality.** For example, when graphing $x + y < 3$, test (0, 0) in $x + y < 3$, not $x + y = 3$. Since (0, 0) is a solution of $x + y < 3$, shade the half-plane containing (0, 0) as shown.

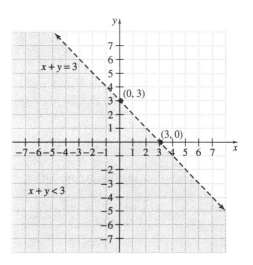

EXAMPLE 4 Graph $x > 2y$.

Solution: Find the boundary line by graphing $x = 2y$. The boundary line is a dashed line since the inequality symbol is $>$. We cannot use (0, 0) as a test point because it is a point on the boundary line. Choose (0, 2) as the test point.

$x > 2y$
$0 > 2(2)$ Let $x = 0$ and $y = 2$.
$0 > 4$ False.

Since the statement is false, shade the half-plane that does not contain the test point (0, 2).

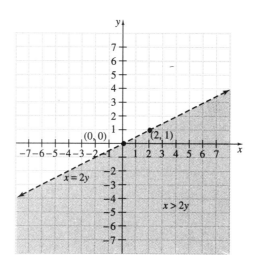

EXAMPLE 5 Graph $5x + 4y \leq 20$.

Solution: Graph the solid boundary line $5x + 4y = 20$. Choose $(0, 0)$ as the test point.

$$5x + 4y \leq 20$$
$$5(0) + 4(0) \leq 20 \quad \text{Let } x = 0 \text{ and } y = 0.$$
$$0 \leq 20 \quad \text{True.}$$

Shade the half-plane that contains $(0, 0)$. The graph of $5x + 4y \leq 20$ is given next.

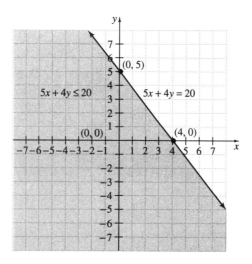

EXAMPLE 6 Graph $y > 3$.

Solution: Graph the dashed boundary line $y = 3$. Recall that the graph of $y = 3$ is a horizontal line with y-intercept 3. Choose $(0, 0)$ as the test point.

$y > 3$
$0 > 3$ Let $y = 0$.
$0 > 3$ False.

Shade the half-plane that does not contain $(0, 0)$. The graph of $y > 3$ is shown next.

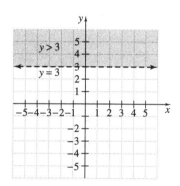

MENTAL MATH

State whether the graph of each inequality includes its corresponding boundary line.

1. $y \geq x + 4$ **2.** $x - y > -7$ **3.** $y \geq x$ **4.** $x > 0$

Decide whether $(0, 0)$ is a solution of each given inequality.

5. $x + y > -5$ **6.** $2x + 3y < 10$ **7.** $x - y \leq -1$ **8.** $\frac{2}{3}x + \frac{5}{6}y > 4$

EXERCISE SET 3.5

Determine which ordered pairs given are solutions of the linear inequality in two variables. See Example 1.

1. $x - y > 3$; $(0, 3), (2, -1), (5, 1)$
2. $y - x < -2$; $(2, 1), (5, -1), (3, 7)$
3. $3x - 5y \leq -4$; $(2, 3), (-1, -1), (4, 0)$
4. $2x + y \geq 10$; $(0, 11), (-1, -4), (5, 0)$
5. $x < -y$; $(6, 6), (0, 2), (-5, 1)$
6. $y > 3x$; $(0, 0), (1, 4), (-1, -4)$

Graph each inequality. See Examples 2 through 6.

7. $x + y \leq 1$
8. $x + y \geq -2$
9. $2x + y > -4$
10. $x + 3y \leq 3$
11. $x + 6y \leq -6$
12. $7x + y > -14$
13. $2x + 5y > -10$
14. $5x + 2y \leq 10$
15. $x + 2y \leq 3$
16. $2x + 3y > -5$
17. $2x + 7y > 5$
18. $3x + 5y \leq -2$
19. $x - 2y \geq 3$
20. $4x + y \leq 2$

216 CHAPTER 3   GRAPHING

21. $5x + y < 3$
22. $x + 2y > -7$
23. $4x + y < 8$
24. $9x + 2y \geq -9$
25. $y \geq 2x$
26. $x < 5y$
27. $x \geq 0$
28. $y \leq 0$
29. $y \leq -3$
30. $x > -\dfrac{2}{3}$
31. $2x - 7y > 0$
32. $5x + 2y \leq 0$
33. $3x - 7y \geq 0$
34. $-2x - 9y > 0$
35. $x > y$
36. $x \leq -y$
37. $x - y \leq 6$
38. $x - y > 10$
39. $-\dfrac{1}{4}y + \dfrac{1}{3}x > 1$
40. $\dfrac{1}{2}x - \dfrac{1}{3}y \leq -1$
41. $-x < 0.4y$
42. $0.3x \geq 0.1y$

 43. Write an inequality whose solutions are all pairs of numbers x and y whose sum is at least 13. Graph the inequality.

 44. Write an inequality whose solutions are all the pairs of numbers x and y whose sum is at most -4. Graph the inequality.

Match each inequality with its graph.

  a. $x > 2$ **b.** $y < 2$ **c.** $y \leq 2x$ **d.** $y \leq -3x$
  e. $2x + 3y < 6$ **f.** $3x + 2y > 6$

45.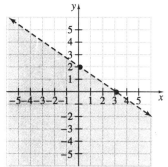

46.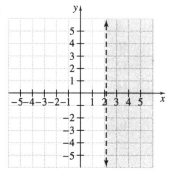

47.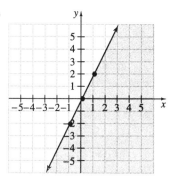

48.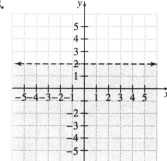

49.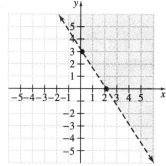

50.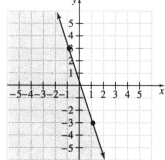

51. Explain why a point on the boundary line should not be chosen as the test point.

52. Describe the graph of a linear inequality.

Review Exercises

Evaluate. See Section 1.3.

53. 2^3

54. 3^4

55. $(-2)^5$

56. -2^5

57. $3 \cdot 4^2$

58. $4 \cdot 3^3$

Evaluate each expression for the given replacement value. See Section 1.7.

59. x^2 if x is -5

60. x^3 if x is -5

61. $2x^3$ if x is -1

62. $3x^2$ if x is -1

GROUP ACTIVITY

FINANCIAL ANALYSIS

OPTIONAL MATERIALS:
- Financial magazines
- Annual reports

The table below gives the sales in millions of dollars for the leading U.S. businesses in the aerospace industry for the years 1993 and 1994. You have been asked to analyze the performances of these companies and, based on this information alone, make an investment recommendation.

AEROSPACE SALES (IN MILLIONS OF DOLLARS)

COMPANY	1993	1994
Boeing	25,300	21,900
United Technologies	20,700	21,200
McDonnell-Douglas	14,500	13,200
Lockheed	13,100	13,100
Allied-Signal	11,800	12,800
Martin Marietta	9,400	9,900
Textron	8,700	9,700
General Dynamics	4,700	3,700

Source: *FORTUNE* magazine

1. Write the data for each company as two ordered pairs in the form (year, sales).

2. Assuming that the trends in the sales are linear, graph the line represented by the ordered pairs for each company. Describe the trend shown by each graph.

2. Find the slope of the line for each company.

3. Which of the lines, if any, have negative slopes? What does that mean in this context? Which of the lines, if any, have zero slopes? What does that mean in this context? Which of the lines, if any, are parallel? What does that mean in this context? Which, if any, of the lines are perpendicular? What does that mean in this context?

4. Of these aerospace industry companies, which one(s) would you recommend as an investment choice? Why?

5. Do you think it is wise to make a decision after looking at only 2 years of sales? What other factors do you think should be taken into consideration when making an investment choice?

6. (Optional) Use financial magazines and/or company annual reports to find sales or revenue information for two different years for two to four companies in the same industry. Analyze the sales and make an investment recommendation.

Chapter 3 Highlights

Definitions and Concepts	Examples
Section 3.1 The Rectangular Coordinate System	
The **rectangular coordinate system** consists of a plane and a vertical and a horizontal number line intersecting at their 0 coordinate. The vertical number line is called the **y-axis** and the horizontal number line is called the **x-axis.** The point of intersection of the axes is called the **origin.** To **plot** or **graph** an ordered pair means to find its corresponding point on a rectangular coordinate system. To plot or graph an ordered pair such as $(3, -2)$, start at the origin. Move 3 units to the right and from there, 2 units down. To plot or graph $(-3, 4)$ start at the origin. Move 3 units to the left and from there, 4 units up.	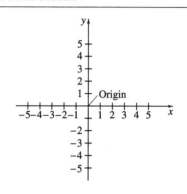 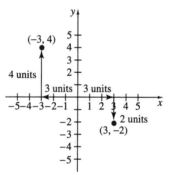
An ordered pair is a **solution** of an equation in two variables if replacing the variables by the coordinates of the ordered pair results in a true statement.	Determine whether $(-1, 5)$ is a solution to $2x + 3y = 13$. $2x + 3y = 13$ $2(-1) + 3 \cdot 5 = 13$ Let $x = -1, y = 5$ $-2 + 15 = 13$ $13 = 13$ True.
If one coordinate of an ordered pair solution is known, the other value can be determined by substitution.	Complete the ordered pair $(0,)$ for the equation $x - 6y = 12$. $x - 6y = 12$ $0 - 6y = 12$ Let $x = 0$. $\dfrac{-6y}{-6} = \dfrac{12}{-6}$ Divide by -6. $y = -2$ The ordered pair solution is $(0, -2)$.

DEFINITIONS AND CONCEPTS	EXAMPLES

SECTION 3.2 GRAPHING LINEAR EQUATIONS

A **linear equation in two variables** is an equation that can be written in the form $Ax + By = C$ where A and B are not both 0. The form $Ax + By = C$ is called **standard form.**

To graph a linear equation in two variables, find three ordered pair solutions. Plot the solution points and draw the line connecting the points.

LINEAR EQUATIONS

$3x + 2y = -6$ $x = -5$
$y = 3$ $y = -x + 10$

$x + y = 10$ is in standard form.

Graph $x - 2y = 5$

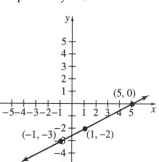

x	y
5	0
1	−2
−1	−3

SECTION 3.3 INTERCEPTS

An **intercept point** of a graph is a point where the graph intersects an axis. If a graph intersects the x-axis at a, then a is the **x-intercept** and the corresponding intercept point is $(a, 0)$. If a graph intersects the y-axis at b, then b is the **y-intercept** and the corresponding intercept point is $(0, b)$.

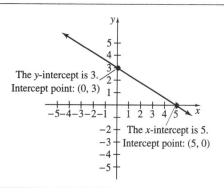

To find the x-intercept, let $y = 0$ and solve for x.
To find the y-intercept, let $x = 0$ and solve for y.

Graph $2x - 5y = -10$ by finding intercepts.

If $y = 0$, then
$2x - 5 \cdot 0 = -10$
$2x = -10$
$\dfrac{2x}{2} = \dfrac{-10}{2}$
$x = -5$

The x-intercept is -5.
Intercept point: $(-5, 0)$.

If $x = 0$, then
$2 \cdot 0 - 5y = -10$
$-5y = -10$
$\dfrac{-5y}{-5} = \dfrac{-10}{-5}$
$y = 2$

The y-intercept is 2.
Intercept point: $(0, 2)$.

(continued)

DEFINITIONS AND CONCEPTS	EXAMPLES
SECTION 3.3 INTERCEPTS	
	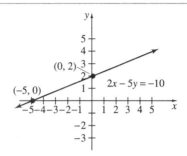
The graph of $x = c$ is a vertical line with x-intercept c.	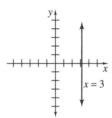
The graph of $y = c$ is a horizontal line with y-intercept c.	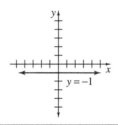
SECTION 3.4 SLOPE	
The **slope** m of the line through points (x_1, y_1) and (x_2, y_2) is given by $$m = \frac{y_2 - y_1}{x_2 - x_1} \quad \text{as long as } x_2 \neq x_1$$	The slope of the line through points $(-1, 6)$ and $(-5, 8)$ is $$m = \frac{y_2 - y_1}{x_2 - x_1} = \frac{8 - 6}{-5 - (-1)} = \frac{2}{-4} = -\frac{1}{2}$$
A horizontal line has slope 0.	The slope of the line $y = -5$ is 0.
The slope of a vertical line is undefined.	The line $x = 3$ has undefined slope.
Nonvertical parallel lines have the same slope.	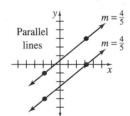

(continued)

DEFINITIONS AND CONCEPTS	EXAMPLES
SECTION 3.4 SLOPE	
Two nonvertical lines are perpendicular if the slope of one is the negative reciprocal of the slope of the other.	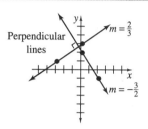 Perpendicular lines
SECTION 3.5 GRAPHING LINEAR INEQUALITIES	
A **linear inequality in two variables** is an inequality that can be written in one of the forms: $Ax + By < C \qquad Ax + By \leq C$ $Ax + By > C \qquad Ax + By \geq C$	LINEAR INEQUALITIES $2x - 5y < 6 \qquad x \geq -5$ $y > -8x \qquad y \leq 2$
To graph a linear inequality 1. Graph the boundary line by graphing the related equation. Draw the line solid if the inequality symbol is $\leq$ or $\geq$. Draw the line dashed if the inequality symbol is $<$ or $>$. 2. Choose a test point not on the line. Substitute its coordinates into the original inequality. 3. If the resulting inequality is true, shade the half-plane that contains the test point. If the inequality is not true, shade the half-plane that does not contain the test point.	Graph $2x - y \leq 4$. 1. Graph $2x - y = 4$. Draw a solid line because the inequality symbol is $\leq$. 2. Check the test point $(0, 0)$ in the inequality $2x - y \leq 4$. $$2 \cdot 0 - 0 \leq 4 \qquad \text{Let } x = 0 \text{ and } y = 0.$$ $$0 \leq 4 \qquad \text{True.}$$ 3. The inequality is true so we shade the half-plane containing $(0, 0)$. 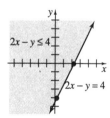

CHAPTER 3 REVIEW

(3.1) *Plot the following ordered pairs on a Cartesian coordinate system.*

1. $(-7, 0)$
2. $\left(0, 4\frac{4}{5}\right)$
3. $(-2, -5)$
4. $(1, -3)$
5. $(0.7, 0.7)$
6. $(-6, 4)$

Determine whether each ordered pair is a solution of the given equation.

7. $7x - 8y = 56$; $(0, 56)$, $(8, 0)$
8. $-2x + 5y = 10$; $(-5, 0)$, $(1, 1)$
9. $x = 13$; $(13, 5)$, $(13, 13)$
10. $y = 2$; $(7, 2)$, $(2, 7)$

Complete the ordered pairs so that each is a solution of the given equation.

11. $-2 + y = 6x$; $(7,)$
12. $y = 3x + 5$; $(, -8)$

Complete the table of values for each given equation; then plot the ordered pairs. Use a single coordinate system for each exercise.

13. $9 = -3x + 4y$

x	y
	0
	3
9	

14. $y = 5$

x	y
7	
-7	
0	

15. $x = 2y$

x	y
	0
	5
	-5

16. The cost in dollars of producing x compact disk holders is given by $y = 5x + 2000$.

   a. Complete the following table.

x	1	100	1000
y			

   b. Find the number of compact disk holders that can be produced for $6430.

(3.2) *Graph each linear equation.*

17. $x - y = 1$
18. $x + y = 6$
19. $x - 3y = 12$
20. $5x - y = -8$
21. $x = 3y$
22. $y = -2x$
23. $2x - 3y = 6$
24. $4x - 3y = 12$

(3.3) *Identify the intercepts and intercept points.*

25.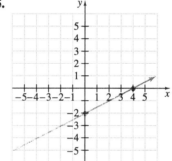

26.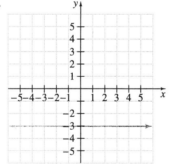

27.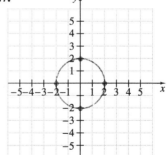

28.

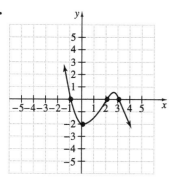

Graph each linear equation by finding its intercepts.

29. $x - 3y = 12$ **30.** $-4x + y = 8$
31. $y = -3$ **32.** $x = 5$
33. $y = -3x$ **34.** $x = 5y$
35. $x - 2 = 0$ **36.** $y + 6 = 0$

(3.4) *Find the slope of each line.*

37.

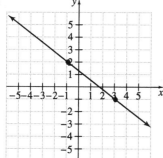

38.

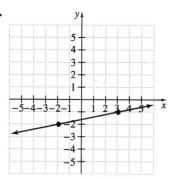

Match each line with its slope.

a.
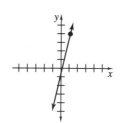

b.
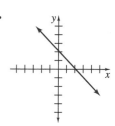

c.
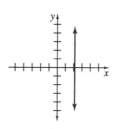

d.
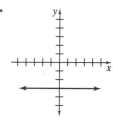

e.
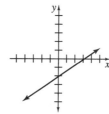

39. $m = 0$ **40.** $m = -1$
41. no slope   **42.** $m = 4$
43. $m = \dfrac{2}{3}$

Find the slope of the line that goes through the given points.

44. $(2, 5)$ and $(6, 8)$ **45.** $(4, 7)$ and $(1, 2)$
46. $(1, 3)$ and $(-2, -9)$ **47.** $(-4, 1)$ and $(3, -6)$

Find the slope of each line.

48. $y = 3x + 7$ **49.** $x - 2y = 4$
50. $y = -2$ **51.** $x = 0$

Graph each line passing through the given point with the given slope.

52. through $(0, 3)$ with slope $\frac{1}{4}$
53. through $(-5, 3)$ with slope -3

Determine whether the lines through the pairs of points are parallel, perpendicular, or neither.

54. $(-3, 1)$ and $(1, -2)$ **55.** $(-7, 6)$ and $(0, 4)$
     $(2, 4)$ and $(6, 1)$ $(-9, -3)$ and $(1, 5)$

56. (9, 10) and (8, −7)
(−1, −3) and (2, −8)

57. (−1, 3) and (3, −2)
(−2, −2) and (3, 2)

(3.5) *Graph the following inequalities.*

58. $3x - 4y \leq 0$

59. $3x - 4y \geq 0$

60. $x + 6y < 6$

61. $x + y > -2$

62. $y \geq -7$

63. $y \leq -4$

64. $-x \leq y$

65. $x \geq -y$

CHAPTER 3 TEST

Determine whether the ordered pairs are solutions of the equations.

1. $x - 2y = 3$; (1, 1)

2. $2x + 3y = 6$; (0, −2)

Complete the ordered pair solution for the following equations.

3. $12y - 7x = 5$; (1,)

4. $y = 17$; (−4,)

Find the slopes of the following lines.

5.
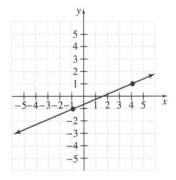

6.
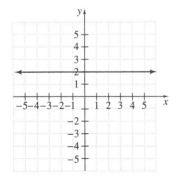

7. Through (6, −5) and (−1, 2)

8. Through (0, −8) and (−1, −1)

9. $-3x + y = 5$

10. $x = 6$

Determine whether the lines through the pairs of points are parallel, perpendicular, or neither.

11. (−1, 3) and (1, −3)
(2, −1) and (4, −7)

12. (−6, −6) and (−1, −2)
(−4, 3) and (3, −3)

Graph the following.

13. $2x + y = 8$

14. $-x + 4y = 5$

15. $x - y \geq -2$

16. $y \geq -4x$

17. $5x - 7y = 10$

18. $2x - 3y > -6$

19. $6x + y > -1$

20. $y = -1$

21. $x - 3 = 0$

22. $5x - 3y = 15$

Write each statement as an equation in two variables. Then graph the equation.

23. The *y*-value is 1 more than twice the *x*-value.

24. The *x*-value added to four times the *y*-value is less than −4.

25. The perimeter of the parallelogram below is 42 meters. Write a linear equation in two variables for the perimeter. Use this equation to find *x* when *y* is 8.

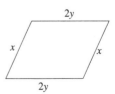

Chapter 3 Cumulative Review

1. Insert <, >, or = in the space between the paired numbers to make each statement true.

   a. −1 0   b. 7 $\frac{14}{2}$ c. −5 −6

2. Write each fraction in lowest terms.

   a. $\frac{42}{49}$ b. $\frac{11}{27}$ c. $\frac{88}{20}$

3. Simplify $\frac{8 + 2 \cdot 3}{2^2 - 1}$.

4. Write an algebraic expression that represents each of the following phrases. Let the variable x represent the unknown number.

   a. The sum of a number and 3
   b. The product of 3 and a number
   c. Twice a number
   d. 10 decreased by a number
   e. 5 times a number increased by 7

5. If $x = 2$ and $y = -5$, evaluate the following expressions.

   a. $\frac{x - y}{12 + x}$ b. $x^2 - y$

6. Find each quotient.

   a. $\frac{-24}{-4}$ b. $\frac{-36}{3}$ c. $\frac{2}{3} \div \left(-\frac{5}{4}\right)$

7. Find each product by using the distributive property to remove parentheses.

   a. $5(x + 2)$ b. $-2(y + 0.3z - 1)$
   c. $-(x + y - 2z + 6)$

8. Solve $-5(2a - 1) - (-11a + 6) = 7$ for a.

9. Solve $\frac{y}{7} = 20$ for y.

10. Solve $.25x + .10(x - 3) = .05(x + 18)$.

11. Twice the sum of a number and 4 is the same as four times the number decreased by 12. Find the number.

12. Charles Pecot can afford enough fencing to enclose a rectangular garden with a perimeter of 140 feet. If the width of his garden is to be 30 feet, find the length.

30 feet

13. Find 72% of 200.

14. Solve $-4x + 7 \geq -9$, and graph the solution set.

15. On a single coordinate system, plot the ordered pairs. State in which quadrant, if any, each point lies.

    a. $(3, 2)$ b. $(-2, -4)$ c. $(1, -2)$ d. $(-5, 3)$
    e. $(0, 0)$ f. $(0, 2)$ g. $(-5, 0)$ h. $\left(0, -1\frac{1}{2}\right)$

16. Graph the linear equation $2x + y = 5$.

17. Complete the table for the equation $y = 3x$.

x	y
−1	
	0
	−9

18. Graph $x = 2$.

19. Find the slope of the line through $(-1, 5)$ and $(2, -3)$. Graph the line.

20. Graph the line through $(-1, 5)$ with slope -2.

21. Find the slope of the line $y = -1$

22. Graph $x + y < 7$.

CHAPTER

4

4.1 Exponents
4.2 Adding and Subtracting Polynomials
4.3 Multiplying Polynomials
4.4 Special Products
4.5 Negative Exponents and Scientific Notation
4.6 Division of Polynomials

Exponents and Polynomials

Making Predictions Based on Historical Data

Accidental deaths include those due to motor vehicle accidents, poisonings, drownings, and fires. To help arrange for equipment/personnel resources and program funding for 911 service, it would be helpful for planners to be able to predict the number of accidental deaths that might occur in a given year. The first step in predicting such information is using the trends in historical data to model what the future might bring.

In the Chapter Group Activity on page 273, you will have the opportunity to investigate trends in polynomials representing historical data on the annual numbers of accidental deaths and then use that data and related polynomials to make predictions for the future.

228 CHAPTER 4 EXPONENTS AND POLYNOMIALS

Recall from Chapter 1 that an exponent is a shorthand notation for repeated factors. This chapter explores additional concepts about exponents and exponential expressions. An especially useful type of exponential expression is a polynomial. Polynomials model many real-world phenomena. Our goal in this chapter is to become proficient with operations on polynomials.

4.1 EXPONENTS

TAPE BA 4.1

OBJECTIVES

1. Evaluate exponential expressions.
2. Use the product rule for exponents.
3. Use the power rule for exponents.
4. Use the power rules for products and quotients.
5. Use the quotient rule for exponents, and define a number raised to the 0 power.

As we reviewed in Section 1.3, an exponent is a shorthand notation for repeated factors. For example, $2 \cdot 2 \cdot 2 \cdot 2 \cdot 2$ can be written as 2^5. The expression 2^5 is called an **exponential expression.** It is also called the fifth **power** of 2, or we say that 2 is **raised** to the fifth power.

$$5^6 = \underbrace{5 \cdot 5 \cdot 5 \cdot 5 \cdot 5 \cdot 5}_{\text{6 factors; each factor is 5}} \quad \text{and} \quad (-3)^4 = \underbrace{(-3) \cdot (-3) \cdot (-3) \cdot (-3)}_{\text{4 factors; each factor is } -3}$$

The **base** of an exponential expression is the repeated factor. The **exponent** is the number of times that the base is used as a factor.

DEFINITION OF a^n

If a is a real number and n is a positive integer, then **a raised to the nth power,** written a^n, is the product of n factors, each of which is a.

$$a^n = \underbrace{a \cdot a \cdot a \cdot a \cdot a \ldots a}_{n \text{ factors of } a}$$

EXAMPLE 1 Evaluate each expression.

a. 2^3 b. 3^1 c. $(-4)^2$ d. -4^2 e. $\left(\dfrac{1}{2}\right)^4$ f. $4 \cdot 3^2$

Solution: **a.** $2^3 = 2 \cdot 2 \cdot 2 = 8$

b. To raise 3 to the first power means to use 3 as a factor only once. Therefore, $3^1 = 3$. Also, when no exponent is shown, the exponent is assumed to be 1.

c. $(-4)^2 = (-4)(-4) = 16$

d. $-4^2 = -(4 \cdot 4) = -16$

e. $\left(\dfrac{1}{2}\right)^4 = \dfrac{1}{2} \cdot \dfrac{1}{2} \cdot \dfrac{1}{2} \cdot \dfrac{1}{2} = \dfrac{1}{16}$

f. $4 \cdot 3^2 = 4 \cdot 9 = 36$

Notice how similar -4^2 is to $(-4)^2$ in the example above. The difference between the two is the parentheses. In $(-4)^2$, the parentheses tell us that the base, or repeated factor, is -4. In -4^2, only 4 is the base.

REMINDER Be careful when identifying the base of an exponential expression.

$(-3)^2$	-3^2	$2 \cdot 3^2$
Base is -3	Base is 3	Base is 3
$(-3)^2 = (-3)(-3) = 9$	$-3^2 = -(3 \cdot 3) = -9$	$2 \cdot 3^2 = 2 \cdot 3 \cdot 3 = 18$

An exponent has the same meaning whether the base is a number or a variable. If x is a real number and n is a positive integer, then x^n is the product of n factors, each of which is x.

$$x^n = \underbrace{x \cdot x \cdot x \cdot x \cdot x \ldots x}_{n \text{ factors of } x}$$

EXAMPLE 2 Evaluate for the given value of x.

a. $2x^3$; x is 5

b. $\dfrac{9}{x^2}$; x is -3

Solution: **a.** If x is 5, $2x^3 = 2 \cdot 5^3$

$= 2 \cdot (5 \cdot 5 \cdot 5)$

$= 2 \cdot 125$

$= 250$

b. If x is -3, $\dfrac{9}{x^2} = \dfrac{9}{(-3)^2}$

$= \dfrac{9}{(-3)(-3)}$

$= \dfrac{9}{9} = 1$

Exponential expressions can be multiplied, divided, added, subtracted, and themselves raised to powers. By our definition of an exponent,

$$5^4 \cdot 5^3 = \underbrace{(5 \cdot 5 \cdot 5 \cdot 5)}_{\text{4 factors of 5}} \cdot \underbrace{(5 \cdot 5 \cdot 5)}_{\text{3 factors of 5}}$$

$$= \underbrace{5 \cdot 5 \cdot 5 \cdot 5 \cdot 5 \cdot 5 \cdot 5}_{\text{7 factors of 5}}$$

$$= 5^7$$

Also,

$$x^2 \cdot x^3 = (x \cdot x) \cdot (x \cdot x \cdot x)$$
$$= x \cdot x \cdot x \cdot x \cdot x$$
$$= x^5$$

In both cases, notice that the result is exactly the same if the exponents are added.

$$5^4 \cdot 5^3 = 5^{4+3} = 5^7 \quad \text{and} \quad x^2 \cdot x^3 = x^{2+3} = x^5$$

This suggests the following **product rule for exponents.**

> **PRODUCT RULE FOR EXPONENTS**
> If m and n are positive integers and a is a real number, then
> $$a^m \cdot a^n = a^{m+n}$$

In other words, to multiply two exponential expressions with a **common base,** keep the base and add the exponents.

EXAMPLE 3 Use the product rule to simplify.

a. $4^2 \cdot 4^5$ **b.** $x^2 \cdot x^5$ **c.** $y^3 \cdot y$ **d.** $y^3 \cdot y^2 \cdot y^7$ **e.** $(-5)^7 \cdot (-5)^8$

Solution: **a.** $4^2 \cdot 4^5 = 4^{2+5} = 4^7$
b. $x^2 \cdot x^5 = x^{2+5} = x^7$
c. $y^3 \cdot y = y^3 \cdot y^1$ Recall that if no exponent is written, it is assumed to be 1.
   $= y^{3+1}$
   $= y^4$
d. $y^3 \cdot y^2 \cdot y^7 = y^{3+2+7} = y^{12}$
e. $(-5)^7 \cdot (-5)^8 = (-5)^{7+8} = (-5)^{15}$.

We can simplify this expression further. Because $(-5)^{15}$ is the product of an odd number of negative numbers, the product is negative, so that

$(-5)^{15}$ ← odd number   can also be written as -5^{15}.

Both expressions have the same value.

EXAMPLE 4 Use the product rule to simplify $(2x^2)(-3x^5)$.

Solution: Recall that $2x^2$ means $2 \cdot x^2$ and $-3x^5$ means $-3 \cdot x^5$.

$(2x^2)(-3x^5) = 2 \cdot x^2 \cdot -3 \cdot x^5$ Remove parentheses.
$\qquad = 2 \cdot -3 \cdot x^2 \cdot x^5$ Group factors with common bases.
$\qquad = -6x^7$ Simplify.

3 Exponential expressions can themselves be raised to powers. Let's try to discover a rule that simplifies an expression like $(x^2)^3$. By the definition of a^n,

$$(x^2)^3 = \underbrace{(x^2)(x^2)(x^2)}_{\text{3 factors of } x^2}$$

which can be simplified by the product rule for exponents.

$$(x^2)^3 = (x^2)(x^2)(x^2) = x^{2+2+2} = x^6$$

Notice that the result is exactly the same if we multiply the exponents.

$$(x^2)^3 = x^{2 \cdot 3} = x^6$$

The following property states this result.

> **POWER OF A POWER RULE FOR EXPONENTS**
> If m and n are positive integers and a is a real number, then
> $$(a^m)^n = a^{mn}$$

To raise a power to a power, keep the base and multiply the exponents.

EXAMPLE 5 Simplify each of the following expressions.

    a. $(x^2)^5$ **b.** $(y^8)^2$ **c.** $[(-5)^3]^4$

Solution: **a.** $(x^2)^5 = x^{2 \cdot 5} = x^{10}$

b. $(y^8)^2 = y^{8 \cdot 2} = y^{16}$

c. $[(-5)^3]^4 = (-5)^{12}$. Because $(-5)^{12}$ is the product of an even number of negative numbers, their product is a positive number, so that

$(-5)^{12}$ ←— even number   can be written as 5^{12}

Both expressions have the same value.

4 When the base of an exponential expression is a product or quotient, the definition of a^n still applies. To simplify $(xy)^3$, for example,

$$(xy)^3 = \underbrace{(xy)(xy)(xy)}_{3 \text{ factors of } xy}$$

$= x \cdot x \cdot x \cdot y \cdot y \cdot y$ Group factors with common bases.

$= x^3 y^3$ Simplify.

Similarly, to simplify $\left(\dfrac{x}{y}\right)^3$:

$$\left(\dfrac{x}{y}\right)^3 = \underbrace{\left(\dfrac{x}{y}\right)\left(\dfrac{x}{y}\right)\left(\dfrac{x}{y}\right)}_{3 \text{ factors of } \frac{x}{y}}$$

$= \dfrac{x \cdot x \cdot x}{y \cdot y \cdot y}$ Multiply fractions.

$= \dfrac{x^3}{y^3}$ Simplify.

Notice that the power of a product (or quotient) can be written as a product (or quotient) of powers.

$$(xy)^3 = x^3 y^3 \quad \text{and} \quad \left(\dfrac{x}{y}\right)^3 = \dfrac{x^3}{y^3}$$

In general, we have the following.

POWER OF A PRODUCT OR QUOTIENT RULE

If n is a positive integer and a, b, and c are real numbers, then

$$(ab)^n = a^n b^n \quad \text{and} \quad \left(\dfrac{a}{c}\right)^n = \dfrac{a^n}{c^n}$$

as long as c is not 0.

In other words, to raise a product to a power, raise each factor to the power. Also, to raise a quotient to a power, raise both the numerator and the denominator to the power.

EXAMPLE 6 Simplify each expression.

a. $(st)^4$ **b.** $\left(\dfrac{m}{n}\right)^7$ **c.** $(2a)^3$ **d.** $(-5x^2y^3z)^2$ **e.** $\left(\dfrac{2x^4}{3y^5}\right)^4$

Solution:

a. $(st)^4 = s^4 \cdot t^4 = s^4 t^4$ Use the power of a product rule.

b. $\left(\dfrac{m}{n}\right)^7 = \dfrac{m^7}{n^7}, n \neq 0$ Use the power of a quotient rule.

c. $(2a)^3 = 2^3 \cdot a^3 = 8a^3$ Use the power of a product rule.

d. $(-5x^2y^3z)^2 = (-5)^2 \cdot (x^2)^2 \cdot (y^3)^2 \cdot (z^1)^2$ Use the power of a product rule.
$= 25x^4y^6z^2$ Use the power rule for exponents.

e. $\left(\dfrac{2x^4}{3y^5}\right)^4 = \dfrac{2^4 \cdot (x^4)^4}{3^4 \cdot (y^5)^4}$ Use the power of a product or quotient rule.
$= \dfrac{16x^{16}}{81y^{20}}, y \neq 0$ Use the power rule for exponents.

5 Another pattern for simplifying exponential expressions involves quotients.

To simplify an expression like $\dfrac{x^5}{x^3}$, in which the numerator and the denominator have a common base, we can apply the fundamental principle of fractions and divide the numerator and the denominator by the common base factors. Assume for the remainder of this section that denominators are not 0.

$$\dfrac{x^5}{x^3} = \dfrac{x \cdot x \cdot x \cdot x \cdot x}{x \cdot x \cdot x}$$

$$= \dfrac{\boxed{x} \cdot \boxed{x} \cdot \boxed{x} \cdot x \cdot x}{\boxed{x} \cdot \boxed{x} \cdot \boxed{x}}$$

$$= x \cdot x$$

$$= x^2$$

Notice that the result is exactly the same if we subtract exponents of the common bases.

$$\dfrac{x^5}{x^3} = x^{5-3} = x^2$$

The quotient rule for exponents states this result in a general way.

> **QUOTIENT RULE FOR EXPONENTS**
>
> If m and n are positive integers and a is a real number, then
>
> $$\frac{a^m}{a^n} = a^{m-n}$$
>
> as long as a is not 0.

In other words, to divide one exponential expression by another with a common base, keep the base and subtract exponents.

EXAMPLE 7 Simplify each quotient.

a. $\dfrac{x^5}{x^2}$ b. $\dfrac{4^7}{4^3}$ c. $\dfrac{(-3)^5}{(-3)^2}$ d. $\dfrac{2x^5y^2}{xy}$

Solution:
a. $\dfrac{x^5}{x^2} = x^{5-2} = x^3$ Use the quotient rule.

b. $\dfrac{4^7}{4^3} = 4^{7-3} = 4^4 = 256$ Use the quotient rule.

c. $\dfrac{(-3)^5}{(-3)^2} = (-3)^3 = -27$

d. Begin by grouping common bases.

$\dfrac{2x^5y^2}{xy} = 2 \cdot \dfrac{x^5}{x^1} \cdot \dfrac{y^2}{y^1}$

$= 2 \cdot (x^{5-1}) \cdot (y^{2-1})$ Use the quotient rule.

$= 2x^4y^1$ or $2x^4y$

Let's look at one more case. To simplify $\dfrac{x^3}{x^3}$, we use the quotient rule and subtract exponents.

$$\dfrac{x^3}{x^3} = x^{3-3} = x^0$$

But our definition of a^n does not include the possibility that n might be 0, as in x^0. What is the meaning when 0 is an exponent? To find out, use another method to simplify $\dfrac{x^3}{x^3}$. Divide the numerator and denominator by common factors, by applying the fundamental principle.

$$\dfrac{x^3}{x^3} = \dfrac{x \cdot x \cdot x}{x \cdot x \cdot x} = 1$$

Since $\dfrac{x^3}{x^3} = x^0$ and $\dfrac{x^3}{x^3} = 1$, we conclude that $x^0 = 1$ as long as x is not 0.

EXPONENTS SECTION 4.1 **235**

> **ZERO EXPONENT**
>
> $a^0 = 1$, as long as a is not 0.

EXAMPLE 8 Simplify the following expressions.

    a. 3^0 **b.** $(ab)^0$ **c.** $(-5)^0$ **d.** -5^0

Solution: **a.** $3^0 = 1$

    b. Assume that neither a nor b is zero.

$$(ab)^0 = a^0 \cdot b^0 = 1 \cdot 1 = 1$$

    c. $(-5)^0 = 1$ **d.** $-5^0 = -1 \cdot 5^0 = -1 \cdot 1 = -1$

> **REMINDER** These examples will remind you of the difference between adding and multiplying terms.
>
> | Addition | Multiplication |
> |---|---|
> | $5x^3 + 3x^3 = (5 + 3)x^3 = 8x^3$ | $(5x^3)(3x^3) = 5 \cdot 3 \cdot x^3 \cdot x^3 = 15x^{3+3} = 15x^6$ |
> | $7x + 4x^2 = 7x + 4x^2$ | $(7x)(4x^2) = 7 \cdot 4 \cdot x \cdot x^2 = 28x^{1+2} = 28x^3$ |

In the next example, exponential expressions are simplified using two or more of the exponent rules presented in this section.

EXAMPLE 9 Simplify the following.

    a. $\left(\dfrac{-5x^2}{y^3}\right)^2$ **b.** $\dfrac{(x^3)^4 x}{x^7}$ **c.** $\dfrac{(2x)^5}{x^3}$ **d.** $\dfrac{(a^2 b)^3}{a^3 b^2}$

Solution: **a.** Use the power of a product or quotient rule; then use the power of a power rule for exponents.

$$\left(\frac{-5x^2}{y^3}\right)^2 = \frac{(-5)^2 (x^2)^2}{(y^3)^2} = \frac{25x^4}{y^6}$$

    b. $\dfrac{(x^3)^4 x}{x^7} = \dfrac{x^{12} \cdot x}{x^7} = \dfrac{x^{12+1}}{x^7} = \dfrac{x^{13}}{x^7} = x^{13-7} = x^6$

    c. Use the power of a product or quotient rule; then use the quotient rule.

$$\frac{(2x)^5}{x^3} = \frac{2^5 \cdot x^5}{x^3} = 2^5 \cdot x^{5-3} = 32x^2$$

d. Begin by applying the power of a product rule to the numerator.

$$\frac{(a^2 b)^3}{a^3 b^2} = \frac{(a^2)^3 \cdot b^3}{a^3 \cdot b^2}$$

$$= \frac{a^6 b^3}{a^3 b^2} \quad \text{Use the power of a power rule for exponents.}$$

$$= a^{6-3} b^{3-2} \quad \text{Use the quotient rule.}$$

$$= a^3 b^1 \quad \text{or} \quad a^3 b$$

Mental Math

State the bases and the exponents for each of the following expressions.

1. 3^2
2. 5^4
3. $(-3)^6$
4. -3^7
5. -4^2
6. $(-4)^3$
7. $5 \cdot 3^4$
8. $9 \cdot 7^6$
9. $5x^2$
10. $(5x)^2$

Exercise Set 4.1

Evaluate each expression. See Example 1.

1. 7^2
2. -3^2
3. $(-5)^1$
4. $(-3)^2$
5. -2^4
6. -4^3
7. $(-2)^4$
8. $(-4)^3$
9. $\left(\dfrac{1}{3}\right)^3$
10. $\left(-\dfrac{1}{9}\right)^2$
11. $7 \cdot 2^4$
12. $9 \cdot 1^2$

13. Explain why $(-5)^4 = 625$, while $-5^4 = -625$.
14. Explain why $5 \cdot 4^2 = 80$, while $(5 \cdot 4)^2 = 400$.

Evaluate each expression given the replacement values for x. See Example 2.

15. $x^2; x = -2$
16. $x^3; x = -2$
17. $5x^3; x = 3$
18. $4x^2; x = -1$
19. $2xy^2; x = 3$ and $y = 5$
20. $-4x^2 y^3; x = 2$ and $y = -1$
21. $\dfrac{2z^4}{5}; z = -2$
22. $\dfrac{10}{3y^3}; y = 5$

23. The formula $V = x^3$ can be used to find the volume V of a cube with side length x. Find the volume of a cube with side length 7 meters. (Volume is measured in cubic units.)

24. The formula $S = 6x^2$ can be used to find the surface area S of a cube with side length x. Find the surface area of the cube with side length 5 meters. (Surface area is measured in square units.)

25. To find the amount of water that a swimming pool in the shape of a cube can hold, do we use the formula for volume of the cube or surface area of the cube? (See Exercises 23 and 24.)

26. To find the amount of material needed to cover an ottoman in the shape of a cube, do we use the formula for volume of the cube or surface area of the cube? (See Exercises 23 and 24.)

Use the product rule to simplify each expression. Write the results using exponents. See Examples 3 and 4.

27. $x^2 \cdot x^5$ **28.** $y^2 \cdot y$

29. $(-3)^3 \cdot (-3)^9$ **30.** $(-5)^7 \cdot (-5)^6$

31. $(5y^4)(3y)$ **32.** $(-2z^3)(-2z^2)$

33. $(4z^{10})(-6z^7)(z^3)$ **34.** $(12x^5)(-x^6)(x^4)$

35. The following rectangle has width $4x^2$ feet and length $5x^3$ feet. Find its area.

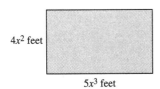

36. The following parallelogram has base length $9y^7$ meters and height $2y^{10}$ meters. Find its area.

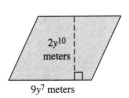

Use the power of a power rule and the power of a product or quotient rule to simplify each expression. See Examples 5 and 6.

37. $(pq)^7$ **38.** $(4s)^3$

39. $\left(\dfrac{m}{n}\right)^9$ **40.** $\left(\dfrac{xy}{7}\right)^2$

41. $(x^2y^3)^5$ **42.** $(a^4b)^7$

43. $\left(\dfrac{-2xz}{y^5}\right)^2$ **44.** $\left(\dfrac{y^4}{-3z^3}\right)^3$

45. The square shown has sides of length $8z^5$ decimeters. Find its area.

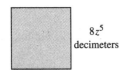

46. Given the following circle with radius $5y$ centimeters, find its area. Do not approximate π.

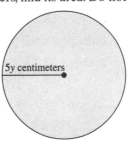

47. The following vault is in the shape of a cube. If each side is $3y^4$ feet, find its volume.

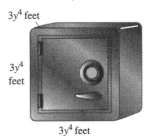

48. The silo shown is in the shape of a cylinder. If its radius is $4x$ meters and its height is $5x^3$ meters, find its volume. Do not approximate π.

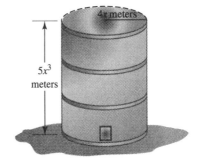

Use the quotient rule and simplify each expression. See Example 7.

49. $\dfrac{x^3}{x}$ **50.** $\dfrac{y^{10}}{y^9}$ **51.** $\dfrac{(-2)^5}{(-2)^3}$

52. $\dfrac{(-5)^{14}}{(-5)^{11}}$ **53.** $\dfrac{p^7q^{20}}{pq^{15}}$ **54.** $\dfrac{x^8y^6}{y^5}$

55. $\dfrac{7x^2y^6}{14x^2y^3}$ **56.** $\dfrac{9a^4b^7}{3ab^2}$

Simplify the following. See Example 8.

57. $(2x)^0$
58. $-4x^0$
59. $-2x^0$
60. $(4y)^0$
61. $5^0 + y^0$
62. $-3^0 + 4^0$

63. In your own words, explain why $5^0 = 1$.
64. In your own words, explain when $(-3)^n$ is positive and when it is negative.

Simplify the following. See Example 9.

65. $\left(\dfrac{-3a^2}{b^3}\right)^3$
66. $\left(\dfrac{q^7}{-2p^5}\right)^5$
67. $\dfrac{(x^5)^7 \cdot x^8}{x^4}$
68. $\dfrac{y^{20}}{(y^2)^3 \cdot y^9}$
69. $\dfrac{(z^3)^6}{(5z)^4}$
70. $\dfrac{(3x)^4}{(x^2)^2}$
71. $\dfrac{(6mn)^5}{mn^2}$
72. $\dfrac{(6xy)^2}{9x^2y^2}$

Simplify the following.

73. -5^2
74. $(-5)^2$
75. $\left(\dfrac{1}{4}\right)^3$
76. $\left(\dfrac{2}{3}\right)^3$
77. $(9xy)^2$
78. $(2ab)^5$
79. $(6b)^0$
80. $(5ab)^0$
81. $2^3 + 2^5$
82. $7^2 - 7^0$
83. $b^4 b^2$
84. $y^4 y^1$
85. $a^2 a^3 a^4$
86. $x^2 x^{15} x^9$
87. $(2x^3)(-8x^4)$
88. $(3y^4)(-5y)$
89. $(4a)^3$
90. $(2ab)^4$
91. $(-6xyz^3)^2$
92. $(-3xy^2a^3b)^3$
93. $\left(\dfrac{3y^5}{6x^4}\right)^3$
94. $\left(\dfrac{2ab}{6yz}\right)^4$

95. $\dfrac{x^5}{x^4}$
96. $\dfrac{5x^9}{x^3}$
97. $\dfrac{2x^3y^2z}{xyz}$
98. $\dfrac{x^{12}y^{13}}{x^5y^7}$
99. $\dfrac{(3x^2y^5)^5}{x^3y}$
100. $\dfrac{(4a^2)^4}{a^4b}$

Review Exercises

Simplify each expression by combining any like terms. Use the distributive property to remove any parenthesis. See Section 2.1.

101. $3x - 5x + 7$
102. $7w + w - 2w$
103. $y - 10 + y$
104. $-6z + 20 - 3z$
105. $7x + 2 - 8x - 6$
106. $10y - 14 - y - 14$
107. $2(x - 5) + 3(5 - x)$
108. $-3(w + 7) + 5(w + 1)$

A Look Ahead

EXAMPLE

Simplify $x^a \cdot x^{3a}$.

Solution:

Like bases, so add exponents.

$$x^a \cdot x^{3a} = x^{a+3a} = x^{4a}$$

Simplify each expression. Assume that variables represent positive integers. See the example.

109. $x^{5a} x^{4a}$
110. $b^{9a} b^{4a}$
111. $(a^b)^5$
112. $(2a^{4b})^4$
113. $\dfrac{x^{9a}}{x^{4a}}$
114. $\dfrac{y^{15b}}{y^{6b}}$
115. $(x^a y^b z^c)^{5a}$
116. $(9a^2 b^3 c^4 d^5)^{ab}$

4.2 ADDING AND SUBTRACTING POLYNOMIALS

OBJECTIVES

1. Define monomial, binomial, trinomial, polynomial, and degree.
2. Find the value of a polynomial given replacement values for the variables.
3. Combine like terms.
4. Add and subtract polynomials.

TAPE BA 4.2

1

In this section, we introduce a special algebraic expression called a polynomial. Let's first review some definitions presented in Section 2.1.

Recall that a term is a number or the product of a number and variables raised to powers. The terms of the expression $4x^2 + 3x$ are $4x^2$ and $3x$. The terms of the expression $9x^4 - 7x - 1$ are $9x^4$, $-7x$, and -1.

Expression	Terms
$4x^2 + 3x$	$4x^2, 3x$
$9x^4 - 7x - 1$	$9x^4, -7x, -1$
$7y^3$	$7y^3$

The **numerical coefficient** of a term, or simply the **coefficient,** is the numerical factor of each term. If no numerical factor appears in the term, then the coefficient is understood to be 1. If the term is a number only, it is called a **constant** term or simply a constant.

Term	Coefficient
x^5	1
$3x^2$	3
$-4x$	-4
$-x^2y$	-1
3 (constant)	3

A **polynomial in x** is a finite sum of terms of the form ax^n, where a is a real number and n is a whole number. For example,

$$x^5 - 3x^3 + 2x^2 - 5x + 1$$

is a polynomial. Notice that this polynomial is written in **descending powers** of x because the powers of x decrease from left to right. (Recall that the term 1 can be thought of as $1x^0$.)

On the other hand,

$$x^{-5} + 2x - 3$$

is **not** a polynomial because it contains an exponent, -5, that is not a whole number. (We study negative exponents in Section 5 of this chapter.)

A **monomial** is a polynomial with exactly one term.

A **binomial** is a polynomial with exactly two terms.

A **trinomial** is a polynomial with exactly three terms.

The following are examples of monomials, binomials, and trinomials. Each of these examples is also a polynomial.

Monomials	Binomials	Trinomials
ax^2	$x + y$	$x^2 + 4xy + y^2$
$-3z$	$3p + 2$	$x^5 + 7x^2 - x$
4	$4x^2 - 7$	$-q^4 + q^3r - 2q$

Each term of a polynomial has a **degree.**

> **DEGREE OF A TERM**
> The degree of a term is the sum of the exponents on the variables contained in the term.

EXAMPLE 1 Find the degree of each term.

  a. $-3x^2$ **b.** $5x^3yz$ **c.** 2

Solution: **a.** The exponent on x is 2, so the degree of the term is 2.
  b. $5x^3yz$ can be written as $5x^3y^1z^1$. The degree of the term is the sum of its exponents, so the degree is $3 + 1 + 1$ or 5.
  c. The constant, 2, can be written as $2x^0$ (since $x^0 = 1$). The degree of 2 or $2x^0$ is 0.

From the preceding, we can say that **the degree of a constant is 0.**
Each polynomial also has a degree.

> **DEGREE OF A POLYNOMIAL**
> The degree of a polynomial is the greatest degree of any term of the polynomial.

EXAMPLE 2 Find the degree of each polynomial and tell whether the polynomial is a monomial, binomial, trinomial, or none of these.

  a. $-2t^2 + 3t + 6$ **b.** $15x - 10$ **c.** $7x + 3x^3 + 2x^2 - 1$ **d.** $7x^2y - 6xy$

Solution: **a.** The degree of the trinomial $-2t^2 + 3t + 6$ is 2, the greatest degree of any of its terms.
  b. The degree of the binomial $15x - 10$ is 1.
  c. The degree of the polynomial $7x + 3x^3 + 2x^2 - 1$ is 3.
  d. The degree of the binomial $7x^2y - 6xy$ is 3.

2 Polynomials have different values depending on replacement values for the variables.

EXAMPLE 3 Find the value of the polynomial $3x^2 - 2x + 1$ when $x = -2$.

Solution: Replace x with -2 and simplify.

$$3x^2 - 2x + 1 = 3(-2)^2 - 2(-2) + 1$$
$$= 3(4) + 4 + 1$$
$$= 12 + 4 + 1$$
$$= 17$$

Many physical phenomena can be modeled by polynomials.

EXAMPLE 4 The CN Tower in Toronto, Ontario, is 1821 feet tall and is the world's tallest self-supporting structure. An object is dropped from the top of this building. Neglecting air resistance, the height of the object at time t seconds is given by the polynomial $-16t^2 + 1821$. Find the height of the object when $t = 1$ second and when $t = 10$ seconds.

Solution: To find each height, we evaluate the polynomial when $t = 1$ and when $t = 10$.

$$-16t^2 + 1821 = -16(1)^2 + 1821 \quad \text{Replace } t \text{ with 1.}$$
$$= -16(1) + 1821$$
$$= -16 + 1821$$
$$= 1805$$

The height of the object at 1 second is 1805 feet.

$$-16t^2 + 1821 = -16(10)^2 + 1821 \quad \text{Replace } t \text{ with 10.}$$
$$= -16(100) + 1821$$
$$= -1600 + 1821$$
$$= 221$$

The height of the object at 10 seconds is 221 feet.

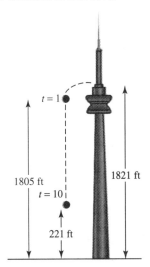

3 Polynomials with like terms can be simplified by combining like terms. Recall that like terms are terms that contain exactly the same variables raised to exactly the same powers.

$$\begin{array}{c} \text{Like Terms} \\ 5x^2, -7x^2 \\ y, 2y \\ \frac{1}{2}a^2b, -a^2b \end{array}$$

Only like terms can be combined. We combine like terms by applying the distributive property.

EXAMPLE 5 Combine like terms.

a. $-3x + 7x$ **b.** $11x^2 + 5 + 2x^2 - 7$

Solution: **a.** $-3x + 7x = (-3 + 7)x = 4x$
b. $11x^2 + 5 + 2x^2 - 7 = 11x^2 + 2x^2 + 5 - 7$
    $= 13x^2 - 2$ Combine like terms.

4 We now practice adding and subtracting polynomials.

> **To Add Polynomials**
> To add polynomials, combine all like terms.

EXAMPLE 6 Add $(-2x^2 + 5x - 1)$ and $(-2x^2 + x + 3)$.

Solution: $(-2x^2 + 5x - 1) + (-2x^2 + x + 3) = -2x^2 + 5x - 1 - 2x^2 + x + 3$
$= (-2x^2 - 2x^2) + (5x + 1x) + (-1 + 3)$
$= -4x^2 + 6x + 2$

EXAMPLE 7 Add: $(4x^3 - 6x^2 + 2x + 7) + (5x^2 - 2x)$.

Solution: $(4x^3 - 6x^2 + 2x + 7) + (5x^2 - 2x) = 4x^3 - 6x^2 + 2x + 7 + 5x^2 - 2x$
$= 4x^3 + (-6x^2 + 5x^2) + (2x - 2x) + 7$
$= 4x^3 - x^2 + 7$

Polynomials can be added vertically if we line up like terms underneath one another.

EXAMPLE 8 Add $(7y^3 - 2y^2 + 7)$ and $(6y^2 + 1)$ using the vertical format.

Solution: Vertically line up like terms and add.

$$\begin{array}{r} 7y^3 - 2y^2 + 7 \\ 6y^2 + 1 \\ \hline 7y^3 + 4y^2 + 8 \end{array}$$

To subtract one polynomial from another, recall the definition of subtraction. To subtract a number, we add its opposite: $a - b = a + (-b)$. To subtract a polynomial, we also add its opposite. Just as $-b$ is the opposite of b, $-(x^2 + 5)$ is the opposite of $(x^2 + 5)$.

EXAMPLE 9 Subtract: $(5x - 3) - (2x - 11)$.

Solution: From the definition of subtraction, we have

$$(5x - 3) - (2x - 11) = (5x - 3) + [-(2x - 11)] \quad \text{Add the opposite.}$$
$$= (5x - 3) + (-2x + 11) \quad \text{Apply the distributive property.}$$
$$= 3x + 8$$

> **TO SUBTRACT POLYNOMIALS**
> To subtract two polynomials, change the signs of the terms of the polynomial being subtracted and then add.

EXAMPLE 10 Subtract: $(2x^3 + 8x^2 - 6x) - (2x^3 - x^2 + 1)$.

Solution: First, change the sign of each term of the second polynomial and then add.

$$(2x^3 + 8x^2 - 6x) - (2x^3 - x^2 + 1) = (2x^3 + 8x^2 - 6x) + (-2x^3 + x^2 - 1)$$
$$= 2x^3 - 2x^3 + 8x^2 + x^2 - 6x - 1$$
$$= 9x^2 - 6x - 1 \quad \text{Combine like terms.}$$

EXAMPLE 11 Subtract $(5y^2 + 2y - 6)$ from $(-3y^2 - 2y + 11)$ using the vertical format.

Solution: Arrange the polynomials in vertical format, lining up like terms.

$$\begin{array}{r} -3y^2 - 2y + 11 \\ -(5y^2 + 2y - 6) \\ \hline \end{array} \qquad \begin{array}{r} -3y^2 - 2y + 11 \\ -5y^2 - 2y + 6 \\ \hline -8y^2 - 4y + 17 \end{array}$$

EXAMPLE 12   Subtract $(5z - 7)$ from the sum of $(8z + 11)$ and $(9z - 2)$.

Solution: Notice that $(5z - 7)$ is to be subtracted **from** a sum. The translation is

$$[(8z + 11) + (9z - 2)] - (5z - 7)$$
$$= 8z + 11 + 9z - 2 - 5z + 7 \quad \text{Remove grouping symbols.}$$
$$= 8z + 9z - 5z + 11 - 2 + 7 \quad \text{Group like terms.}$$
$$= 12z + 16 \quad \text{Combine like terms.}$$

MENTAL MATH

Combine like terms.

1. $-9y - 5y$
2. $6m^5 + 7m^5$
3. $4y^3 + 3y^3$
4. $21y^5 - 19y^5$
5. $x + 6x$
6. $7z - z$

EXERCISE SET 4.2

Find the degree of each of the following polynomials and determine whether it is a monomial, binomial, trinomial, or none of these. See Examples 1 and 2.

1. $x + 2$
2. $-6y + y^2 + 4$
3. $9m^3 - 5m^2 + 4m - 8$
4. $5a^2 + 3a^3 - 4a^4$
5. $12x^4y - x^2y^2 - 12x^2y^4$
6. $7r^2s^2 + 2r - 3s^5$
7. $3zx - 5x^2$
8. $5y + 2$
9. Describe how to find the degree of a term.
10. Describe how to find the degree of a polynomial.
11. Explain why xyz is a monomial while $x + y + z$ is a trinomial.
12. Explain why the degree of the term $5y^3$ is 3 and the degree of the polynomial $2y + y + 2y$ is 1.

Find the value of each polynomial when **(a)** $x = 0$ *and* **(b)** $x = -1$. *See Examples 3 and 4.*

13. $x + 6$
14. $2x - 10$
15. $x^2 - 5x - 2$
16. $x^2 - 4$
17. $x^3 - 15$
18. $-2x^3 + 3x^2 - 6$
19. Find the height of the object in Example 4 when t is 10.8 seconds. Explain your result.
20. Approximate how long (to the nearest tenth of a second) before the object in Example 4 hits the ground.

Simplify each of the following by combining like terms. See Example 5.

21. $14x^2 + 9x^2$
22. $18x^3 - 4x^3$
23. $15x^2 - 3x^2 - y$
24. $12k^3 - 9k^3 + 11$
25. $8s - 5s + 4s$
26. $5y + 7y - 6y$
27. $0.1y^2 - 1.2y^2 + 6.7 - 1.9$
28. $7.6y + 3.2y^2 - 8y - 2.5y^2$

Perform the indicated operations. See Examples 6, 7, 9, and 10.

29. $(3x + 7) + (9x + 5)$
30. $(3x^2 + 7) + (3x^2 + 9)$
31. $(-7x + 5) + (-3x^2 + 7x + 5)$
32. $(3x - 8) + (4x^2 - 3x + 3)$
33. $(2x + 5) - (3x - 9)$
34. $(5x^2 + 4) - (-2y^2 + 4)$
35. $3x - (5x - 9)$
36. $4 - (-y - 4)$

37. $(-5x^2 + 3) + (2x^2 + 1)$
38. $(-y - 2) + (3y + 5)$
39. $(2x^2 + 3x - 9) - (-4x + 7)$
40. $(-7x^2 + 4x + 7) - (-8x + 2)$
41. Given the following triangle, find its perimeter.

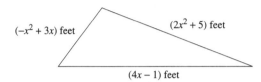

42. Given the following quadrilateral, find its perimeter.

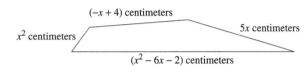

43. A wooden beam is $(4y^2 + 4y + 1)$ meters long. If a piece $(y^2 - 10)$ meters is cut, express the length of the remaining piece of beam as a polynomial in y.

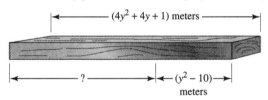

44. A piece of quarter-round molding is $(13x - 7)$ inches long. If a piece $(2x + 2)$ inches is removed, express the length of the remaining piece of molding as a polynomial in x.

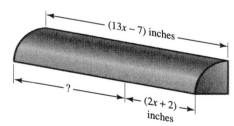

Perform the indicated operations. See Examples 8 and 11.

45. $3t^2 + 4$
    $+ 5t^2 - 8$

46. $7x^3 + 3$
    $+ 2x^3 + 1$

47. $4z^2 - 8z + 3$
    $- (6z^2 + 8z - 3)$

48. $5u^5 - 4u^2 + 3u - 7$
    $- (3u^5 + 6u^2 - 8u + 2)$

49. $5x^3 - 4x^2 + 6x - 2$
    $- (3x^3 - 2x^2 - x - 4)$

50. $7a^2 - 9a + 6$
    $- (11a^2 - 4a + 2)$

51. $10a^3 - 8a^2 + 9$
    $+ 5a^3 + 9a^2 + 7$

52. $2x^3 - 3x^2 + x - 4$
    $+ 5x^3 + 2x^2 - 3x + 2$

Perform the indicated operations. See Example 12.

53. Subtract $(19x^2 + 5)$ from $(81x^2 + 10)$.
54. Subtract $(2x + xy)$ from $(3x - 9xy)$.
55. Subtract $(2x + 2)$ from the sum of $(8x + 1)$ and $(6x + 3)$.
56. Subtract $(-12x - 3)$ from the sum of $(-5x - 7)$ and $(12x + 3)$.
57. Subtract $(8x + 9)$ from $(9xy^2 + 7x - 18)$.
58. Subtract $(4x^2 + 7)$ from $(9x^3 + 9x^2 - 9)$.

Perform the indicated operations.

59. $-15x - (-4x)$
60. $16y - (-4y)$
61. $2x - 5 + 5x - 8$
62. $x - 3 + 8x + 10$
63. $(-3y^2 - 4y) + (2y^2 + y - 1)$
64. $(7x^2 + 2x - 9) + (-3^2 + 5)$
65. $(-7y^2 + 5) - (-8y^2 + 12)$
66. $(4 + 5a) - (-a - 5)$
67. $(5x + 8) - (-2x^2 - 6x + 8)$
68. $(-6y^2 + 3y - 4) - (9y^2 - 3y)$
69. $(-8x^4 + 7x) + (-8x^4 + x + 9)$
70. $(6y^5 - 6y^3 + 4) + (-2y^5 - 8y^3 - 7)$
71. $(3x^2 + 5x - 8) + (5x^2 + 9x + 12) - (x^2 - 14)$
72. $(-a^2 + 1) - (a^2 - 3) + (5a^2 - 6a + 7)$
73. Subtract $4x$ from $7x - 3$.
74. Subtract y from $y^2 - 4y + 1$.
75. Subtract $(5x + 7)$ from $(7x^2 + 3x + 9)$.
76. Subtract $(5y^2 + 8y + 2)$ from $(7y^2 + 9y - 8)$.
77. Subtract $(4y^2 - 6y - 3)$ from the sum of $(8y^2 + 7)$ and $(6y + 9)$.

246 CHAPTER 4 EXPONENTS AND POLYNOMIALS

78. Subtract $(5y + 7x^2)$ from the sum of $(8y - x)$ and $(3 + 8x^2)$.
79. Subtract $(-2x^2 + 4x - 12)$ from the sum of $(-x^2 - 2x)$ and $(5x^2 + x + 9)$.
80. Subtract $(4x^2 - 2x + 2)$ from the sum of $(x^2 + 7x + 1)$ and $(7x + 5)$.
81. $[(1.2x^2 - 3x + 9.1) - (7.8x^2 - 3.1 + 8)] + (1.2x - 6)$
82. $[(7.9y^4 - 6.8y^3 + 3.3y) + (6.1y^3 - 5)] - (4.2y^4 + 1.1y - 1)$
83. A rocket is fired upward from the ground with an initial velocity of 200 feet per second. Neglecting air resistance, the height of the rocket at any time t can be described by the polynomial $-16t^2 + 200t$. Find the height of the rocket at the given times.

   a. $t = 1$ second
   b. $t = 5$ seconds
   c. $t = 7.6$ seconds
   d. $t = 10.3$ seconds.

84. Explain why the height of the rocket in Exercise 83 increases and then decreases as time passes.
85. Approximate (to the nearest tenth of a second) how long before the rocket in Exercise 83 hits the ground.
86. Write a polynomial that describes the surface area of the given figure in terms of its sides x and y. (Recall that the surface area of a solid is the sum of the areas of the faces or sides of the solid.)

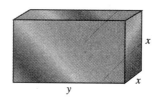

Review Exercises

Multiply. See Section 4.1.

87. $3x(2x)$
88. $-7x(x)$
89. $(12x^3)(-x^5)$
90. $6r^3(7r^{10})$
91. $10x^2(20xy^2)$
92. $-z^2y(11zy)$

Graph each linear equation. See Sections 3.2 and 3.3
93. $2x - y = 6$
94. $3x - y = -6$
95. $x = -4$
96. $y = 2$

4.3 Multiplying Polynomials

OBJECTIVES

1. Use the distributive property to multiply polynomials.
2. Multiply polynomials vertically.

TAPE BA 4.3

1 To multiply polynomials, we apply our knowledge of the rules and definitions of exponents.
To multiply two monomials such as $(-5x^3)$ and $(-2x^4)$, use the associative and commutative properties and regroup. Remember that to multiply exponential expressions with a common base we add exponents.

$$(-5x^3)(-2x^4) = (-5)(-2)(x^3)(x^4) = 10x^7$$

To multiply polynomials that are not monomials, use the distributive property.

EXAMPLE 1 Use the distributive property to find each product.

a. $5x(2x^3 + 6)$ **b.** $-3x^2(5x^2 + 6x - 1)$ **c.** $(3n^2 - 5n + 4)(2n)$

Solution:

a. $5x(2x^3 + 6) = 5x(2x^3) + 5x(6)$ Use the distributive property.
$= 10x^4 + 30x$ Multiply.

b. $-3x^2(5x^2 + 6x - 1)$
$= (-3x^2)(5x^2) + (-3x^2)(6x) + (-3x^2)(-1)$ Use the distributive property.
$= -15x^4 - 18x^3 + 3x^2$ Multiply.

c. $(3n^2 - 5n + 4)(2n)$
$= (3n^2)(2n) + (-5n)(2n) + 4(2n)$ Use the distributive property.
$= 6n^3 - 10n^2 + 8n$ Multiply.

We also use the distributive property to multiply two binomials. To multiply $(x + 3)$ by $(x + 1)$, distribute the factor $(x + 1)$ first.

$(x + 3)(x + 1) = x(x + 1) + 3(x + 1)$ Distribute $(x + 1)$.
$= x(x) + x(1) + 3(x) + 3(1)$ Apply distributive property a second time.
$= x^2 + x + 3x + 3$ Multiply.
$= x^2 + 4x + 3$ Combine like terms.

This idea can be expanded so that we can multiply any two polynomials.

TO MULTIPLY TWO POLYNOMIALS

Multiply each term of the first polynomial by each term of the second polynomial, and then combine like terms.

EXAMPLE 2 Find the product: $(3x + 2)(2x - 5)$.

Solution: Multiply each term of the first binomial by each term of the second.

$(3x + 2)(2x - 5) = 3x(2x) + 3x(-5) + 2(2x) + 2(-5)$
$= 6x^2 - 15x + 4x - 10$ Multiply.
$= 6x^2 - 11x - 10$ Combine like terms.

EXAMPLE 3 Multiply: $(2x - y)^2$.

Solution: Recall that $a^2 = a \cdot a$, so $(2x - y)^2 = (2x - y)(2x - y)$. Multiply each term of the first polynomial by each term of the second.

$$(2x - y)(2x - y) = 2x(2x) + 2x(-y) + (-y)(2x) + (-y)(-y)$$
$$= 4x^2 - 2xy - 2xy + y^2 \quad \text{Multiply.}$$
$$= 4x^2 - 4xy + y^2 \quad \text{Combine like terms.}$$

EXAMPLE 4 Multiply: $(t + 2)$ by $(3t^2 - 4t + 2)$.

Solution: Multiply each term of the first polynomial by each term of the second.

$$(t + 2)(3t^2 - 4t + 2) = t(3t^2) + t(-4t) + t(2) + 2(3t^2) + 2(-4t) + 2(2)$$
$$= 3t^3 - 4t^2 + 2t + 6t^2 - 8t + 4$$
$$= 3t^3 + 2t^2 - 6t + 4 \quad \text{Combine like terms.}$$

EXAMPLE 5 Multiply: $(3a + b)^3$.

Solution: Write $(3a + b)^3$ as $(3a + b)(3a + b)(3a + b)$.

$$(3a + b)(3a + b)(3a + b) = (9a^2 + 3ab + 3ab + b^2)(3a + b)$$
$$= (9a^2 + 6ab + b^2)(3a + b)$$
$$= (9a^2 + 6ab + b^2)3a + (9a^2 + 6ab + b^2)b$$
$$= 27a^3 + 18a^2b + 3ab^2 + 9a^2b + 6ab^2 + b^3$$
$$= 27a^3 + 27a^2b + 9ab^2 + b^3$$

Another convenient method for multiplying polynomials is to use a vertical format similar to the format used to multiply real numbers. We demonstrate this method by multiplying $(3y^2 - 4y + 1)$ by $(y + 2)$.

Step 1. Write the polynomials in a vertical format.

$$\begin{array}{r} 3y^2 - 4y + 1 \\ \times \quad y + 2 \\ \hline \end{array}$$

Step 2. Multiply 2 by each term of the top polynomial. Write the first **partial product** below the line.

$$3y^2 - 4y + 1$$
$$\times \quad\quad y + 2$$
$$\overline{6y^2 - 8y + 2} \quad \text{Partial product.}$$

Step 3. Multiply y by each term of the top polynomial. Write this partial product underneath the previous one, being careful to line up like terms.

$$3y^2 - 4y + 1$$
$$\times \quad\quad y + 2$$
$$\overline{6y^2 - 8y + 2} \quad \text{Partial product.}$$
$$3y^3 - 4y^2 + y \quad\quad \text{Partial product.}$$

Step 4. Combine like terms of the partial products.

$$3y^2 - 4y + 1$$
$$\times \quad\quad y + 2$$
$$\overline{6y^2 - 8y + 2}$$
$$3y^3 - 4y^2 + y$$
$$\overline{3y^3 + 2y^2 - 7y + 2}$$

Thus, $(y + 2)(3y^2 - 4y + 1) = 3y^3 + 2y^2 - 7y + 2$.

EXAMPLE 6 Find the product of $(2x^2 - 3x + 4)$ and $(x^2 + 5x - 2)$ using the vertical format.

Solution: Multiply each term of the second polynomial by each term of the first polynomial.

$$2x^2 - 3x + 4$$
$$\times \quad\quad x^2 + 5x - 2$$
$$\overline{\quad\quad\quad -4x^2 + 6x - 8} \quad \text{Multiply } 2x^2 - 3x + 4 \text{ by } -2.$$
$$10x^3 - 15x^2 + 20x \quad\quad\quad \text{Multiply } 2x^2 - 3x + 4 \text{ by } 5x.$$
$$2x^4 - 3x^3 + 4x^2 \quad\quad\quad\quad\quad \text{Multiply } 2x^2 - 3x + 4 \text{ by } x^2.$$
$$\overline{2x^4 + 7x^3 - 15x^2 + 26x - 8} \quad \text{Combine like terms.}$$

MENTAL MATH

Find the following products mentally.

1. $5x(2y)$
2. $7a(4b)$
3. $x^2 \cdot x^5$
4. $z \cdot z^4$
5. $6x(3x^2)$
6. $5a^2(3a^2)$

EXERCISE SET 4.3

Find the following products. See Example 1.
1. $2a(2a - 4)$
2. $3a(2a + 7)$
3. $7x(x^2 + 2x - 1)$
4. $-5y(y^2 + y - 10)$
5. $3x^2(2x^2 - x)$
6. $-4y^2(5y - 6y^2)$

7. The area of the larger rectangle below is $x(x + 3)$. Find another expression for this area by finding the sum of the areas of the smaller rectangles.

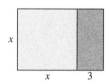

8. Write an expression for the area of the larger rectangle below in two different ways.

Find the following products. See Examples 2 and 3.
9. $(a + 7)(a - 2)$
10. $(y + 5)(y + 7)$
11. $(2y - 4)^2$
12. $(6x - 7)^2$
13. $(5x - 9y)(6x - 5y)$
14. $(3x - 7y)(7x + 2y)$
15. $(2x^2 - 5)^2$
16. $(x^2 - 4)^2$

17. The area of the figure below is $(x + z)(x + 3)$. Find another expression for this area by finding the sum of the areas of the smaller rectangles.

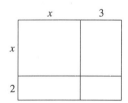

18. Write an expression for the area of the figure below in two different ways.

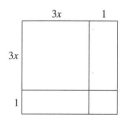

Find the following products. See Example 4.
19. $(x - 2)(x^2 - 3x + 7)$
20. $(x + 3)(x^2 + 5x - 8)$
21. $(x + 5)(x^3 - 3x + 4)$
22. $(a + 2)(a^3 - 3a^2 + 7)$
23. $(2a - 3)(5a^2 - 6a + 4)$
24. $(3 + b)(2 - 5b - 3b^2)$

Find the following products. See Example 5.
25. $(x + 2)^3$
26. $(y - 1)^3$
27. $(2y - 3)^3$
28. $(3x + 4)^3$

Find the following products. Use the vertical multiplication method. See Example 6.
29. $(x + 3)(2x^2 + 4x - 1)$
30. $(2x - 5)(3x^2 - 4x + 7)$
31. $(x^2 + 5x - 7)(x^2 - 7x - 9)$
32. $(3x^2 - x + 2)(x^2 + 2x + 1)$

33. Evaluate each of the following.
   a. $(2 + 3)^2$; $2^2 + 3^2$
   b. $(8 + 10)^2$; $8^2 + 10^2$
   Does $(a + b)^2 = a^2 + b^2$ no matter what the values of a and b are? Why or why not?

34. Perform the indicated operations. Explain the difference between the two expressions.
   a. $(3x + 5) + (3x + 7)$
   b. $(3x + 5)(3x + 7)$

Find the following products.
35. $2a(a + 4)$
36. $-3a(2a + 7)$
37. $3x(2x^2 - 3x + 4)$
38. $-4x(5x^2 - 6x - 10)$
39. $(5x + 9y)(3x + 2y)$
40. $(5x - 5y)(2x - y)$
41. $(x + 2)(x^2 + 5x + 6)$
42. $(x - 7)(x^2 - 15x + 56)$
43. $(7x + 4)^2$
44. $(3x - 2)^2$
45. $-2a^2(3a^2 - 2a + 3)$
46. $-4b^2(3b^3 - 12b^2 - 6)$
47. $(x + 3)(x^2 + 7x + 12)$
48. $(n + 1)(n^2 - 7n - 9)$
49. $(a + 1)^3$
50. $(x - y)^3$
51. $(x + y)(x + y)$
52. $(x + 3)(7x + 1)$

53. $(x - 7)(x - 6)$
54. $(4x + 5)(-3x + 2)$
55. $3a(a^2 + 2)$
56. $x^3(x + 12)$
57. $-4y(y^2 + 3y - 11)$
58. $-2x(5x^2 - 6x + 1)$
59. $(5x + 1)(5x - 1)$
60. $(2x + y)(3x - y)$
61. $(5x + 4)(x^2 - x + 4)$
62. $(x - 2)(x^2 - x + 3)$
63. $(2x - 5)^3$
64. $(3y - 1)^3$
65. $(4x + 5)(8x^2 + 2x - 4)$
66. $(x + 7)(x^2 - 7x - 8)$
67. $(7xy - y)^2$
68. $(x + y)^2$
69. $(5y^2 - y + 3)(y^2 - 3y - 2)$
70. $(2x^2 + x - 1)(x^2 + 3x + 4)$
71. $(3x^2 + 2x - 4)(2x^2 - 4x + 3)$
72. $(a^2 + 3a - 2)(2a^2 - 5a - 1)$

Express each of the following as polynomials.

73. Find the area of the following rectangle.

74. Find the area of the square field.

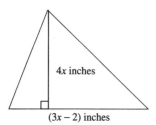

75. Find the area of the following triangle.

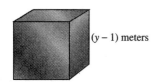

76. Find the volume of the cube-shaped glass block.

77. Multiply the following polynomials.
   a. $(a + b)(a - b)$
   b. $(2x + 3y)(2x - 3y)$
   c. $(4x + 7)(4x - 7)$

   Can you make a general statement about all products of the form $(x + y)(x - y)$?

Review Exercises

Perform the indicated operation. See Section 4.1.

78. $(5x)^2$
79. $(4p)^2$
80. $(-3y^3)^2$
81. $(-7m^2)^2$

For income tax purposes, Rob Calcutta, the owner of Copy Services, uses a method called **straight-line depreciation** *to show the depreciated (or decreased) value of a copy machine he recently purchased. Rob assumes that he can use the machine for 7 years. The graph below shows the depreciated value of the machine over the years. See Sections 1.9 and 3.1.*

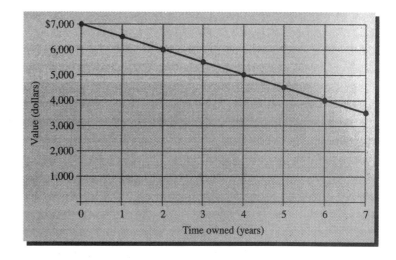

82. What was the purchase price of the copy machine?
83. What is the depreciated value of the machine in 7 years?
84. What loss in value occurred during the first year?
85. What loss in value occurred during the second year?
86. Why do you think this method of depreciating is called straight-line depreciation?
87. Why is the line tilted downward?

4.5 NEGATIVE EXPONENTS AND SCIENTIFIC NOTATION

OBJECTIVES

1. Evaluate numbers raised to negative integer powers.
2. Use all the rules and definitions for exponents to simplify exponential expressions.
3. Write numbers in scientific notation.
4. Convert numbers from scientific notation to standard form.

Tape BA 4.5

Our work with exponential expressions so far has been limited to exponents that are positive integers or 0. Here we expand to give meaning to an expression like x^{-3}.

Suppose that we wish to simplify the expression $\dfrac{x^2}{x^5}$. If we use the quotient rule for exponents, we subtract exponents:

$$\dfrac{x^2}{x^5} = x^{2-5} = x^{-3}, \quad x \neq 0$$

But what does x^{-3} mean? Let's simplify $\dfrac{x^2}{x^5}$ using the definition of a^n.

$$\dfrac{x^2}{x^5} = \dfrac{x \cdot x}{x \cdot x \cdot x \cdot x \cdot x}$$

$$= \dfrac{\boxed{x} \cdot \boxed{x}}{\boxed{x} \cdot \boxed{x} \cdot x \cdot x \cdot x} \quad \text{Divide numerator and denominator by common factors by applying the fundamental principle.}$$

$$= \dfrac{1}{x^3}$$

If the quotient rule is to hold true for negative exponents, then x^{-3} must equal $\dfrac{1}{x^3}$.

From this example, we state the definition for negative exponents.

> **NEGATIVE EXPONENTS**
> If a is a real number other than 0 and n is an integer, then
> $$a^{-n} = \dfrac{1}{a^n}$$

EXAMPLE 1 Simplify by writing each expression with positive exponents only.

    a. 3^{-2} **b.** $2x^{-3}$ **c.** $\dfrac{1}{2^{-5}}$

Solution: **a.** $3^{-2} = \dfrac{1}{3^2} = \dfrac{1}{9}$ Use the definition of negative exponent.

           b. $2x^{-3} = 2 \cdot \dfrac{1}{x^3} = \dfrac{2}{x^3}$ Use the definition of negative exponent.

Since there are no parentheses, notice that only x is the base for the exponent -3.

           c. $\dfrac{1}{2^{-5}} = \dfrac{1}{\dfrac{1}{2^5}}$ Use the definition of negative exponent.

              $= 1 \div \dfrac{1}{2^5}$

              $= 1 \cdot \dfrac{2^5}{1} = 2^5 \quad \text{or} \quad 32$

> **REMINDER** A negative exponent does not affect the sign of its base.
> Remember: a negative exponent means a reciprocal. $a^{-n} = \frac{1}{a^n}$. For example,
>
> $$x^{-2} = \frac{1}{x^2}, \qquad 2^{-3} = \frac{1}{2^3} \text{ or } \frac{1}{8}$$
>
> $$\frac{1}{y^{-4}} = \frac{1}{\frac{1}{y^4}} = y^4, \qquad \frac{1}{5^{-2}} = 5^2 \text{ or } 25$$

EXAMPLE 2 Simplify each expression. Write results with positive exponents.

a. $\left(\frac{2}{3}\right)^{-4}$ b. $2^{-1} + 4^{-1}$ c. $(-2)^{-4}$

Solution:
a. $\left(\frac{2}{3}\right)^{-4} = \frac{1}{\left(\frac{2}{3}\right)^4} = \frac{1}{\frac{2^4}{3^4}} = \frac{3^4}{2^4} = \frac{81}{16}$ b. $2^{-1} + 4^{-1} = \frac{1}{2} + \frac{1}{4} = \frac{2}{4} + \frac{1}{4} = \frac{3}{4}$

c. $(-2)^{-4} = \frac{1}{(-2)^4} = \frac{1}{(-2)(-2)(-2)(-2)} = \frac{1}{16}$

2 All the previously stated rules for exponents apply for negative exponents also. Here is a summary of the rules and definitions for exponents.

> **SUMMARY OF EXPONENT RULES**
> If m and n are integers and a, b, and c are real numbers, then:
> Product rule for exponents: $a^m \cdot a^n = a^{m+n}$
> Power of a power rule for exponents: $(a^m)^n = a^{m \cdot n}$
> Power of a product or quotient: $(ab)^n = a^n b^n$ and
> $$\left(\frac{a}{c}\right)^n = \frac{a^n}{c^n}, c \neq 0$$
> Quotient rule for exponents: $\frac{a^m}{a^n} = a^{m-n}, a \neq 0$
> Zero exponent: $a^0 = 1, a \neq 0$
> Negative exponent: $a^{-n} = \frac{1}{a^n}, a \neq 0$

EXAMPLE 3 Simplify each expression. Write answers with positive exponents.

a. $\frac{y}{y^{-2}}$ b. $\frac{p^{-4}}{q^{-9}}$ c. $\frac{x^{-5}}{x^7}$

Solution: a. $\dfrac{y}{y^{-2}} = \dfrac{y^1}{y^{-2}} = y^{1-(-2)} = y^3$ b. $\dfrac{p^{-4}}{q^{-9}} = \dfrac{q^9}{p^4}$ c. $\dfrac{x^{-5}}{x^7} = x^{-5-7} = x^{-12} = \dfrac{1}{x^{12}}$

EXAMPLE 4 Simplify the following expressions. Write each result using positive exponents only.

a. $(2x^3)(5x)^{-2}$ b. $\left(\dfrac{3a^2}{b}\right)^{-3}$ c. $\dfrac{4^{-1}x^{-3}y}{4^{-3}x^2y^{-6}}$ d. $(y^{-3}z^{-6})^{-6}$ e. $\left(\dfrac{-2x^3y}{xy^{-1}}\right)^3$

Solution: a. $(2x^3)(5x)^{-2} = 2x^3 \cdot 5^{-2}x^{-2}$ Use the power of a power rule.

$= \dfrac{2x^{3+(-2)}}{5^2}$ Use the product and quotient rules and definition of negative exponent.

$= \dfrac{2x}{25}$

b. $\left(\dfrac{3a^2}{b}\right)^{-3} = \dfrac{3^{-3}a^{-6}}{b^{-3}} = \dfrac{b^3}{3^3a^6} = \dfrac{b^3}{27a^6}$

c. $\dfrac{4^{-1}x^{-3}y}{4^{-3}x^2y^{-6}} = 4^{-1-(-3)}x^{-3-2}y^{1-(-6)} = 4^2x^{-5}y^7 = \dfrac{4^2y^7}{x^5} = \dfrac{16y^7}{x^5}$

d. $(y^{-3}z^6)^{-6} = y^{18} \cdot z^{-36} = \dfrac{y^{18}}{z^{36}}$

e. $\left(\dfrac{-2x^3y}{xy^{-1}}\right)^3 = \dfrac{(-2)^3x^9y^3}{x^3y^{-3}} = \dfrac{-8x^9y^3}{x^3y^{-3}} = -8x^{9-3}y^{3-(-3)} = -8x^6y^6$

3 Both very large and very small numbers frequently occur in many fields of science. For example, the distance between the sun and the planet Pluto is approximately 5,906,000,000 kilometers, and the mass of a proton is approximately 0.00000000000000000000000165 gram. It can be tedious to write these numbers in this standard decimal notation, so **scientific notation** is used as a convenient shorthand for expressing very large and very small numbers.

> **SCIENTIFIC NOTATION**
> A positive number is written in scientific notation if it is written as the product of a number a, where $1 \leq a < 10$, and an integer power r of 10:
> $a \times 10^r$

The following numbers are written in scientific notation. The $\times$ sign for multiplication is used as part of the notation.

$2.03 \times 10^2 \qquad 7.362 \times 10^7$
$1 \times 10^{-3} \qquad 8.1 \times 10^{-5}$

To write the distance between the sun and Pluto in scientific notation, begin by moving the decimal point to the left until we have a number between 1 and 10.

5906000000.

Next, count the number of places the decimal point is moved.

5906000000.

9 decimal places

We moved the decimal point 9 places **to the left.** This count is used as the power of 10.

$5{,}906{,}000{,}000 = 5.906 \times 10^9$

To express the mass of a proton in scientific notation, move the decimal until the number is between 1 and 10.

0.00000000000000000000000165

The decimal point was moved 24 places **to the right,** so the exponent on 10 is negative 24.

$0.000\ 000\ 000\ 000\ 000\ 000\ 000\ 001\ 65 = 1.65 \times 10^{-24}$

To Write a Number in Scientific Notation

Step 1. Move the decimal point in the original number so that the new number has a value between 1 and 10.

Step 2. Count the number of decimal places the decimal point is moved in step 1. If the decimal point is moved to the left, the count is positive. If the decimal point is moved to the right, the count is negative.

Step 3. Multiply the new number in step 1 by 10 raised to an exponent equal to the count found in step 2.

EXAMPLE 5 Write the following numbers in scientific notation.

a. 367,000,000 **b.** 0.000003 **c.** 20,520,000,000 **d.** 0.00085

Solution: **a.** *Step 1* Move the decimal point until the number is between 1 and 10.

367,000,000.

Step 2 The decimal point is moved to the left 8 places, so the count is positive 8.

Step 3 $367{,}000{,}000 = 3.67 \times 10^8$.

b. *Step 1* Move the decimal point until the number is between 1 and 10.

0.000003

Step 2 The decimal point is moved 6 places to the right, so the count is -6.
Step 3 $0.000003 = 3.0 \times 10^{-6}$.
c. $20{,}520{,}000{,}000 = 2.052 \times 10^{10}$
d. $0.00085 = 8.5 \times 10^{-4}$

A number written in scientific notation can be rewritten in standard form. To write 8.63×10^3 in standard form, recall that $10^3 = 1000$.

$$8.63 \times 10^3 = 8.63(1000) = 8630$$

Notice that the exponent on the 10 is positive three and we moved the decimal point three places to the right.
To write 8.63×10^{-3} in standard form, recall that $10^{-3} = \dfrac{1}{10^3} = \dfrac{1}{1000}$.

$$8.63 \times 10^{-3} = 8.63\left(\dfrac{1}{1000}\right) = \dfrac{8.63}{1000} = 0.00863$$

The exponent on the 10 is negative three, and we moved the decimal to the left three places.

In general, **to write a scientific notation number in standard form,** move the decimal point the same number of places as the exponent on 10. If the exponent is positive, move the decimal point to the right; if the exponent is negative, move the decimal point to the left.

EXAMPLE 6 Write the following numbers in standard notation, without exponents.

a. 1.02×10^5 b. 7.358×10^{-3} c. 8.4×10^7 d. 3.007×10^{-5}

Solution: a. Move the decimal point 5 places to the right and

$$1.02 \times 10^5 = 102{,}000.$$

b. Move the decimal point 3 places to the left and

$$7.358 \times 10^{-3} = 0.007358$$

c. $8.4 \times 10^7 = 84{,}000{,}000.$

7 places to the right

d. $3.007 \times 10^{-5} = 0.00003007$

5 places to the left

Performing operations on numbers written in scientific notation makes use of the rules and definitions for exponents.

EXAMPLE 7 Perform the indicated operations. Write each answer in standard decimal notation.

a. $(8 \times 10^{-6})(7 \times 10^3)$ **b.** $\dfrac{12 \times 10^2}{6 \times 10^{-3}}$

Solution: **a.** $(8 \times 10^{-6})(7 \times 10^3) = 8 \cdot 7 \cdot 10^{-6} \cdot 10^3$
$= 56 \times 10^{-3}$
$= 0.056$

b. $\dfrac{12 \times 10^2}{6 \times 10^{-3}} = \dfrac{12}{6} \times 10^{2-(-3)} = 2 \times 10^5 = 200,000$

SCIENTIFIC CALCULATOR EXPLORATIONS

SCIENTIFIC NOTATION

To enter a number written in scientific notation on a scientific calculator, locate the key marked $\boxed{\text{EE}}$.

To enter 3.1×10^7, press $\boxed{3.1}$ $\boxed{\text{EE}}$ $\boxed{7}$. The display should read $\boxed{3.1 \quad 07}$.

Enter the following numbers written in scientific notation on your calculator.

1. 5.31×10^3
2. -4.8×10^{14}
3. 6.6×10^{-9}
4. -9.9811×10^{-2}

Multiply the following on your calculator. Notice the form of the result.

5. $3,000,000 \times 5,000,000$
6. $230,000 \times 1,000$

Multiply the following on your calculator. Write the product in scientific notation.

7. $(3.26 \times 10^6)(2.5 \times 10^{13})$
8. $(8.76 \times 10^{-4})(1.237 \times 10^9)$

MENTAL MATH

State each expression using positive exponents.

1. $5x^{-2}$
2. $3x^{-3}$
3. $\dfrac{1}{y^{-6}}$
4. $\dfrac{1}{x^{-3}}$
5. $\dfrac{4}{y^{-3}}$
6. $\dfrac{16}{y^{-7}}$

EXERCISE SET 4.5

Simplify each expression. Write results with positive exponents. See Examples 1 and 2.

1. 4^{-3}
2. 6^{-2}
3. $7x^{-3}$
4. $(7x)^{-3}$
5. $\left(-\dfrac{1}{4}\right)^{-3}$
6. $\left(-\dfrac{1}{8}\right)^{-2}$
7. $3^{-1} + 2^{-1}$
8. $4^{-1} + 4^{-2}$
9. $\dfrac{1}{p^{-3}}$
10. $\dfrac{1}{q^{-5}}$

Simplify each expression. Write results with positive exponents. See Example 3.

11. $\dfrac{p^{-5}}{q^{-4}}$
12. $\dfrac{r^{-5}}{s^{-2}}$
13. $\dfrac{x^{-2}}{x}$
14. $\dfrac{y}{y^{-3}}$
15. $\dfrac{z^{-4}}{z^{-7}}$
16. $\dfrac{x^{-4}}{x^{-1}}$

17. It was stated earlier that, for an integer n,
$$x^{-n} = \dfrac{1}{x^n}, \quad x \neq 0$$
Explain why x may not equal 0.

18. If $a = \dfrac{1}{10}$, then find the value of a^{-2}.

Simplify the following. Write results with positive exponents only. See Example 4.

19. $(a^{-5}b^2)^{-6}$
20. $(4^{-1}x^5)^{-2}$
21. $\left(\dfrac{x^{-2}y^4}{x^3y^7}\right)^2$
22. $\left(\dfrac{a^5b}{a^7b^{-2}}\right)^{-3}$
23. $\dfrac{4^2z^{-3}}{4^3z^{-5}}$
24. $\dfrac{3^{-1}x^4}{3^3x^{-7}}$

25. Explain why $(a^{-1})^3$ has the same value as $(a^3)^{-1}$.
26. Determine whether each statement is true or false.
    a. $5^{-1} < 5^{-2}$
    b. $\left(\dfrac{1}{5}\right)^{-1} < \left(\dfrac{1}{5}\right)^{-2}$
    c. $a^{-1} < a^{-2}$ for all nonzero numbers.

Simplify the following. Write results with positive exponents.

27. $(-3)^{-2}$
28. $(-2)^{-4}$
29. $\dfrac{-1}{p^{-4}}$
30. $\dfrac{-1}{y^{-6}}$
31. $-2^0 - 3^0$
32. $5^0 + (-5)^0$
33. $\dfrac{r}{r^{-3}r^{-2}}$
34. $\dfrac{p}{p^{-3}q^{-5}}$
35. $(x^5y^3)^{-3}$
36. $(z^5x^5)^{-3}$
37. $2^0 + 3^{-1}$
38. $4^{-2} - 4^{-3}$
39. $\dfrac{2^{-3}x^{-4}}{2^2x}$
40. $\dfrac{5^{-1}z^7}{5^{-2}z^9}$
41. $\dfrac{7ab^{-4}}{7^{-1}a^{-3}b^2}$
42. $\dfrac{6^{-5}x^{-1}y^2}{6^{-2}x^{-4}y^4}$
43. $\left(\dfrac{a^{-5}b}{ab^3}\right)^{-4}$
44. $\left(\dfrac{r^{-2}s^{-3}}{r^{-4}s^{-3}}\right)^{-3}$
45. $\dfrac{(xy^3)^5}{(xy)^{-4}}$
46. $\dfrac{(rs)^{-3}}{(r^2s^3)^2}$
47. $\dfrac{(-2xy^{-3})^{-3}}{(xy^{-1})^{-1}}$
48. $\dfrac{(-3x^2y^2)^{-2}}{(xyz)^{-2}}$

49. Find the volume of the cube.

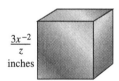

$\dfrac{3x^{-2}}{z}$ inches

50. Find the area of the triangle.

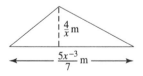

$\dfrac{4}{x}$ m

$\dfrac{5x^{-3}}{7}$ m

51. The product of a monomial and $7x^{-2}$ is $14x^5$. Find the monomial.

52. The product of an exponential expression and $-5y^3$ is $15y$. Find the expression.

Write each number in scientific notation. See Example 5.

53. 78,000
54. 9,300,000,000

55. 0.00000167
56. 0.00000017
57. 0.00635
58. 0.00194
59. 1,160,000
60. 700,000

61. The temperature at the interior of the Earth is 20,000,000 degrees Celsius. Write 20,000,000 in scientific notation.

62. The half-life of a carbon isotope is 5000 years. Write 5000 in scientific notation.

63. The distance between the earth and the sun is 93,000,000 miles. Write 93,000,000 in scientific notation.

64. The population of the world is 5,506,000,000. Write 5,506,000,000 in scientific notation.

Write each number in standard notation. See Example 6.

65. 7.86×10^8
66. 1.43×10^7
67. 8.673×10^{-10}
68. 9.056×10^{-4}
69. 3.3×10^{-2}
70. 4.8×10^{-6}
71. 2.032×10^4
72. 9.07×10^{10}

73. One coulomb of electricity is 6.25×10^{18}. Write this number in standard notation.

74. The mass of a hydrogen atom is 1.7×10^{-24} grams. Write this number in standard notation.

75. The distance light travels in 1 year is 9.460×10^{12} kilometers. Write this number in standard notation.

76. The population of the United States is 2.48×10^8. Write this number in standard notation.

Evaluate the following expressions using exponential rules. Write the results in standard notation. See Example 7.

77. $(1.2 \times 10^{-3})(3 \times 10^{-2})$
78. $(2.5 \times 10^6)(2 \times 10^{-6})$

79. $(4 \times 10^{-10})(7 \times 10^{-9})$
80. $(5 \times 10^6)(4 \times 10^{-8})$
81. $\dfrac{8 \times 10^{-1}}{16 \times 10^5}$
82. $\dfrac{25 \times 10^{-4}}{5 \times 10^{-9}}$
83. $\dfrac{1.4 \times 10^{-2}}{7 \times 10^{-8}}$
84. $\dfrac{0.4 \times 10^5}{0.2 \times 10^{11}}$

85. The average amount of water flowing past the mouth of the Amazon River is 4.2×10^6 cubic feet per second. How much water flows past in an hour? (1 hour equals 3600 seconds.) Write the result in scientific notation.

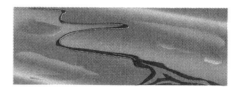

86. A beam of light travels 9.460×10^{12} kilometers per year. How far does light travel in 10,000 years? Write the result in scientific notation.

87. The total force (F) against the face of a dam that is 100 feet long by 20 feet high is given by the following:

$$F = \dfrac{(6.24 \times 10)(4 \times 10^4)}{2}$$

Compute the force and express it in scientific notation.

88. Suppose $1000 is invested at a rate of 9% and compounded monthly. The amount (A) after one year is given by

$$A = (1 \times 10^3)(1.09381)$$

Compute this amount in standard notation.

Simplify. Write results in standard notation.

89. $(2.63 \times 10^{12})(-1.5 \times 10^{-10})$
90. $(6.785 \times 10^{-4})(4.68 \times 10^{10})$

Light travels at a rate of 1.86×10^5 miles per second. Use this information and the distance formula $d = r \cdot t$ to answer Exercises 91 and 92.

91. If the distance from the moon to the Earth is 238,857 miles, find how long it takes the reflected light of the moon to reach the Earth. (Round to the nearest tenth of a second.)

92. If the distance from the sun to the earth is 93,000,000 miles, find how long it takes the light of the sun to reach the earth. (Round to the nearest tenth of a second.)

Review Exercises

Simplify the following. See Section 4.1.

93. $\dfrac{5x^7}{3x^4}$

94. $\dfrac{27y^{14}}{3y^7}$

95. $\dfrac{15z^4 y^3}{21zy}$

96. $\dfrac{18a^7 b^{17}}{30a^7 b}$

Use the distributive property and multiply. See Section 4.3.

97. $\dfrac{1}{y}(5y^2 - 6y + 5)$

98. $\dfrac{2}{x}(3x^5 + x^4 - 2)$

99. $2x^2\left(10x - 6 + \dfrac{1}{x}\right)$

100. $-5y^3\left(2y^2 - 4y + 2 - \dfrac{3}{y}\right)$

A Look Ahead

EXAMPLE

Simplify the following expressions. Assume that the variable in the exponent represents an integer value.

a. $x^{m+1} \cdot x^m$ **b.** $(z^{2x+1})^x$ **c.** $\dfrac{y^{6a}}{y^{4a}}$

Solution:

a. $x^{m+1} \cdot x^m = x^{(m+1)+m} = x^{2m+1}$

b. $(z^{2x+1})^x = z^{(2x+1)x} = z^{2x^2+x}$

c. $\dfrac{y^{6a}}{y^{4a}} = y^{6a-4a} = y^{2a}$

Simplify each expression. Assume that variables represent positive integers. See the example.

101. $a^{-4m} \cdot a^{5m}$

102. $(x^{-3s})^3$

103. $(3y^{2z})^3$

104. $a^{4m+1} \cdot a^4$

105. $\dfrac{y^{4a}}{y^{-a}}$

106. $\dfrac{y^{-6a}}{zy^{6a}}$

107. $(z^{3a+2})^{-2}$

108. $(a^{4x-1})^{-1}$

4.6 DIVISION OF POLYNOMIALS

OBJECTIVES

1. Divide a polynomial by a monomial.
2. Use long division to divide a polynomial by another polynomial.

TAPE BA 4.6

Although we didn't use the word monomial, in Sections 4.1 and 4.5 we divided monomials in developing rules and definitions for exponents. For example, $\dfrac{36a^2 b}{6ab^3}$ can be thought of as the monomial $36a^2 b$ divided by the monomial $6ab^3$.

EXAMPLE 1 Simplify by performing the indicated division of monomials. Write each answer with positive exponents.

a. $\dfrac{36a^2 b}{6ab^3}$

b. $\dfrac{3x^4 y^5}{9x^4 y^2}$

Solution: To divide the monomials, use rules for exponents.

a. $\dfrac{36a^2 b}{6ab^3} = \dfrac{6a}{b^2}$

b. $\dfrac{3x^4 y^5}{9x^4 y^2} = \dfrac{y^3}{3}$

To divide a polynomial by a monomial, recall addition of fractions. Fractions that have a common denominator are added by adding the numerators:

$$\frac{a}{c} + \frac{b}{c} = \frac{a+b}{c}$$

If we read this equation from right to left and let a, b, and c be monomials, $c \neq 0$, the following emerges.

TO DIVIDE A POLYNOMIAL BY A MONOMIAL

Divide each term of the polynomial by the monomial.

$$\frac{a+b}{c} = \frac{a}{c} + \frac{b}{c}, \quad c \neq 0$$

Throughout this section, we assume that denominators are not 0.

EXAMPLE 2 Divide $(6m^2 + 2m)$ by $2m$.

Solution: Begin by writing the quotient in fraction form. Then divide each term of the polynomial $6m^2 + 2m$ by the monomial $2m$.

$$\frac{6m^2 + 2m}{2m} = \frac{6m^2}{2m} + \frac{2m}{2m} = 3m + 1$$

To check, multiply the quotient $(3m + 1)$ by the divisor $2m$ and see that the product is the dividend $(6m^2 + 2m)$.

$$2m(3m + 1) = 2m(3m) + 2m(1) = 6m^2 + 2m$$

The quotient $(3m + 1)$ is correct.

EXAMPLE 3 Simplify the quotient $\dfrac{8x^2y^2 - 16xy + 2x}{4xy}$.

Solution: $\dfrac{8x^2y^2 - 16xy + 2x}{4xy} = \dfrac{8x^2y^2}{4xy} - \dfrac{16xy}{4xy} + \dfrac{2x}{4xy}$ Divide each term by $4xy$.

$$= 2xy - 4 + \frac{1}{2y}$$

Notice that the quotient is not a polynomial because of the term $\frac{1}{2y}$. This expression is called a rational expression and we will study rational expressions further in Chapter 6. Although the quotient of two polynomials is not always a polynomial, we may still check by multiplying.

$$4xy\left(2xy - 4 + \frac{1}{2y}\right) = 4xy(2xy) + 4xy(-4) + 4xy\left(\frac{1}{2y}\right)$$
$$= 8x^2y^2 - 16xy + 2x$$

EXAMPLE 4 Simplify $\dfrac{12x^5y^6 - 6x^2y^2 + 9x^3y^4 + 3}{6x^2y^2}$.

Solution: Divide each term by $6x^2y^2$.

$$\dfrac{12x^5y^6 - 6x^2y^2 + 9x^3y^4 + 3}{6x^2y^2} = \dfrac{12x^5y^6}{6x^2y^2} - \dfrac{6x^2y^2}{6x^2y^2} + \dfrac{9x^3y^4}{6x^2y^2} + \dfrac{3}{6x^2y^2}$$

$$= 2x^3y^4 - 1 + \dfrac{3xy^2}{2} + \dfrac{1}{2x^2y^2}$$

Check this result by multiplying.

2 To divide a polynomial by a polynomial other than a monomial, we use a process known as long division. Polynomial long division is similar to number long division, so we review long division by dividing 13 into 3660.

$$\begin{array}{r} 281 \\ 13\overline{)3660} \end{array}$$

26↓	$2 \cdot 13 = 26$
106	Subtract and bring down the next digit in the dividend.
104↓	$8 \cdot 13 = 104$
20	Subtract and bring down the next digit in the dividend.
13	$1 \cdot 13 = 13$
7	Subtract. There are no more digits to bring down, so the remainder is 7.

The quotient is 281 R 7, which can be written as $281\,\dfrac{7}{13} \begin{array}{l}\leftarrow \text{remainder}\\ \leftarrow \text{divisor}\end{array}$. Recall that division can be checked by multiplication. To check a division problem such as this one, we see that

$$13 \cdot 281 + 7 = 3660$$

Now we demonstrate long division of polynomials.

EXAMPLE 5 Divide $(x^2 + 7x + 12)$ by $(x + 3)$.

Solution: $(x^2 + 7x + 12)$ is the **dividend polynomial** and $(x + 3)$ is the **divisor polynomial.**

$$\begin{array}{r} x \\ x + 3 \overline{)x^2 + 7x + 12} \\ \underline{x^2 + 3x} \\ 4x + 12 \end{array}$$

How many times does x divide x^2? $\dfrac{x^2}{x} = x$.

Multiply: $x(x + 3)$.

Subtract and bring down the next term.

To subtract, change the signs of these terms and add.

Next, repeat this process.

$$\begin{array}{r} x + 4 \\ x+3\overline{\smash{)}x^2 + 7x + 12} \\ \underline{x^2 + 3x} \\ 4x + 12 \\ \underline{4x + 12} \\ 0 \end{array}$$

How many times does x divide $4x$? $\dfrac{4x}{x} = 4$.

To subtract, change the signs of these terms and add. ⟶

Multiply: $4(x + 3)$.
Subtract. The remainder is 0.

Then $(x^2 + 7x + 12)$ divided by $(x + 3)$ is $(x + 4)$, the **quotient polynomial.** To check, see that

divisor · quotient + remainder = dividend

or

$(x + 3) \cdot (x + 4) + \quad 0 \quad = x^2 + 7x + 12$, the dividend, so the division checks.

EXAMPLE 6 Divide $6x^2 + 10x - 5$ by $3x - 1$.

Solution: The divisor polynomial is $(3x - 1)$ and the dividend polynomial is $(6x^2 + 10x - 5)$.

$$\begin{array}{r} 2x + 4 \\ 3x-1\overline{\smash{)}6x^2 + 10x - 5} \\ \underline{6x^2 - 2x}\downarrow \\ 12x - 5 \\ \underline{12x - 4} \\ -1 \end{array}$$

$\dfrac{6x^2}{3x} = 2x$, so $2x$ is a term of the quotient.
$2x(3x - 1)$
Subtract and bring down the next term.
$\dfrac{12x}{3x} = 4$, $4(3x - 1)$
Subtract. The remainder is -1.

Then $(6x^2 + 10x - 5)$ divided by $(3x - 1)$ is $(2x + 4)$ with a remainder of -1. This can be written as

$$\dfrac{6x^2 + 10x - 5}{3x - 1} = 2x + 4 + \dfrac{-1}{3x - 1} \begin{pmatrix} \text{remainder} \\ \text{divisor} \end{pmatrix}$$

We call $(2x + 4)$ the quotient polynomial and -1 the **remainder polynomial.** To check, see that divisor · quotient + remainder = dividend.

$$(3x - 1)(2x + 4) + (-1) = (6x^2 + 12x - 2x - 4) - 1$$
$$= 6x^2 + 10x - 5$$

The division checks.

Notice that the division process is continued until the degree of the remainder polynomial is less than the degree of the divisor polynomial.

270 CHAPTER 4 EXPONENTS AND POLYNOMIALS

EXAMPLE 7 Divide: $\dfrac{4x^2 + 7 + 8x^3}{2x + 3}$.

Solution: Before we begin the division process, the dividend polynomial and the divisor polynomial should be written in descending order of exponents. Any missing powers are represented by a term whose coefficient is zero.

$$\frac{4x^2 + 7 + 8x^3}{2x + 3} = \frac{8x^3 + 4x^2 + 0x + 7}{2x + 3} \quad \text{There is no } x \text{ term, so include } 0x \text{ as the missing power.}$$

$$\begin{array}{r}
4x^2 - 4x + 6 \\
2x + 3 \overline{) 8x^3 + 4x^2 + 0x + 7} \\
\underline{8x^3 + 12x^2} \\
-8x^2 + 0x \\
\underline{-8x^2 - 12x} \\
12x + 7 \\
\underline{12x + 18} \\
-11 \quad \text{Remainder polynomial.}
\end{array}$$

Thus, $\dfrac{4x^2 + 7 + 8x^3}{2x + 3} = 4x^2 - 4x + 6 + \dfrac{-11}{2x + 3}$.

EXAMPLE 8 Divide: $\dfrac{2x^4 - x^3 + 3x^2 + x - 1}{x^2 + 1}$.

Solution: Before dividing, rewrite the divisor polynomial $(x^2 + 1)$ as $(x^2 + 0x + 1)$. The $0x$ term represents the missing x^1 term in the divisor.

$$\begin{array}{r}
2x^2 - x + 1 \\
x^2 + 0x + 1 \overline{) 2x^4 - x^3 + 3x^2 + x - 1} \\
\underline{2x^4 + 0x^3 + 2x^2} \\
-x^3 + x^2 + x \\
\underline{-x^3 - 0x^2 - x} \\
x^2 + 2x - 1 \\
\underline{x^2 + 0x + 1} \\
2x - 2 \quad \text{Remainder polynomial.}
\end{array}$$

Thus, $\dfrac{2x^4 - x^3 + 3x^2 + x - 1}{x^2 + 1} = 2x^2 - x + 1 + \dfrac{2x - 2}{x^2 + 1}$.

MENTAL MATH

Simplify each expression mentally.

1. $\dfrac{a^6}{a^4}$
2. $\dfrac{y^2}{y}$
3. $\dfrac{a^3}{a}$

4. $\dfrac{p^8}{p^3}$

5. $\dfrac{k^5}{k^2}$

6. $\dfrac{k^7}{k^5}$

EXERCISE SET 4.6

Simplify each expression. See Example 1.

1. $\dfrac{8k^4}{2k}$
2. $\dfrac{27r^4}{3r^6}$
3. $\dfrac{-6m^4}{-2m^3}$
4. $\dfrac{15a^4}{-15a^5}$
5. $\dfrac{-24a^6b}{6ab^2}$
6. $\dfrac{-5x^4y^5}{15x^4y^2}$
7. $\dfrac{6x^2y^3}{-7xy^5}$
8. $\dfrac{-8xa^2b}{-5xa^5b}$

Perform each division. See Examples 2 through 4.

9. $\dfrac{15p^3 + 18p^2}{3p}$
10. $\dfrac{14m^2 - 27m^3}{7m}$
11. $\dfrac{-9x^4 + 18x^5}{6x^5}$
12. $\dfrac{6x^5 + 3x^4}{3x^4}$
13. $\dfrac{-9x^5 + 3x^4 - 12}{3x^3}$
14. $\dfrac{6a^2 - 4a + 12}{2a^2}$
15. $\dfrac{4x^4 - 6x^3 + 7}{-4x^4}$
16. $\dfrac{-12a^3 + 36a - 15}{3a}$
17. $\dfrac{25x^5 - 15x^3 + 5}{5x^2}$
18. $\dfrac{-4y^2 + 4y + 6}{2y}$

19. The perimeter of a square is $(12x^3 + 4x - 16)$ feet. Find the length of its side.

20. The volume of the swimming pool shown is $(36x^5 - 12x^3 + 6x^2)$ cubic feet. If its height is $2x$ feet and its width is $3x$ feet, find its length.

Perform each division. See Examples 5 through 8.

21. $\dfrac{x^2 + 4x + 3}{x + 3}$
22. $\dfrac{x^2 + 7x + 10}{x + 5}$
23. $\dfrac{2x^2 + 13x + 15}{x + 5}$
24. $\dfrac{3x^2 + 8x + 4}{x + 2}$
25. $\dfrac{2x^2 - 7x + 3}{x - 4}$
26. $\dfrac{3x^2 - x - 4}{x - 1}$
27. $\dfrac{8x^2 + 6x - 27}{2x - 3}$
28. $\dfrac{18w^2 + 18w - 8}{3w + 4}$
29. $\dfrac{9a^3 - 3a^2 - 3a + 4}{3a + 2}$
30. $\dfrac{-x^3 - 6x^2 + 2x - 3}{x - 1}$
31. $\dfrac{2b^3 + 9b^2 + 6b - 4}{b + 4}$
32. $\dfrac{2x^3 + 3x^2 - 3x + 4}{x + 2}$

33. Explain how to check a polynomial long division result when the remainder is 0.

34. Explain how to check a polynomial long division result when the remainder is not 0.

35. The area of the following parallelogram is $(10x^2 + 31x + 15)$ square meters. If its base is $(5x + 3)$ meters, find its height.

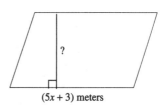

36. The area of the top of the Ping-Pong table is $(49x^2 + 70x - 200)$ square inches. If its length is $(7x + 20)$ inches, find its width.

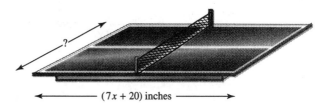

Perform each division.

37. $\dfrac{20x^2 + 5x + 9}{5x^3}$

38. $\dfrac{8x^3 - 4x^2 + 6x + 2}{2x^2}$

39. $\dfrac{5x^2 + 28x - 10}{x + 6}$

40. $\dfrac{2x^2 + x - 15}{x + 3}$

41. $\dfrac{10x^3 - 24x^2 - 10x}{10x}$

42. $\dfrac{2x^3 + 12x^2 + 16}{4x^2}$

43. $\dfrac{6x^2 + 17x - 4}{x + 3}$

44. $\dfrac{2x^2 - 9x + 15}{x - 6}$

45. $\dfrac{12x^4 + 3x^2}{3x^2}$

46. $\dfrac{15x^2 - 9x^5}{9x^5}$

47. $\dfrac{2x^3 + 2x^2 - 17x + 8}{x - 2}$

48. $\dfrac{4x^3 + 11x^2 - 8x - 10}{x + 3}$

49. $\dfrac{30x^2 - 17x + 2}{5x - 2}$

50. $\dfrac{4x^2 - 13x - 12}{4x + 3}$

51. $\dfrac{3x^4 - 9x^3 + 12}{-3x}$

52. $\dfrac{8y^6 - 3y^2 - 4y}{4y}$

53. $\dfrac{8x^2 + 10x + 1}{2x + 1}$

54. $\dfrac{3x^2 + 17x + 7}{3x + 2}$

55. $\dfrac{4x^2 - 81}{2x - 9}$

56. $\dfrac{16x^2 - 36}{4x + 6}$

57. $\dfrac{4x^3 + 12x^2 + x - 12}{2x + 3}$

58. $\dfrac{6x^2 + 11x - 10}{3x - 2}$

59. $\dfrac{x^3 - 27}{x - 3}$

60. $\dfrac{x^3 + 64}{x + 4}$

61. $\dfrac{x^3 + 1}{x + 1}$

62. $\dfrac{x^5 + x^2}{x^2 + x}$

63. $\dfrac{1 - 3x^2}{x + 2}$

64. $\dfrac{7 - 5x^2}{x + 3}$

65. $\dfrac{-4b + 4b^2 - 5}{2b - 1}$

66. $\dfrac{-3y + 2y^2 - 15}{2y + 5}$

Review Exercises

Multiply each expression. See Section 4.3.

67. $2a(a^2 + 1)$
68. $-4a(3a^2 - 4)$
69. $2x(x^2 + 7x - 5)$
70. $4y(y^2 - 8y - 4)$
71. $-3xy(xy^2 + 7x^2y + 8)$
72. $-9xy(4xyz + 7xy^2z + 2)$
73. $9ab(ab^2c + 4bc - 8)$
74. $-7sr(6s^2r + 9sr^2 + 9rs + 8)$

Use the bar graph below to answer exercises 75 through 78. See Section 1.9.

75. Which album has sold the most copies?
76. Estimate how many more copies the album *Rumours* has sold than the album *Boston*.
77. Which album(s) shown has sold the least copies?
78. If Michael Jackson made $7 for every *Thriller* album sold, what was his income for this album?

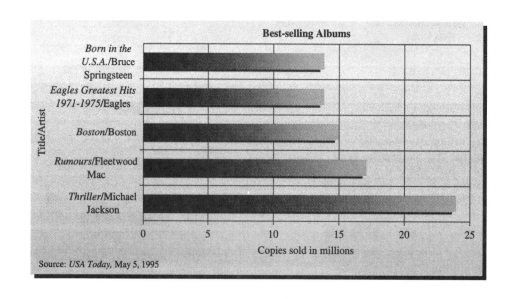

GROUP ACTIVITY

MAKING PREDICTIONS BASED ON HISTORICAL DATA

MATERIALS:
- Calculator
- Grapher with bar graph capabilities (optional)

According to data from the National Safety Council, a polynomial that represents the annual number of deaths due to motor vehicle accidents over the period 1990–1994 is

$$46{,}426 - 4381x + 893x^2,$$

where $x = 0$ represents 1990. A polynomial, also based on data from the National Safety Council, that represents the annual number of deaths due to all other types of accidents over the period 1990–1994 is

$$33{,}139 - 2540x + 730x^2,$$

where $x = 0$ represents 1990.

1. Use the polynomials to complete the following table showing the number of accidental deaths per year over the period 1990–1994 by evaluating each polynomial at the given values of x. What trends do you notice in the data?

YEAR	x	NUMBER OF ACCIDENTAL DEATHS DUE TO MOTOR VEHICLE ACCIDENTS	NUMBER OF ACCIDENTAL DEATHS DUE TO ALL OTHER TYPES OF ACCIDENTS
1990	0		
1991	1		
1992	2		
1993	3		
1994	4		

2. Use the given polynomials to find a polynomial that represents the total number of accidental deaths per year. Evaluate this new polynomial to find the total number of accidental deaths for the years 1990–1994.

3. Numerically verify the values you calculated in Question 2 by adding a new column, titled "Total Number of Accidental Deaths Due to All Causes," to the table in Question 1 and combining values from the table to complete the new column.

4. Use the new polynomial representing total number of accidental deaths per year to predict the number of accidental deaths in the years 2000 and 2002.

5. Create a bar graph that represents the data for total accidental deaths per year for the years 1990–1994 and your predictions for 2000 and 2002. Study your bar graph. Discuss what the graph implies about the future.

6. (Optional) Use a grapher to create the bar graph in Question 5. Compare it to your own graph. Can you draw the same conclusions? Explain.

Chapter 4 Highlights

Definitions and Concepts	Examples
Section 4.1 Exponents	
a^n means the product of n factors, each of which is a.	$3^2 = 3 \cdot 3 = 9$ $(-5)^3 = (-5)(-5)(-5) = -125$ $\left(\dfrac{1}{2}\right)^4 = \dfrac{1}{2} \cdot \dfrac{1}{2} \cdot \dfrac{1}{2} \cdot \dfrac{1}{2} = \dfrac{1}{16}$
If m and n are integers and no denominators are 0, Product Rule: $a^m \cdot a^n = a^{m+n}$ Power of a Power Rule: $(a^m)^n = a^{mn}$ Power of a Product Rule: $(ab)^n = a^n b^n$ Power of a Quotient Rule: $\left(\dfrac{a}{b}\right)^n = \dfrac{a^n}{b^n}$ Quotient Rule: $\dfrac{a^m}{a^n} = a^{m-n}$ Zero Exponent: $a^0 = 1, a \ne 0$.	$x^2 \cdot x^7 = x^{2+7} = x^9$ $(5^3)^8 = 5^{3 \cdot 8} = 5^{24}$ $(7y)^4 = 7^4 y^4$ $\left(\dfrac{x}{8}\right)^3 = \dfrac{x^3}{8^3}$ $\dfrac{x^9}{x^4} = x^{9-4} = x^5$ $5^0 = 1, x^0 = 1, x \ne 0$.
Section 4.2 Adding and Subtracting Polynomials	
A **term** is a number or the product of numbers and variables raised to powers.	Terms $-5x, 7a^2 b, \dfrac{1}{4} y^4, 0.2$
The **numerical coefficient** or **coefficient** of a term is its numerical factor.	Term Coefficient $7x^2$ 7 y 1 $-a^2 b$ -1
A **polynomial** is a term or a finite sum of terms in which variables may appear in the numerator raised to whole number powers only. A **monomial** is a polynomial with exactly 1 term. A **binomial** is a polynomial with exactly 2 terms. A **trinomial** is a polynomial with exactly 3 terms.	Polynomials $3x^2 - 2x + 1$ (Trinomial) $-0.2a^2 b - 5b^2$ (Binomial) $\dfrac{5}{6} y^3$ (Monomial)
The **degree of a term** is the sum of the exponents on the variables in the term.	Term Degree $-5x^3$ 3 3 (or $3x^0$) 0 $2a^2 b^2 c$ 5

(continued)

DEFINITIONS AND CONCEPTS	EXAMPLES
SECTION 4.2 ADDING AND SUBTRACTING POLYNOMIALS	
The **degree of a polynomial** is the greatest degree of any term of the polynomial.	Polynomial Degree $5x^2 - 3x + 2$ 2 $7y + 8y^2z^3 - 12$ $2 + 3 = 5$
To add polynomials, add like terms.	Add: $(7x^2 - 3x + 2) + (-5x - 6) = 7x^2 - 3x + 2 - 5x - 6$ $= 7x^2 - 8x - 4$
To subtract two polynomials, change the signs of the terms of the second polynomial, then add.	Subtract: $(17y^2 - 2y + 1) - (-3y^3 + 5y - 6)$ $= (17y^2 - 2y + 1) + (3y^3 - 5y + 6)$ $= 17y^2 - 2y + 1 + 3y^3 - 5y + 6$ $= 3y^3 + 17y^2 - 7y + 7$
SECTION 4.3 MULTIPLYING POLYNOMIALS	
To multiply two polynomials, multiply each term of one polynomial by each term of the other polynomial, and then combine like terms.	Multiply: $(2x + 1)(5x^2 - 6x + 2)$ $= 2x(5x^2 - 6x + 2) + 1(5x^2 - 6x + 2)$ $= 10x^3 - 12x^2 + 4x + 5x^2 - 6x + 2$ $= 10x^3 - 7x^2 - 2x + 2$
SECTION 4.4 SPECIAL PRODUCTS	
The **FOIL method** may be used when multiplying two binomials.	Multiply: $(5x - 3)(2x + 3)$ F O I L $(5x - 3)(2x + 3) = (5x)(2x) + (5x)(3) + (-3)(2x) + (-3)(3)$ $= 10x^2 + 15x - 6x - 9$ $= 10x^2 + 9x - 9$
Squaring a Binomial $(a + b)^2 = a^2 + 2ab + b^2$ $(a - b)^2 = a^2 - 2ab + b^2$	Square each binomial. $(x + 5)^2 = x^2 + 2(x)(5) + 5^2$ $= x^2 + 10x + 25$ $(3x - 2y)^2 = (3x)^2 - 2(3x)(2y) + (2y)^2$ $= 9x^2 - 12xy + 4y^2$
Multiplying the Sum and Difference of Two Terms $(a + b)(a - b) = a^2 - b^2$	Multiply. $(6y + 5)(6y - 5) = (6y)^2 - 5^2$ $= 36y^2 - 25$

(continued)

DEFINITIONS AND CONCEPTS	EXAMPLES
SECTION 4.5 NEGATIVE EXPONENTS AND SCIENTIFIC NOTATION	
If $a \neq 0$ and n is an integer, $$a^{-n} = \frac{1}{a^n}$$ Rules for exponents are true for positive and negative integers.	$3^{-2} = \frac{1}{3^2} = \frac{1}{9}$; $5x^{-2} = \frac{5}{x^2}$ Simplify: $\left(\frac{x^{-2}y}{x^5}\right)^{-2} = \frac{x^4 y^{-2}}{x^{-10}}$ $= x^{4-(-10)}y^{-2}$ $= \frac{x^{14}}{y^2}$
A positive number is written in scientific notation if it is as the product of a number a, $1 \leq a < 10$, and an integer power r of 10. $$a \times 10^r$$	Numbers Written in Scientific Notation $12{,}000 = 1.2 \times 10^4$ $0.00000568 = 5.68 \times 10^{-6}$
SECTION 4.6 DIVISION OF POLYNOMIALS	
To divide a polynomial by a monomial: $$\frac{a+b}{c} = \frac{a}{c} + \frac{b}{c}$$ To divide a polynomial by a polynomial other than a monomial, use long division.	Divide: $$\frac{15x^5 - 10x^3 + 5x^2 - 2x}{5x^2} = \frac{15x^5}{5x^2} - \frac{10x^3}{5x^2} + \frac{5x^2}{5x^2} - \frac{2x}{5x^2}$$ $$= 3x^3 - 2x + 1 - \frac{2}{5x}$$ $$5x - 1 + \frac{-4}{2x+3}$$ $2x+3\overline{)10x^2 + 13x - 7}$ $\underline{10x^2 + 15x}$ $-2x - 7$ $\underline{-2x - 3}$ -4

CHAPTER 4 REVIEW

(4.1) State the base and the exponent for each expression.

1. 3^2 **2.** $(-5)^4$

3. -5^4

Evaluate each expression.

4. 8^3 **5.** $(-6)^2$

6. -6^2 **7.** $-4^3 - 4^0$

8. $(3b)^0$ **9.** $\dfrac{8b}{8b}$

Simplify each expression.

10. $5b^3 b^5 a^6$ **11.** $2^3 \cdot x^0$

12. $[(-3)^2]^3$ **13.** $(2x^3)(-5x^2)$

14. $\left(\dfrac{mn}{q}\right)^2 \cdot \left(\dfrac{mn}{q}\right)$ **15.** $\left(\dfrac{3ab^2}{6ab}\right)^4$

16. $\dfrac{x^9}{x^4}$ **17.** $\dfrac{2x^7 y^8}{8xy^2}$

18. $\dfrac{12xy^6}{3x^4 y^{10}}$ **19.** $5a^7(2a^4)^3$

20. $(2x)^2(9x)$

21. $\dfrac{(-4)^2(3^3)}{(4^5)(3^2)}$

22. $\dfrac{(-7)^2(3^5)}{(-7)^3(3^4)}$

23. $\dfrac{(2x)^0(-4)^2}{16x}$

24. $\dfrac{(8xy)(3xy)}{18x^2y^2}$

25. $m^0 + p^0 + 3q^0$

26. $(-5a)^0 + 7^0 + 8^0$

27. $(3xy^2 + 8x + 9)^0$

28. $8x^0 + 9^0$

29. $6(a^2b^3)^3$

30. $\dfrac{(x^3z)^a}{x^2z^2}$

(4.2) *Find the degree of each term.*

31. $-5x^4y^3$

32. $10x^3y^2z$

33. $35a^5bc^2$

34. $95xyz$

Find the degree of each polynomial.

35. $y^5 + 7x - 8x^4$

36. $9y^2 + 30y + 25$

37. $-14x^2yb - 28x^2y^3b - 42x^2y^2$

38. $6x^2y^2z^2 + 5x^2y^3 - 12xyz$

39. The surface area of a box with a square base and a height of 5 units is given by the polynomial $2x^2 + 20x$. Fill in the table below by evaluating $2x^2 + 20x$ for the given values of x.

x	1	3	5.1	10
$2x^2 + 20x$				

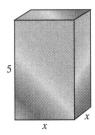

Combine like terms.

40. $6a^2b^2 + 4ab + 9a^2b^2$

41. $21x^2y^3 + 3xy + x^2y^3 + 6$

42. $4a^2b - 3b^2 - 8q^2 - 10a^2b + 7q^2$

43. $2s^{14} + 3s^{13} + 12s^{12} - s^{10}$

Add or subtract as indicated.

44. $(3k^2 + 2k + 6) + (5k^2 + k)$

45. $(2s^5 + 3s^4 + 4s^3 + 5s^2) - (4s^2 + 7s + 6)$

46. $(2m^7 + 3x^4 + 7m^6) - (8m^7 + 4m^2 + 6x^4)$

47. Subtract $(4x^2 + 8x - 7)$ from the sum of $(x^2 + 7x + 9)$ and $(x^2 + 4)$.

(4.3) *Multiply each expression.*

48. $9x(x^2y)$

49. $-7(8xz^2)$

50. $(6xa^2)(xya^3)$

51. $(4xy)(-3xa^2y^3)$

52. $6(x + 5)$

53. $9(x - 7)$

54. $4(2a + 7)$

55. $9(6a - 3)$

56. $-7x(x^2 + 5)$

57. $-8y(4y^2 - 6)$

58. $-2(x^3 - 9x^2 + x)$

59. $-3a(a^2b + ab + b^2)$

60. $(3a^3 - 4a + 1)(-2a)$

61. $(6b^3 - 4b + 2)(7b)$

62. $(2x + 2)(x - 7)$

63. $(2x - 5)(3x + 2)$

64. $(4a - 1)(a + 7)$

65. $(6a - 1)(7a + 3)$

66. $(x + 7)(x^3 + 4x - 5)$

67. $(x + 2)(x^5 + x + 1)$

68. $(x^2 + 2x + 4)(x^2 + 2x - 4)$

69. $(x^3 + 4x + 4)(x^3 + 4x - 4)$

70. $(x + 7)^3$

71. $(2x - 5)^3$

(4.4) *Use the special product rules to find each product.*

72. $(x + 7)^2$

73. $(x - 5)^2$

74. $(3x - 7)^2$

75. $(4x + 2)^2$

76. $(5x - 9)^2$

77. $(5x + 1)(5x - 1)$

78. $(7x + 4)(7x - 4)$

79. $(a + 2b)(a - 2b)$

80. $(2x - 6)(2x + 6)$

81. $(4a^2 - 2b)(4a^2 + 2b)$

(4.5) *Simplify each expression.*

82. 7^{-2}

83. -7^{-2}

84. $2x^{-4}$

85. $(2x)^{-4}$

86. $\left(\dfrac{1}{5}\right)^{-3}$

87. $\left(\dfrac{-2}{3}\right)^{-2}$

88. $2^0 + 2^{-4}$

89. $6^{-1} - 7^{-1}$

Simplify each expression. Assume that variables in an exponent represent positive integers only. Write each answer using positive exponents.

90. $\dfrac{1}{(2q)^{-3}}$

91. $\dfrac{-1}{(qr)^{-3}}$

92. $\dfrac{r^{-3}}{s^{-4}}$

93. $\dfrac{rs^{-3}}{r^{-4}}$

94. $\dfrac{-6}{8x^{-3}r^4}$

95. $\dfrac{-4s}{16s^{-3}}$

96. $(2x^{-5})^{-3}$

97. $(3y^{-6})^{-1}$

278 CHAPTER 4 EXPONENTS AND POLYNOMIALS

98. $(3a^{-1}b^{-1}c^{-2})^{-2}$
99. $(4x^{-2}y^{-3}z)^{-3}$
100. $\dfrac{5^{-2}x^8}{5^{-3}x^{11}}$
101. $\dfrac{7^5y^{-2}}{7^7y^{-10}}$
102. $\left(\dfrac{bc^{-2}}{bc^{-3}}\right)^4$
103. $\left(\dfrac{x^{-3}y^{-4}}{x^{-2}y^{-5}}\right)^{-3}$
104. $\dfrac{x^{-4}y^{-6}}{x^2y^7}$
105. $\dfrac{a^5b^{-5}}{a^{-5}b^5}$
106. $-2^0 + 2^{-4}$
107. $-3^{-2} - 3^{-3}$
108. $a^{6m}a^{5m}$
109. $\dfrac{(x^{5+h})^3}{x^5}$
110. $(3xy^{2z})^3$
111. $a^{m+2}a^{m+3}$

Write each number in scientific notation.

112. 0.00027
113. 0.8868
114. 80,800,000
115. −868,000
116. The population of California is 29,760,000. Write 29,760,000 in scientific notation.
117. The radius of the earth is 4000 miles. Write 4000 in scientific notation.

Write each number in standard form.

118. 8.67×10^5
119. 3.86×10^{-3}
120. 8.6×10^{-4}
121. 8.936×10^5
122. The number of photons of light emitted by a 100-watt bulb every second is 1×10^{20}. Write 1×10^{20} in standard notation.

123. The real mass of all the galaxies in the constellation of Virgo is 3×10^{-25}. Write 3×10^{-25} in standard notation.

Simplify. Express each result in standard form.

124. $(8 \times 10^4)(2 \times 10^{-7})$
125. $\dfrac{8 \times 10^4}{2 \times 10^{-7}}$

(4.6) *Perform each division.*

126. $\dfrac{4xy^2}{3xz^2y^3}$
127. $\dfrac{4xy^3}{32xy^2z}$
128. $\dfrac{x^2 + 21x + 49}{7x^2}$
129. $\dfrac{5a^3b - 15ab^2 + 20ab}{-5ab}$
130. $(a^2 - a + 4) \div (a - 2)$
131. $(4x^2 + 20x + 7) \div (x + 5)$
132. $\dfrac{a^3 + a^2 + 2a + 6}{a - 2}$
133. $\dfrac{9b^3 - 18b^2 + 8b - 1}{3b - 2}$
134. $\dfrac{4x^4 - 4x^3 + x^2 + 4x - 3}{2x - 1}$
135. $\dfrac{-10x^2 - x^3 - 21x + 18}{x - 6}$

CHAPTER 4 TEST

Evaluate each expression.

1. 2^5
2. $(-3)^4$
3. -3^4
4. 4^{-3}

Simplify each exponential expression.

5. $\left(\dfrac{5x^6y^3}{35x^7y}\right)^2$
6. $\dfrac{7(xy)^4}{(xy)^2}$
7. $4(x^2y^3)^{-3}$

Simplify each expression. Write the result using only positive exponents.

8. $\left(\dfrac{x^2y^3}{x^3y^{-4}}\right)^{-2}$
9. $\dfrac{6^2x^{-4}y^{-1}}{6^3x^{-3}y^7}$

Express each number in scientific notation.

10. 563,000
11. 0.0000863

Write each number in standard form.

12. 1.5×10^{-3}
13. 6.23×10^4
14. Simplify. Write the answer in standard form.
$(1.2 \times 10^5)(3 \times 10^{-7})$
15. Find the degree of the following polynomial.
$4xy^2 + 7xyz + 9x^3yz$
16. Simplify by combining like terms.
$6xyz + 9x^2y - 3xyz + 9x^2y$

Perform the indicated operations.

17. $(8x^3 + 7x^2 + 4x - 7) + (8x^3 - 7x - 6)$

18. $\quad 5x^3 + x^2 + 5x - 2$
$\quad -(8x^3 - 4x^2 + x - 7)$

t	0 seconds	1 second	3 seconds	5 seconds
$-16t^2 + 1001$				

19. Subtract $(4x + 2)$ from the sum of $(8x^2 + 7x + 5)$ and $(x^3 - 8)$.

20. Multiply: $(3x + 7)(x^2 + 5x + 2)$.

21. Multiply $x^3 - x^2 + x + 1$ by $2x^2 - 3x + 7$ using the vertical format.

22. Use the FOIL method to multiply $(x + 7)(3x - 5)$.

Use special products to multiply each of the following.

23. $(3x - 7)(3x + 7)$ **24.** $(4x - 2)^2$

25. $(8x + 3)^2$ **26.** $(x^2 - 9b)(x^2 + 9b)$

27. The height of the Bank of China in Hong Kong is 1001 feet. Neglecting air resistance, the height of an object dropped from this building at time t seconds is given by the polynomial $-16t^2 + 1001$. Find the height of the object at the given times below.

Divide.

28. $\dfrac{8xy^2}{4x^3y^3z}$

29. $\dfrac{4x^2 + 2xy - 7x}{8xy}$

30. $(x^2 + 7x + 10) \div (x + 5)$

31. $\dfrac{27x^3 - 8}{3x + 2}$

Chapter 4 Cumulative Review

1. Translate each sentence into a mathematical statement.
   a. Nine is less than or equal to eleven.
   b. Eight is greater than one.
   c. Three is not equal to four.

2. Write each of the following numbers as a product of primes.
   a. 40 **b.** 63

3. Add or subtract as indicated. Write each result in lowest terms.
   a. $\dfrac{2}{7} + \dfrac{4}{7}$ **b.** $\dfrac{3}{10} + \dfrac{2}{10}$
   c. $\dfrac{9}{7} - \dfrac{2}{7}$ **d.** $\dfrac{5}{3} - \dfrac{1}{3}$

4. Decide whether 2 is a solution of $3x + 10 = 8x$.

5. Subtract 8 from -4.

6. If $x = -2$ and $y = -4$, evaluate each expression.
   a. $5x - y$ **b.** $x^3 - y^2$ **c.** $\dfrac{3x}{2y}$

7. Simplify each expression by combining like terms.
   a. $2x + 3x + 5 + 2$
   b. $-5a - 3 + a + 2$
   c. $4y - 3y^2$
   d. $2.3x + 5x - 6$

8. Solve $-3x = 33$ for x.

9. Solve $3(x - 4) = 3x - 12$.

10. Solve $V = lwh$ for l.

11. The cost of a large hand-tossed pepperoni pizza at Domino's recently increased from $5.80 to $7.03. Find the percent increase.

12. Rajiv Puri invested part of his $20,000 inheritance in a mutual funds account that pays 7% simple interest yearly and the rest in a certificate of deposit that pays 9% simple interest yearly. At the end of one year, Rajiv's investments earned $1550. Find the amount he invested at each rate.

13. Graph the linear equation $-5x + 3y = 15$.

14. Graph the line with x-intercept 1 and y-intercept -1.

15. Find the slope of the line through $(-1, -2)$ and $(2, 4)$. Graph the line.

16. Simplify each of the following expressions.
    a. $(x^2)^5$
    b. $(y^8)^2$
    c. $[(-5)^3]^4$

17. Combine like terms.
    a. $-3x + 7x$
    b. $11x^2 + 5 + 2x^2 - 7$

18. Multiply: $(2x - y)^2$.

19. Use special products to square the following binomials.
    a. $(t + 2)^2$
    b. $(p - q)^2$
    c. $(2x + 3y)^2$
    d. $(5r - 7s)^2$

20. Simplify the following expressions. Write each result using positive exponents only.
    a. $(2x^3)(5x)^{-2}$
    b. $\left(\dfrac{3a^2}{b}\right)^{-3}$
    c. $\dfrac{4^{-1}x^{-3}y}{4^{-3}x^2y^{-6}}$
    d. $(y^{-3}z^{-6})^{-6}$
    e. $\left(\dfrac{-2x^3y}{xy^{-1}}\right)^3$

21. Simplify: $\dfrac{8x^2y^2 - 16xy + 2x}{4xy}$.

CHAPTER

5

FACTORING POLYNOMIALS

5.1 THE GREATEST COMMON FACTOR AND FACTORING BY GROUPING

5.2 FACTORING TRINOMIALS OF THE FORM $x^2 + bx + c$

5.3 FACTORING TRINOMIALS OF THE FORM $ax^2 + bx + c$

5.4 FACTORING BINOMIALS

5.5 CHOOSING A FACTORING STRATEGY

5.6 SOLVING QUADRATIC EQUATIONS BY FACTORING

5.7 QUADRATIC EQUATIONS AND PROBLEM SOLVING

CHOOSING AMONG BUILDING OPTIONS

Whether putting in a new floor, hanging new wallpaper, or retiling a bathroom, it is usually necessary to choose among several different building or decorating materials with different pricing schemes. Because it is often the case that only a fixed amount of money is available for such projects, it can be helpful to compare the choices by calculating how much area can be covered by a fixed dollar-value of the material.

IN THE CHAPTER GROUP ACTIVITY ON PAGE 333, YOU WILL HAVE THE OPPORTUNITY TO HELP SAM CHOOSE AMONG THREE DIFFERENT CHOICES OF MATERIALS FOR BUILDING A PATIO AROUND HIS NEW IN-GROUND SWIMMING POOL.

282 CHAPTER 5 FACTORING POLYNOMIALS

In Chapter 4, you learned how to multiply polynomials. This chapter deals with an operation that is the reverse process of multiplying, called factoring. Factoring is an important algebraic skill because this process allows us to write a sum as a product.

At the end of this chapter, we use factoring to help us solve equations other than linear equations, and in Chapter 6 we use factoring to simplify and perform arithmetic operations on rational expressions.

5.1 THE GREATEST COMMON FACTOR AND FACTORING BY GROUPING

TAPE
BA 5.1

OBJECTIVES

1. Find the greatest common factor of a list of integers.
2. Find the greatest common factor of a list of terms.
3. Factor out the greatest common factor from a polynomial.
4. Factor a polynomial by grouping.

When an integer is written as the product of two or more other integers, each of these integers is called a **factor** of the product. This is true for polynomials, also. When a polynomial is written as the product of two or more other polynomials, each of these polynomials is called a factor of the product. The process of writing a polynomial as a product is called **factoring.**

$$2 \cdot 3 = 6 \qquad x^2 \cdot x^3 = x^5 \qquad (x+2)(x+3) = x^2 + 5x + 6$$
factor factor product $\quad$ factor factor product $\quad$ factor factor product

Notice that factoring is the reverse process of multiplying.

$$\overset{\text{factoring}}{x^2 + 5x + 6 = (x+2)(x+3)}$$
$$\underset{\text{multiplying}}{}$$

We begin our study of factoring by reviewing the **greatest common factor (GCF).** The GCF of a list of integers is the largest integer that is a factor of all the integers in the list. For example, the GCF of 12 and 20 is 4 because 4 is the largest integer that is a factor of both 12 and 20. With large integers, the GCF may not be easily found by inspection. When this happens, use the following steps.

TO FIND THE GCF OF A LIST OF INTEGERS

Step 1. Write each number as a product of prime numbers.

Step 2. Identify the common prime factors.

Step 3. The product of all common prime factors found in step 2 is the greatest common factor. If there are no common prime factors, the greatest common factor is 1.

Recall from Section 1.2 that a prime number is a whole number other than 1, whose only factors are 1 and itself.

EXAMPLE 1 Find the GCF of each list of numbers.

a. 28 and 40 **b.** 55 and 21 **c.** 15, 18, and 66

Solution: **a.** Write each number as a product of primes.

$$28 = 2 \cdot 2 \cdot 7 = 2^2 \cdot 7$$
$$40 = 2 \cdot 2 \cdot 2 \cdot 5 = 2^3 \cdot 5$$

There are two common factors, each of which is 2, so the GCF is

$$\text{GCF} = 2 \cdot 2 = 4$$

b. $55 = 5 \cdot 11$
$21 = 3 \cdot 7$

There are no common prime factors; thus, the GCF is 1.

c. $15 = 3 \cdot 5$
$18 = 2 \cdot 3 \cdot 3 = 2 \cdot 3^2$
$66 = 2 \cdot 3 \cdot 11$

The only prime factor common to all three numbers is 3, so the GCF is

$$\text{GCF} = 3$$

The greatest common factor of a list of variables raised to powers is found in a similar way. For example, the GCF of x^2, x^3, and x^5 is x^2 because each term contains a factor of x^2 and no higher power of x is a factor of each term.

$$x^2 = x^2$$
$$x^3 = x^2 \cdot x$$
$$x^5 = x^2 \cdot x^3$$

From this example, we see that **the GCF of a list of common variables raised to powers is the variable raised to the smallest exponent in the list.**

EXAMPLE 2 Find the GCF of each list of terms.

a. x^3, x^7, and x^3 **b.** y, y^4, and y^7

Solution: **a.** The GCF is x^3, since 3 is the smallest exponent to which x is raised.
b. The GCF is y^1 or y, since 1 is the smallest exponent on y.

In general, **the GCF of a list of terms is the product of all common factors.**

EXAMPLE 3 Find the greatest common factor of each list of terms.

a. $6x^2, 10x^3$, and $-8x$ b. $8y^2, y^3$, and y^5 c. a^3b^2, a^5b, and a^6b^2

Solution: a. The GCF of the numerical coefficients 6, 10, and -8 is 2.
The GCF of variable factors x^2, x^3, and x is x.
Thus, the GCF of the terms $6x^2, 10x^3$, and $-8x$ is $2x$.

b. The GCF of the numerical coefficients 8, 1, and 1 is 1.
The GCF of variable factors y^2, y^3, and y^5 is y^2.
Thus, the GCF of terms $8y^2, y^3$, and y^5 is $1y^2$ or y^2.

c. The GCF of a^3, a^5, and a^6 is a^3.
The GCF of b^2, b, and b^2 is b. Thus, the GCF of the terms is a^3b.

The first step in factoring a polynomial is to find the GCF of its terms. Once we do so, we can write the polynomial as a product by **factoring out** the GCF.

The polynomial $8x + 14$, for example, contains two terms: $8x$ and 14. The GCF of these terms is 2. We factor out 2 from each term by writing each term as a product of 2 and the term's remaining factors.

$$8x + 14 = 2 \cdot 4x + 2 \cdot 7$$

Using the distributive property, we can write

$$8x + 14 = 2 \cdot 4x + 2 \cdot 7$$
$$= 2(4x + 7)$$

Thus, a factored form of $8x + 14$ is $2(4x + 7)$.

EXAMPLE 4 Factor each polynomial by factoring out the GCF.

a. $6t + 18$ b. $y^5 - y^7$

Solution: a. The GCF of terms $6t$ and 18 is 6.

$$6t + 18 = 6 \cdot t + 6 \cdot 3$$
$$= 6(t + 3) \quad \text{Apply the distributive property.}$$

Our work can be checked by multiplying 6 and $(t + 3)$.

$6(t + 3) = 6 \cdot t + 6 \cdot 3 = 6t + 18$, the original polynomial.

b. The GCF of y^5 and y^7 is y^5. Thus,

$$y^5 - y^7 = (y^5)\,1 - (y^5)\,y^2$$
$$= y^5(1 - y^2)$$

EXAMPLE 5 Factor $-9a^5 + 18a^2 - 3a$.

Solution: $-9a^5 + 18a^2 - 3a = (3a)(-3a^4) + (3a)(6a) + (3a)(-1)$
$ = 3a(-3a^4 + 6a - 1)$

In Example 5 we could have chosen to factor out a $-3a$ instead of $3a$. If we factor out a $-3a$, we have

$$-9a^5 + 18a^2 - 3a = (-3a)(3a^4) + (-3a)(-6a) + (-3a)(1)$$
$$= -3a(3a^4 - 6a + 1)$$

EXAMPLE 6 Factor $25x^4z + 15x^3z + 5x^2z$.

Solution: The greatest common factor is $5x^2z$.

$$25x^4z + 15x^3z + 5x^2z = 5x^2z(5x^2 + 3x + 1)$$

> **REMINDER** Be careful when the GCF of the terms is the same as one of the terms in the polynomial. The greatest common factor of the terms of $8x^2 - 6x^3 + 2x$ is $2x$. When factoring out $2x$ from the terms of $8x^2 - 6x^3 + 2x$, don't forget a term of 1.
>
> $$8x^2 - 6x^3 + 2x = 2x(4x) - 2x(3x^2) + 2x(1)$$
> $$= 2x(4x - 3x^2 + 1)$$
>
> Check by multiplying.
>
> $$2x(4x - 3x^2 + 1) = 8x^2 - 6x^3 + 2x$$

EXAMPLE 7 Factor $5(x + 3) + y(x + 3)$.

Solution: The binomial $(x + 3)$ is the greatest common factor. Use the distributive property to factor out $(x + 3)$.

$$5(x + 3) + y(x + 3) = (x + 3)(5 + y)$$

EXAMPLE 8 Factor $3m^2n(a + b) - (a + b)$.

Solution: The greatest common factor is $(a + b)$.

$$3m^2n(a + b) - 1(a + b) = (a + b)(3m^2n - 1)$$

4 Once the GCF is factored out, we can often continue to factor the polynomial, using a variety of techniques. We discuss here a technique for factoring polynomials called **grouping.**

EXAMPLE 9 Factor $xy + 2x + 3y + 6$ by grouping. Check by multiplying.

Solution: The GCF of the first two terms is x, and the GCF of the last two terms is 3.

$$xy + 2x + 3y + 6 = x(y + 2) + 3(y + 2)$$

Notice that $x(y + 2) + 3(y + 2)$ is not a factored form of the original polynomial because it is a sum and not a product.

Next, factor out the common binomial factor of $(y + 2)$.
$$x(y + 2) + 3(y + 2) = (y + 2)(x + 3)$$
To check, multiply $(y + 2)$ by $(x + 3)$.
$$(y + 2)(x + 3) = xy + 2x + 3y + 6, \text{ the original polynomial.}$$
Thus, the factored form of $xy + 2x + 3y + 6$ is $(y + 2)(x + 3)$.

TO FACTOR A FOUR-TERM POLYNOMIAL BY GROUPING

Step 1. Arrange the terms so that the first two terms have a common factor and the last two terms have a common factor.

Step 2. For each pair of terms, use the distributive property to factor out the pair's greatest common factor.

Step 3. If there is now a common binomial factor, factor it out.

Step 4. If there is no common binomial factor in step 3, begin again, rearranging the terms differently. If no rearrangement leads to a common binomial factor, the polynomial cannot be factored.

EXAMPLE 10 Factor $3x^2 + 4xy - 3x - 4y$ by grouping.

Solution: The first two terms have a common factor x. Factor -1 from the last two terms so that the common binomial factor of $(3x + 4y)$ appears.
$$3x^2 + 4xy - 3x - 4y = x(3x + 4y) - 1(3x + 4y)$$
Next, factor out the common factor $(3x + 4y)$.
$$= (3x + 4y)(x - 1)$$

R E M I N D E R When **factoring** a polynomial, make sure the polynomial is written as a **product.** For example, it is true that
$$xy + 2x + 3y + 6 = x(y + 2) + 3(y + 2)$$
but $x(y + 2) + 3(y + 2)$ is not a **factored form** of the original polynomial since it is a **sum,** not a **product.** The factored form of $xy + 2x + 3y + 6$ is $(y + 2)(x + 3)$.

Factoring out a greatest common factor first makes factoring by any method easier, as we see in the next example.

EXAMPLE 11 Factor $4ax - 4ab - 2bx + 2b^2$.

THE GREATEST COMMON FACTOR AND FACTORING BY GROUPING SECTION 5.1 287

Solution: First, factor out the common factor 2 from all four terms.

$$4ax - 4ab - 2bx + 2b^2$$
$$= 2(2ax - 2ab - bx + b^2) \quad \text{Factor out 2 from all four terms.}$$
$$= 2[2a(x - b) - b(x - b)] \quad \text{Factor out common factors from each pair of terms.}$$
$$= 2(x - b)(2a - b) \quad \text{Factor out the common binomial.}$$

Notice that we factored out $-b$ instead of b from the second pair of terms so that the binomial factor of each pair is the same.

MENTAL MATH

Find the prime factorization of the following integers mentally.

1. 14 **2.** 15 **3.** 10 **4.** 70

Find the GCF of the following pairs of integers mentally.

5. 6, 15 **6.** 20, 15 **7.** 3, 18 **8.** 14, 35

EXERCISE SET 5.1

Find the GCF for each list. See Examples 1 through 3.

1. 32, 36
2. 36, 90
3. 12, 18, 36
4. 24, 14, 21
5. y^2, y^4, y^7
6. x^3, x^2, x^3
7. $x^{10}y^2, xy^2, x^3y^3$
8. p^7q, p^8q^2, p^9q^3
9. $8x, 4$
10. $9y, y$
11. $12y^4, 20y^3$
12. $32x, 18x^2$
13. $12x^3, 6x^4, 3x^5$
14. $15y^2, 5y^7, 20y^3$
15. $18x^2y, 9x^3y^3, 36x^3y$
16. $7x, 21x^2y^2, 14xy$

Factor out the GCF from each polynomial. See Examples 4 through 6.

17. $3a + 6$
18. $18a - 12$
19. $30x - 15$
20. $42x - 7$
21. $24cd^3 - 18c^2d$
22. $25x^4y^3 - 15x^2y^2$
23. $-24a^4x + 18a^3x$
24. $-15a^2x + 9ax$
25. $12x^3 + 16x^2 - 8x$
26. $6x^3 - 9x^2 + 12x$
27. $5x^3y - 15x^2y + 10xy$
28. $14x^3y + 7x^2y - 7xy$

29. Construct a binomial whose greatest common factor is $5a^3$.

30. Construct a trinomial whose greatest common factor is $2x^2$.

Factor out the GCF from each polynomial. See Examples 7 and 8.

31. $y(x + 2) + 3(x + 2)$
32. $z(y + 4) + 3(y + 4)$
33. $x(y - 3) - 4(y - 3)$
34. $6(x + 2) - y(x + 2)$
35. $2x(x + y) - (x + y)$
36. $xy(y + 1) - (y + 1)$

Factor the following four-term polynomials by grouping. See Examples 9 through 11.

37. $5x + 15 + xy + 3y$
38. $xy + y + 2x + 2$
39. $2y - 8 + xy - 4x$
40. $6x - 42 + xy - 7y$
41. $3xy - 6x + 8y - 16$
42. $xy - 2yz + 5x - 10z$
43. $y^3 + 3y^2 + y + 3$
44. $x^3 + 4x + x^2 + 4$

Write an expression for the area of each shaded region. Then write the expression as a factored polynomial.

45.
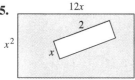

46.
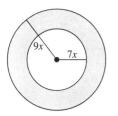

47.
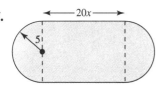

48.
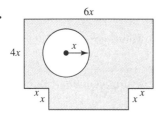

Factor the following polynomials.

49. $3x - 6$
50. $4x - 16$
51. $-8x - 18$
52. $-6x - 40$
53. $32xy - 18x^2$
54. $10xy - 15x^2$
55. $4x - 8y + 4$
56. $7x + 21y - 7$
57. $8(x + 2) - y(x + 2)$
58. $x(y^2 + 1) - 3(y^2 + 1)$
59. $-40x^8y^6 - 16x^9y^5$
60. $-21x^3y - 49x^2y^2$
61. $5x + 10$
62. $7x + 35$
63. $-3x + 12$
64. $-10x + 20$
65. $18x^3y^3 - 12x^3y^2 + 6x^5y^2$
66. $32x^3y^3 - 24x^2y^3 + 8x^2y^4$
67. $-2a^3 - 6a^2b$
68. $-3b^3c - 9b^2$
69. $y^2(x - 2) + (x - 2)$
70. $x(y + 4) + (y + 4)$
71. $5xy + 15x + 6y + 18$
72. $2x^3 + x^2 + 8x + 4$
73. $4x^2 - 8xy - 3x + 6y$
74. $2x^3 - x^2 - 10x + 5$
75. $126x^3yz + 210y^4z^3$
76. $231x^3y^2z - 143yz^2$
77. $4y^2 - 12y + 4yz - 12z$
78. $5x^2 - 20x^2y + 5z - 20zy$
79. $3y - 5x + 15 - xy$
80. $2x - 9y + 18 - xy$
81. $36x + 15y + 30 + 18xy$
82. $21y - 15x + 15xy - 21$
83. $12x^2y - 42x^2 - 4y + 14$
84. $90 + 15y^2 - 18x - 3xy^2$

85. Explain how you can tell whether a polynomial is written in factored form.
86. Construct a 4-term polynomial that can be factored by grouping.

Which of the following expressions is factored?

87. $(a + 6)(b - 2)$
88. $(x + 5)(x + y)$
89. $5(2y + z) - b(2y + z)$
90. $3x(a + 2b) + 2(a + 2b)$

Write an expression for the length of each rectangle.

91.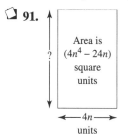

92.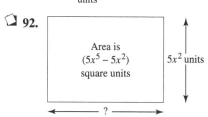

Review Exercises

Multiply. See Section 4.4.

93. $(x + 2)(x + 5)$
94. $(y + 3)(y + 6)$
95. $(a - 7)(a - 8)$
96. $(z - 4)(z - 4)$
97. $(b + 1)(b - 4)$
98. $(x - 5)(x + 10)$
99. $(y - 9)(y + 2)$
100. $(a + 3)(a - 1)$

5.2 FACTORING TRINOMIALS OF THE FORM $x^2 + bx + c$

TAPE BA 5.2

OBJECTIVES

1. Factor trinomials of the form $x^2 + bx + c$.
2. Factor out the greatest common factor and then factor a trinomial of the form $x^2 + bx + c$.

In this section, we factor trinomials of the form $x^2 + bx + c$, where the numerical coefficient of the squared variable is 1. Recall that factoring a polynomial is the process of writing the polynomial as a product. For example, since $(x + 3)(x + 1) = x^2 + 4x + 3$, we say that the factored form of $x^2 + 4x + 3$ is

$$x^2 + 4x + 3 = (x + 3)(x + 1)$$

Notice that the product of the first terms of the binomials is $x \cdot x = x^2$, the first term of the trinomial. Also, the product of the last two terms of the binomials is $3 \cdot 1 = 3$, the third term of the trinomial. The sum of these same terms is $3 + 1 = 4$, the coefficient of the x term of the trinomial.

Many trinomials, such as the preceding, factor into two binomials. To factor $x^2 + 7x + 10$, assume that it factors into two binomials and begin by writing two pairs of parentheses. The first term of the trinomial is x^2, so we use x and x as the first terms of the binomial factors.

$$(x +)(x +)$$

To determine the last term of each binomial factor, we look for two integers whose product is 10 and whose sum is 7. Since our numbers must have a positive product and a positive sum, we list pairs of positive integer factors of 10 only.

Positive Factors of 10	Sum of Factors
1, 10	$1 + 10 = 11$
2, 5	$2 + 5 = 7$

The correct pair of numbers is 2 and 5 because their product is 10 and their sum is 7. Now we can fill in the last terms of the binomial factors.

$$x^2 + 7x + 10 = (x + 2)(x + 5)$$

To see if we have factored correctly, multiply.

$$(x + 2)(x + 5) = x^2 + 2x + 5x + 10$$
$$= x^2 + 7x + 10 \qquad \text{Combine like terms.}$$

Since multiplication is commutative, the factored form of $x^2 + 7x + 10$ can also be written as $(x + 5)(x + 2)$.

To Factor a Trinomial of the Form $x^2 + bx + c$

To factor a trinomial of the form $x^2 + bx + c$, look for two numbers whose product is c and whose sum is b. The factored form of $x^2 + bx + c$ is

$$(x + \text{one number})(x + \text{other number})$$

EXAMPLE 1 Factor $x^2 + 7x + 12$.

Solution: Begin by writing the first terms of the binomial factors.

$$(x +)(x +)$$

Next, look for two numbers whose product is 12 and whose sum is 7. Since our numbers must have a positive product and a positive sum, we look at positive pairs of factors of 12 only.

Positive Factors of 12	Sum of Factors
1, 12	$1 + 12 = 13$
2, 6	$2 + 6 = 8$
3, 4	$3 + 4 = 7$

The correct pair of numbers is 3 and 4 because their product is 12 and their sum is 7. Use these numbers as the last terms of the binomial factors.

$$x^2 + 7x + 12 = (x + 3)(x + 4)$$

To check, multiply $(x + 3)$ by $(x + 4)$.

EXAMPLE 2 Factor $x^2 - 8x + 15$.

Solution: Begin by writing the first terms of the binomials.

$$(x +)(x +)$$

Now look for two numbers whose product is 15 and whose sum is -8. Since our numbers must have a positive product and a negative sum, we look at negative factors of 15 only.

Negative Factors of 15	Sum of Factors
$-1, -15$	$-1 + (-15) = -16$
$-3, -5$	$-3 + (-5) = -8$

The correct pair of numbers is -3 and -5 because their product is 15 and their sum is -8. Then

$$x^2 - 8x + 15 = (x - 3)(x - 5)$$

EXAMPLE 3 Factor $x^2 + 4x - 12$.

Solution: $x^2 + 4x - 12 = (x +)(x +)$

Look for two numbers whose product is -12 and whose sum is 4.

Factors of -12	Sum of Factors
$-1, 12$	$-1 + 12 = 11$
$1, -12$	$1 + (-12) = -11$
$-2, 6$	$-2 + 6 = 4$
$2, -6$	$2 + (-6) = -4$
$-3, 4$	$-3 + 4 = 1$
$3, -4$	$3 + (-4) = -1$

The correct pair of numbers is -2 and 6 since their product is -12 and their sum is 4. Hence

$$x^2 + 4x - 12 = (x - 2)(x + 6)$$

EXAMPLE 4 Factor $r^2 - r - 42$.

Solution: Because the variable in this trinomial is r, the first term of each binomial factor is r.

$$r^2 - r - 42 = (r +)(r +)$$

Find two numbers whose product is -42 and whose sum is -1, the numerical coefficient of r. The numbers are 6 and -7. Therefore,

$$r^2 - r - 42 = (r + 6)(r - 7)$$

EXAMPLE 5 Factor $a^2 + 2a + 10$.

Solution: Look for two numbers whose product is 10 and whose sum is 2. Neither 1 and 10 nor 2 and 5 give the required sum, 2. We conclude that $a^2 + 2a + 10$ is not factorable with integers. The polynomial $a^2 + 2a + 10$ is called a **prime polynomial.**

EXAMPLE 6 Factor $x^2 + 5yx + 6y^2$.

Solution: $x^2 + 5yx + 6y^2 = (x +)(x +)$

Look for two terms whose product is $6y^2$ and whose sum is $5y$, the coefficient of x in the middle term of the trinomial. The terms are $2y$ and $3y$ because $2y \cdot 3y = 6y^2$ and $2y + 3y = 5y$. Therefore,

$$x^2 + 5yx + 6y^2 = (x + 2y)(x + 3y)$$

> **2** Remember that the first step in factoring any polynomial is to factor out the greatest common factor (if there is one other than 1 or -1).

EXAMPLE 7 Factor $3m^2 - 24m - 60$.

Solution: First factor out the greatest common factor, 3, from each term.

$$3m^2 - 24m - 60 = 3(m^2 - 8m - 20)$$

Next, factor $m^2 - 8m - 20$ by looking for two factors of -20 whose sum is -8. The factors are -10 and 2.

$$3m^2 - 24m - 60 = 3(m + 2)(m - 10)$$

Remember to write the common factor 3 as part of the answer.

Check by multiplying.

$$3(m + 2)(m - 10) = 3(m^2 - 8m - 20)$$
$$= 3m^2 - 24m - 60$$

> **REMINDER** When factoring a polynomial, remember that factored out common factors are part of the final factored form. For example,
> $$5x^2 - 15x - 50 = 5(x^2 - 3x - 10)$$
> $$= 5(x + 2)(x - 5)$$
> Thus, $5x^2 - 15x - 50$ **factored completely** is $5(x + 2)(x - 5)$.

MENTAL MATH

Complete the following.

1. $x^2 + 9x + 20 = (x + 4)(x \quad\quad)$
2. $x^2 + 12x + 35 = (x + 5)(x \quad\quad)$
3. $x^2 - 7x + 12 = (x - 4)(x \quad\quad)$
4. $x^2 - 13x + 22 = (x - 2)(x \quad\quad)$
5. $x^2 + 4x + 4 = (x + 2)(x \quad\quad)$
6. $x^2 + 10x + 24 = (x + 6)(x \quad\quad)$

EXERCISE SET 5.2

Factor each trinomial completely. See Examples 1 through 5.

1. $x^2 + 7x + 6$
2. $x^2 + 6x + 8$
3. $x^2 + 9x + 20$
4. $x^2 + 13x + 30$
5. $x^2 - 8x + 15$
6. $x^2 - 9x + 14$
7. $x^2 - 10x + 9$
8. $x^2 - 6x + 9$
9. $x^2 - 15x + 5$
10. $x^2 - 13x + 30$
11. $x^2 - 3x - 18$
12. $x^2 - x - 30$
13. $x^2 + 5x + 2$
14. $x^2 - 7x + 5$

Factor each trinomial completely. See Example 6.

15. $x^2 + 8xy + 15y^2$
16. $x^2 + 6xy + 8y^2$
17. $x^2 - 2xy + y^2$
18. $x^2 - 11xy + 30y^2$
19. $x^2 - 3xy - 4y^2$
20. $x^2 - 4xy - 77y^2$

Factor each trinomial completely. See Example 7.

21. $2z^2 + 20z + 32$
22. $3x^2 + 30x + 63$
23. $2x^3 - 18x^2 + 40x$
24. $x^3 - x^2 - 56x$
25. $7x^2 + 14xy - 21y^2$
26. $6r^2 - 3rs - 3s^2$

Factor each trinomial completely.

27. $x^2 + 15x + 36$
28. $x^2 + 19x + 60$
29. $x^2 - x - 2$
30. $x^2 - 5x - 14$
31. $r^2 - 16r + 48$
32. $r^2 - 10r + 21$
33. $x^2 - 4x - 21$
34. $x^2 - 4x - 32$
35. $x^2 + 7xy + 10y^2$
36. $x^2 - 3xy - 4y^2$
37. $r^2 - 3r + 6$
38. $x^2 + 4x - 10$
39. $2t^2 + 24t + 64$
40. $2t^2 + 20t + 50$
41. $x^3 - 2x^2 - 24x$
42. $x^3 - 3x^2 - 28x$
43. $x^2 - 16x + 63$
44. $x^2 - 19x + 88$
45. $x^2 + xy - 2y^2$
46. $x^2 - xy - 6y^2$
47. $3x^2 + 9x - 30$
48. $4x^2 - 4x - 48$
49. $3x^2 - 60x + 108$
50. $2x^2 - 24x + 70$
51. $x^2 - 18x - 144$
52. $x^2 + x - 42$
53. $6x^3 + 54x^2 + 120x$
54. $3x^3 + 3x^2 - 126x$
55. $2t^5 - 14t^4 + 24t^3$
56. $3x^6 + 30x^5 + 72x^4$
57. $5x^3y - 25x^2y^2 - 120xy^3$
58. $3x^2 - 6xy - 72y^2$
59. $4x^2y + 4xy - 12y$
60. $3x^2y - 9xy + 45y$
61. $2a^2b - 20ab^2 + 42b^3$
62. $-1x^2z + 14xz^2 - 28z^3$

Find a positive value of b so that each trinomial is factorable.

63. $x^2 + bx + 15$
64. $y^2 + by + 20$
65. $m^2 + bm - 27$
66. $x^2 + bx - 14$

Find a positive value of c so that each trinomial is factorable.

67. $x^2 + 6x + c$
68. $t^2 + 8t + c$
69. $y^2 - 4y + c$
70. $n^2 - 16n + c$

Complete the following sentences in your own words.

71. If $x^2 + bx + c$ is factorable and c is negative, then the signs of the last term factors of the binomial are opposite because . . . .

72. If $x^2 + bx + c$ is factorable and c is positive, then the signs of the last term factors of the binomials are the same because . . . .

Review Exercises

Multiply. See Section 4.4.

73. $(2x + 1)(x + 5)$
74. $(3x + 2)(x + 4)$
75. $(5y - 4)(3y - 1)$
76. $(4z - 7)(7z - 1)$
77. $(a + 3)(9a - 4)$
78. $(y - 5)(6y + 5)$

Graph each linear equation. See Section 3.2.

79. $y = -3x$
80. $y = 5x$
81. $y = 2x - 7$
82. $y = -x + 4$

A Look Ahead

EXAMPLE

Factor $2t^5y^2 - 22t^4y^2 + 56t^3y^2$.

Solution:

First, factor out the greatest common factor of $2t^3y^2$.

$$2t^5y^2 - 22t^4y^2 + 56t^3y^2 = 2t^3y^2(t^2 - 11t + 28)$$
$$= 2t^3y^2(t - 4)(t - 7)$$

Factor each trinomial completely.

83. $2x^2y + 30xy + 100y$

84. $3x^2z^2 + 9xz^2 + 6z^2$

85. $-12x^2y^3 - 24xy^3 - 36y^3$

86. $-4x^2t^4 + 4xt^4 + 24t^4$

87. $y^2(x+1) - 2y(x+1) - 15(x+1)$

88. $z^2(x+1) - 3z(x+1) - 70(x+1)$

5.4 FACTORING BINOMIALS

OBJECTIVES

1. Factor the difference of two squares.
2. Factor the sum or difference of two cubes.

TAPE
BA 5.4

When learning to multiply binomials in Chapter 4, we studied a special product, the product of the sum and difference of two terms, a and b

$$(a + b)(a - b) = a^2 - b^2$$

For example, the product of $x + 3$ and $x - 3$ is

$$(x + 3)(x - 3) = x^2 - 9$$

The binomial $x^2 - 9$ is called a **difference of squares.** In this section, we use the pattern for the product of a sum and difference to factor the binomial difference of squares.

To use this pattern to help us factor, we must be able to recognize a difference of squares. A binomial is a difference of squares when it is the difference of the square of some expression a and the square of some expression b.

> **DIFFERENCE OF TWO SQUARES**
> $$a^2 - b^2 = (a + b)(a - b)$$

EXAMPLE 1   Factor $4x^2 - 1$.

Solution: $4x^2 - 1$ is the difference of two squares since $4x^2 = (2x)^2$ and $1 = (1)^2$; therefore,

$$4x^2 - 1 = (2x)^2 - 1^2 = (2x + 1)(2x - 1)$$

Multiply to check.

EXAMPLE 2   Factor $25a^2 - 9b^2$.

Solution: $25a^2 - 9b^2 = (5a)^2 - (3b)^2 = (5a + 3b)(5a - 3b)$

EXAMPLE 3   Factor $9x^2 - 36$.

Solution: Remember when factoring always to check first for common factors. If there are common factors, factor out the GCF and then factor the resulting polynomial.

$$\begin{aligned} 9x^2 - 36 &= 9(x^2 - 4) \quad \text{Factor out the GCF 9.} \\ &= 9(x^2 - 2^2) \\ &= 9(x + 2)(x - 2) \end{aligned}$$

In this example, if we forget to factor out the GCF first, we still have the difference of two squares.

$$9x^2 - 36 = (3x)^2 - (6)^2 = (3x + 6)(3x - 6)$$

This binomial has not been factored completely since both terms of both binomial factors have a common factor of 3.

$$3x + 6 = 3(x + 2) \quad \text{and} \quad 3x - 6 = 3(x - 2)$$

Then

$$9x^2 - 36 = (3x + 6)(3x - 6) = 3(x + 2)3(x - 2) = 9(x + 2)(x - 2)$$

Factoring is easier if the GCF is factored out first before using other methods.

EXAMPLE 4   Factor $x^2 + 4$.

Solution: The binomial $x^2 + 4$ is the **sum** of squares since we can write $x^2 + 4$ as $x^2 + 2^2$. We might try to factor using $(x + 2)(x + 2)$ or $(x - 2)(x - 2)$. But when multiplying to check, neither factoring is correct.

$$(x + 2)(x + 2) = x^2 + 4x + 4$$
$$(x - 2)(x - 2) = x^2 - 4x + 4$$

In both cases, the product is a trinomial, not the required binomial. In fact, $x^2 + 4$ is a prime polynomial.

Although the sum of two squares usually does not factor, the sum or difference of two cubes can be factored and reveals factoring patterns. The pattern for the sum of cubes is illustrated by multiplying the binomial $x + y$ and the trinomial $x^2 - xy + y^2$.

$$\begin{array}{r} x^2 - xy + y^2 \\ x + y \\ \hline x^2y - xy^2 + y^3 \\ x^3 - x^2y + xy^2 \\ \hline x^3 + y^3 \end{array}$$

$(x + y)(x^2 - xy + y^2) = x^3 + y^3$ Sum of cubes.

The pattern for the difference of two cubes is illustrated by multiplying the binomial $x - y$ by the trinomial $x^2 + xy + y^2$. The result is

$(x - y)(x^2 + xy + y^2) = x^3 - y^3$ Difference of cubes.

> **SUM OR DIFFERENCE OF TWO CUBES**
> $a^3 + b^3 = (a + b)(a^2 - ab + b^2)$
> $a^3 - b^3 = (a - b)(a^2 + ab + b^2)$

EXAMPLE 5   Factor $x^3 + 8$.

Solution: First, write the binomial in the form $a^3 + b^3$.

$x^3 + 8 = x^3 + 2^3$ Write in the form $a^3 + b^3$.

If we replace a with x and b with 2 in the formula above, we have

$x^3 + 2^3 = (x + 2)(x^2 - (x)(2) + 2^2)$
$ = (x + 2)(x^2 - 2x + 4)$

> R E M I N D E R   When factoring sums or differences of cubes, notice the sign patterns.
>
> same sign
> $$x^3 + y^3 = (x + y)(x^2 - xy + y^2)$$
> opposite sign   always positive
>
> same sign
> $$x^3 - y^3 = (x - y)(x^2 + xy + y^2)$$
> opposite sign   always positive

EXAMPLE 6 Factor $y^3 - 27$.

Solution: $y^3 - 27 = y^3 - 3^3$ Write in the form $a^3 - b^3$.
$= (y - 3)[y^2 + (y)(3) + 3^2]$
$= (y - 3)(y^2 + 3y + 9)$

EXAMPLE 7 Factor $64x^3 + 1$.

Solution: $64x^3 + 1 = (4x)^3 + 1^3$
$= (4x + 1)[(4x)^2 - (4x)(1) + 1^2]$
$= (4x + 1)(16x^2 - 4x + 1)$

EXAMPLE 8 Factor $54a^3 - 16b^3$.

Solution: Remember to factor out common factors first before using other factoring methods.

$54a^3 - 16b^3 = 2(27a^3 - 8b^3)$ Factor out the GCF 2.
$= 2[(3a)^3 - (2b)^3]$ Difference of two cubes.
$= 2(3a - 2b)[(3a)^2 + (3a)(2b) + (2b)^2]$
$= 2(3a - 2b)(9a^2 + 6ab + 4b^2)$

MENTAL MATH

State each number as a square.

1. 1   **2.** 25   **3.** 81   **4.** 64
5. 9   **6.** 100

State each number as a cube.

7. 1   **8.** 64   **9.** 8   **10.** 27

EXERCISE SET 5.4

Factor the difference of two squares. See Examples 1 through 3.

1. $25y^2 - 9$
2. $49a^2 - 16$
3. $121 - 100x^2$
4. $144 - 81x^2$
5. $12x^2 - 27$
6. $36x^2 - 64$
7. $169a^2 - 49b^2$
8. $225a^2 - 81b^2$
9. $x^2y^2 - 1$
10. $16 - a^2b^2$
11. What binomial multiplied by $(x - 6)$ gives the difference of two squares?

12. What binomial multiplied by $(5 + y)$ gives the difference of two squares?

Factor the sum or difference of two cubes. See Examples 5 through 8.

13. $a^3 + 27$
14. $b^3 - 8$
15. $8a^3 + 1$
16. $64x^3 - 1$
17. $5k^3 + 40$
18. $6r^3 - 162$
19. $x^3y^3 - 64$
20. $8x^3 - y^3$
21. $x^3 + 125$
22. $a^3 - 216$
23. $24x^4 - 81xy^3$
24. $375y^6 - 24y^3$
25. What binomial multiplied by $(4x^2 - 2xy + y^2)$ gives the sum or difference of two cubes?
26. What binomial multiplied by $(1 + 4y + 16y^2)$ gives the sum or difference of two cubes?

Factor the binomials completely.

27. $x^2 - 4$
28. $x^2 - 36$
29. $81 - p^2$
30. $100 - t^2$
31. $4r^2 - 1$
32. $9t^2 - 1$
33. $9x^2 - 16$
34. $36y^2 - 25$
35. $16r^2 + 1$
36. $49y^2 + 1$
37. $27 - t^3$
38. $125 + r^3$
39. $8r^3 - 64$
40. $54r^3 + 2$
41. $t^3 - 343$
42. $s^3 + 216$
43. $x^2 - 169y^2$
44. $x^2 - 225y^2$
45. $x^2y^2 - z^2$
46. $x^3y^3 - z^3$
47. $x^3y^3 + 1$
48. $x^2y^2 + z^2$
49. $s^3 - 64t^3$
50. $8t^3 + s^3$
51. $18r^2 - 8$
52. $32t^2 - 50$
53. $9xy^2 - 4x$
54. $16xy^2 - 64x$
55. $25y^4 - 100y^2$
56. $xy^3 - 9xyz^2$
57. $x^3y - 4xy^3$
58. $12s^3t^3 + 192s^5t$
59. $8s^6t^3 + 100s^3t^6$
60. $25x^5y + 121x^3y$
61. $27x^2y^3 - xy^2$
62. $8x^3y^3 + x^3y$

Review Exercises

Divide the following. See Section 4.6.

63. $\dfrac{8x^4 + 4x^3 - 2x + 6}{2x}$

64. $\dfrac{3y^4 + 9y^2 - 6y + 1}{3y^2}$

Use long division to divide the following. See Section 4.6.

65. $\dfrac{2x^2 - 3x - 2}{x - 2}$

66. $\dfrac{4x^2 - 21x + 21}{x - 3}$

67. $\dfrac{3x^2 + 13x + 10}{x + 3}$

68. $\dfrac{5x^2 + 14x + 12}{x + 2}$

A Look Ahead

EXAMPLE:

Factor $(x + y)^2 - (x - y)^2$.

Solution:

Use the method for factoring the difference of squares.

$(x + y)^2 - (x - y)^2 = [(x + y) + (x - y)]$
$\qquad\qquad\qquad\qquad [(x + y) - (x - y)]$
$= (x + y + x - y)(x + y - x + y)$
$= (2x)(2y)$
$= 4xy$

Factor each difference of squares. See the example.

69. $x^4 - 16$

70. $x^6 - 1$

71. $a^2 - (2 + b)^2$

72. $(x + 3)^2 - y^2$

73. $(x^2 - 4)^2 - (x - 2)^2$

74. $(x^2 - 9) - (3 - x)$

5.5 CHOOSING A FACTORING STRATEGY

TAPE
BA 5.5

OBJECTIVE

 Factor polynomials completely.

 A polynomial is factored completely when it is written as the product of prime polynomials. This section uses the various methods of factoring polynomials that have been discussed in earlier sections. Since these methods are applied throughout the remainder of this text, as well as in later courses, it is important to master the skills of factoring. The following is a set of guidelines for factoring polynomials.

TO FACTOR A POLYNOMIAL

Step 1. Are there any common factors? If so, factor out the GCF.

Step 2. How many terms are in the polynomial?
  a. If there are **two** terms, decide if one of the following can be applied.
   i. Difference of two squares: $a^2 - b^2 = (a - b)(a + b)$.
   ii. Difference of two cubes:
    $a^3 - b^3 = (a - b)(a^2 + ab + b^2)$.
   iii. Sum of two cubes: $a^3 + b^3 = (a + b)(a^2 - ab + b^2)$.
  b. If there are **three** terms, try one of the following.
   i. Perfect square trinomial: $a^2 + 2ab + b^2 = (a + b)^2$.
   ii. If not a perfect square trinomial, factor using the methods presented in Sections 5.2 and 5.3.
  c. If there are **four** or more terms, try factoring by grouping.

Step 3. See if any factors in the factored polynomial can be factored further.

Step 4. Check by multiplying.

EXAMPLE 1 Factor $10t^2 - 17t + 3$.

Solution: *Step 1.* The terms of this polynomial have no common factor (other than 1).

Step 2. There are three terms, so this polynomial is a trinomial. This trinomial is not a perfect square trinomial, so factor using methods from earlier sections.

$$\text{Factors of } 10t^2: \quad 10t^2 = 2t \cdot 5t, \quad 10t^2 = t \cdot 10t$$

Since the middle term, $-17t$, has a negative numerical coefficient, find negative factors of 3.

$$\text{Factors of 3:} \quad 3 = -1 \cdot -3$$

The correct combination is

$$(2t - 3)(5t - 1) = 10t^2 - 17t + 3$$

$$\underbrace{}_{-15t}$$

$$\underbrace{}_{-2t}$$

$$\overline{-17t} \quad \text{Correct middle term.}$$

Step 3. No factor can be factored further, so we have factored completely.

Step 4. To check, multiply $2t - 3$ and $5t - 1$.

$$(2t - 3)(5t - 1) = 10t^2 - 2t - 15t + 3 = 10t^2 - 17t + 3$$

The factored form of $10t^2 - 17t + 3$ is $(2t - 3)(5t - 1)$.

EXAMPLE 2 Factor $2x^3 + 3x^2 - 2x - 3$.

Solution: *Step 1.* There are no factors common to all terms.

Step 2. Try factoring by grouping since this polynomial has four terms.

$$2x^3 + 3x^2 - 2x - 3 = x^2(2x + 3) - (2x + 3) \quad \text{Factor out the greatest common factor for each pair of terms.}$$

$$= (2x + 3)(x^2 - 1) \quad \text{Factor out } 2x + 3.$$

Step 3. The binomial $x^2 - 1$ can be factored further.

$$= (2x + 3)(x + 1)(x - 1) \quad \text{Factor } x^2 - 1 \text{ as a difference of squares.}$$

Step 4. Check by finding the product of the three binomials.

The polynomial factored completely is $(2x + 3)(x + 1)(x - 1)$.

EXAMPLE 3   Factor $12m^2 - 3n^2$.

Solution: **Step 1.**   The terms of this binomial contain a greatest common factor of 3.

$$12m^2 - 3n^2 = 3(4m^2 - n^2) \quad \text{Factor out the greatest common factor.}$$

Step 2.   The binomial $4m^2 - n^2$ is a difference of squares.

$$= 3(2m + n)(2m - n) \quad \text{Factor the difference of squares.}$$

Step 3.   No factor can be factored further.

Step 4.   We check by multiplying.

$$3(2m + n)(2m - n) = 3(4m^2 - n^2) = 12m^2 - 3n^2$$

The factored form of $12m^2 - 3n^2$ is $3(2m + n)(2m - n)$.

EXAMPLE 4   Factor $x^3 + 27y^3$.

Solution: **Step 1.**   The terms of this binomial contain no common factor (other than 1).

Step 2.   This binomial is the sum of two cubes.

$$\begin{aligned} x^3 + 27y^3 &= (x)^3 + (3y)^3 \\ &= (x + 3y)[x^2 - 3xy + (3y)^2] \\ &= (x + 3y)(x^2 - 3xy + 9y^2) \end{aligned}$$

Step 3.   No factor can be factored further.

Step 4.   We check by multiplying.

$$\begin{aligned} (x + 3y)(x^2 - 3xy + 9y^2) &= x(x^2 - 3xy + 9y^2) \\ &\quad + 3y(x^2 - 3xy + 9y^2) \\ &= x^3 - 3x^2y + 9xy^2 + 3x^2y \\ &\quad - 9xy^2 + 27y^3 = x^3 + 27y^3 \end{aligned}$$

Thus, $x^3 + 27y^3$ factored completely is $(x + 3y)(x^2 - 3xy + 9y^2)$.

EXAMPLE 5   Factor $30a^2b^3 + 55a^2b^2 - 35a^2b$.

Solution: **Step 1.** $30a^2b^3 + 55a^2b^2 - 35a^2b = 5a^2b(6b^2 + 11b - 7)$ Factor out the GCF.

Step 2. $= 5a^2b(2b - 1)(3b + 7)$ Factor the resulting trinomial.

Step 3. No factor can be factored further.
Step 4. Check by multiplying.
The trinomial factored completely is $5a^2b(2b - 1)(3b + 7)$.

EXERCISE SET 5.5

Factor the following completely. See Examples 1 through 5.

1. $a^2 + 2ab + b^2$
2. $a^2 - 2ab + b^2$
3. $a^2 + a - 12$
4. $a^2 - 7a + 10$
5. $a^2 - a - 6$
6. $a^2 + 2a + 1$
7. $x^2 + 2x + 1$
8. $x^2 + x - 2$
9. $x^2 + 4x + 3$
10. $x^2 + x - 6$
11. $x^2 + 7x + 12$
12. $x^2 + x - 12$
13. $x^2 + 3x - 4$
14. $x^2 - 7x + 10$
15. $x^2 + 2x - 15$
16. $x^2 + 11x + 30$
17. $x^2 - x - 30$
18. $x^2 + 11x + 24$
19. $2x^2 - 98$
20. $3x^2 - 75$
21. $x^2 + 3x + xy + 3y$
22. $3y - 21 + xy - 7x$
23. $x^2 + 6x - 16$
24. $x^2 - 3x - 28$
25. $4x^3 + 20x^2 - 56x$
26. $6x^3 - 6x^2 - 120x$
27. $12x^2 + 34x + 24$
28. $8a^2 + 6ab - 5b^2$
29. $4a^2 - b^2$
30. $28 - 13x - 6x^2$
31. $20 - 3x - 2x^2$
32. $x^2 - 2x + 4$
33. $a^2 + a - 3$
34. $6y^2 + y - 15$
35. $4x^2 - x - 5$
36. $x^2y - y^3$
37. $4t^2 + 36$
38. $x^2 + x + xy + y$
39. $ax + 2x + a + 2$
40. $18x^3 - 63x^2 + 9x$
41. $12a^3 - 24a^2 + 4a$
42. $x^2 + 14x - 32$
43. $x^2 - 14x - 48$
44. $16a^2 - 56ab + 49b^2$
45. $25p^2 - 70pq + 49q^2$
46. $7x^2 + 24xy + 9y^2$
47. $125 - 8y^3$
48. $64x^3 + 27$
49. $-x^2 - x + 30$
50. $-x^2 + 6x - 8$
51. $14 + 5x - x^2$
52. $3 - 2x - x^2$
53. $3x^4y + 6x^3y - 72x^2y$
54. $2x^3y + 8x^2y^2 - 10xy^3$
55. $5x^3y^2 - 40x^2y^3 + 35xy^4$
56. $4x^4y - 8x^3y - 60x^2y$
57. $12x^3y + 243xy$
58. $6x^3y^2 + 8xy^2$
59. $(x - y)^2 - z^2$
60. $(x + 2y)^2 - 9$

61. $3rs - s + 12r - 4$
62. $x^3 - 2x^2 + 3x - 6$
63. $4x^2 - 8xy - 3x + 6y$
64. $4x^2 - 2xy - 7yz + 14xz$
65. $6x^2 + 18xy + 12y^2$
66. $12x^2 + 46xy - 8y^2$
67. $xy^2 - 4x + 3y^2 - 12$
68. $x^2y^2 - 9x^2 + 3y^2 - 27$
69. $5(x + y) + x(x + y)$
70. $7(x - y) + y(x - y)$
71. $14t^2 - 9t + 1$
72. $3t^2 - 5t + 1$
73. $3x^2 + 2x - 5$
74. $7x^2 + 19x - 6$
75. $x^2 + 9xy - 36y^2$
76. $3x^2 + 10xy - 8y^2$
77. $1 - 8ab - 20a^2b^2$
78. $1 - 7ab - 60a^2b^2$
79. $x^4 - 10x^2 + 9$
80. $x^4 - 13x^2 + 36$
81. $x^4 - 14x^2 - 32$
82. $x^4 - 22x^2 - 75$
83. $x^2 - 23x + 120$
84. $y^2 + 22y + 96$
85. $6x^3 - 28x^2 + 16x$
86. $6y^3 - 8y^2 - 30y$
87. $27x^3 - 125y^3$
88. $216y^3 - z^3$
89. $x^3y^3 + 8z^3$
90. $27a^3b^3 + 8$
91. $2xy - 72x^3y$
92. $2x^3 - 18x$
93. $x^3 + 6x^2 - 4x - 24$
94. $x^3 - 2x^2 - 36x + 72$
95. $6a^3 + 10a^2$
96. $4n^2 - 6n$
97. $a^2(a + 2) + 2(a + 2)$
98. $a - b + x(a - b)$

99. $x^3 - 28 + 7x^2 - 4x$
100. $a^3 - 45 - 9a + 5a^2$
101. Explain why it makes good sense to factor out the GCF first, before using other methods of factoring.
102. The sum of two squares usually does not factor. Is the sum of two squares $9x^2 + 81y^2$ factorable?

Review Exercises

Solve each equation. See Section 2.4.

103. $x - 6 = 0$
104. $y + 5 = 0$
105. $2m + 4 = 0$
106. $3x - 9 = 0$
107. $5z - 1 = 0$
108. $4a + 2 = 0$

Solve the following. See Section 2.6.

109. A suitcase has a volume of 960 cubic inches. Find x.

110. The sail shown has an area of 25 square feet. Find its height, x.

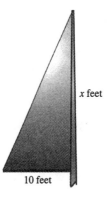

List the x- and y-intercept points for each graph. See Section 3.3.

111.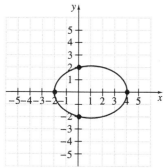

112.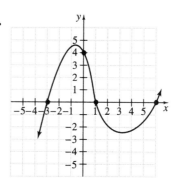

113.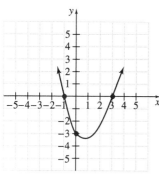

114.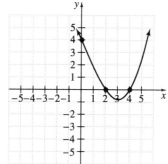

5.6 SOLVING QUADRATIC EQUATIONS BY FACTORING

OBJECTIVES

1. Define quadratic equation.
2. Solve quadratic equations by factoring.
3. Solve equations with degree greater than 2 by factoring.

TAPE
BA 5.6

1. Linear equations, while versatile, are not versatile enough to model many real-life phenomena. For example, let's suppose an object is dropped from the top of a 256-foot cliff and we want to know how long before the object strikes the ground. The answer to this question is found by solving the equation $-16t^2 + 256 = 0$. (See Example 1 in the next section.) This equation is called a **quadratic equation** because it contains a variable with an exponent of 2 and no other variable in the equation contains an exponent greater than 2. In this section, we solve quadratic equations by factoring.

> **QUADRATIC EQUATION**
>
> A quadratic equation is one that can be written in the form $ax^2 + bx + c = 0$, where a, b, and c are real numbers, and $a \neq 0$.

Notice that the degree of the polynomial $ax^2 + bx + c$ is 2. Here are a few more examples of quadratic equations.

QUADRATIC EQUATIONS

$$3x^2 + 5x + 6 = 0 \qquad x^2 = 9 \qquad y^2 + y = 1$$

The form $ax^2 + bx + c = 0$ is called the **standard form** of a quadratic equation. The quadratic equations $3x^2 + 5x + 6 = 0$ and $-16t^2 + 256 = 0$ are in standard form. One side of the equation is 0 and the other side is a polynomial of degree 2 written in descending powers of the variable.

2 Some quadratic equations can be solved by making use of factoring and the **zero factor theorem.**

> **ZERO FACTOR THEOREM**
> If a and b are real numbers and if $ab = 0$, then $a = 0$ or $b = 0$.

This theorem states that if the product of two numbers is 0 then at least one of the numbers must be 0. If the equation

$$(x - 3)(x + 1) = 0$$

is a true statement, then either the factor $x - 3$ must be 0 or the factor $x + 1$ must be 0. In other words,

$$x - 3 = 0 \quad \text{or} \quad x + 1 = 0$$
$$x = 3 \quad \text{or} \quad x = -1$$

Thus, 3 and -1 are both solutions of the equation $(x - 3)(x + 1) = 0$. To check, replace x with 3 in the original equation. Then replace x with -1 in the original equation.

$$(x - 3)(x + 1) = 0$$
$$(3 - 3)(3 + 1) = 0 \qquad \text{Replace } x \text{ with 3.}$$
$$0(4) = 0 \qquad \text{True.}$$
$$(x - 3)(x + 1) = 0$$
$$(-1 - 3)(-1 + 1) = 0 \qquad \text{Replace } x \text{ with } -1.$$
$$(-4)(0) = 0 \qquad \text{True.}$$

EXAMPLE 1 Solve $(x - 5)(2x + 7) = 0$.

Solution: Use the zero factor theorem; set each factor equal to 0 and solve the resulting linear equations.

$$(x - 5)(2x + 7) = 0$$
$$x - 5 = 0 \quad \text{or} \quad 2x + 7 = 0$$
$$x = 5 \quad \text{or} \quad 2x = -7$$
$$x = -\frac{7}{2}$$

If x is either 5 or $-\frac{7}{2}$, the product $(x-5)(2x+7)$ is 0. Check by replacing x with 5 in the original equation; then replace x with $-\frac{7}{2}$ in the original equation. Both 5 and $-\frac{7}{2}$ are solutions and the solution set is $\{5, -\frac{7}{2}\}$.

To use the zero factor theorem, one side of the quadratic equation must be 0 and the other side must be in factored form as in Example 1. If a quadratic equation is not in this form, first write the equation in standard form, then factor the polynomial.

EXAMPLE 2 Solve $x^2 - 9x = -20$.

Solution: First, write the equation in standard form; then factor.

$$x^2 - 9x = -20$$
$$x^2 - 9x + 20 = 0 \qquad \text{Write in standard form by adding 20 to both sides.}$$
$$(x-4)(x-5) = 0 \qquad \text{Factor.}$$

Next, use the zero factor theorem and set each factor equal to 0.

$$x - 4 = 0 \quad \text{or} \quad x - 5 = 0 \qquad \text{Set each factor equal to 0.}$$
$$x = 4 \quad \text{or} \quad x = 5 \qquad \text{Solve.}$$

Check the solutions by replacing x with each value in the original equation. The solution set is $\{4, 5\}$.

The following steps may be used to solve a quadratic equation by factoring.

TO SOLVE QUADRATIC EQUATIONS BY FACTORING

Step 1. Write the equation in standard form: $ax^2 + bx + c = 0$.
Step 2. Factor the quadratic completely.
Step 3. Set each factor containing a variable equal to 0.
Step 4. Solve the resulting equations.
Step 5. Check each solution in the original equation.

Since it is not always possible to factor a quadratic polynomial, not all quadratic equations can be solved by factoring. Other methods of solving quadratic equations are presented in Chapter 10.

EXAMPLE 3 Solve $x(2x - 7) = 4$.

Solution: First, write the equation in standard form; then factor.

$$x(2x - 7) = 4$$
$$2x^2 - 7x = 4 \quad \text{Multiply.}$$
$$2x^2 - 7x - 4 = 0 \quad \text{Write in standard form.}$$
$$(2x + 1)(x - 4) = 0 \quad \text{Factor.}$$
$$2x + 1 = 0 \quad \text{or} \quad x - 4 = 0 \quad \text{Set each factor equal to zero.}$$
$$2x = -1 \quad \text{or} \quad x = 4 \quad \text{Solve.}$$
$$x = -\frac{1}{2}$$

Check both solutions $-\frac{1}{2}$ and 4. The solution set is $\left\{-\frac{1}{2}, 4\right\}$.

REMINDER To apply the zero factor theorem, one side of the equation must be 0 and the other side of the equation must be factored. To solve the equation $x(2x - 7) = 4$, for example, you may **not** set each factor equal to 4.

EXAMPLE 4 Solve $-2x^2 - 4x + 30 = 0$.

Solution: The equation is in standard form so we begin by factoring out a common factor of -2.

$$-2x^2 - 4x + 30 = 0$$
$$-2(x^2 + 2x - 15) = 0 \quad \text{Factor out } -2.$$
$$-2(x + 5)(x - 3) = 0 \quad \text{Factor the quadratic.}$$

Next, set each factor **containing a variable** equal to 0.

$$x + 5 = 0 \quad \text{or} \quad x - 3 = 0 \quad \text{Set each factor containing a variable equal to 0.}$$
$$x = -5 \quad \text{or} \quad x = 3 \quad \text{Solve.}$$

Note that the factor -2 is a constant term containing no variables and can never equal 0. The solution set is $\{-5, 3\}$.

Some equations involving polynomials of degree higher than 2 may also be solved by factoring and then applying the zero factor theorem.

EXAMPLE 5 Solve $3x^3 - 12x = 0$.

Solution: Factor the left side of the equation. Begin by factoring out the common factor of $3x$.

$$3x^3 - 12x = 0$$
$$3x(x^2 - 4) = 0 \quad \text{Factor out the GCF } 3x.$$
$$3x(x + 2)(x - 2) = 0 \quad \text{Factor } x^2 - 4, \text{ a difference of squares.}$$
$$3x = 0 \quad \text{or} \quad x + 2 = 0 \quad \text{or} \quad x - 2 = 0 \quad \text{Set each factor equal to 0.}$$

$x = 0$ or $\quad x = -2$ or $\quad x = 2 \quad$ Solve.

Thus, the equation $3x^3 - 12x = 0$ has three solutions: 0, -2, and 2. To check, replace x with each solution in the original equation.

Let $x = 0$	Let $x = -2$	Let $x = 2$
$3(0)^3 - 12(0) = 0$	$3(-2)^3 - 12(-2) = 0$	$3(2)^3 - 12(2) = 0$
$0 = 0$	$3(-8) + 24 = 0$	$3(8) - 24 = 0$
	$0 = 0$	$0 = 0$

Substituting 0, -2, or 2 into the original equation results each time in a true equation. The solution set is $\{-2, 0, 2\}$.

EXAMPLE 6 Solve $(5x - 1)(2x^2 + 15x + 18) = 0$.

Solution:

$(5x - 1)(2x^2 + 15x + 18) = 0$
$(5x - 1)(2x + 3)(x + 6) = 0 \qquad$ Factor the trinomial.
$5x - 1 = 0 \quad$ or $\quad 2x + 3 = 0 \quad$ or $\quad x + 6 = 0 \qquad$ Set each factor equal to 0.
$5x = 1 \quad$ or $\quad 2x = -3 \quad$ or $\quad x = -6 \qquad$ Solve.
$x = \dfrac{1}{5} \quad$ or $\quad x = -\dfrac{3}{2}$

The solutions are $\dfrac{1}{5}$, $-\dfrac{3}{2}$, and -6. Check by replacing x with each solution in the original equation. The solution set is $\left\{-6, -\dfrac{3}{2}, \dfrac{1}{5}\right\}$.

EXAMPLE 7 Solve $2x^3 - 4x^2 - 30x = 0$.

Solution: Begin by factoring out the GCF $2x$.

$2x^3 - 4x^2 - 30x = 0$
$2x(x^2 - 2x - 15) = 0 \qquad$ Factor out the GCF $2x$.
$2x(x - 5)(x + 3) = 0 \qquad$ Factor the quadratic.
$2x = 0 \quad$ or $\quad x - 5 = 0 \quad$ or $\quad x + 3 = 0 \qquad$ Set each factor containing a variable equal to 0.
$x = 0 \quad$ or $\quad x = 5 \quad$ or $\quad x = -3 \qquad$ Solve.

Check by replacing x with each solution in the cubic equation. The solution set is $\{-3, 0, 5\}$.

In Chapter 3, we graphed linear equations in two variables, such as $y = 5x - 6$. Recall that to find the x-intercept of the graph of a linear equation, let $y = 0$ and solve for x. This is also how to find the x-intercepts of the graph of a **quadratic equation in two variables,** such as $y = x^2 - 5x + 4$.

EXAMPLE 8 Find the x-intercepts of the graph of $y = x^2 - 5x + 4$.

Solution: Let $y = 0$ and solve for x.

$$y = x^2 - 5x + 4$$
$$0 = x^2 - 5x + 4 \quad \text{Let } y = 0$$
$$0 = (x - 1)(x - 4) \quad \text{Factor.}$$
$$x - 1 = 0 \quad \text{or} \quad x - 4 = 0 \quad \text{Set each factor equal to 0.}$$
$$x = 1 \quad \text{or} \quad x = 4 \quad \text{Solve.}$$

The x-intercepts of the graph of $y = x^2 - 5x + 4$ are 1 and also 4. The intercept points are $(1, 0)$ and $(4, 0)$.

The graph of $y = x^2 - 5x + 4$ is shown below.

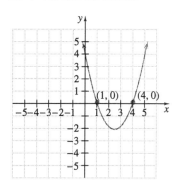

In general, a quadratic equation in two variables is one that can be written in the form $y = ax^2 + bx + c$ where $a \neq 0$. The graph of such an equation is called a **parabola** and will open up or down depending on the value of a.

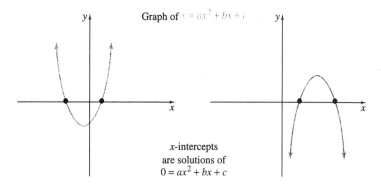

Graph of $y = ax^2 + bx + c$

x-intercepts are solutions of $0 = ax^2 + bx + c$

Notice that the x-intercepts of the graph of $y = ax^2 + bx + c$ are the real number solutions of $0 = ax^2 + bx + c$. Also, the real number solutions of $0 = ax^2 + bx + c$ are the x-intercepts of the graph of $y = ax^2 + bx + c$. We study more about graphs of quadratic equations in two variables in Chapter 10.

GRAPHING CALCULATOR EXPLORATIONS

A grapher may be used to find solutions of a quadratic equation whether the related quadratic polynomial is factorable or not. For example, let's use a grapher to approximate the solutions of $0 = x^2 + 4x - 3$. To do so, graph $y_1 = x^2 + 4x - 3$. Recall that the x-intercepts of this graph are the solutions of $0 = x^2 + 4x - 3$.

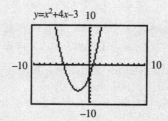

Notice that the graph appears to have an x-intercept between -5 and -4 and one between 0 and 1. Many graphers contain a TRACE feature. This feature activates a graph cursor that can be used to *trace* along a graph while the corresponding x- and y-coordinates are shown on the screen. Use the TRACE feature to confirm that x-intercepts lie between -5 and -4 and also 0 and 1. To approximate the x-intercepts to the nearest tenth, use a ROOT or a ZOOM feature on your grapher or redefine the viewing window. (A ROOT feature calculates the x-intercept. A ZOOM feature magnifies the viewing window around a specific location such as the graph cursor.) If we redefine the window to [0, 1] on the x-axis and [-1, 1] on the y-axis, the following graph is generated.

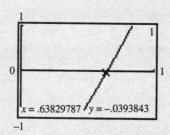

By using the TRACE feature, we can conclude that one x-intercept is approximately 0.6 to the nearest tenth. By repeating these steps for the other x-intercept, we find that it is approximately -4.6.

Use a grapher to approximate the real number solutions. If an equation has no real number solution, state so.

1. $3x^2 - 4x - 6 = 0$
2. $x^2 - x - 9 = 0$
3. $2x^2 + x + 2 = 0$
4. $-4x^2 - 5x - 4 = 0$
5. $-x^2 + x + 5 = 0$
6. $10x^2 + 6x - 3 = 0$

MENTAL MATH

Solve each equation by inspection.

1. $(a - 3)(a - 7) = 0$
2. $(a - 5)(a - 2) = 0$
3. $(x + 8)(x + 6) = 0$
4. $(x + 2)(x + 3) = 0$
5. $(x + 1)(x - 3) = 0$
6. $(x - 1)(x + 2) = 0$

EXERCISE SET 5.6

Solve each equation. See Example 1.

1. $(x - 2)(x + 1) = 0$
2. $(x + 3)(x + 2) = 0$
3. $x(x + 6) = 0$
4. $2x(x - 7) = 0$
5. $(2x + 3)(4x - 5) = 0$
6. $(3x - 2)(5x + 1) = 0$
7. $(2x - 7)(7x + 2) = 0$
8. $(9x + 1)(4x - 3) = 0$
9. Write a quadratic equation that has two solutions, 6 and −1. Leave the polynomial in the equation in factored form.
10. Write a quadratic equation that has two solutions, 0 and −2. Leave the polynomial in the equation in factored form.

Solve each equation. See Examples 2 through 4.

11. $x^2 - 13x + 36 = 0$
12. $x^2 + 2x - 63 = 0$
13. $x^2 + 2x - 8 = 0$
14. $x^2 - 5x + 6 = 0$
15. $x^2 - 4x = 32$
16. $x^2 - 5x = 24$
17. $x(3x - 1) = 14$
18. $x(4x - 11) = 3$
19. $3x^2 + 19x - 72 = 0$
20. $36x^2 + x - 21 = 0$
21. Write a quadratic equation in standard form that has two solutions, 5 and 7.
22. Write an equation that has three solutions, 0, 1, and 2.

Solve each equation. See Examples 5 through 7.

23. $x^3 - 12x^2 + 32x = 0$
24. $x^3 - 14x^2 + 49x = 0$
25. $(4x - 3)(16x^2 - 24x + 9) = 0$
26. $(2x + 5)(4x^2 - 10x + 25) = 0$
27. $4x^3 - x = 0$
28. $4y^3 - 36y = 0$
29. $32x^3 - 4x^2 - 6x = 0$
30. $15x^3 + 24x^2 - 63x = 0$

Find the x-intercepts of the graph of each equation. See Example 8.

31. $y = (3x + 4)(x - 1)$
32. $y = (5x - 3)(x - 4)$
33. $y = x^2 - 3x - 10$
34. $y = x^2 + 7x + 6$
35. $y = 2x^2 + 11x - 6$
36. $y = 4x^2 + 11x + 6$

For Exercises 37 through 42, match each equation with its graph.

A
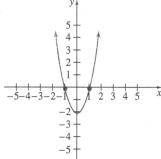

B
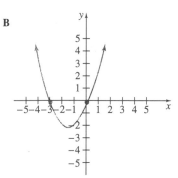

C
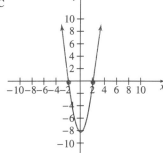

D
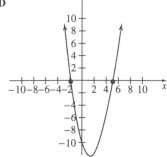

E
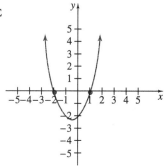

F
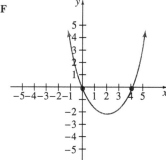

37. $y = (x + 2)(x - 1)$
38. $y = (x - 5)(x + 2)$
39. $y = x(x + 3)$
40. $y = x(x - 4)$
41. $y = 2x^2 - 8$
42. $y = 2x^2 - 2$

Solve each equation. Be careful. Some of the equations are quadratic and higher degree and some are linear.

43. $x(x + 7) = 0$
44. $y(6 - y) = 0$
45. $(x + 5)(x - 4) = 0$
46. $(x - 8)(x - 1) = 0$
47. $x^2 - x = 30$
48. $x^2 + 13x = -36$
49. $6y^2 - 22y - 40 = 0$
50. $3x^2 - 6x - 9 = 0$

51. $(2x + 3)(2x^2 - 5x - 3) = 0$
52. $(2x - 9)(x^2 + 5x - 36) = 0$
53. $x^2 - 15 = -2x$
54. $x^2 - 26 = -11x$
55. $x^2 - 16x = 0$
56. $x^2 + 5x = 0$
57. $-18y^2 - 33y + 216 = 0$
58. $-20y^2 + 145y - 35 = 0$
59. $12x^2 - 59x + 55 = 0$
60. $30x^2 - 97x + 60 = 0$
61. $18x^2 + 9x - 2 = 0$
62. $28x^2 - 27x - 10 = 0$
63. $x(6x + 7) = 5$
64. $4x(8x + 9) = 5$
65. $4(x - 7) = 6$
66. $5(3 - 4x) = 9$
67. $5x^2 - 6x - 8 = 0$
68. $9x^2 + 6x + 2 = 0$
69. $(y - 2)(y + 3) = 6$
70. $(y - 5)(y - 2) = 28$
71. $4y^2 - 1 = 0$
72. $4y^2 - 81 = 0$
73. $t^2 + 13t + 22 = 0$
74. $x^2 - 9x + 18 = 0$
75. $5t - 3 = 12$
76. $9 - t = -1$
77. $x^2 + 6x - 17 = -26$
78. $x^2 - 8x - 4 = -20$
79. $12x^2 + 7x - 12 = 0$
80. $30x^2 - 11x - 30 = 0$
81. $10t^3 - 25t - 15t^2 = 0$
82. $36t^3 - 48t - 12t^2 = 0$

83. A compass is accidentally thrown upward and out of an air balloon at a height of 300 feet. The height, y, of the compass at time x is given by the equation
$$y = -16x^2 + 20x + 300$$

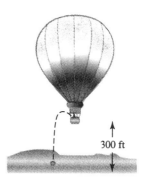

a. Find the height of the compass at the given times by filling in the table below.

time x	0	1	2	3	4	5	6
height y							

b. Use the table to determine when the compass strikes the ground.

c. Use the table to approximate the maximum height of the compass.

d. Plot the points (x, y) on a rectangular coordinate system and connect them with a smooth curve. Explain your results.

84. A rocket is fired upward from the ground with an initial velocity of 100 feet per second. The height, y, of the rocket at any time x is given by the equation
$$y = -16x^2 + 100x$$

a. Find the height of the rocket at the given times by filling in the table below.

time x	0	1	2	3	4	5	6	7
height y								

b. Use the table to approximate when the rocket strikes the ground to the nearest tenth of a second.

c. Use the table to approximate the maximum height of the rocket.

d. Plot the points (x, y) on a rectangular coordinate system and connect them with a smooth curve. Explain your results.

Review Exercises

Perform the following operations. Write all results in lowest terms. See Section 1.2.

85. $\dfrac{3}{5} + \dfrac{4}{9}$

86. $\dfrac{2}{3} + \dfrac{3}{7}$

87. $\dfrac{7}{10} - \dfrac{5}{12}$

88. $\dfrac{5}{9} - \dfrac{5}{12}$

89. $\dfrac{7}{8} \div \dfrac{7}{15}$

90. $\dfrac{5}{12} - \dfrac{3}{10}$

91. $\dfrac{4}{5} \cdot \dfrac{7}{8}$

92. $\dfrac{3}{7} \cdot \dfrac{12}{17}$

A Look Ahead

EXAMPLE Solve $(x - 6)(2x - 3) = (x + 2)(x + 9)$.

Solution:
$$(x - 6)(2x - 3) = (x + 2)(x + 9)$$
$$2x^2 - 15x + 18 = x^2 + 11x + 18$$
$$x^2 - 26x = 0$$
$$x(x - 26) = 0$$
$$x = 0 \quad \text{or} \quad x - 26 = 0$$
$$x = 26$$

Solve each equation. See the example.

93. $(x - 3)(3x + 4) = (x + 2)(x - 6)$

94. $(2x - 3)(x + 6) = (x - 9)(x + 2)$

95. $(2x - 3)(x + 8) = (x - 6)(x + 4)$

96. $(x + 6)(x - 6) = (2x - 9)(x + 4)$

97. $(4x - 1)(x - 8) = (x + 2)(x + 4)$

98. $(5x - 2)(x + 3) = (2x - 3)(x + 2)$

5.7 QUADRATIC EQUATIONS AND PROBLEM SOLVING

TAPE
BA 5.7

OBJECTIVE

1. Solve problems that can be modeled by quadratic equations.

Some problems may be modeled by quadratic equations. To solve these problems, we use the same problem-solving steps that were introduced in Section 2.5. When solving these problems, keep in mind that a solution of an equation that models a problem may not be a solution to the problem. For example, a person's age or the length of a rectangle is always a positive number. Discard solutions that do not make sense as solutions of the problem.

EXAMPLE 1 For a TV commercial, a piece of luggage is dropped from a cliff 256 feet above the ground to show the durability of the luggage. Neglecting air resistance, the height h in feet of the luggage above the ground after t seconds is given by the quadratic equation

$$h = -16t^2 + 256$$

Find how long it takes for the luggage to hit the ground.

Solution: **1. UNDERSTAND.** Read and reread the problem. Then draw a picture of the problem.

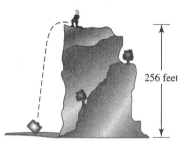

The equation $h = -16t^2 + 256$ models the height of the falling luggage at time t. Familiarize yourself with this equation by finding the height of the luggage at $t = 1$ second and $t = 2$ seconds.

When $t = 1$ second, the height of the suitcase is $h = -16(1)^2 + 256 = 240$ feet

When $t = 2$ seconds, the height of the suitcase is $h = -16(2)^2 + 256 = 192$ feet

Since we have been given the needed equation, we proceed to step 4.

4. TRANSLATE. To find how long it takes the luggage to hit the ground, we want to know the value of t for which the height $h = 0$.

$$0 = -16t^2 + 256$$

5. COMPLETE. Solve the quadratic equation by factoring.

$$0 = -16t^2 + 256$$
$$0 = -16(t^2 - 16)$$
$$0 = -16(t - 4)(t + 4)$$

$t - 4 = 0$ or $t + 4 = 0$
$t = 4$ or $t = -4$

6. INTERPRET. Since the time t cannot be negative, the proposed solution is 4 seconds. To *check*, see that the height of the luggage when t is 4 seconds is 0.

When $t = 4$ seconds $h = -16(4)^2 + 256 = -256 + 256 = 0$ feet

State: The solution checks and the luggage hits the ground 4 seconds after it is dropped.

EXAMPLE 2 In May 1995, 24 inches of rain fell on Slidell, Louisiana, in just two days, causing approximately one-third of the houses in that community to flood. When a home floods, all flooring in the home must be removed and replaced. The Callacs' home flooded and the soiled rug in their den was removed and now needs to be replaced. The length of their den is 4 feet more than the width. If the area of the floor is 117 square feet, find its length and width.

Solution: **1. UNDERSTAND.** Read and reread the problem. Propose and check a solution.

2. ASSIGN. Let

   x = the width of the floor; then

   $x + 4$ = the length of the floor since it is 4 feet longer.

3. ILLUSTRATE. An illustration is shown to the left.
4. TRANSLATE. Here, we use the formula for the area of a rectangle.

   In words: width · length = area

   Translate: $x \cdot (x + 4) = 117$

5. COMPLETE. $x(x + 4) = 117$

   $x^2 + 4x = 117$ Multiply.
   $x^2 + 4x - 117 = 0$ Write in standard form.
   $(x + 13)(x - 9) = 0$ Factor.

   Next, set each factor equal to 0.

   $x + 13 = 0$ or $x - 9 = 0$
   $x = -13$ or $x = 9$ Solve.

6. INTERPRET. The solutions are -13 and 9. Since x represents the width of the room, the solution -13 must be discarded. The proposed width is 9 feet and the proposed length is $x + 4$ or $9 + 4$ or 13 feet.

   Check: The area of a 9-foot by 13-foot room is (9 feet)(13 feet) = 117 square feet. The proposed solution checks.

   State: The floor is 9 feet by 13 feet.

EXAMPLE 3 The height of a triangular sail is 2 meters less than twice the length of the base. If the sail has an area of 30 square meters, find the length of its base and the height.

Solution:
1. UNDERSTAND. Read and reread the problem. Propose and check a solution.
2. ASSIGN. Since we are finding the length of the base and the height, let

   x = the length of the base and since the height is 2 meters less than twice the base,

   $2x - 2$ = the height.

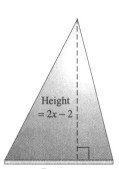

3. ILLUSTRATE. An illustration is shown to the left.
4. TRANSLATE. We are given that the area of the triangle is 30 square meters, so we use the formula for area of a triangle.

   In words: area of triangle = $\frac{1}{2}$ · base · height

   Translate: $30 = \frac{1}{2} \cdot x \cdot (2x - 2)$

5. COMPLETE. Here we solve the quadratic equation.

$$30 = \frac{1}{2}x(2x - 2)$$

$$30 = x^2 - x \quad \text{Multiply.}$$

$$x^2 - x - 30 = 0 \quad \text{Write in standard form.}$$

$$(x - 6)(x + 5) = 0 \quad \text{Factor.}$$

$$x - 6 = 0 \quad \text{or} \quad x + 5 = 0 \quad \text{Set each factor equal to 0.}$$

$$x = 6 \quad \text{or} \quad x = -5$$

6. INTERPRET. Since x represents the length of the base, discard the solution -5. The base of a triangle cannot be negative. The base is then 6 feet and the height is $2(6) - 2 = 10$ feet.

Check: To check this problem, recall that $\frac{1}{2}$ base · height = area, or

$$\frac{1}{2}(6)(10) = 30, \quad \text{the required area.}$$

State: The base of the triangular sail is 6 meters and the height is 10 meters.

The next example makes use of the **Pythagorean theorem** and consecutive integers. Before we review this theorem, recall that a **right triangle** is a triangle that contains a 90° or right angle. The **hypotenuse** of a right triangle is the side opposite the right angle and is the longest side of the triangle. The **legs** of a right triangle are the other sides of the triangle.

PYTHAGOREAN THEOREM

In a right triangle, the sum of the squares of the lengths of the two legs is equal to the square of the length of the hypotenuse.

$$(\text{leg})^2 + (\text{leg})^2 = (\text{hypotenuse})^2 \quad \text{or} \quad a^2 + b^2 = c^2$$

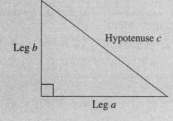

Study the following diagrams for a review of consecutive integers.

Consecutive integers: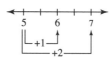

$x \quad x+1 \quad x+2$

Consecutive even integers: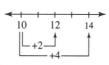

$x \quad x+2 \quad x+4$

Consecutive odd integers: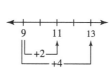

$x \quad x+2 \quad x+4$

EXAMPLE 4 Find the lengths of the sides of a right triangle if the lengths can be expressed by three consecutive even integers.

Solution: **1.** UNDERSTAND. Read and reread the problem. Let's propose and check a solution. If the length of one leg of the right triangle is 4 units, then the other leg is the next even integer, or 6 units, and the hypotenuse of the triangle is the next even integer, or 8 units. Remember that the hypotenuse is the longest side. Let's see if a triangle with sides of these lengths forms a right triangle. To do this, check to see whether the Pythagorean theorem holds true.

$$4^2 + 6^2 = 8^2$$
$$16 + 36 = 64$$
$$52 = 64 \quad \text{False.}$$

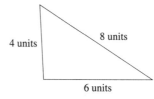

Our guess does not check, but we now have a better understanding of the problem.

2. ASSIGN. Let x, $x + 2$, and $x + 4$ be three consecutive even integers. Since these integers represent lengths of the sides of a right triangle, we have

$$x = \text{one leg}$$
$$x + 2 = \text{other leg}$$
$$x + 4 = \text{hypotenuse (longest side)}$$

3. ILLUSTRATE.

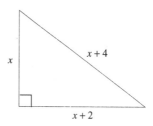

4. TRANSLATE. By the Pythagorean theorem, we have that
   In words: $(\text{hypotenuse})^2 = (\text{leg})^2 + (\text{leg})^2$
   Translate: $(x + 4)^2 = (x)^2 + (x + 2)^2$

5. COMPLETE. Now solve the equation.

$$(x + 4)^2 = x^2 + (x + 2)^2$$
$$x^2 + 8x + 16 = x^2 + x^2 + 4x + 4 \quad \text{Multiply.}$$
$$x^2 + 8x + 16 = 2x^2 + 4x + 4$$
$$x^2 - 4x - 12 = 0 \quad \text{Write in standard form.}$$
$$(x - 6)(x + 2) = 0$$
$$x - 6 = 0 \quad \text{or} \quad x + 2 = 0$$
$$x = 6 \quad \text{or} \quad x = -2$$

6. INTERPRET. Discard $x = -2$ since length cannot be negative. If $x = 6$, then $x + 2 = 8$ and $x + 4 = 10$. To *check*, see that $(\text{hypotenuse})^2 = (\text{leg})^2 + (\text{leg})^2$, or $10^2 = 6^2 + 8^2$, or $100 = 36 + 64$.

State: The sides of the right triangle have lengths 6 units, 8 units, and 10 units.

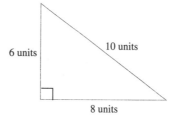

EXERCISE SET 5.7

Represent the given conditions using a single variable, x.

1. Two numbers whose sum is 36.
2. Two numbers whose sum is 10.
3. Two consecutive odd integers.
4. Two consecutive even integers.
5. The length and width of a rectangle whose length is 4 centimeters less than three times the width.
6. The length and width of a rectangle whose length is twice its width.
7. A woman's age now and her age 10 years ago.

8. The age of a man and the age of his son if the man is 5 years older than twice his son's age.

9. Three consecutive integers.

10. Three consecutive odd integers.

11. The three sides of a triangle if the first side is 2 inches less than twice the second side. The third side is 10 inches longer than the second side.

12. The three sides of a triangle if the first side is twice the second side and the third side is 5 more than three times the second side.

Solve. See Example 1.

13. An object is thrown upward from the top of an 80-foot building with an initial velocity of 64 feet per second. The height h of the object after t seconds is given by the quadratic

$$h = -16t^2 + 64t + 80$$

When will the object hit the ground?

14. A hang-glider pilot accidentally drops her compass from the top of a 400-foot cliff. The height h of the compass after t seconds is given by the quadratic equation

$$h = -16t^2 + 400$$

When will the compass hit the ground?

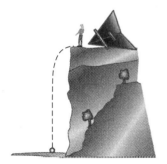

Solve the following. See Examples 2 and 3.

15. The length of a rectangle is 7 centimeters less than twice the width. Its area is 30 square centimeters. Find the dimensions of the rectangle.

16. The length of a rectangle is 9 inches more than its width. Its area is 112 square inches. Find the dimensions of the rectangle.

17. The altitude of a triangle is 8 centimeters more than twice the length of the base. If the triangle has an area of 96 square centimeters, find the length of its base and altitude.

18. The base of a triangle is 4 meters less than twice the length of the altitude. If the triangle has an area of 15 square meters, find the length of its base and altitude.

19. If the sides of a square are increased by 3 inches, the area becomes 64 square inches. Find the length of the sides of the original square.

20. If the sides of a square are increased by 5 meters, the area becomes 100 square meters. Find the length of the sides of the original square.

Solve. See Example 4.

21. Find the lengths of the sides of a right triangle if the hypotenuse is 10 centimeters longer than the short leg and 5 centimeters longer than the long leg.

22. Find the lengths of the sides of a right triangle if the length of the hypotenuse is 12 kilometers longer than the short leg and 6 kilometers longer than the long leg.

23. Find the length of the short leg of a right triangle if the long leg is 12 feet more than the short leg and the hypotenuse is 12 feet less than twice the short leg.

24. Find the length of the short leg of a right triangle if the long leg is 10 miles more than the short leg and the hypotenuse is 10 miles less than twice the short leg.

Solve.

25. The sum of a number and its square is 132. Find the number.

26. The sum of a number and its square is 182. Find the number.

27. The sum of two numbers is 20, and the sum of their squares is 218. Find the numbers.

28. The sum of two numbers is 25, and the sum of their squares is 325. Find the numbers.

29. If $D = \frac{n(n-3)}{2}$ is the formula for the number of diagonals D of a polygon with n sides, find the number of sides for a polygon with five diagonals.

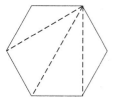

30. If $D = \frac{n(n-3)}{2}$ is the formula for the number of diagonals D of a polygon with n sides, find the number of sides for a polygon with 35 diagonals.

31. A rectangle has a perimeter of 42 miles and an area of 104 square miles. Find the dimensions of the rectangle.

32. A rectangle has a perimeter of 60 meters and an area of 209 square meters. Find the dimensions of the rectangle.

33. The sum of the squares of two consecutive integers is 9 greater than 8 times the smaller integer. Find the integers.

34. The square of the largest of three consecutive integers is equal to the sum of the squares of the other two. Find the integers.

35. A rectangular pool is surrounded by a walk 4 meters wide. The pool is 6 meters longer than its width. If the total area is 576 square meters more than the area of the pool, find the dimensions of the pool.

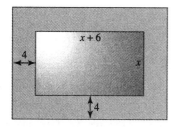

36. A rectangular garden is surrounded by a walk of uniform width. The area of the garden is 180 square yards. If the dimensions of the garden plus the walk are 16 yards by 24 yards, find the width of the walk. An illustration is at the top of the next column.

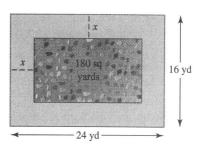

37. Find two consecutive even numbers whose product is 624.

38. Find two consecutive odd numbers whose product is 399.

39. One leg of a right triangle is 4 millimeters more than the smaller leg and the hypotenuse is 8 millimeters more than the smaller leg. Find the lengths of the sides of the triangle.

40. One leg of a right triangle is 9 centimeters longer than the other leg and the hypotenuse is 45 centimeters. Find the lengths of the legs of the triangle.

41. The length of the base of a triangle is twice its altitude. If the area of the triangle is 100 square kilometers, find the altitude.

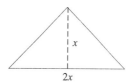

42. The altitude of a triangle is 2 millimeters less than the base. If the area is 60 square millimeters, find the base.

43. The sum of the squares of two consecutive negative integers is 221. Find the integers.

44. The sum of the squares of two consecutive even positive integers is 100. Find the integers.

45. Find the dimensions of a rectangle whose length is 2 yards more than twice its width and whose area is 60 square yards.

46. Find the dimensions of a rectangle whose length is 9 centimeters more than its width and whose area is 112 square centimeters.

47. Find the dimensions of a rectangle whose width is 7 miles less than its length and whose area is 120 square miles.

48. Find the dimensions of a rectangle whose width is 2 inches less than half its length and whose area is 160 square inches.

49. At the end of 2 years, P dollars invested at an interest rate r compounded annually increases to an amount, A dollars, given by
$$A = P(1 + r)^2$$
Find the interest rate if \$100 increased to \$144 in 2 years.

50. At the end of 2 years, P dollars invested at an interest rate r compounded annually increases to an amount, A dollars, given by
$$A = P(1 + r)^2$$
Find the interest rate if \$2000 increased to \$2420 in 2 years.

51. If the cost, C, for manufacturing x units of a certain product is given by $C = x^2 - 15x + 50$, find the number of units manufactured at a cost of \$9500.

52. If a switchboard handles n telephones, the number C of telephone connections it can make simultaneously is given by the equation $C = \frac{n(n-1)}{2}$. Find how many telephones are handled by a switchboard making 120 telephone connections simultaneously.

53. Two boats travel at a right angle to each other after leaving the same dock at the same time. One hour later the boats are 17 miles apart. If one boat travels 7 miles per hour faster than the other boat, find the rate of each boat.

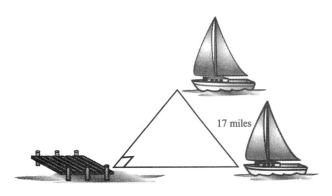

17 miles

54. The side of a square equals the width of a rectangle. The length of the rectangle is 6 meters longer than its width. The sum of the areas of the square and the rectangle is 176 square meters. Find the side of the square.

55. A rectangle has a perimeter of 42 yards and an area of 104 square yards. Find the dimensions of the rectangle.

56. Describe the kind of applied problem you find easiest to solve and why.

57. Describe the kind of applied problem you find most difficult to solve and why.

Review Exercises

The following double line graph shows a comparison of the number of farms in the United States and the size of the average farm. Use this graph for Exercises 58 through 63. See Sections 1.9 and 3.1.

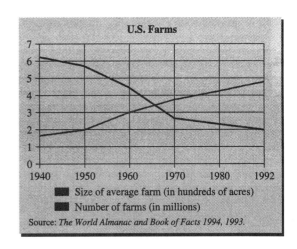

58. Approximate the size of the average farm in 1940.

59. Approximate the size of the average farm in 1992.

60. Approximate the number of farms in 1940.

61. Approximate the number of farms in 1992.

62. Approximate the year that the broken lines in this graph intersect.

63. In your own words, explain the meaning of the point of intersection in the graph.

64. Describe the trends shown in this graph and speculate as to why these trends have occurred.

Write each fraction in simplest form. See Section 1.2.

65. $\dfrac{20}{35}$ 66. $\dfrac{24}{32}$ 67. $\dfrac{27}{18}$

68. $\dfrac{15}{27}$ 69. $\dfrac{14}{42}$ 70. $\dfrac{45}{50}$

GROUP ACTIVITY

CHOOSING AMONG BUILDING OPTIONS

Sam has just had a 10-feet-by-15-feet, in-ground swimming pool installed in his backyard. He has $3000 left from the building project that he would like to spend on surrounding the pool with a patio, equally wide on all sides. He has talked to several local suppliers about options for building this patio and must choose among the following.

OPTION	MATERIAL	PRICE
A	Poured cement	$5 per square foot
B	Brick	$7.50 per square foot plus a $30 flat fee for delivering the bricks
C	Outdoor carpeting	$4.50 per square foot plus $10.86 per foot of the pool's perimeter to install an edging

1. Draw a diagram to represent the problem.
2. Write an algebraic expression that represents the total cost of the patio for each option.
3. If Sam plans to spend the entire $3000 he has saved for the patio, how wide would the patio be in each option? For each option, draw the final dimensions of the pool and patio to scale.
4. Which option should Sam choose? Why? Discuss the pros and cons of each option.
5. Summarize your findings and recommendations that could be used in a presentation to Sam.

Chapter 5 Highlights

Definitions and Concepts	Examples
Section 5.1 The Greatest Common Factor and Factoring by Grouping	
Factoring is the process of writing an expression as a product.	Factor: $6 = 2 \cdot 3$ $x^2 + 5x + 6 = (x+2)(x+3)$
To find the GCF of a list of integers, *Step 1.* Write each number as a product of primes. *Step 2.* Identify the common prime factors. *Step 3.* The product of all common factors is the greatest common factor. If there are no common prime factors, the GCF is 1.	Find the GCF of 12, 36, and 48. $12 = 2 \cdot 2 \cdot 3$ $36 = 2 \cdot 2 \cdot 3 \cdot 3$ $48 = 2 \cdot 2 \cdot 2 \cdot 2 \cdot 3$ GCF $= 2 \cdot 2 \cdot 3 = 12$.
The GCF of a list of common variables raised to powers is the variable raised to the smallest exponent in the list.	The GCF of z^5, z^3, and z^{10} is z^3.
The GCF of a list of terms is the product of all common factors.	Find the GCF of $8x^2y$, $10x^3y^2$, and $25x^2y^3$. The GCF of 8, 10, and 25 is 2. The GCF of x^2, x^3, and x^2 is x^2. The GCF of y, y^2, and y^3 is y. The GCF of the terms is $2x^2y$.
To Factor by Grouping *Step 1.* Arrange the terms so that the first two terms have a common factor and the last two have a common factor. *Step 2.* For each pair of terms, factor out the pair's GCF. *Step 3.* If there is now a common binomial factor, factor it out. *Step 4.* If there is no common binomial factor, begin again, rearranging the terms differently. If no rearrangement leads to a common binomial factor, the polynomial cannot be factored.	Factor $10ax + 15a - 6xy - 9y$. *Step 1.* $10ax + 15a - 6xy - 9y$ *Step 2.* $5a(2x+3) - 3y(2x+3)$ *Step 3.* $(2x+3)(5a-3y)$
Section 5.2 Factoring Trinomials of the Form $x^2 + bx + c$	
To factor a trinomial of the form $x^2 + bx + c$, look for two numbers whose product is c and whose sum is b. The factored form is $(x + \text{one number})(x + \text{other number})$	Factor: $x^2 + 7x + 12$ $3 + 4 = 7 \qquad 3 \cdot 4 = 12$ $(x+3)(x+4)$

(continued)

Definitions and Concepts	Examples
Section 5.3 Factoring Trinomials of the Form $ax^2 + bx + c$	
Method 1: To factor $ax^2 + bx + c$, try various combinations of factors of ax^2 and c until a middle term of bx is obtained when checking.	Factor: $3x^2 + 14x - 5$ Factors of $3x^2$: $3x, x$ Factors of -5: $-1, 5$ and $1, -5$. $(3x - 1)(x + 5)$ $\underbrace{}_{-1x}$ $\overline{15x}$ $\overline{14x}$ **correct** middle term
Method 2: Factor $ax^2 + bx + c$ by grouping. *Step 1.* Find two numbers whose product is $a \cdot c$ and whose sum is b. *Step 2.* Rewrite bx, using the factors found in step 1. *Step 3.* Factor by grouping.	Factor: $3x^2 + 14x - 5$ *Step 1.* Find two numbers whose product is $3 \cdot (-5)$ or -15 and whose sum is 14. They are 15 and -1. *Step 2* $3x^2 + 14x - 5$ $\qquad = 3x^2 + 15x - 1x - 5$ *Step 3* $\qquad = 3x(x + 5) - 1(x + 5)$ $\qquad = (x + 5)(3x - 1)$
A **perfect square trinomial** is a trinomial that is the square of some binomial.	Perfect Square Trinomial = square of binomial $x^2 + 4x + 4 \qquad = (x + 2)^2$ $25x^2 - 10x + 1 \qquad = (5x - 1)^2$
Factoring perfect square trinomials: $a^2 + 2ab + b^2 = (a + b)^2$ $a^2 - 2ab + b^2 = (a - b)^2$	Factor: $x^2 + 6x + 9 = x^2 + 2(x \cdot 3) + 3^2 = (x + 3)^2$ $4x^2 - 12x + 9 = (2x)^2 - 2(2x \cdot 3) + 3^2 = (2x - 3)^2$
Section 5.4 Factoring Binomials	
Difference of Squares $a^2 - b^2 = (a + b)(a - b)$ Sum or Difference of Cubes $a^3 + b^3 = (a + b)(a^2 - ab + b^2)$ $a^3 - b^3 = (a - b)(a^2 + ab + b^2)$	Factor: $x^2 - 9 = x^2 - 3^2 = (x + 3)(x - 3)$ $y^3 + 8 = y^3 + 2^3 = (y + 2)(y^2 - 2y + 4)$ $125z^3 - 1 = (5z)^3 - 1^3 = (5z - 1)(25z^2 + 5z + 1)$
Section 5.5 Choosing a Factoring Strategy	

To factor a polynomial,
Step 1. Factor out the GCF.
Step 2.   **a.** If two terms,
    i. $a^2 - b^2 = (a - b)(a + b)$
    ii. $a^3 - b^3 = (a - b)(a^2 + ab + b^2)$
    iii. $a^3 + b^3 = (a + b)(a^2 - ab + b^2)$
	Factor: $2x^4 - 6x^2 - 8$
	Step 1. $2x^4 - 6x^2 - 8 = 2(x^4 - 3x^2 - 4)$
	Step 2. **b. ii.** $\qquad = 2(x^2 + 1)(x^2 - 4)$

(continued)

DEFINITIONS AND CONCEPTS	EXAMPLES
SECTION 5.5 CHOOSING A FACTORING STRATEGY	

b. If three terms,
    i. $a^2 + 2ab + b^2 = (a + b)^2$
    ii. Methods in Sections 5.2 and 5.3
 c. If four or more terms, try factoring by grouping.	

Step 3. See if any factors can be factored further.	*Step 3.* $\qquad = 2(x^2 + 1)(x + 2)(x - 2)$
Step 4. Check by multiplying.	*Step 4.* Check by multiplying. $2(x + 2)(x - 2)(x^2 + 1) = 2(x^2 - 4)(x^2 + 1)$ $\qquad = 2(x^4 - 3x^2 - 4)$ $\qquad = 2x^4 - 6x^2 - 8$
SECTION 5.6 SOLVING QUADRATIC EQUATIONS BY FACTORING	
A **quadratic equation** is an equation that can be written in the form $ax^2 + bx + c = 0$ with a not 0. The form $ax^2 + bx + c = 0$ is called the **standard form** of a quadratic equation.	Quadratic Equation Standard Form $\quad x^2 = 16 \qquad\qquad\quad x^2 - 16 = 0$ $\quad y = -2y^2 + 5 \qquad 2y^2 + y - 5 = 0$
Zero Factor Theorem If a and b are real numbers and if $ab = 0$, then $a = 0$ or $b = 0$.	If $(x + 3)(x - 1) = 0$, then $x + 3 = 0$ or $x - 1 = 0$
To solve quadratic equations by factoring,	Solve: $3x^2 = 13x - 4$
Step 1. Write the equation in standard form: $ax^2 + bx + c = 0$.	*Step 1.* $\quad 3x^2 - 13x + 4 = 0$
Step 2. Factor the quadratic.	*Step 2.* $\quad (3x - 1)(x - 4) = 0$
Step 3. Set each factor containing a variable equal to 0.	*Step 3.* $\quad 3x - 1 = 0 \quad$ or $\quad x - 4 = 0$
Step 4. Solve the equations.	*Step 4.* $\quad 3x = 1 \quad$ or $\quad x = 4$ $\qquad\qquad x = \dfrac{1}{3}$
Step 5. Check in the original equation.	*Step 5.* Check both $\tfrac{1}{3}$ and 4 in the original equation.
SECTION 5.7 QUADRATIC EQUATIONS AND PROBLEM SOLVING	
Problem-Solving Steps	A garden is in the shape of a rectangle whose length is two feet more than its width. If the area of the garden is 35 square feet, find its dimensions.
1. UNDERSTAND the problem.	**1.** Read and reread the problem. Guess a solution and check your guess.
2. ASSIGN a variable.	**2.** Let x be the width of the rectangular garden. Then $x + 2$ is the length.

(continued)

DEFINITIONS AND CONCEPTS	EXAMPLES
SECTION 5.7 QUADRATIC EQUATIONS AND PROBLEM SOLVING	
3. ILLUSTRATE.	3. x [rectangle] $x+2$
4. TRANSLATE.	4. In words: length · width = area Translate: $(x+2) \cdot x = 35$
5. COMPLETE by solving.	5. $(x+2)x = 35$ $x^2 + 2x - 35 = 0$ $(x-5)(x+7) = 0$ $x - 5 = 0$ or $x + 7 = 0$ $x = 5$ or $x = -7$
6. INTERPRET.	6. Discard the solution of -7 since x represents width. *Check:* If x is 5 feet then $x + 2 = 5 + 2 = 7$ feet. The area of a rectangle whose width is 5 feet and whose length is 7 feet is (5 feet)(7 feet) or 35 square feet. *State:* The garden is 5 feet by 7 feet.

CHAPTER 5 REVIEW

(5.1) *Complete the factoring.*

1. $6x^2 - 15x = 3x()$
2. $2x^3y - 6x^2y^2 - 8xy^3 = 2xy()$

Factor the GCF from each polynomial.

3. $20x^2 + 12x$
4. $6x^2y^2 - 3xy^3$
5. $-8x^3y + 6x^2y^2$
6. $3x(2x + 3) - 5(2x + 3)$
7. $5x(x + 1) - (x + 1)$

Factor.

8. $3x^2 - 3x + 2x - 2$
9. $6x^2 + 10x - 3x - 5$
10. $3a^2 + 9ab + 3b^2 + ab$

(5.2) *Factor each trinomial.*

11. $x^2 + 6x + 8$
12. $x^2 - 11x + 24$
13. $x^2 + x + 2$
14. $x^2 - 5x - 6$
15. $x^2 + 2x - 8$
16. $x^2 + 4xy - 12y^2$
17. $x^2 + 8xy + 15y^2$
18. $3x^2y + 6xy^2 + 3y^3$
19. $72 - 18x - 2x^2$
20. $32 + 12x - 4x^2$

(5.3) *Factor each trinomial.*

21. $2x^2 + 11x - 6$
22. $4x^2 - 7x + 4$
23. $4x^2 + 4x - 3$
24. $6x^2 + 5xy - 4y^2$
25. $6x^2 - 25xy + 4y^2$
26. $18x^2 - 60x + 50$
27. $2x^2 - 23xy - 39y^2$

338 CHAPTER 5 FACTORING POLYNOMIALS

28. $4x^2 - 28xy + 49y^2$
29. $18x^2 - 9xy - 20y^2$
30. $36x^3y + 24x^2y^2 - 45xy^3$

(5.4) *Factor each binomial.*
31. $4x^2 - 9$
32. $9t^2 - 25s^2$
33. $16x^2 + y^2$
34. $x^3 - 8y^3$
35. $8x^3 + 27$
36. $2x^3 + 8x$
37. $54 - 2x^3y^3$
38. $9x^2 - 4y^2$
39. $16x^4 - 1$
40. $x^4 + 16$

(5.5) *Factor.*
41. $2x^2 + 5x - 12$
42. $3x^2 - 12$
43. $x(x - 1) + 3(x - 1)$
44. $x^2 + xy - 3x - 3y$
45. $4x^2y - 6xy^2$
46. $8x^2 - 15x - x^3$
47. $125x^3 + 27$
48. $24x^2 - 3x - 18$
49. $(x + 7)^2 - y^2$
50. $x^2(x + 3) - 4(x + 3)$

Solve the following equations.
51. $(x + 6)(x - 2) = 0$
52. $3x(x + 1)(7x - 2) = 0$
53. $4(5x + 1)(x + 3) = 0$
54. $x^2 + 8x + 7 = 0$
55. $x^2 - 2x - 24 = 0$
56. $x^2 + 10x = -25$
57. $x(x - 10) = -16$
58. $(3x - 1)(9x^2 + 3x + 1) = 0$
59. $56x^2 - 5x - 6 = 0$

60. $20x^2 - 7x - 6 = 0$
61. $5(3x + 2) = 4$
62. $6x^2 - 3x + 8 = 0$
63. $12 - 5t = -3$
64. $5x^3 + 20x^2 + 20x = 0$
65. $4t^3 - 5t^2 - 21t = 0$

(5.7) *Solve the following problems.*

66. A flag for a local organization is in the shape of a rectangle whose length is 15 inches less than twice its width. If the area of the flag is 500 square inches, find its dimensions.

67. The base of a triangular sail is four times its height. If the area of the triangle is 162 square yards, find the base.

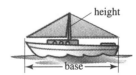

68. Find two consecutive positive integers whose product is 380.

69. A rocket is fired from the ground with an initial velocity of 440 feet per second. Its height h after t seconds is given by the equation
$$h = -16t^2 + 440t$$

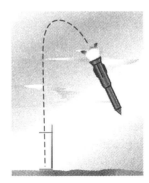

a. Find how many seconds pass before the rocket reaches a height of 2800 feet. Explain why two answers are obtained.

b. Find how many seconds pass before the rocket reaches the ground again.

70. An architect's squaring instrument is in the shape of a right triangle. Find the length of the long leg of the right triangle if the hypotenuse is 8 centimeters longer than the long leg and the short leg is 8 centimeters shorter than the long leg.

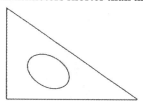

CHAPTER 5 TEST

Factor each polynomial completely. If a polynomial cannot be factored, write "prime."

1. $9x^3 + 39x^2 + 12x$
2. $x^2 + x - 10$
3. $x^2 + 4$
4. $y^2 - 8y - 48$
5. $3a^2 + 3ab - 7a - 7b$
6. $3x^2 - 5x + 2$
7. $x^2 + 20x + 90$
8. $x^2 + 14xy + 24y^2$
9. $26x^6 - x^4$
10. $50x^3 + 10x^2 - 35x$
11. $180 - 5x^2$
12. $64x^3 - 1$
13. $6t^2 - t - 5$
14. $xy^2 - 7y^2 - 4x + 28$
15. $x - x^5$
16. $-xy^3 - x^3y$

Solve each equation.

17. $x^2 + 5x = 14$
18. $(x + 3)^2 = 16$
19. $3x(2x - 3)(3x + 4) = 0$
20. $5t^3 - 45t = 0$
21. $3x^2 = -12x$
22. $t^2 - 2t - 15 = 0$
23. $7x^2 = 168 + 35x$
24. $6x^2 = 15x$

Solve each problem.

25. Find the dimensions of a rectangular garden whose length is 5 feet longer than its width and whose area is 66 square feet.

26. A deck for a home is in the shape of a triangle. The length of the base of the triangle is 9 feet longer than its altitude. If the area of the triangle is 68 square feet, find the length of the base.

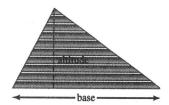

27. The sum of two numbers is 17, and the sum of their squares is 145. Find the numbers.

28. An object is dropped from the top of the Woolworth Building on Broadway in New York City. The height h of the object after t seconds is given by the equation

$$h = -16t^2 + 784$$

Find how many seconds pass before the object reaches the ground.

CHAPTER 5 CUMULATIVE REVIEW

1. Find the absolute value of each number.
   a. $|4|$ b. $|-5|$ c. $|0|$

2. Evaluate each expression if $x = 3$ and $y = 2$.
   a. $2x - y$ b. $\dfrac{3x}{2y}$ c. $\dfrac{x}{y} + \dfrac{y}{2}$ d. $x^2 - y^2$

3. The lowest point in North America is in Death Valley, at an elevation of 282 feet below sea level. Nearby, Mount Whitney reaches 14,494 feet, the highest point in the United States outside Alaska. How much of a variation in elevation is there between these two extremes?

4. Solve $5t - 5 = 6t + 2$ for t.

5. Solve $4(2x - 3) + 7 = 3x + 5$.

6. A local cellular phone company charges Elaine Chapoton $50 per month and $0.36 per minute of phone use in her usage category. If Elaine was charged $99.68 for a month's cellular phone use, determine the number of whole minutes of phone use.

7. Solve $-2x \leq -4$, and graph the solution set.

8. Determine whether each ordered pair is a solution of the equation $x - 2y = 6$.
   a. $(6, 0)$
   b. $(0, 3)$
   c. $(2, -2)$

9. Graph the linear equation $y = -\dfrac{1}{3}x$.

10. Identify the x- and y-intercepts and the intercept points.

A.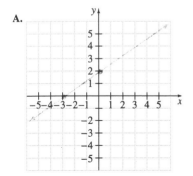

B.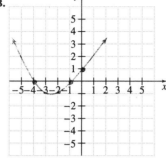

C.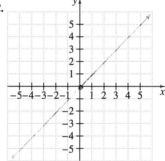

D.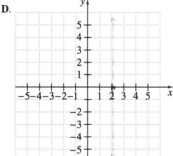

E.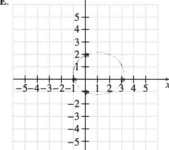

11. Find the slope of the line whose equation is $-2x + 3y = 12$.
12. Graph $2x - y \geq 3$.
13. Simplify each expression.
    a. $(st)^4$
    b. $\left(\dfrac{m}{n}\right)^7$
    c. $(2a)^3$
    d. $(-5x^2y^3z)^2$
    e. $\left(\dfrac{2x^4}{3y^5}\right)^4$
14. Add: $(4x^3 - 6x^2 + 2x + 7) + (5x^2 - 2x)$.
15. Divide: $(x^2 + 7x + 12)$ by $(x + 3)$.
16. Factor each polynomial by factoring out the GCF.
    a. $6t + 18$
    b. $y^5 - y^7$
17. Factor $x^2 - 8x + 15$.
18. Factor $3x^2 + 11x + 6$.
19. Factor $9x^2 - 36$.
20. Solve $(x - 5)(2x + 7) = 0$.
21. The height of a triangular sail is 2 meters less than twice the length of the base. If the sail has an area of 30 square meters, find the length of its base and the height.

CHAPTER

6

RATIONAL EXPRESSIONS

6.1 SIMPLIFYING RATIONAL EXPRESSIONS

6.2 MULTIPLYING AND DIVIDING RATIONAL EXPRESSIONS

6.3 ADDING AND SUBTRACTING RATIONAL EXPRESSIONS WITH COMMON DENOMINATORS AND LEAST COMMON DENOMINATOR

6.4 ADDING AND SUBTRACTING RATIONAL EXPRESSIONS WITH UNLIKE DENOMINATORS

6.5 SIMPLIFYING COMPLEX FRACTIONS

6.6 SOLVING EQUATIONS CONTAINING RATIONAL EXPRESSIONS

6.7 RATIO AND PROPORTION

6.8 RATIONAL EQUATIONS AND PROBLEM SOLVING

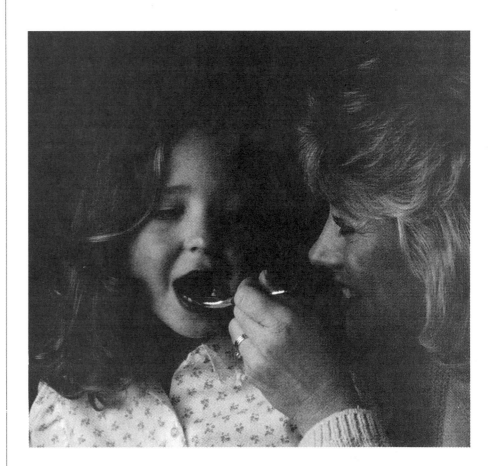

COMPARING FORMULAS FOR DOSES OF MEDICATION

Depending on the medicine, too large a dose can be extremely dangerous and even fatal. Particularly for children, gauging the proper dose is critical. Mathematical models predicting the correct doses are available, but are, at best, approximations.

IN THE CHAPTER GROUP ACTIVITY ON PAGE 399, YOU WILL HAVE THE OPPORTUNITY TO INVESTIGATE TWO WELL-KNOWN FORMULAS FOR PREDICTING THE CORRECT DOSAGES FOR CHILDREN.

344 CHAPTER 6 RATIONAL EXPRESSIONS

In this chapter, we expand our knowledge of algebraic expressions to include another category called **rational expressions,** such as $\frac{x+1}{x}$. We explore the operations of addition, subtraction, multiplication, and division for these algebraic fractions, using principles similar to the principles for number fractions. Thus, the material in this chapter will make full use of your knowledge of number fractions.

6.1 SIMPLIFYING RATIONAL EXPRESSIONS

OBJECTIVES

1. Find the value of a rational expression given a replacement number.
2. Identify values for which a rational expression is undefined.
3. Write rational expressions in lowest terms.

TAPE
BA 6.1

1 As we reviewed in chapter 1, a rational number is a number that can be written as a quotient of integers. A **rational expression** is also a quotient; it is a quotient of polynomials.

> **RATIONAL EXPRESSION**
>
> A rational expression is an expression that can be written in the form $\frac{P}{Q}$, where P and Q are polynomials and Q does not equal 0.

RATIONAL EXPRESSIONS

$$\frac{3y^3}{8} \qquad \frac{-4p}{p^3 + 2p + 1} \qquad \frac{5x^2 - 3x + 2}{3x + 7}$$

Rational expressions have different values depending on what value replaces the variable. Next, we review the standard order of operations by finding values of rational expressions at given replacement values of the variable.

EXAMPLE 1 Find the value of $\frac{x+4}{2x-3}$ for the given replacement values.

a. $x = 5$ **b.** $x = -2$

Solution: **a.** Replace each x in the expression with 5 and then simplify.

$$\frac{x+4}{2x-3} = \frac{5+4}{2(5)-3} = \frac{9}{10-3} = \frac{9}{7}.$$

b. Replace each x in the expression with -2 and then simplify.

$$\frac{x+4}{2x-3} = \frac{-2+4}{2(-2)-3} = \frac{2}{-7} \quad \text{or} \quad -\frac{2}{7}.$$

For a negative fraction such as $\frac{2}{-7}$, recall from Chapter 1 that

$$\frac{2}{-7} = \frac{-2}{7} = -\frac{2}{7}$$

In general, for any fraction

$$\frac{-a}{b} = \frac{a}{-b} = -\frac{a}{b}, \quad b \neq 0$$

This is also true for rational expressions. For example,

$$\underbrace{\frac{-(x+2)}{x}}_{\uparrow} = \frac{x+2}{-x} = -\frac{x+2}{x}$$

Notice the parentheses.

2 In the preceding box, notice that we wrote $b \neq 0$ for the denominator b. This is because the denominator of a rational expression must not equal 0 since division by 0 is not defined. This means we must be careful when replacing the variable in a rational expression by a number. For example, suppose we replace x with 5 in the rational expression $\frac{2+x}{x-5}$. The expression becomes

$$\frac{2+x}{x-5} = \frac{2+5}{5-5} = \frac{7}{0}$$

But division by 0 is undefined. Therefore, in this expression we can allow x to be any real number *except* 5. A rational expression is undefined for values that make the denominator 0.

EXAMPLE 2 Are there any values for x for which each expression is undefined?

a. $\frac{x}{x-3}$ **b.** $\frac{x^2+2}{x^2-3x+2}$ **c.** $\frac{x^3-6x^2-10x}{3}$ **d.** $\frac{2}{x^2+1}$

Solution: To find values for which a rational expression is undefined, find values that make the denominator 0.

a. The denominator of $\frac{x}{x-3}$ is 0 when $x - 3 = 0$ or when $x = 3$. Thus, when $x = 3$, the expression $\frac{x}{x-3}$ is undefined.

b. Set the denominator equal to zero.

$$x^2 - 3x + 2 = 0$$
$$(x-2)(x-1) = 0 \qquad \text{Factor.}$$
$$x - 2 = 0 \quad \text{or} \quad x - 1 = 0 \qquad \text{Set each factor equal to zero.}$$
$$x = 2 \quad \text{or} \quad x = 1 \qquad \text{Solve.}$$

Thus, when $x = 2$ or $x = 1$, the denominator $x^2 - 3x + 2$ is 0. So the rational expression $\dfrac{x^2 + 2}{x^2 - 3x + 2}$ is undefined when $x = 2$ or when $x = 1$.

c. The denominator of $\dfrac{x^3 - 6x^2 - 10x}{3}$ is never zero, so there are no values of x for which this expression is undefined.

d. No matter which real number x is replaced by, the denominator $x^2 + 1$ does not equal 0, so there are no real numbers for which this expression is undefined.

3 A fraction is said to be written in lowest terms or simplest form when the numerator and denominator have no common factors other than 1 (or -1). For example, the fraction $\dfrac{7}{10}$ is in lowest terms since the numerator and denominator have no common factors other than 1 (or -1).

The process of writing a rational expression in lowest terms or simplest form is called **simplifying** a rational expression. The following fundamental principle of rational expressions is used to simplify a rational expression.

FUNDAMENTAL PRINCIPLE OF RATIONAL EXPRESSIONS

If P, Q, and R are polynomials, and Q and R are not 0,

$$\frac{PR}{QR} = \frac{P}{Q}$$

Simplifying a rational expression is similar to simplifying a fraction. To simplify the fraction $\dfrac{15}{20}$, we factor the numerator and the denominator, look for common factors in both, and then use the fundamental principle.

$$\frac{15}{20} = \frac{3 \cdot 5}{2 \cdot 2 \cdot 5} = \frac{3}{2 \cdot 2} = \frac{3}{4}$$

To simplify the rational expression $\dfrac{x^2 - 9}{x^2 + x - 6}$, we also factor the numerator and denominator, look for common factors in both, and then use the fundamental principle of rational expressions.

$$\frac{x^2 - 9}{x^2 + x - 6} = \frac{(x - 3)(x + 3)}{(x - 2)(x + 3)} = \frac{x - 3}{x - 2}$$

This means that the rational expression $\dfrac{x^2 - 9}{x^2 + x - 6}$ has the same value as the rational expression $\dfrac{x - 3}{x - 2}$ for all values of x except 2 and -3. (Remember that when x is 2, the denominator of both rational expressions is 0 and when x is -3, the original rational expression has a denominator of 0.) As we simplify rational expressions, we will assume that the simplified rational expression is equal to the original rational expression for all real numbers except those for which either denominator is 0. The following steps may be used to simplify rational expressions.

To Simplify a Rational Expression

Step 1. Completely factor the numerator and denominator.

Step 2. Apply the fundamental principle of rational expressions to divide out common factors.

EXAMPLE 3 Write $\dfrac{21a^2b}{3a^5b}$ in simplest form.

Solution: Factor the numerator and denominator. Then apply the fundamental principle.

$$\frac{21a^2b}{3a^5b} = \frac{7 \cdot 3 \cdot a^2 \cdot b}{3 \cdot a^3 \cdot a^2 \cdot b} = \frac{7}{a^3}$$

EXAMPLE 4 Simplify: $\dfrac{5x - 5}{x^3 - x^2}$.

Solution: Factor the numerator and denominator if possible and then apply the fundamental principle.

$$\frac{5x - 5}{x^3 - x^2} = \frac{5(x - 1)}{x^2(x - 1)} = \frac{5}{x^2}$$

EXAMPLE 5 Write $\dfrac{x^2 + 8x + 7}{x^2 - 4x - 5}$ in lowest terms.

Solution: Factor the numerator and denominator and apply the fundamental principle.

$$\frac{x^2 + 8x + 7}{x^2 - 4x - 5} = \frac{(x + 7)(x + 1)}{(x - 5)(x + 1)} = \frac{x + 7}{x - 5}$$

EXAMPLE 6 Simplify: $\dfrac{x^2 + 4x + 4}{x^2 + 2x}$.

Solution: Factor the numerator and denominator and apply the fundamental principle.

$$\frac{x^2 + 4x + 4}{x^2 + 2x} = \frac{(x + 2)(x + 2)}{x(x + 2)} = \frac{x + 2}{x}$$

> **REMINDER** When simplifying a rational expression, the fundamental principle applies to **common factors, not their common terms.** For example, $\dfrac{x + 2}{x}$ cannot be simplified any further because the numerator and denominator have no **common factors.**

EXAMPLE 7
Simplify each rational expression.

a. $\dfrac{x+y}{y+x}$ b. $\dfrac{x-y}{y-x}$

Solution: a. The expression $\dfrac{x+y}{y+x}$ can be simplified by using the commutative property of addition to rewrite the denominator $y+x$ as $x+y$.

$$\dfrac{x+y}{y+x} = \dfrac{x+y}{x+y} = 1$$

b. The expression $\dfrac{x-y}{y-x}$ can be simplified by recognizing that $y-x$ and $x-y$ are opposites. In other words, $y-x = -1(x-y)$. Proceed as follows:

$$\dfrac{x-y}{y-x} = \dfrac{1 \cdot (x-y)}{(-1)(x-y)} = \dfrac{1}{-1} = -1$$

EXAMPLE 8
Simplify: $\dfrac{4-x^2}{3x^2-5x-2}$.

Solution:
$$\dfrac{4-x^2}{3x^2-5x-2} = \dfrac{(2-x)(2+x)}{(x-2)(3x+1)} \quad \text{Factor.}$$

$$= \dfrac{(-1)(x-2)(2+x)}{(x-2)(3x+1)} \quad \text{Write } 2-x \text{ as } -1(x-2).$$

$$= \dfrac{(-1)(2+x)}{3x+1} \quad \text{or} \quad \dfrac{-2-x}{3x+1} \quad \text{Simplify.}$$

EXAMPLE 9
Simplify: $\dfrac{2x^2-2xy+3x-3y}{2x+3}$.

Solution: First, factor the four-term numerator by grouping.

$$= \dfrac{2x^2-2xy+3x-3y}{2x+3} = \dfrac{2x(x-y)+3(x-y)}{2x+3}$$

$$= \dfrac{(2x+3)(x-y)}{2x+3} \quad \text{Factor.}$$

$$= \dfrac{x-y}{1} \quad \text{or} \quad x-y \quad \text{Simplify.}$$

MENTAL MATH

Find any real numbers for which each equal rational expression is undefined. See Example 2.

1. $\dfrac{x+5}{x}$ 2. $\dfrac{x^2-5x}{x-3}$ 3. $\dfrac{x^2+4x-2}{x(x-1)}$ 4. $\dfrac{x+2}{(x-5)(x-6)}$

EXERCISE SET 6.1

Find the value of the following expressions when $x = 2$, $y = -2$, and $z = -5$. See Example 1.

1. $\dfrac{x + 5}{x + 2}$

2. $\dfrac{x + 8}{2x + 5}$

3. $\dfrac{z - 8}{z + 2}$

4. $\dfrac{y - 2}{-5 + y}$

5. $\dfrac{x^2 + 8x + 2}{x^2 - x - 6}$

6. $\dfrac{z^2 + 8}{z^3 - 25z}$

7. $\dfrac{x + 5}{x^2 + 4x - 8}$

8. $\dfrac{z^3 + 1}{z^2 + 1}$

9. $\dfrac{y^3}{y^2 - 1}$

10. $\dfrac{z}{z^2 - 5}$

11. The total revenue R from the sale of a popular music compact disc is approximately given by the equation

$$R = \dfrac{150x^2}{x^2 + 3}$$

where x is the number of years since the CD has been released and revenue R is in millions of dollars.

a. Find the total revenue generated by the end of the first year.

b. Find the total revenue generated by the end of the second year.

c. Find the total revenue generated in the second year only.

12. For a certain model fax machine, the manufacturing cost C per machine is given by the equation

$$C = \dfrac{250x + 10{,}000}{x}$$

where x is the number of fax machines manufactured and cost C is in dollars per machine.

a. Find the cost per fax machine when manufacturing 100 fax machines.

b. Find the cost per fax machine when manufacturing 1000 fax machines.

c. Does the cost per machine decrease or increase when more machines are manufactured? Explain why this is so.

Find any real numbers for which each rational expression is undefined. See Example 2.

13. $\dfrac{x + 3}{x + 2}$

14. $\dfrac{5x + 1}{x - 3}$

15. $\dfrac{4x^2 + 9}{2x - 8}$

16. $\dfrac{9x^3 + 4x}{15x + 45}$

17. $\dfrac{9x^3 + 4}{15x + 30}$

18. $\dfrac{19x^3 + 2}{x^3 - x}$

19. $\dfrac{x^2 - 5x - 2}{x^2 + 4}$

20. $\dfrac{9y^5 + y^3}{x^2 + 9}$

21. Explain why the denominator of a fraction or a rational expression must not equal zero.

22. Does $\dfrac{(x - 3)(x + 3)}{x - 3}$ have the same value as $x + 3$ for all real numbers? Explain why or why not.

Simplify each expression. See Examples 3 through 6.

23. $\dfrac{8x^5}{4x^9}$

24. $\dfrac{12y^7}{-2y^6}$

25. $\dfrac{5(x - 2)}{(x - 2)(x + 1)}$

26. $\dfrac{9(x - 7)(x + 7)}{3(x - 7)}$

27. $\dfrac{-5a - 5b}{a + b}$

28. $\dfrac{7x + 35}{x^2 + 5x}$

29. $\dfrac{x + 5}{x^2 - 4x - 45}$

30. $\dfrac{x - 3}{x^2 - 6x + 9}$

31. $\dfrac{5x^2 + 11x + 2}{x + 2}$

32. $\dfrac{12x^2 + 4x - 1}{2x + 1}$

33. $\dfrac{x^2 + x - 12}{2x^2 - 5x - 3}$

34. $\dfrac{x^2 + 3x - 4}{x^2 - x - 20}$

35. Explain how to write a fraction in lowest terms.

36. Explain how to write a rational expression in lowest terms.

Simplify each expression. See Examples 7 through 9.

37. $\dfrac{x - 7}{7 - x}$

38. $\dfrac{y - z}{z - y}$

39. $\dfrac{y^2 - 2y}{4 - 2y}$

40. $\dfrac{x^2 + 5x}{20 + 4x}$

41. $\dfrac{x^2 - 4x + 4}{4 - x^2}$

42. $\dfrac{x^2 + 10x + 21}{-2x - 14}$

43. $\dfrac{x^2 + xy + 2x + 2y}{x + 2}$

44. $\dfrac{ab + ac + b^2 + bc}{b + c}$

45. $\dfrac{5x + 15 - xy - 3y}{2x + 6}$

46. $\dfrac{xy - 6x + 2y - 12}{y^2 - 6y}$

Simplify each expression.

47. $\dfrac{15x^4y^8}{-5x^8y^3}$

48. $\dfrac{24a^3b^3}{6a^2b^4}$

49. $\dfrac{(x - 2)(x + 3)}{5(x + 3)}$

50. $\dfrac{-2(y - 9)}{(y - 9)^2}$

51. $\dfrac{-6a - 6b}{a + b}$

52. $\dfrac{4a - 4y}{4y - 4a}$

53. $\dfrac{2x^2 - 8}{4x - 8}$

54. $\dfrac{5x^2 - 500}{35x + 350}$

55. $\dfrac{11x^2 - 22x^3}{6x - 12x^2}$

56. $\dfrac{16r^2 - 4s^2}{4r - 2s}$

57. $\dfrac{x + 7}{x^2 + 5x - 14}$

58. $\dfrac{x - 10}{x^2 - 17x + 70}$

59. $\dfrac{2x^2 + 3x - 2}{2x - 1}$

60. $\dfrac{4x^2 + 24x}{x + 6}$

61. $\dfrac{x^2 - 1}{x^2 - 2x + 1}$

62. $\dfrac{x^2 - 16}{x^2 - 8x + 16}$

63. $\dfrac{m^2 - 6m + 9}{m^2 - 9}$

64. $\dfrac{m^2 - 4m + 4}{m^2 + m - 6}$

65. $\dfrac{-2a^2 + 12a - 18}{9 - a^2}$

66. $\dfrac{-4a^2 + 8a - 4}{2a^2 - 2}$

67. $\dfrac{2 - x}{x - 2}$

68. $\dfrac{7 - y}{y - 7}$

69. $\dfrac{x^2 - 1}{1 - x}$

70. $\dfrac{x^2 - xy}{2y - 2x}$

71. $\dfrac{x^2 + 7x + 10}{x^2 - 3x - 10}$

72. $\dfrac{2x^2 + 7x - 4}{x^2 + 3x - 4}$

73. $\dfrac{3x^2 + 7x + 2}{3x^2 + 13x + 4}$

74. $\dfrac{4x^2 - 4x + 1}{2x^2 + 9x - 5}$

75. $\dfrac{x^2 + 3x - 2x - 6}{x^2 - 2x}$

76. $\dfrac{ax + ay - bx - by}{x^2 - y^2}$

77. $\dfrac{x^3 + 8}{x + 2}$

78. $\dfrac{x^3 + 64}{x + 4}$

79. $\dfrac{x^2 + xy + 5x + 5y}{3x + 3y}$

80. $\dfrac{x^2 + 2x + xy + 2y}{x^2 - 4}$

81. $\dfrac{x^3 - 1}{1 - x}$

82. $\dfrac{3 - x}{x^3 - 27}$

How does the graph of $y = \dfrac{x^2 - 9}{x - 3}$ compare to the graph of $y = x + 3$?

Recall that $\dfrac{x^2 - 9}{x - 3} = \dfrac{(x + 3)(x - 3)}{x - 3} = x + 3$ as long as x is not 3.

This means that the graph of $y = \dfrac{x^2 - 9}{x - 3}$ is the same as the graph of $y = x + 3$ with $x \neq 3$. To graph $y = \dfrac{x^2 - 9}{x - 3}$, then, graph the linear equation $y = x + 3$ and place an open dot on the graph at 3. This open dot or interruption of the line at 3 means $x \neq 3$.

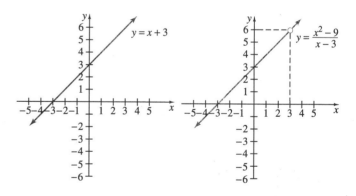

83. Graph $y = \dfrac{x^2 - 25}{x + 5}$.

84. Graph $y = \dfrac{x^2 - 16}{x - 4}$.

85. Graph $y = \dfrac{x^2 + x - 12}{x + 4}$.

86. Graph $y = \dfrac{x^2 - 6x + 8}{x - 2}$.

Review Exercises

Perform the indicated operations. See Section 1.2.

87. $\dfrac{1}{3} \cdot \dfrac{9}{11}$ **88.** $\dfrac{5}{27} \cdot \dfrac{2}{5}$ **91.** $\dfrac{5}{6} \cdot \dfrac{10}{11} \cdot \dfrac{2}{3}$ **92.** $\dfrac{4}{3} \cdot \dfrac{1}{7} \cdot \dfrac{10}{13}$

89. $\dfrac{1}{3} \div \dfrac{1}{4}$ **90.** $\dfrac{7}{8} \div \dfrac{1}{2}$ **93.** $\dfrac{13}{20} \div \dfrac{2}{9}$ **94.** $\dfrac{8}{15} \div \dfrac{5}{8}$

6.2 Multiplying and Dividing Rational Expressions

OBJECTIVES

1. Multiply rational expressions.
2. Divide rational expressions.

TAPE BA 6.2

1 Just as simplifying rational expressions is similar to simplifying number fractions, multiplying and dividing rational expressions is similar to multiplying and dividing number fractions. To find the product of $\frac{3}{5}$ and $\frac{1}{4}$, multiply the numerators and then multiply the denominators of both fractions.

$$\frac{3}{5} \cdot \frac{1}{4} = \frac{3 \cdot 1}{5 \cdot 4} = \frac{3}{20}$$

Use this same procedure to multiply rational expressions.

MULTIPLYING RATIONAL EXPRESSIONS

Let P, Q, R, and S be polynomials. Then

$$\frac{P}{Q} \cdot \frac{R}{S} = \frac{PR}{QS}$$

as long as $Q \neq 0$ and $S \neq 0$.

EXAMPLE 1 Multiply.

a. $\dfrac{25x}{2} \cdot \dfrac{1}{y^3}$ **b.** $\dfrac{-7x^2}{5y} \cdot \dfrac{3y^5}{14x^2}$

Solution: To multiply rational expressions, multiply the numerators and then multiply the denominators of both expressions. Then write in lowest terms.

a. $\dfrac{25x}{2} \cdot \dfrac{1}{y^3} = \dfrac{25x \cdot 1}{2 \cdot y^3} = \dfrac{25x}{2y^3}$

The expression $\dfrac{25x}{2y^3}$ is in lowest terms.

b. $\dfrac{-7x^2}{5y} \cdot \dfrac{3y^5}{14x^2} = \dfrac{-7x^2 \cdot 3y^5}{5y \cdot 14x^2}$ Multiply.

The expression $\dfrac{-7x^2 \cdot 3y^5}{5y \cdot 14x^2}$ is not in lowest terms, so we factor the numerator and the denominator and apply the fundamental principle.

$$= \dfrac{-1 \cdot \boxed{7} \cdot 3 \cdot \boxed{x^2} \cdot \boxed{y} \cdot y^4}{5 \cdot 2 \cdot \boxed{7} \cdot \boxed{x^2} \cdot \boxed{y}}$$

$$= -\dfrac{3y^4}{10}$$

When multiplying rational expressions, it is usually best to factor each numerator and denominator. This will help us when we apply the fundamental principle to write the product in lowest terms.

EXAMPLE 2 Multiply: $\dfrac{x^2 + x}{3x} \cdot \dfrac{6}{5x + 5}$.

Solution: $\dfrac{x^2 + x}{3x} \cdot \dfrac{6}{5x + 5} = \dfrac{x(x+1)}{3x} \cdot \dfrac{2 \cdot 3}{5(x+1)}$ Factor numerators and denominators.

$$= \dfrac{x\,(x+1) \cdot 2 \cdot \boxed{3}}{3x \cdot 5\,\boxed{(x+1)}}$$ Multiply.

$$= \dfrac{2}{5}$$ Simplify.

The following steps may be used to multiply rational expressions.

TO MULTIPLY RATIONAL EXPRESSIONS

Step 1.   Completely factor numerators and denominators.

Step 2.   Multiply numerators and multiply denominators.

Step 3.   Simplify or write the product in lowest terms by applying the fundamental principle to all common factors.

EXAMPLE 3 Multiply $\dfrac{3x + 3}{5x - 5x^2} \cdot \dfrac{2x^2 + x - 3}{4x^2 - 9}$.

Solution: $\dfrac{3x + 3}{5x - 5x^2} \cdot \dfrac{2x^2 + x - 3}{4x^2 - 9} = \dfrac{3(x + 1)}{5x(1 - x)} \cdot \dfrac{(2x + 3)(x - 1)}{(2x - 3)(2x + 3)}$ Factor.

$$= \dfrac{3(x + 1)\,\boxed{(2x + 3)}(x - 1)}{5x(1 - x)(2x - 3)\,\boxed{(2x + 3)}}$$ Multiply.

$$= \dfrac{3(x + 1)(x - 1)}{5x(1 - x)(2x - 3)}$$ Apply the fundamental principle.

Next, recall that $x - 1$ and $1 - x$ are opposites so that $x - 1 = -1(1 - x)$.

$$= \frac{3(x+1)(-1)(1-x)}{5x(1-x)(2x-3)} \quad \text{Write } x-1 \text{ as } -1(1-x).$$

$$= \frac{-3(x+1)}{5x(2x-3)} \quad \text{Apply the fundamental principle.}$$

2 We can divide by a rational expression in the same way we divide by a fraction. To divide by a fraction, multiply by its reciprocal.

For example, to divide $\frac{3}{2}$ by $\frac{7}{8}$, multiply $\frac{3}{2}$ by $\frac{8}{7}$.

$$\frac{3}{2} \div \frac{7}{8} = \frac{3}{2} \cdot \frac{8}{7} = \frac{3 \cdot 4 \cdot 2}{2 \cdot 7} = \frac{12}{7}$$

DIVIDING RATIONAL EXPRESSIONS

Let P, Q, R, and S be polynomials. Then,

$$\frac{P}{Q} \div \frac{R}{S} = \frac{P}{Q} \cdot \frac{S}{R} = \frac{PS}{QR}$$

as long as $Q \neq 0$, $S \neq 0$, and $R \neq 0$.

EXAMPLE 4 Divide: $\dfrac{3x^3y^7}{40} \div \dfrac{4x^3}{y^2}$.

Solution:
$$\frac{3x^3y^7}{40} \div \frac{4x^3}{y^2} = \frac{3x^3y^7}{40} \cdot \frac{y^2}{4x^3} \quad \text{Multiply by the reciprocal of } \frac{4x^3}{y^2}.$$

$$= \frac{3 x^3 y^9}{160 x^3}$$

$$= \frac{3y^9}{160} \quad \text{Simplify.}$$

EXAMPLE 5 Divide $\dfrac{(x-1)(x+2)}{10}$ by $\dfrac{2x+4}{5}$.

Solution:
$$\frac{(x-1)(x+2)}{10} \div \frac{2x+4}{5} = \frac{(x-1)(x+2)}{10} \cdot \frac{5}{2x+4} \quad \text{Multiply by the reciprocal of } \frac{2x+4}{5}.$$

$$= \frac{(x-1)(x+2) \cdot 5}{5 \cdot 2 \cdot 2 \cdot (x+2)} \quad \text{Factor and multiply.}$$

$$= \frac{x-1}{4} \quad \text{Simplify.}$$

The following may be used to divide by a rational expression.

CHAPTER 6 RATIONAL EXPRESSIONS

> **TO DIVIDE BY A RATIONAL EXPRESSION**
> Multiply by its reciprocal.

EXAMPLE 6 Divide: $\dfrac{6x+2}{x^2-1} \div \dfrac{3x^2+x}{x-1}$.

Solution:
$$\dfrac{6x+2}{x^2-1} \div \dfrac{3x^2+x}{x-1} = \dfrac{6x+2}{x^2-1} \cdot \dfrac{x-1}{3x^2+x}$$ Multiply by the reciprocal.

$$= \dfrac{2(3x+1)(x-1)}{(x+1)(x-1) \cdot x(3x+1)}$$ Factor and multiply.

$$= \dfrac{2}{x(x+1)}$$ Simplify.

EXAMPLE 7 Divide: $\dfrac{2x^2-11x+5}{5x-25} \div \dfrac{4x-2}{10}$.

Solution:
$$\dfrac{2x^2-11x+5}{5x-25} \div \dfrac{4x-2}{10} = \dfrac{2x^2-11x+5}{5x-25} \cdot \dfrac{10}{4x-2}$$ Multiply by the reciprocal.

$$= \dfrac{(2x-1)(x-5) \cdot 2 \cdot 5}{5(x-5) \cdot 2(2x-1)}$$ Factor and multiply.

$$= \dfrac{1}{1} \text{ or } 1$$ Simplify.

MENTAL MATH

Find the following products. See Example 1.

1. $\dfrac{2}{y} \cdot \dfrac{x}{3}$
2. $\dfrac{3x}{4} \cdot \dfrac{1}{y}$
3. $\dfrac{5}{7} \cdot \dfrac{y^2}{x^2}$
4. $\dfrac{x^5}{11} \cdot \dfrac{4}{z^3}$
5. $\dfrac{9}{x} \cdot \dfrac{x}{5}$
6. $\dfrac{y}{7} \cdot \dfrac{3}{y}$

EXERCISE SET 6.2

Find each product and simplify if possible. See Examples 1 through 3.

1. $\dfrac{3x}{y^2} \cdot \dfrac{7y}{4x}$
2. $\dfrac{9x^2}{y} \cdot \dfrac{4y}{3x^2}$
3. $\dfrac{8x}{2} \cdot \dfrac{x^5}{4x^2}$
4. $\dfrac{6x^2}{10x^3} \cdot \dfrac{5x}{12}$
5. $-\dfrac{5a^2b}{30a^2b^2} \cdot b^3$
6. $-\dfrac{9x^3y^2}{18xy^5} \cdot y^3$
7. $\dfrac{x}{2x-14} \cdot \dfrac{x^2-7x}{5}$
8. $\dfrac{4x-24}{20x} \cdot \dfrac{5}{x-6}$
9. $\dfrac{6x+6}{5} \cdot \dfrac{10}{36x+36}$
10. $\dfrac{x^2+x}{8} \cdot \dfrac{16}{x+1}$

11. $\dfrac{m^2 - n^2}{m + n} \cdot \dfrac{m}{m^2 - mn}$ 12. $\dfrac{(m - n)^2}{m + n} \cdot \dfrac{m}{m^2 - mn}$

13. $\dfrac{x^2 - 25}{x^2 - 3x - 10} \cdot \dfrac{x + 2}{x}$ 14. $\dfrac{a^2 + 6a + 9}{a^2 - 4} \cdot \dfrac{a + 3}{a - 2}$

15. Find the area of the following rectangle.

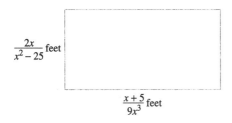

16. Find the area of the following square.

Find each quotient and simplify. See Examples 4 through 7.

17. $\dfrac{5x^7}{2x^5} \div \dfrac{10x}{4x^3}$ 18. $\dfrac{9y^4}{6y} \div \dfrac{y^2}{3}$

19. $\dfrac{8x^2}{y^3} \div \dfrac{4x^2y^3}{6}$ 20. $\dfrac{7a^2b}{3ab^2} \div \dfrac{21a^2b^2}{14ab}$

21. $\dfrac{(x - 6)(x + 4)}{4x} \div \dfrac{2x - 12}{8x^2}$

22. $\dfrac{(x + 3)^2}{5} \div \dfrac{5x + 15}{25}$ 23. $\dfrac{3x^2}{x^2 - 1} \div \dfrac{x^5}{(x + 1)^2}$

24. $\dfrac{(x + 1)}{(x + 1)(2x + 3)} \div \dfrac{20}{2x + 3}$

25. $\dfrac{m^2 - n^2}{m + n} \div \dfrac{m}{m^2 + nm}$ 26. $\dfrac{(m - n)^2}{m + n} \div \dfrac{m^2 - mn}{m}$

27. $\dfrac{x + 2}{7 - x} \div \dfrac{x^2 - 5x + 6}{x^2 - 9x + 14}$ 28. $(x - 3) \div \dfrac{x^2 + 3x - 18}{x}$

29. $\dfrac{x^2 + 7x + 10}{1 - x} \div \dfrac{x^2 + 2x - 15}{x - 1}$

30. $\dfrac{a^2 - b^2}{9} \cdot \dfrac{27x^2}{3b - 3a}$

31. Explain how to multiply rational expressions.

32. Explain how to divide rational expressions.

Perform the indicated operations.

33. $\dfrac{5a^2b}{30a^2b^2} \cdot \dfrac{1}{b^3}$ 34. $\dfrac{9x^2y^2}{42xy^5} \cdot \dfrac{6}{x^5}$

35. $\dfrac{12x^3y}{8xy^7} \div \dfrac{7x^5y}{6x}$ 36. $\dfrac{4y^2z}{3y^7z^7} \div \dfrac{12y}{6z}$

37. $\dfrac{5x - 10}{12} \div \dfrac{4x - 8}{8}$ 38. $\dfrac{6x + 6}{5} \div \dfrac{3x + 3}{10}$

39. $\dfrac{x^2 + 5x}{8} \cdot \dfrac{9}{3x + 15}$ 40. $\dfrac{3x^2 + 12x}{6} \cdot \dfrac{9}{2x + 8}$

41. $\dfrac{7}{6p^2 + q} \div \dfrac{14}{18p^2 + 3q}$ 42. $\dfrac{5x - 10}{12} \div \dfrac{4x - 8}{8}$

43. $\dfrac{3x + 4y}{x^2 + 4xy + 4y^2} \cdot \dfrac{x + 2y}{2}$

44. $\dfrac{2a + 2b}{3} \div \dfrac{a^2 - b^2}{a - b}$ 45. $\dfrac{x^2 - 9}{x^2 + 8} \div \dfrac{3 - x}{2x^2 + 16}$

46. $\dfrac{x^2 - y^2}{3x^2 + 3xy} \cdot \dfrac{3x^2 + 6x}{3x^2 - 2xy - y^2}$

47. $\dfrac{(x + 2)^2}{x - 2} \div \dfrac{x^2 - 4}{2x - 4}$ 48. $\dfrac{x^2 - 4}{2y} \div \dfrac{2 - x}{6xy}$

49. $\dfrac{a^2 + 7a + 12}{a^2 + 5a + 6} \cdot \dfrac{a^2 + 8a + 15}{a^2 + 5a + 4}$

50. $\dfrac{b^2 + 2b - 3}{b^2 + b - 2} \cdot \dfrac{b^2 - 4}{b^2 + 6b + 8}$

51. $\dfrac{1}{-x - 4} \div \dfrac{x^2 - 7x}{x^2 - 3x - 28}$

52. $\dfrac{x^2 - 10x + 21}{7 - x} \div (x + 3)$

53. $\dfrac{x^2 - 5x - 24}{2x^2 - 2x - 24} \cdot \dfrac{4x^2 + 4x - 24}{x^2 - 10x + 16}$

54. $\dfrac{a^2 - b^2}{a} \cdot \dfrac{a + b}{a^2 + ab}$ 55. $(x - 5) \div \dfrac{5 - x}{x^2 + 2}$

56. $\dfrac{2x^2 + 3xy + y^2}{x^2 - y^2} \div \dfrac{1}{2x + 2y}$

57. $\dfrac{x^2 - y^2}{x^2 - 2xy + y^2} \cdot \dfrac{y - x}{x + y}$ 58. $\dfrac{x + 3}{x^2 - 9} \cdot \dfrac{x^2 - 8x + 15}{5x}$

59. $\dfrac{a^2 + ac + ba + bc}{a - b} \div \dfrac{a + c}{a + b}$

60. $\dfrac{x^2 + 2x - xy - 2y}{x^2 - y^2} \div \dfrac{2x + 4}{x + y}$

61. $\dfrac{3x^2 + 8x + 5}{x^2 + 8x + 7} \cdot \dfrac{x + 7}{x^2 + 4}$

62. $\dfrac{16x^2 + 2x}{16x^2 + 10x + 1} \cdot \dfrac{1}{4x^2 + 2x}$

63. Find the quotient of $\dfrac{x^2 - 9}{2x}$ and $\dfrac{x + 3}{8x^4}$.

64. Find the quotient of $\dfrac{4x^2 + 4x + 1}{4x + 2}$ and $\dfrac{4x + 2}{16}$.

65. $\dfrac{x^3 + 8}{x^2 - 2x + 4} \cdot \dfrac{4}{x^2 - 4}$

66. $\dfrac{9y}{3y - 3} \cdot \dfrac{y^3 - 1}{y^3 + y^2 + y}$

67. $\dfrac{a^2 - ab}{6a^2 + 6ab} \div \dfrac{a^3 - b^3}{a^2 - b^2}$

68. $\dfrac{x^3 + 27y^3}{6x} \div \dfrac{x^2 - 9y^2}{x^2 - 3xy}$

A Look Ahead

EXAMPLE Perform the indicated operations.

$$\dfrac{15x^2 - x - 6}{12x^3} \cdot \dfrac{4x}{9 - 25x^2} \div \dfrac{x}{3x - 2}$$

Solution:

$$\dfrac{15x^2 - x - 6}{12x^3} \cdot \dfrac{4x}{9 - 25x^2} \div \dfrac{x}{3x - 2}$$

$$= \left(\dfrac{15x^2 - x - 6}{12x^3} \cdot \dfrac{4x}{9 - 25x^2}\right) \cdot \dfrac{3x - 2}{x}$$

$$= \dfrac{(3x - 2)(5x + 3) \cdot 4x(3x - 2)}{12x^3(3 - 5x)(3 + 5x) \cdot x}$$

$$= \dfrac{(3x - 2)^2}{3x^3(3 - 5x)}$$

Perform the following operations.

77. $\left(\dfrac{x^2 - y^2}{x^2 + y^2} \div \dfrac{x^2 - y^2}{3x}\right) \cdot \dfrac{x^2 + y^2}{6}$

78. $\left(\dfrac{x^2 - 9}{x^2 - 1} \cdot \dfrac{x^2 + 2x + 1}{2x^2 + 9x + 9}\right) \div \dfrac{2x + 3}{1 - x}$

79. $\left(\dfrac{2a + b}{b^2} \cdot \dfrac{3a^2 - 2ab}{ab + 2b^2}\right) \div \dfrac{a^2 - 3ab + 2b^2}{5ab - 10b^2}$

80. $\left(\dfrac{x^2y^2 - xy}{4x - 4y} \div \dfrac{3y - 3x}{8x - 8y}\right) \cdot \dfrac{y - x}{8}$

Review Exercises

Perform each operation. See Section 1.2.

69. $\dfrac{1}{5} + \dfrac{4}{5}$

70. $\dfrac{3}{15} + \dfrac{6}{15}$

71. $\dfrac{9}{9} - \dfrac{19}{9}$

72. $\dfrac{4}{3} - \dfrac{8}{3}$

73. $\dfrac{6}{5} + \left(\dfrac{1}{5} - \dfrac{8}{5}\right)$

74. $-\dfrac{3}{2} + \left(\dfrac{1}{2} - \dfrac{3}{2}\right)$

See Section 3.2.

75. Graph the linear equation $x - 2y = 6$.

76. Graph the linear equation $5x + y = 10$.

6.3 ADDING AND SUBTRACTING RATIONAL EXPRESSIONS WITH COMMON DENOMINATORS AND LEAST COMMON DENOMINATOR

TAPE
BA 6.3

OBJECTIVES

1. Add and subtract rational expressions with common denominators.
2. Find the least common denominator of a list of rational expressions.
3. Write a rational expression as an equivalent expression whose denominator is given.

 Like multiplication and division, addition and subtraction of rational expressions is similar to addition and subtraction of rational numbers. For example, to add fractions with common denominators such as $\frac{6}{5}$ and $\frac{2}{5}$, add the numerators and write the sum over the common denominator 5.

$$\frac{6}{5} + \frac{2}{5} = \frac{6+2}{5} = \frac{8}{5}$$

To subtract two fractions with common denominators, subtract the numerators and write the result over the common denominator.

$$\frac{6}{13} - \frac{5}{13} = \frac{6-5}{13} = \frac{1}{13}$$

Rational expressions with common denominators are added and subtracted in the same manner.

ADDING AND SUBTRACTING RATIONAL EXPRESSIONS WITH COMMON DENOMINATORS

If P, Q, and R are polynomials, then

$$\frac{P}{R} + \frac{Q}{R} = \frac{P+Q}{R} \quad \text{and} \quad \frac{P}{R} - \frac{Q}{R} = \frac{P-Q}{R}$$

as long as $R \neq 0$.

EXAMPLE 1 Add: $\dfrac{5m}{2n} + \dfrac{m}{2n}$.

Solution:
$\dfrac{5m}{2n} + \dfrac{m}{2n} = \dfrac{5m + m}{2n}$ Add the numerators.

$= \dfrac{6m}{2n}$ Simplify the numerator by combining like terms.

$= \dfrac{3m}{n}$ Simplify the rational expression by applying the fundamental principle.

EXAMPLE 2 Subtract: $\dfrac{2y}{2y-7} - \dfrac{7}{2y-7}$.

Solution:
$\dfrac{2y}{2y-7} - \dfrac{7}{2y-7} = \dfrac{2y-7}{2y-7}$ Subtract the numerators.

$= \dfrac{1}{1}$ or 1 Simplify.

EXAMPLE 3 Subtract: $\dfrac{3x^2 + 2x}{x - 1} - \dfrac{10x - 5}{x - 1}$.

Solution:
$\dfrac{3x^2 + 2x}{x - 1} - \dfrac{10x - 5}{x - 1} = \dfrac{3x^2 + 2x - (10x - 5)}{x - 1}$ Subtract the numerators. Notice the parentheses.

$= \dfrac{3x^2 + 2x - 10x + 5}{x - 1}$ Use the distributive property.

$= \dfrac{3x^2 - 8x + 5}{x - 1}$ Combine like terms.

$= \dfrac{(x - 1)(3x - 5)}{x - 1}$ Factor.

$= 3x - 5$ Simplify.

> **REMINDER** Notice how the numerator $10x - 5$ has been subtracted in Example 3.
>
> This $-$ sign applies to the entire numerator of $10x - 5$.   So parentheses are inserted here to indicate this.
>
> $\downarrow$ $\downarrow$ $\downarrow$
>
> $\dfrac{3x^2 + 2x}{x - 1} - \dfrac{10x - 5}{x - 1} = \dfrac{3x^2 + 2x - (10x - 5)}{x - 1}$

2 To add and subtract fractions with **unlike** denominators, first find a least common denominator (LCD), and then write all fractions as equivalent fractions with the LCD.

For example, suppose we add $\tfrac{8}{3}$ and $\tfrac{2}{5}$. The LCD of denominators 3 and 5 is 15, since 15 is the smallest number that both 3 and 5 divide into evenly. Rewrite each fraction so that its denominator is 15. (Notice how we apply the fundamental principle.)

$$\dfrac{8}{3} + \dfrac{2}{5} = \dfrac{8(5)}{3(5)} + \dfrac{2(3)}{5(3)} = \dfrac{40}{15} + \dfrac{6}{15} = \dfrac{40 + 6}{15} = \dfrac{46}{15}$$

To add or subtract rational expressions with unlike denominators, we also first find an LCD and then write all rational expressions as equivalent expressions with the LCD. The **least common denominator LCD of a list of rational expressions** is a polynomial of least degree whose factors include all the factors of the denominators in the list.

> **TO FIND THE LEAST COMMON DENOMINATOR (LCD)**
>
> *Step 1.*   Factor each denominator completely.
>
> *Step 2.*   The least common denominator LCD is the product of all unique factors found in step 1, each raised to a power equal to the greatest number of times that the factor appears in any one factored denominator.

EXAMPLE 4 Find the LCD for each pair.

a. $\dfrac{1}{8}, \dfrac{3}{22}$

b. $\dfrac{7}{5x}, \dfrac{6}{15x^2}$

Solution: **a.** Start by finding the prime factorization of each denominator.

$$8 = 2 \cdot 2 \cdot 2 = 2^3 \quad \text{and} \quad 22 = 2 \cdot 11$$

Next, write the product of all the unique factors, each raised to a power equal to the greatest number of times that the factor appears.

The greatest number of times that the factor 2 appears is 3.
The greatest number of times that the factor 11 appears is 1.

$$\text{LCD} = 2^3 \cdot 11^1 = 8 \cdot 11 = 88$$

b. Factor each denominator.

$$5x = 5 \cdot x \quad \text{and} \quad 15x^2 = 3 \cdot 5 \cdot x^2$$

The greatest number of times that the factor 5 appears is 1.
The greatest number of times that the factor 3 appears is 1.
The greatest number of times that the factor x appears is 2.

$$\text{LCD} = 3^1 \cdot 5^1 \cdot x^2 = 15x^2$$

EXAMPLE 5 Find the LCD of $\dfrac{7x}{x+2}$ and $\dfrac{5x^2}{x-2}$.

Solution: The denominators $x + 2$ and $x - 2$ are completely factored already. The factor $x + 2$ appears once and the factor $x - 2$ appears once.

$$\text{LCD} = (x+2)(x-2)$$

EXAMPLE 6 Find the LCD of $\dfrac{6m^2}{3m+15}$ and $\dfrac{2}{(m+5)^2}$.

Solution: Factor each denominator.

$$3m + 15 = 3(m + 5)$$
$(m + 5)^2$ is already factored.

The greatest number of times that the factor 3 appears is 1.
The greatest number of times that the factor $m + 5$ appears *in any one denominator* is 2.

$$\text{LCD} = 3(m+5)^2$$

EXAMPLE 7 Find the LCD of $\dfrac{t-10}{t^2-t-6}$ and $\dfrac{t+5}{t^2+3t+2}$.

Solution: Start by factoring each denominator.

$$t^2 - t - 6 = (t-3)(t+2)$$
$$t^2 + 3t + 2 = (t+1)(t+2)$$
$$\text{LCD} = (t-3)(t+2)(t+1)$$

EXAMPLE 8 Find the LCD of $\dfrac{2}{x-2}$ and $\dfrac{10}{2-x}$.

Solution: The denominators $x - 2$ and $2 - x$ are opposites. That is, $2 - x = -1(x - 2)$. Use $x - 2$ or $2 - x$ as the LCD.

$$\text{LCD} = x - 2 \quad \text{or} \quad \text{LCD} = 2 - x$$

3 Next we practice writing a rational expression as an equivalent rational expression with a given denominator. To do this, we apply the fundamental principle, which says that $\dfrac{PR}{QR} = \dfrac{P}{Q}$, or equivalently that $\dfrac{P}{Q} = \dfrac{PR}{QR}$. This can be seen by recalling that multiplying an expression by 1 produces an equivalent expression. In other words,

$$\dfrac{P}{Q} = \dfrac{P}{Q} \cdot 1 = \dfrac{P}{Q} \cdot \dfrac{R}{R} = \dfrac{PR}{QR}.$$

EXAMPLE 9 Write $\dfrac{4b}{9a}$ as an equivalent fraction with the given denominator.

$$\dfrac{4b}{9a} = \dfrac{}{27a^2b}$$

Solution: Ask yourself: "What do we multiply $9a$ by to get $27a^2b$?" The answer is $3ab$, since $9a(3ab) = 27a^2b$. Multiply the numerator and denominator by $3ab$.

$$\dfrac{4b}{9a} = \dfrac{4b(3ab)}{9a(3ab)} = \dfrac{12ab^2}{27a^2b}$$

EXAMPLE 10 Write the rational expression as an equivalent rational expression with the given denominator.

$$\dfrac{5}{x^2-4} = \dfrac{}{(x-2)(x+2)(x-4)}$$

Solution: First, factor the denominator $x^2 - 4$.

$$\frac{5}{x^2 - 4} = \frac{5}{(x+2)(x-2)}$$

If we multiply the original denominator $(x + 2)(x - 2)$ by $x - 4$, the result is the new denominator $(x + 2)(x - 2)(x - 4)$. Thus, multiply the numerator and the denominator by $x - 4$.

$$\frac{5}{x^2 - 4} = \frac{5}{(x-2)(x+2)} = \frac{5(x-4)}{(x-2)(x+2)(x-4)}$$
$$= \frac{5x - 20}{(x-2)(x+2)(x-4)}$$

MENTAL MATH

Perform the indicated operation.

1. $\dfrac{2}{3} + \dfrac{1}{3}$
2. $\dfrac{5}{11} + \dfrac{1}{11}$
3. $\dfrac{3x}{9} + \dfrac{4x}{9}$
4. $\dfrac{3y}{8} + \dfrac{2y}{8}$
5. $\dfrac{8}{9} - \dfrac{7}{9}$
6. $-\dfrac{4}{12} - \dfrac{3}{12}$
7. $\dfrac{7}{5} - \dfrac{10y}{5}$
8. $\dfrac{12x}{7} - \dfrac{4x}{7}$

EXERCISE SET 6.3

Add or subtract as indicated. Simplify the result if possible. See examples 1 through 3.

1. $\dfrac{a}{13} + \dfrac{9}{13}$
2. $\dfrac{x+1}{7} + \dfrac{6}{7}$
3. $\dfrac{9}{3+y} + \dfrac{y+1}{3+y}$
4. $\dfrac{9}{y+9} + \dfrac{y}{y+9}$
5. $\dfrac{4m}{3n} + \dfrac{5m}{3n}$
6. $\dfrac{3p}{2} + \dfrac{11p}{2}$
7. $\dfrac{2x+1}{x-3} + \dfrac{3x+6}{x-3}$
8. $\dfrac{4p-3}{2p+7} + \dfrac{3p+8}{2p+7}$
9. $\dfrac{7}{8} - \dfrac{3}{8}$
10. $\dfrac{4}{5} - \dfrac{13}{5}$
11. $\dfrac{4m}{m-6} - \dfrac{24}{m-6}$
12. $\dfrac{8y}{y-2} - \dfrac{16}{y-2}$
13. $\dfrac{2x^2}{x-5} - \dfrac{25+x^2}{x-5}$
14. $\dfrac{6x^2}{2x-5} - \dfrac{25+2x^2}{2x-5}$
15. $\dfrac{-3x^2-4}{x-4} - \dfrac{12-4x^2}{x-4}$
16. $\dfrac{7x^2-9}{2x-5} - \dfrac{16+3x^2}{2x-5}$
17. $\dfrac{2x+3}{x+1} - \dfrac{x+2}{x+1}$
18. $\dfrac{1}{x^2-2x-15} - \dfrac{4-x}{x^2-2x-15}$
19. $\dfrac{3}{x^3} + \dfrac{9}{x^3}$
20. $\dfrac{5}{xy} + \dfrac{8}{xy}$
21. $\dfrac{5}{x+4} - \dfrac{10}{x+4}$
22. $\dfrac{4}{2x+1} - \dfrac{8}{2x+1}$
23. $\dfrac{x}{x+y} - \dfrac{2}{x+y}$
24. $\dfrac{y+1}{y+2} - \dfrac{3}{y+2}$
25. $\dfrac{8x}{2x+5} + \dfrac{20}{2x+5}$
26. $\dfrac{12y-5}{3y-1} + \dfrac{1}{3y-1}$

362 CHAPTER 6 RATIONAL EXPRESSIONS

27. $\dfrac{5x+4}{x-1} - \dfrac{2x+7}{x-1}$

28. $\dfrac{x^2+9x}{x+7} - \dfrac{4x+14}{x+7}$

29. $\dfrac{a}{a^2+2a-15} - \dfrac{3}{a^2+2a-15}$

30. $\dfrac{3y}{y^2+3y-10} - \dfrac{6}{y^2+3y-10}$

31. $\dfrac{2x+3}{x^2-x-30} - \dfrac{x-2}{x^2-x-30}$

32. $\dfrac{3x-1}{x^2+5x-6} - \dfrac{2x-7}{x^2+5x-6}$

33. A square-shaped pasture has a side of length $\dfrac{5}{x-2}$ meters. Express its perimeter as a rational expression.

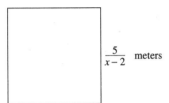

34. The following trapezoid has sides of indicated length. Find its perimeter.

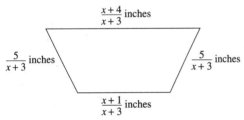

35. Describe the process for adding and subtracting two rational expressions with the same denominators.

36. Explain the similarities between subtracting $\tfrac{3}{8}$ from $\tfrac{7}{8}$ and subtracting $\tfrac{6}{x+3}$ from $\tfrac{9}{x+3}$.

Find the LCD for the following lists of rational expressions. See examples 4 through 8.

37. $\dfrac{2}{3}, \dfrac{4}{33}$

38. $\dfrac{8}{20}, \dfrac{4}{15}$

39. $\dfrac{19}{2x}, \dfrac{5}{4x^3}$

40. $\dfrac{17x}{4y^5}, \dfrac{2}{8y}$

41. $\dfrac{9}{8x}, \dfrac{3}{2x+4}$

42. $\dfrac{1}{6y}, \dfrac{3x}{4y+12}$

43. $\dfrac{1}{3x+3}, \dfrac{8}{2x^2+4x+2}$

44. $\dfrac{19x+5}{4x-12}, \dfrac{3}{2x^2-12x+18}$

45. $\dfrac{5}{x-8}, \dfrac{3}{8-x}$

46. $\dfrac{2x+5}{3x-7}, \dfrac{5}{7-3x}$

47. $\dfrac{4+x}{8x^2(x-1)^2}, \dfrac{17}{10x^3(x-1)}$

48. $\dfrac{2x+3}{9x(x+2)}, \dfrac{9x+5}{12(x+2)^2}$

49. $\dfrac{9x+1}{2x+1}, \dfrac{3x-5}{2x-1}$

50. $\dfrac{5}{4x-2}, \dfrac{7}{4x+2}$

51. $\dfrac{5x+1}{2x^2+7x-4}, \dfrac{3x}{2x^2+5x-3}$

52. $\dfrac{4}{x^2+4x+3}, \dfrac{4x-2}{x^2+10x+21}$

53. Write some instructions to help a friend who is having difficulty finding the LCD of two rational expressions.

54. Explain why the LCD of rational expressions $\dfrac{7}{x+1}$ and $\dfrac{9x}{(x+1)^2}$ is $(x+1)^2$ and not $(x+1)^3$.

Rewrite each rational expression as an equivalent rational expression whose denominator is the given polynomial. See Examples 9 and 10.

55. $\dfrac{3}{2x}; 4x^2$

56. $\dfrac{3}{9y^5}; 72y^9$

57. $\dfrac{6}{3a}; 12ab^2$

58. $\dfrac{17a}{4y^2x}; 32y^3x^2z$

59. $\dfrac{9}{x+3}; 2(x+3)$

60. $\dfrac{4x+1}{3x+6}; 3y(x+2)$

61. $\dfrac{9a+2}{5a+10}; 5b(a+2)$

62. $\dfrac{5+y}{2x^2+10}; 4(x^2+5)$

63. $\dfrac{x}{x^3+6x^2+8x}; x(x+4)(x+2)(x+1)$

64. $\dfrac{5x}{x^2+2x-3}; (x-1)(x-5)(x+3)$

65. $\dfrac{5}{2x^2-9x-5}; 3x(2x+1)(x-7)(x-5)$

66. $\dfrac{x-9}{3x^2+10x+3}; x(x+3)(x+5)(3x+1)$

67. $\dfrac{9y-1}{15x^2-30}; 30x^2-60$

68. $\dfrac{6}{x^2-9}; (x+3)(x-3)(x+2)$

69. $\dfrac{1}{x^2 - 16}$; $x(x - 4)^2(x + 4)$

70. $\dfrac{-3}{x^2 - 9}$; $(x - 3)(x + 3)^2$

Write each rational expression as an equivalent expression with a denominator of $x - 2$.

71. $\dfrac{5}{2 - x}$

72. $\dfrac{8y}{2 - x}$

73. $-\dfrac{7 + x}{2 - x}$

74. $\dfrac{x - 3}{-(x - 2)}$

Review Exercises

Solve the following quadratic equations by factoring. See Section 5.6.

75. $x(x - 3) = 0$

76. $2x(x + 5) = 0$

77. $x^2 + 6x + 5 = 0$

78. $x^2 - 6x + 5 = 0$

Perform each operation. See Section 1.2.

79. $\dfrac{2}{3} + \dfrac{5}{7}$

80. $\dfrac{9}{10} - \dfrac{3}{5}$

81. $\dfrac{2}{6} - \dfrac{3}{4}$

82. $\dfrac{11}{15} + \dfrac{5}{9}$

6.4 ADDING AND SUBTRACTING RATIONAL EXPRESSIONS WITH UNLIKE DENOMINATORS

OBJECTIVE

1. Add and subtract rational expressions with unlike denominators.

In the previous section, we practiced all the skills we need to add and subtract rational expressions with unlike denominators. The steps are as follows:

TO ADD OR SUBTRACT RATIONAL EXPRESSIONS WITH UNLIKE DENOMINATORS

Step 1. Find the LCD of the rational expressions.

Step 2. Rewrite each rational expression as an equivalent expression whose denominator is the LCD found in *step 1*.

Step 3. Add or subtract numerators and write the sum or difference over the common denominator.

Step 4. Write the rational expression in lowest terms.

EXAMPLE 1 Perform the indicated operation.

a. $\dfrac{a}{4} - \dfrac{2a}{8}$

b. $\dfrac{3}{10x^2} + \dfrac{7}{25x}$

Solution: **a.** First, find the LCD. Since $4 = 2^2$ and $8 = 2^3$, the LCD $= 2^3 = 8$. Write each fraction as an equivalent fraction with the denominator 8; then subtract.

$$\dfrac{a}{4} - \dfrac{2a}{8} = \dfrac{a(2)}{4(2)} - \dfrac{2a}{8} = \dfrac{2a}{8} - \dfrac{2a}{8} = \dfrac{2a - 2a}{8} = \dfrac{0}{8} = 0$$

b. Since $10x^2 = 2 \cdot 5 \cdot x \cdot x$ and $25x = 5 \cdot 5 \cdot x$, the LCD $= 2 \cdot 5^2 \cdot x^2 = 50x^2$. Write each fraction as an equivalent fraction with a denominator of $50x^2$.

$$\frac{3}{10x^2} + \frac{7}{25x} = \frac{3(5)}{10x^2(5)} + \frac{7(2x)}{25x(2x)}$$

$$= \frac{15}{50x^2} + \frac{14x}{50x^2}$$

$$= \frac{15 + 14x}{50x^2} \qquad \text{Add numerators, and use the common denominator.}$$

EXAMPLE 2   Subtract: $\dfrac{6x}{x^2 - 4} - \dfrac{3}{x + 2}$.

Solution: Since $x^2 - 4 = (x + 2)(x - 2)$,

$$\text{LCD} = (x - 2)(x + 2)$$

Write equivalent expressions with the LCD as denominators.

$$\frac{6x}{x^2 - 4} - \frac{3}{x + 2} = \frac{6x}{(x - 2)(x + 2)} - \frac{3(x - 2)}{(x + 2)(x - 2)}$$

$$= \frac{6x - 3(x - 2)}{(x + 2)(x - 2)} \qquad \text{Subtract numerators, and use the common denominator.}$$

$$= \frac{6x - 3x + 6}{(x + 2)(x - 2)} \qquad \text{Apply the distributive property in the numerator.}$$

$$= \frac{3x + 6}{(x + 2)(x - 2)} \qquad \text{Combine like terms in the numerator.}$$

Next, factor the numerator to see if this rational expression can be simplified.

$$= \frac{3x + 6}{(x + 2)(x - 2)}$$

$$= \frac{3\,(x + 2)}{(x + 2)\,(x - 2)} \qquad \text{Factor.}$$

$$= \frac{3}{x - 2} \qquad \text{Apply the fundamental principle to simplify.}$$

EXAMPLE 3   Add: $\dfrac{2}{3t} + \dfrac{5}{t + 1}$.

Solution: The LCD is $3t(t + 1)$. Write each rational expression as an equivalent rational expression with a denominator of $3t(t + 1)$.

$$\frac{2}{3t} + \frac{5}{t + 1} = \frac{2(t + 1)}{3t(t + 1)} + \frac{5(3t)}{(t + 1)(3t)}$$

$$= \frac{2(t+1) + 5(3t)}{3t(t+1)}$$ Add numerators, and use the common denominator.

$$= \frac{2t + 2 + 15t}{3t(t+1)}$$ Apply the distributive property in the numerator.

$$= \frac{17t + 2}{3t(t+1)}$$ Combine like terms in the numerator.

EXAMPLE 4   Subtract: $\dfrac{7}{x-3} - \dfrac{9}{3-x}$.

Solution: To find a common denominator, notice that $x - 3$ and $3 - x$ are opposites. That is, $3 - x = -(x - 3)$. Write the denominator $3 - x$ as $-(x - 3)$ and simplify.

$$\frac{7}{x-3} - \frac{9}{3-x} = \frac{7}{x-3} - \frac{9}{-(x-3)}$$

$$= \frac{7}{x-3} - \frac{-9}{x-3}$$ Apply $\dfrac{a}{-b} = \dfrac{-a}{b}$.

$$= \frac{7 - (-9)}{x-3}$$ Subtract numerators, and use the common denominator.

$$= \frac{16}{x-3}$$

EXAMPLE 5   Add: $1 + \dfrac{m}{m+1}$.

Solution: Recall that 1 is the same as $\dfrac{1}{1}$. The LCD of $\dfrac{1}{1}$ and $\dfrac{m}{m+1}$ is $m + 1$.

$$1 + \frac{m}{m+1} = \frac{1}{1} + \frac{m}{m+1}$$ Write 1 as $\dfrac{1}{1}$.

$$= \frac{1(m+1)}{1(m+1)} + \frac{m}{m+1}$$ Multiply both the numerator and the denominator of $\dfrac{1}{1}$ by $m + 1$.

$$= \frac{m + 1 + m}{m+1}$$ Add numerators, and use the common denominator.

$$= \frac{2m + 1}{m+1}$$ Combine like terms in the numerator.

EXAMPLE 6   Subtract: $\dfrac{3}{2x^2 + x} - \dfrac{2x}{6x + 3}$.

Solution: First, factor the denominators.

$$\frac{3}{2x^2 + x} - \frac{2x}{6x + 3} = \frac{3}{x(2x+1)} - \frac{2x}{3(2x+1)}$$

366 CHAPTER 6 RATIONAL EXPRESSIONS

The LCD is $3x(2x + 1)$. Write equivalent expressions with denominators of $3x(2x + 1)$.

$$= \frac{3(3)}{x(2x + 1)(3)} - \frac{2x(x)}{3(2x + 1)(x)}$$

$$= \frac{9 - 2x^2}{3x(2x + 1)} \quad \text{Subtract numerators, and use the common denominator.}$$

EXAMPLE 7 Add: $\dfrac{2x}{x^2 + 2x + 1} + \dfrac{x}{x^2 - 1}$.

Solution: First, factor the denominators.

$$\frac{2x}{x^2 + 2x + 1} + \frac{x}{x^2 - 1} = \frac{2x}{(x + 1)(x + 1)} + \frac{x}{(x + 1)(x - 1)}$$

Write the rational expressions as equivalent expressions with denominators of $(x + 1)(x + 1)(x - 1)$, the LCD.

$$= \frac{2x(x - 1)}{(x + 1)(x + 1)(x - 1)} + \frac{x(x + 1)}{(x + 1)(x - 1)(x + 1)}$$

$$= \frac{2x(x - 1) + x(x + 1)}{(x + 1)^2(x - 1)} \quad \text{Add numerators, and use the common denominator.}$$

$$= \frac{2x^2 - 2x + x^2 + x}{(x + 1)^2(x - 1)} \quad \text{Apply the distributive property in the numerator.}$$

$$= \frac{3x^2 - x}{(x + 1)^2(x - 1)} \quad \text{or} \quad \frac{x(3x - 1)}{(x + 1)^2(x - 1)}$$

The numerator was factored as a last step to see if the rational expression could be simplified further.

EXERCISE SET 6.4

Perform the indicated operations. See Example 1.

1. $\dfrac{4}{2x} + \dfrac{9}{3x}$

2. $\dfrac{15}{7a} + \dfrac{8}{6a}$

3. $\dfrac{15a}{b} + \dfrac{6b}{5}$

4. $\dfrac{4c}{d} - \dfrac{8x}{5}$

5. $\dfrac{3}{x} + \dfrac{5}{2x^2}$

6. $\dfrac{14}{3x^2} + \dfrac{6}{x}$

Perform the indicated operations. See Examples 2 and 3.

7. $\dfrac{6}{x + 1} + \dfrac{9}{2x + 2}$

8. $\dfrac{8}{x + 4} - \dfrac{3}{3x + 12}$

9. $\dfrac{15}{2x - 4} + \dfrac{x}{x^2 - 4}$

10. $\dfrac{3}{x + 2} - \dfrac{1}{x^2 - 4}$

11. $\dfrac{3}{4x} + \dfrac{8}{x - 2}$

12. $\dfrac{x}{x + 1} + \dfrac{3}{x - 1}$

13. $\dfrac{5}{y^2} - \dfrac{y}{2y+1}$ **14.** $\dfrac{x}{4x-3} - \dfrac{3}{8x-6}$

15. In your own words, explain how to add two rational expressions with different denominators.

16. In your own words, explain how to subtract two rational expressions with different denominators.

Add or subtract as indicated. See Example 4.

17. $\dfrac{6}{x-3} + \dfrac{8}{3-x}$ **18.** $\dfrac{9}{x-3} + \dfrac{9}{3-x}$

19. $\dfrac{-8}{x^2-1} - \dfrac{7}{1-x^2}$ **20.** $\dfrac{-9}{25x^2-1} + \dfrac{7}{1-25x^2}$

21. $\dfrac{x}{x^2-4} - \dfrac{2}{4-x^2}$ **22.** $\dfrac{5}{2x-6} - \dfrac{3}{6-2x}$

Add or subtract as indicated. See Example 5.

23. $\dfrac{5}{x} + 2$ **24.** $\dfrac{7}{x^2} - 5x$

25. $\dfrac{5}{x-2} + 6$ **26.** $\dfrac{6y}{y+5} + 1$

27. $\dfrac{y+2}{y+3} - 2$ **28.** $\dfrac{7}{2x-3} - 3$

29. Two angles are said to be complementary if their sum is 90°. If one angle measures $\dfrac{40}{x}$ degrees, find the measure of its complement.

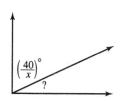

30. Two angles are said to be supplementary if their sum is 180°. If one angle measures $\dfrac{x+2}{x}$ degrees, find the measure of its supplement.

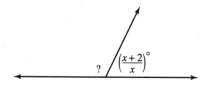

Perform the indicated operations. See Examples 6 and 7.

31. $\dfrac{5x}{x+2} - \dfrac{3x-4}{x+2}$ **32.** $\dfrac{7x}{x-3} - \dfrac{4x+9}{x-3}$

33. $\dfrac{3x^4}{x} - \dfrac{4x^2}{x^2}$ **34.** $\dfrac{5x}{6} + \dfrac{15x^2}{2}$

35. $\dfrac{1}{x+3} - \dfrac{1}{(x+3)^2}$ **36.** $\dfrac{5x}{(x-2)^2} - \dfrac{3}{x-2}$

37. $\dfrac{4}{5b} + \dfrac{1}{b-1}$ **38.** $\dfrac{1}{y+5} + \dfrac{2}{3y}$

39. $\dfrac{2}{m} + 1$ **40.** $\dfrac{6}{x} - 1$

41. $\dfrac{6}{1-2x} - \dfrac{4}{2x-1}$ **42.** $\dfrac{10}{3n-4} - \dfrac{5}{4-3n}$

43. $\dfrac{7}{(x+1)(x-1)} + \dfrac{8}{(x+1)^2}$

44. $\dfrac{5x+2}{(x+1)(x+5)} - \dfrac{2}{x+5}$

45. $\dfrac{x}{x^2-1} - \dfrac{2}{x^2-2x+1}$

46. $\dfrac{x}{x^2-4} - \dfrac{5}{x^2-4x+4}$

47. $\dfrac{3a}{2a+6} - \dfrac{a-1}{a+3}$ **48.** $\dfrac{1}{x+y} - \dfrac{y}{x^2-y^2}$

49. $\dfrac{5}{2-x} + \dfrac{x}{2x-4}$ **50.** $\dfrac{-1}{a-2} + \dfrac{4}{4-2a}$

51. $\dfrac{-7}{y^2-3y+2} - \dfrac{2}{y-1}$ **52.** $\dfrac{2}{x^2+4x+4} + \dfrac{1}{x+2}$

53. $\dfrac{13}{x^2-5x+6} - \dfrac{5}{x-3}$ **54.** $\dfrac{27}{y^2-81} + \dfrac{3}{2(y+9)}$

55. $\dfrac{8}{(x+2)(x-2)} + \dfrac{4}{(x+2)(x-3)}$

56. $\dfrac{5}{6x^2(x+2)} + \dfrac{4x}{x(x+2)^2}$

57. $\dfrac{5}{9x^2-4} + \dfrac{2}{3x-2}$

58. $\dfrac{4}{x^2-x-6} + \dfrac{x}{x^2+5x+6}$

59. $\dfrac{x+8}{x^2-5x-6} + \dfrac{x+1}{x^2-4x-5}$

368 CHAPTER 6 RATIONAL EXPRESSIONS

60. $\dfrac{x}{x^2 + 12x + 20} - \dfrac{1}{x^2 + 8x - 20}$

61. A board of length $\dfrac{3}{x+4}$ inches was cut into two pieces. If one piece is $\dfrac{1}{x-4}$ inches, express the length of the other board as a rational expression.

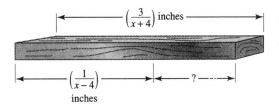

62. The length of the rectangle is $\dfrac{3}{y-5}$ feet, while its width is $\dfrac{2}{y}$ feet. Find its perimeter and then find its area.

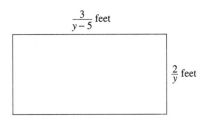

Perform the indicated operations. Addition, subtraction, multiplication, and division of rational expressions are included here.

63. $\dfrac{15x}{x+8} \cdot \dfrac{2x+16}{3x}$

64. $\dfrac{9z+5}{15} \cdot \dfrac{5z}{81z^2 - 25}$

65. $\dfrac{8x+7}{3x+5} - \dfrac{2x-3}{3x+5}$

66. $\dfrac{2z^2}{4z-1} - \dfrac{z - 2z^2}{4z-1}$

67. $\dfrac{5a+10}{18} \div \dfrac{a^2 - 4}{10a}$

68. $\dfrac{9}{x^2 - 1} \div \dfrac{12}{3x+3}$

69. $\dfrac{5}{x^2 - 3x + 2} + \dfrac{1}{x-2}$

70. $\dfrac{4}{2x^2 + 5x - 3} + \dfrac{2}{x+3}$

71. Explain when the LCD is the product of the denominators.

72. Explain when the LCD is the same as one of the denominators of a rational expression to be added or subtracted.

Review Exercises

Factor the following. See Sections 5.1 and 5.4.

73. $x^3 - 1$

74. $8y^3 + 1$

75. $125z^3 + 8$

76. $a^3 - 27$

77. $xy + 2x + 3y + 6$

78. $x^2 - x + xy - y$

Find the slope of each line. See Section 3.4.

79.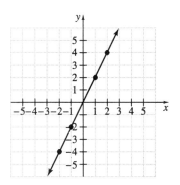

80.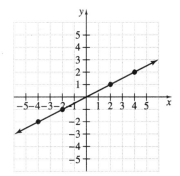

81.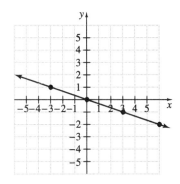

82.

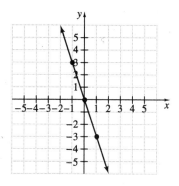

A Look Ahead

EXAMPLE

Perform the indicated operations:

$$\frac{3}{x^2 - 16} + \frac{2}{x^2 - 9x + 20} - \frac{4}{x^2 - x - 20}$$

Solution:

Factor the denominators.

$$= \frac{3}{(x + 4)(x - 4)} + \frac{2}{(x - 4)(x - 5)} - \frac{4}{(x - 5)(x + 4)}$$

Write each expression with an LCD of $(x - 4)(x + 4)(x - 5)$.

$$= \frac{3(x - 5)}{(x - 4)(x + 4)(x - 5)} + \frac{2(x + 4)}{(x - 4)(x + 4)(x - 5)}$$

$$- \frac{4(x - 4)}{(x - 4)(x + 4)(x - 5)}$$

$$= \frac{3(x - 5) + 2(x + 4) - 4(x - 4)}{(x - 4)(x + 4)(x - 5)}$$

$$= \frac{3x - 15 + 2x + 8 - 4x + 16}{(x - 4)(x + 4)(x - 5)}$$

$$= \frac{x + 9}{(x - 4)(x + 4)(x - 5)}$$

Add or subtract as indicated.

83. $\dfrac{5}{x^2 - 4} + \dfrac{2}{x^2 - 4x + 4} - \dfrac{3}{x^2 - x - 6}$

84. $\dfrac{8}{x^2 + 6x + 5} - \dfrac{3x}{x^2 + 4x - 5} + \dfrac{2}{x^2 - 1}$

85. $\dfrac{9}{x^2 + 9x + 14} - \dfrac{3x}{x^2 + 10x + 21} + \dfrac{4}{x^2 + 5x + 6}$

86. $\dfrac{10}{x^2 - 3x - 4} - \dfrac{8}{x^2 + 6x + 5} - \dfrac{9}{x^2 + x - 20}$

87. $\dfrac{5 + x}{x^3 - 27} + \dfrac{x}{x^3 + 3x^2 + 9x}$

88. $\dfrac{x + 5}{x^3 + 1} - \dfrac{3}{2x^2 - 2x + 2}$

6.6 SOLVING EQUATIONS CONTAINING RATIONAL EXPRESSIONS

OBJECTIVES

 Solve equations containing rational expressions.

Understand the difference between solving an equation and performing arithmetic operations on rational expressions.

Solve equations containing rational expressions for a specified variable.

 In Chapter 2, we solved equations containing fractions. In this section, we continue the work we began in Chapter 2 by solving equations containing rational expressions.

EXAMPLES OF EQUATIONS CONTAINING RATIONAL EXPRESSIONS

$$\frac{x}{5} + \frac{x+2}{9} = 8 \quad \text{and} \quad \frac{x+1}{9x-5} = \frac{2}{3x}$$

To solve, use the multiplication property of equality to clear the equation of fractions by multiplying both sides of the equation by the LCD.

TO SOLVE AN EQUATION CONTAINING RATIONAL EXPRESSIONS

Step 1. Multiply both sides of the equation by the LCD of all rational expressions in the equation.

Step 2. Remove any grouping symbols and solve the resulting equation.

Step 3. Check the solution in the original equation.

EXAMPLE 1 Solve: $\dfrac{t-4}{2} - \dfrac{t-3}{9} = \dfrac{5}{18}$.

Solution: The LCD of denominators 2, 9, and 18 is 18. Multiply both sides of the equation by 18.

$$18\left(\frac{t-4}{2} - \frac{t-3}{9}\right) = 18\left(\frac{5}{18}\right)$$

$$18\left(\frac{t-4}{2}\right) - 18\left(\frac{t-3}{9}\right) = 18\left(\frac{5}{18}\right) \qquad \text{Apply the distributive property.}$$

$$9(t-4) - 2(t-3) = 5 \qquad \text{Simplify.}$$

$$9t - 36 - 2t + 6 = 5 \qquad \text{Use the distributive property.}$$

$$7t - 30 = 5 \qquad \text{Combine like terms.}$$

$$7t = 35 \quad \text{or} \quad t = 5 \qquad \text{Solve for } t.$$

To check, substitute 5 for t in the original equation.

$$\frac{t-4}{2} - \frac{t-3}{9} = \frac{5}{18}$$

$$\frac{5-4}{2} - \frac{5-3}{9} = \frac{5}{18} \qquad \text{Replace } t \text{ with 5.}$$

$$\frac{1}{2} - \frac{2}{9} = \frac{5}{18} \qquad \text{Simplify.}$$

Next, subtract the fractions on the left by writing each with a denominator of 18.

$$\frac{1(9)}{2(9)} - \frac{2(2)}{9(2)} = \frac{5}{18}$$

$$\frac{9-4}{18} = \frac{5}{18}$$

$$\frac{5}{18} = \frac{5}{18} \qquad \text{True.}$$

Since the statement is true, 5 is the solution and the solution set is {5}.

Recall from Section 6.1 that a rational expression is defined for all real numbers except those that make the denominator of the expression 0. This means that if an equation contains rational expressions with variables in the denominator, we must be certain that the proposed solution does not make the denominator 0. If replacing the variable with the proposed solution makes the denominator 0, the rational expression is undefined and this proposed solution must be rejected. It is called an **extraneous solution.**

EXAMPLE 2 Solve the equation $3 - \frac{6}{x} = x + 8$.

Solution: In this equation, 0 cannot be a solution because if x is 0, the rational expression $\frac{6}{x}$ is undefined. The LCD is x. Multiply both sides of the equation by x.

$$x\left(3 - \frac{6}{x}\right) = x(x + 8)$$

$$x(3) - x\left(\frac{6}{x}\right) = x \cdot x + x \cdot 8 \qquad \text{Apply the distributive property.}$$

$$3x - 6 = x^2 + 8x \qquad \text{Simplify.}$$

Write the quadratic equation in standard form and solve for x.

$$0 = x^2 + 5x + 6$$

$$0 = (x + 3)(x + 2) \qquad \text{Factor.}$$

$$x + 3 = 0 \quad \text{or} \quad x + 2 = 0 \qquad \text{Set each factor equal to 0 and solve.}$$

$$x = -3 \quad \text{or} \quad x = -2$$

Notice that neither -3 nor -2 makes the denominator in the original equation equal to 0. To check these solutions, replace x in the original equation by -3, and then by −2. This will verify that both −3 and −2 are solutions. The solution set is {−3,−2}.

EXAMPLE 3 Solve the equation $\dfrac{x+1}{x+3} = \dfrac{1}{2x}$.

Solution: Neither −3 nor 0 can be a solution of this equation since each makes a denominator in this equation 0.

Multiply both sides of the equation by the LCD $2x(x+3)$.

$$2x(x+3)\left(\dfrac{x+1}{x+3}\right) = 2x(x+3)\left(\dfrac{1}{2x}\right)$$

$2x(x+1) = (x+3)(1)$	Simplify.
$2x^2 + 2x = x + 3$	Apply the distributive property.
$2x^2 + x - 3 = 0.$	Write the quadratic equation in the standard form.
$(2x+3)(x-1) = 0$	Factor.
$2x + 3 = 0 \quad \text{or} \quad x - 1 = 0$	Set each factor equal to zero.
$x = -\dfrac{3}{2} \quad \text{or} \quad x = 1$	Solve.

Check to see that $-\dfrac{3}{2}$ and 1 are both solutions. The solution set is $\left\{-\dfrac{3}{2}, 1\right\}$.

EXAMPLE 4 Solve the equation $x + \dfrac{14}{x-2} = \dfrac{7x}{x-2} + 1$.

Solution: In this equation, 2 can't be a solution. The LCD is $x - 2$. Multiply both sides of the equation by $x - 2$.

$$(x-2)\left(x + \dfrac{14}{x-2}\right) = (x-2)\left(\dfrac{7x}{x-2} + 1\right)$$

$$(x-2)(x) + (x-2)\left(\dfrac{14}{x-2}\right) = (x-2)\left(\dfrac{7x}{x-2}\right) + (x-2)(1)$$

$x^2 - 2x + 14 = 7x + x - 2$	Simplify.
$x^2 - 2x + 14 = 8x - 2$	Combine like terms.
$x^2 - 10x + 16 = 0$	Write the quadratic equation in standard form.
$(x-8)(x-2) = 0$	Factor.
$x - 8 = 0 \quad \text{or} \quad x - 2 = 0$	Set each factor equal to 0.
$x = 8 \quad \text{or} \quad x = 2$	Solve.

As we stated earlier, 2 can't be a solution of the original equation. Replacing x with 8 in the original equation, we find that 8 is a solution. The only solution is 8 and the solution set is {8}.

If an equation contains rational expressions with variables in the denominator, make sure that you identify the values for which these expressions are not defined. These values cannot be solutions of the equation.

EXAMPLE 5 Solve the equation $\dfrac{3a}{3a-2} - \dfrac{5}{3a^2+7a-6} = 1$.

Solution: Since $3a^2 + 7a - 6 = (3a-2)(a+3)$, both $\dfrac{2}{3}$ and -3 cannot be solutions. The LCD is $(3a-2)(a+3)$, so we multiply both sides of the equation by the LCD.

$$(3a-2)(a+3)\left(\dfrac{3a}{3a-2} - \dfrac{5}{(3a-2)(a+3)}\right) = (3a-2)(a+3)(1)$$

$$(3a-2)(a+3)\left(\dfrac{3a}{3a-2}\right) - (3a-2)(a+3)\left(\dfrac{5}{(3a-2)(a+3)}\right)$$
$$= (3a-2)(a+3)(1)$$

$3a(a+3) - 5 = (3a-2)(a+3)$	Simplify.
$3a^2 + 9a - 5 = 3a^2 + 7a - 6$	Apply the distributive property.
$9a - 5 = 7a - 6$	Subtract $3a^2$ from both sides.
$2a = -1$	Subtract $7a$ from both sides and add 5 to both sides.
$a = -\dfrac{1}{2}$	Divide both sides by 2.

Check that $-\dfrac{1}{2}$ is the solution. The solution set is $\left\{-\dfrac{1}{2}\right\}$.

2 At this point, let's make sure you understand the difference between solving an equation containing rational expressions and performing operations on rational expressions.

EXAMPLE 6 **a.** Solve for x: $\dfrac{x}{4} + 2x = 9$. **b.** Add: $\dfrac{x}{4} + 2x$.

Solution: **a.** This is an equation to solve for x. Begin by multiplying both sides by the LCD, 4.

$$4\left(\dfrac{x}{4} + 2x\right) = 4(9)$$

$4\left(\dfrac{x}{4}\right) + 4(2x) = 4(9)$	Apply the distributive property.
$x + 8x = 36$	Simplify.
$9x = 36$	Combine like terms.
$x = 4$	Solve.

Check to see that 4 is the solution and the solution set is {4}.

b. This example is **not an equation** to solve; it is an addition to perform. To add these rational expressions, find the LCD and write each rational expression as an equivalent expression whose denominator is the LCD. The LCD is 4.

$$\frac{x}{4} + 2x = \frac{x}{4} + \frac{2x(4)}{4}$$

$$= \frac{x + 8x}{4} \qquad \text{Add.}$$

$$= \frac{9x}{4} \qquad \text{Combine like terms in the numerator.}$$

3 The last example in this section is an equation containing several variables. We are directed to solve for one of them. The steps used in the preceeding examples can be applied to solve equations for a specified variable as well.

EXAMPLE 7 Solve for x: $\frac{2x}{a} - 5 = \frac{3x}{b} + a$.

Solution: The LCD is ab. Multiply both sides of the equation by ab.

$$ab\left(\frac{2x}{a} - 5\right) = ab\left(\frac{3x}{b} + a\right)$$

$$ab\left(\frac{2x}{a}\right) - ab(5) = ab\left(\frac{3x}{b}\right) + ab(a) \qquad \text{Apply the distributive property.}$$

$$2xb - 5ab = 3xa + a^2b \qquad \text{Simplify.}$$

Next, write the equation so that all terms containing the variable x appear on one side of the equation. To do this, subtract $3xa$ from both sides and add $5ab$ to both sides.

$$2xb - 3xa = a^2b + 5ab$$

$$x(2b - 3a) = a^2b + 5ab \qquad \text{Factor out } x \text{ from each term on the left side.}$$

$$\frac{x(2b - 3a)}{2b - 3a} = \frac{a^2b + 5ab}{2b - 3a} \qquad \text{Divide both sides by } 2b - 3a.$$

$$x = \frac{a^2b + 5ab}{2b - 3a} \qquad \text{Simplify.}$$

MENTAL MATH

Solve each equation for the variable.

1. $\frac{x}{5} = 2$ 2. $\frac{x}{8} = 4$ 3. $\frac{z}{6} = 6$ 4. $\frac{y}{7} = 8$

EXERCISE SET 6.6

Solve each equation. See Examples 1 and 2.

1. $\frac{x}{5} + 3 = 9$ 2. $\frac{x}{5} - 2 = 9$ 3. $\frac{x}{2} + \frac{5x}{4} = \frac{x}{12}$ 4. $\frac{x}{6} + \frac{4x}{3} = \frac{x}{18}$

5. $2 + \dfrac{10}{x} = x + 5$ **6.** $6 + \dfrac{5}{y} = y - \dfrac{2}{y}$

7. $\dfrac{a}{5} = \dfrac{a-3}{2}$ **8.** $\dfrac{2b}{5} = \dfrac{b+2}{6}$

9. $\dfrac{x-3}{5} + \dfrac{x-2}{2} = \dfrac{1}{2}$ **10.** $\dfrac{a+5}{4} + \dfrac{a+5}{2} = \dfrac{a}{8}$

Recall that two angles are supplementary if the sum of their measures is 180°. Find the measures of the following supplementary angles.

  11.   **12.**

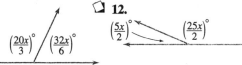

Recall that two angles are complementary if the sum of their measures is 90°. Find the measures of the following complementary angles.

13.   **14.**

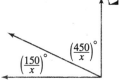 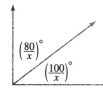

Solve each equation. See Examples 3 through 5.

15. $\dfrac{9}{2a-5} = -2$ **16.** $\dfrac{6}{4-3x} = 3$

17. $\dfrac{y}{y+4} + \dfrac{4}{y+4} = 3$ **18.** $\dfrac{5y}{y+1} - \dfrac{3}{y+1} = 4$

19. $\dfrac{2x}{x+2} - 2 = \dfrac{x-8}{x-2}$ **20.** $\dfrac{4y}{y-3} - 3 = \dfrac{3y-1}{y+3}$

21. $\dfrac{4y}{y-4} + 5 = \dfrac{5y}{y-4}$ **22.** $\dfrac{2a}{a+2} - 5 = \dfrac{7a}{a+2}$

23. $\dfrac{7}{x-2} + 1 = \dfrac{x}{x+2}$ **24.** $1 + \dfrac{3}{x+1} = \dfrac{x}{x-1}$

25. $\dfrac{x+1}{x+3} = \dfrac{2x^2 - 15x}{x^2 + x - 6} - \dfrac{x-3}{x-2}$

26. $\dfrac{3}{x+3} = \dfrac{12x+19}{x^2+7x+12} - \dfrac{5}{x+4}$

27. $\dfrac{y}{2y+2} + \dfrac{2y-16}{4y+4} = \dfrac{2y-3}{y+1}$

28. $\dfrac{1}{x+2} = \dfrac{4}{x^2-4} - \dfrac{1}{x-2}$

Determine whether each of the following is an equation or an expression. If it is an equation, then solve it for its variable. If it is an expression, perform the indicated operation. See Example 6.

29. $\dfrac{1}{x} + \dfrac{2}{3}$ **30.** $\dfrac{3}{a} + \dfrac{5}{6}$

31. $\dfrac{1}{x} + \dfrac{2}{3} = \dfrac{3}{x}$ **32.** $\dfrac{3}{a} + \dfrac{5}{6} = 1$

33. $\dfrac{2}{x+1} - \dfrac{1}{x}$ **34.** $\dfrac{4}{x-3} - \dfrac{1}{x}$

35. $\dfrac{2}{x+1} - \dfrac{1}{x} = 1$ **36.** $\dfrac{4}{x-3} - \dfrac{1}{x} = \dfrac{6}{x(x-3)}$

37. Explain the difference between solving an equation such as $\dfrac{x}{2} + \dfrac{3}{4} = \dfrac{x}{4}$ for x and performing an operation such as adding: $\dfrac{x}{2} + \dfrac{3}{4}$.

38. When solving an equation such as $\dfrac{y}{4} = \dfrac{y}{2} - \dfrac{1}{4}$, we may multiply all terms by 4. When subtracting two rational expressions such as $\dfrac{y}{2} - \dfrac{1}{4}$, we may not. Explain why.

Solve each equation.

39. $\dfrac{2x}{7} - 5x = 9$ **40.** $\dfrac{4x}{8} - 5x = 10$

41. $\dfrac{2}{y} + \dfrac{1}{2} = \dfrac{5}{2y}$ **42.** $\dfrac{6}{3y} + \dfrac{3}{y} = 1$

43. $\dfrac{4x+10}{7} = \dfrac{8}{2}$ **44.** $\dfrac{1}{2} = \dfrac{x+1}{8}$

45. $2 + \dfrac{3}{a-3} = \dfrac{a}{a-3}$

46. $\dfrac{2y}{y-2} - \dfrac{4}{y-2} = 4$ **47.** $\dfrac{5}{x} + \dfrac{2}{3} = \dfrac{7}{2x}$

48. $\dfrac{5}{3} - \dfrac{3}{2x} = \dfrac{5}{4}$ **49.** $\dfrac{2a}{a+4} = \dfrac{3}{a-1}$

50. $\dfrac{5}{3x-8} = \dfrac{x}{x-2}$ **51.** $\dfrac{x+1}{3} - \dfrac{x-1}{6} = \dfrac{1}{6}$

52. $\dfrac{3x}{5} - \dfrac{x-6}{3} = \dfrac{1}{5}$

53. $\dfrac{4r-1}{r^2+5r-14} + \dfrac{2}{r+7} = \dfrac{1}{r-2}$

54. $\dfrac{2t+3}{t-1} - \dfrac{2}{t+3} = \dfrac{5-6t}{t^2+2t-3}$

55. $\dfrac{t}{t-4} = \dfrac{t+4}{6}$ **56.** $\dfrac{15}{x+4} = \dfrac{x-4}{x}$

57. $\dfrac{x}{2x+6} + \dfrac{x+1}{3x+9} = \dfrac{2}{4x+12}$

58. $\dfrac{a}{5a-5} - \dfrac{a-2}{2a-2} = \dfrac{5}{4a-4}$

Solve each equation for the indicated variable. See Example 7.

59. $\dfrac{D}{R} = T$; for R

60. $\dfrac{A}{W} = L$; for W

61. $\dfrac{3}{x} = \dfrac{5y}{x+2}$; for y

62. $\dfrac{7x-1}{2x} = \dfrac{5}{y}$; for y

63. $\dfrac{3a+2}{3b-2} = -\dfrac{4}{2a}$; for b

64. $\dfrac{6x+y}{7x} = \dfrac{3x}{h}$; for h

65. $\dfrac{A}{BH} = \dfrac{1}{2}$; for B

66. $\dfrac{V}{\pi r^2 h} = 1$; for h

67. $\dfrac{C}{\pi r} = 2$; for r

68. $\dfrac{3V}{A} = H$; for V

69. $\dfrac{1}{a} = \dfrac{1}{b} + \dfrac{1}{c}$; for a

70. $\dfrac{1}{2} - \dfrac{1}{x} = \dfrac{1}{y}$; for x

71. $\dfrac{m^2}{6} - \dfrac{n}{3} = \dfrac{p}{2}$; for n

72. $\dfrac{x^2}{r} + \dfrac{y^2}{t} = 1$; for r

Solve each equation.

73. $\dfrac{5}{a^2+4a+3} + \dfrac{2}{a^2+a-6} - \dfrac{3}{a^2-a-2} = 0$

74. $-\dfrac{2}{a^2+2a-8} + \dfrac{1}{a^2+9a+20} = \dfrac{-4}{a^2+3a-10}$

Review Exercises

Graph the line passing through the given point with the given slope. See Section 3.4.

75. $(4, 3)$, $m = -3$

76. $(0, -1)$, $m = \dfrac{1}{2}$

77. $(-5, 2)$, $m = \dfrac{3}{2}$

78. $(-1, -4)$, $m = -\dfrac{1}{5}$

Identify the x- and y-intercepts. Also, write the intercepts as ordered pairs of numbers. See Section 3.3.

79.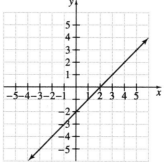

80.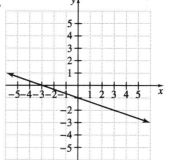

81.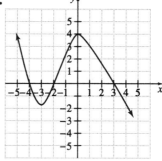

82.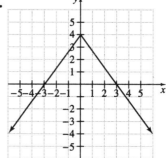

6.7 RATIO AND PROPORTION

OBJECTIVES

1. Use fractional notation to express ratios.
2. Identify and solve proportions.
3. Use proportions to solve problems.
4. Determine unit pricing.

TAPE
BA 6.7

  A **ratio** is the quotient of two numbers or two quantities.

> **RATIO**
> If a and b are two numbers and $b \neq 0$, the **ratio of a to b** is the quotient of a and b. The ratio of a to b can also be written as
>
> $a:b$ or as $\dfrac{a}{b}$

EXAMPLE 1   Write a ratio for each phrase. Use fractional notation.

a. The ratio of 2 parts salt to 5 parts water
b. The ratio of 18 inches to 2 feet

Solution: a. The ratio of 2 parts salt to 5 parts water is $\dfrac{2}{5}$.

b. When comparing measurements, use the same unit of measurement in the numerator as in the denominator. Here, we write 2 feet as $2 \cdot 12$ inches, or 24 inches. The ratio of 18 inches to 2 feet is then $\dfrac{18}{24} = \dfrac{3}{4}$ in lowest terms.

  If two ratios are equal, we say the ratios are **in proportion** to each other. A **proportion** is a mathematical statement that two ratios are equal.

For example, the equation $\dfrac{1}{2} = \dfrac{4}{8}$ is a proportion, as is $\dfrac{x}{5} = \dfrac{8}{10}$, because both sides of the equations are ratios. When we want to emphasize the equation as a proportion, we

read the proportion $\dfrac{1}{2} = \dfrac{4}{8}$ as "one is to two as four is to eight"

In the proportion $\dfrac{1}{2} = \dfrac{4}{8}$, the 1 and the 8 are called the **extremes**; the 2 and the 4 are called the **means**.

> **For any true proportion, the product of the means equals the product of the extremes.**

For example, since the proportion

$$\frac{1}{2} = \frac{4}{8}$$

is true, then it must also be true that

$$1 \cdot 8 = 2 \cdot 4$$
extremes   means
or $8 = 8$

To see why this is so, multiply both sides of the proportion $\frac{a}{b} = \frac{c}{d}$ by the LCD bd.

$$\frac{a}{b} = \frac{c}{d}$$

$$bd\left(\frac{a}{b}\right) = bd\left(\frac{c}{d}\right) \quad \text{Multiply both sides by } bd.$$

$$\underbrace{ad}_{\text{product of extremes}} = \underbrace{bc}_{\text{product of means}} \quad \text{Simplify.}$$

The products ad and bc are also called **cross products.** Equating these cross products is called **cross multiplication.**

$$\frac{a}{b} \diagdown\mkern-12mu\diagup \frac{c}{d} \rightarrow ad = bc$$

> **CROSS MULTIPLICATION**
>
> For any ratios $\frac{a}{b}$ and $\frac{c}{d}$, if
>
> $$\frac{a}{b} = \frac{c}{d}, \text{ then } ad = bc.$$

Cross multiplication can be used to solve a proportion for an unknown.

EXAMPLE 2 Solve for x: $\frac{45}{x} = \frac{5}{7}$.

Solution: To solve, cross multiply.

$$\frac{45}{x} \diagdown\mkern-12mu\diagup \frac{5}{7}$$

RATIO AND PROPORTION   SECTION 6.7   **385**

$$45 \cdot 7 = x \cdot 5 \qquad \text{Cross multiply.}$$
$$\frac{315}{5} = \frac{5x}{5} \qquad \text{Divide both sides by 5.}$$
$$63 = x \qquad \text{Simplify.}$$

To check, substitute 63 for x in the original proportion. The solution set is {63}.

EXAMPLE 3   Solve for x: $\dfrac{x}{x-2} = \dfrac{x-3}{x+1}$.

Solution: To solve, cross multiply.

$$\frac{x}{x-2} = \frac{x-3}{x+1}$$

$$\begin{aligned}
x(x+1) &= (x-2)(x-3) & &\text{Cross multiply.} \\
x^2 + x &= x^2 - 5x + 6 & &\text{Multiply.} \\
x &= -5x + 6 & &\text{Subtract } x^2 \text{ from both sides.} \\
6x &= 6 & &\text{Add } 5x \text{ to both sides.} \\
x &= 1 & &\text{Divide both sides by 6.}
\end{aligned}$$

To check, substitute 1 for x in the original proportion. The solution set is {1}.

Proportions can be used to model and solve many real-life problems. When using proportions in this way, it is important to judge whether the solution is reasonable. Doing so helps us to decide if the proportion has been formed correctly. We use the same problem-solving steps that were introduced in Section 2.5.

EXAMPLE 4   Three boxes of 3.5-inch high-density diskettes cost $37.47. How much should 5 boxes cost?

Solution: **1. UNDERSTAND** the problem. To do so, read and reread the problem. Guess the solution and check the guess. We know that the cost of 5 boxes is more than the cost of 3 boxes, or $37.47, and less than the cost of 6 boxes, which is double the cost of 3 boxes, or 2($37.47) = $74.94. Let's guess that 5 boxes cost $60.00. To check this guess, see if 3 boxes is to 5 boxes as the *price* of 3 boxes is to the *price* of 5 boxes. In other words, see if

$$\frac{3 \text{ boxes}}{5 \text{ boxes}} = \frac{\text{price of 3 boxes}}{\text{price of 5 boxes}}$$

or

$$\frac{3}{5} = \frac{37.47}{60.00}$$

To see if the proportion is true, we cross multiply.

$$3(60.00) = 5(37.47)$$

or

$$180.00 = 187.35 \quad \textbf{Not a true statement.}$$

Thus, $60 is not correct but we now have a better understanding of the problem.

2. ASSIGN a variable. Let x = price of 5 boxes of diskettes.
3. ILLUSTRATE the problem. No illustration is needed.
4. TRANSLATE the problem.

$$\text{In words: } \frac{3 \text{ boxes}}{5 \text{ boxes}} = \frac{\text{price of 3 boxes}}{\text{price of 5 boxes}}$$

$$\text{Translate: } \frac{3}{5} = \frac{37.47}{x}$$

5. COMPLETE the work. Here, we solve the proportion.

$$\frac{3}{5} = \frac{37.47}{x}$$

$3x = 5(37.47)$ Cross multiply.

$3x = 187.35$

$x = 62.45$ Divide both sides by 3.

6. INTERPRET the results. First *check* the solution. To do so, see that 3 boxes is to 5 boxes as $37.47 is to $62.45. Also, notice that our solution is a reasonable one as discussed in step 1.

Next *state* the conclusions. Five boxes of high-density diskettes cost $62.45. ▬

The proportion $\dfrac{5 \text{ boxes}}{3 \text{ boxes}} = \dfrac{\text{price of 5 boxes}}{\text{price of 3 boxes}}$ could also have been used to solve the problem above.

EXAMPLE 5   To estimate the number of people in Jackson, population 50,000, who have no health insurance, 250 people were polled, and 39 of those polled had no insurance. How many people in the city might we expect to be uninsured?

Solution:
1. UNDERSTAND. Read and reread the problem. Guess the solution and check your guess.
2. ASSIGN. Let x = number of people in the city with no health insurance.
3. ILLUSTRATE. No diagram or chart is needed.
4. TRANSLATE.

In words: $\dfrac{\text{total number polled}}{\text{number polled with no insurance}} = \dfrac{\text{total city population}}{\text{number in city with no insurance}}$

Translate: $\dfrac{250}{39} = \dfrac{50{,}000}{x}$

5. COMPLETE. Solve the proportion for x.

$$\dfrac{250}{39} = \dfrac{50{,}000}{x}$$

$250x = 39(50{,}000)$ Cross multiply.

$250x = 1{,}950{,}000$

$x = 7800$ Divide both sides by 250.

6. INTERPRET. *Check* the solution and *state* the conclusion. We expect the city of Jackson to have approximately 7800 citizens with no health insurance.

4 When shopping for an item offered in many different sizes, it is important to be able to determine the best buy, or the best price per unit. To find the unit price of an item, divide the total price of the item by the total number of units.

$$\text{unit price} = \dfrac{\text{total price}}{\text{number of units}}$$

For example, if a 16-ounce can of green beans is priced at $0.88, its unit price is

$$\text{unit price} = \dfrac{\$0.88}{16} = \$0.055$$

EXAMPLE 6 A supermarket offers a 14-ounce box of cereal for $3.79 and an 18-ounce box of the same brand of cereal for $4.99. Which is the better buy?

Solution: To find the better buy, we compare unit prices. The following unit prices were rounded to three decimal places.

Size	Price	Unit Price
14-ounce	$3.79	$\dfrac{\$3.79}{14} \approx \$0.271$
18-ounce	$4.99	$\dfrac{\$4.99}{18} \approx \$0.277$

The 14-ounce box of cereal has the lower unit price so it is the better buy.

EXERCISE SET 6.7

Write each ratio in fractional notation in lowest terms. See Example 1.

1. 2 megabytes to 15 megabytes
2. 18 disks to 41 disks
3. 10 inches to 12 inches
4. 15 miles to 40 miles
5. 5 quarts to 3 gallons
6. 8 inches to 3 feet
7. 4 nickels to 2 dollars
8. 12 quarters to 2 dollars
9. 175 centimeters to 5 meters
10. 90 centimeters to 4 meters
11. 190 minutes to 3 hours
12. 60 hours to 2 days
13. Suppose someone tells you that the ratio of 11 inches to 2 feet is $\frac{11}{2}$. How do you correct that person and explain the error?
14. Write a ratio that can be written in fractional notation as $\frac{3}{2}$.

Solve each proportion. See Examples 2 and 3.

15. $\frac{2}{3} = \frac{x}{6}$
16. $\frac{x}{2} = \frac{16}{6}$
17. $\frac{x}{10} = \frac{5}{9}$
18. $\frac{9}{4x} = \frac{6}{2}$
19. $\frac{4x}{6} = \frac{7}{2}$
20. $\frac{a}{5} = \frac{3}{2}$
21. $\frac{a}{25} = \frac{12}{10}$
22. $\frac{n}{10} = 9$
23. $\frac{x-3}{x} = \frac{4}{7}$
24. $\frac{y}{y-16} = \frac{5}{3}$
25. $\frac{5x+1}{x} = \frac{6}{3}$
26. $\frac{3x-2}{5} = \frac{4x}{1}$
27. $\frac{x+1}{2x+3} = \frac{2}{3}$
28. $\frac{x+1}{x+2} = \frac{5}{3}$
29. $\frac{9}{5} = \frac{12}{3x+2}$
30. $\frac{6}{11} = \frac{27}{3x-2}$
31. $\frac{3}{x+1} = \frac{5}{2x}$
32. $\frac{7}{x-3} = \frac{8}{2x}$
33. $\frac{x+1}{x} = \frac{x+2}{x-2}$
34. $\frac{x-1}{x+3} = \frac{x+3}{x}$
35. If x is 10, is $\frac{2}{x}$ in proportion to $\frac{x}{50}$? Explain why or why not.
36. For what value of x is $\frac{x}{x-1}$ in proportion to $\frac{x+1}{x}$? Explain your result.

Given the following prices charged for various sizes of an item, find the best buy. See Example 6.

37. Laundry detergent
    110 ounces for $5.79
    240 ounces for $13.99
38. Jelly
    10 ounces for $1.14
    15 ounces for $1.69
39. Tuna (in cans)
    6 ounces for $0.69
    8 ounces for $0.90
    16 ounces for $1.89
40. Picante sauce
    10 ounces for $0.99
    16 ounces for $1.69
    30 ounces for $3.29

Solve. See Examples 4 and 5.

41. The ratio of the weight of an object on Earth to the weight of the same object on Pluto is 100 to 3. If an elephant weighs 4100 pounds on Earth, find the elephant's weight on Pluto.
42. If a 170-pound person weighs approximately 65 pounds on Mars, how much does a 9000-pound satellite weigh?
43. In a bag of M&M's, 28 out of 80 M&M's were found to be the color brown. How many M&M's would you expect to be brown from a bag containing 208 M&M's? (Round to the nearest whole.)
44. On an architect's blueprint, 1 inch corresponds to 4 feet. Find the length of a wall represented by a line that is $3\frac{7}{8}$ inches long on the blueprint.
45. There are 110 calories per 28.4 grams of Crispy Rice cereal. Find how many calories are in 42.6 grams of this cereal.

46. A box of flea and tick powder instructs the user to mix 4 tablespoons of powder with 1 gallon of water. Find how much powder should be mixed with 5 gallons of water.

47. Miss Babola's new Mazda gets 35 miles per gallon. Find how far she can drive if the tank contains 13.5 gallons of gas.

48. In a week of city driving, Miss Babola noticed that she was able to drive 418.5 miles on a tank of gas (13.5 gallons). Find how many miles per gallon Miss Babola got in city traffic.

49. Ken Hall, a tailback, holds the high school sports record for total yards rushed in a season. In 1953, he rushed for 4045 total yards in 12 games. Find his average rushing yards per game.

50. A recent headline read, "Women earn bigger check in 1 of every 6 couples." If there are 23,000 couples in a nearby metropolitan area, how many women would you expect to earn bigger paychecks?

51. A human factors expert recommends that there be at least 9 square feet of floor space in a college classroom for every student in the class. Find the minimum floor space that 40 students need.

52. There are 1280 calories in a 14-ounce portion of Eagle Brand Milk. Find how many calories are in 2 ounces of Eagle Brand Milk.

53. Due to space problems at a local university, a 20-foot by 12-foot conference room is converted into a classroom. Find the maximum number of students the room can accommodate. (See Exercise 51.)

54. The manufacturers of cans of salted mixed nuts state that the ratio of peanuts to other nuts is 3 to 2. If 324 peanuts are in a can, find how many other nuts should also be in the can.

55. If Sam Abney can travel 343 miles in 7 hours, find how far he can travel if he maintains the same speed for 5 hours.

56. The instructions on a bottle of plant food read as follows: "Use four tablespoons plant food per 3 gallons of water." Find how many tablespoons of plant food should be mixed into 6 gallons of water.

57. To mix weed killer with water correctly, it is necessary to mix 8 teaspoons of weed killer with 2 gallons of water. Find how many gallons of water are needed to mix with the entire box if it contains 36 teaspoons of weed killer.

58. There are 290 milligrams of sodium per 1-ounce serving of Rice Crispies. Find how many milligrams of sodium are in three 1-ounce servings of the cereal.

59. Mr. Lin's contract states that he will be paid $153 per 8-hour day to teach mathematics. Find how much he earns per hour rounded to the nearest cent.

60. Mr. Gonzales, a pool contractor, bases the cost of labor on the volume of the pool to be constructed. The cost of labor on a wading pool of 803 cubic feet is $750.00. If the customer decided to cut the

volume by a third, find the cost of labor for the smaller pool.

61. An accountant finds that Country Collections earned $35,063 during its first 6 months. Find how much the business earned **each week** on average rounded to the nearest cent.

The following graph shows the capacity of the world to generate electricity from the wind.

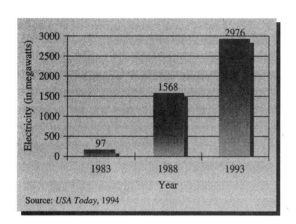

62. Find the increase in megawatt capacity during the 5-year period from 1983 to 1988.

63. Find the increase in megawatt capacity during the 5-year period from 1988 to 1993.

64. If the trend shown on this graph continues, approximate the number of megawatts available from the wind in 1998.

In general, 1000 megawatts will serve the average electricity needs of 560,000 people. Use this fact and the preceding graph to answer the following.

65. In 1993, the number of megawatts that can be generated from wind will serve the electricity needs of how many people?

66. How many megawatts of electricity are needed to serve the city or town in which you live?

67. If $x = ad$, $y = bc$, and $\frac{a}{b} = \frac{c}{d}$, what true statement can you make about the value of x as compared to the value of y?

Review Exercises

Find the slope of the line through each pair of points. Use the slope to determine whether the line is vertical, horizontal, or moves upward or downward from left to right. See Section 3.4.

68. $(-2, 5), (4, -3)$ **69.** $(0, 4)$ $(2, 10)$
70. $(-3, -6)$ $(1, 5)$ **71.** $(-2, 7)$ $(3, -2)$
72. $(3, 7)$ $(3, -2)$ **73.** $(0, -4)$ $(2, -4)$

6.8 RATIONAL EQUATIONS AND PROBLEM SOLVING

OBJECTIVES

1. Translate sentences to equations containing rational expressions.
2. Solve problems involving work.
3. Solve problems involving distance.
4. Solve problems involving similar triangles.

TAPE
BA 6.8

 In this section, we solve problems that can be modeled by equations containing rational expressions. To solve these problems, we use the same problem-solving steps that were first introduced in Section 2.5. In our first example, our goal is to find an unknown number.

EXAMPLE 1 The quotient of a number and 6 minus $\frac{5}{3}$ is the quotient of the number and 2. Find the number.

Solution: **1. UNDERSTAND.** Read and reread the problem, guess a solution, and check your guess. For example, if the unknown number is 2, then we see if the quotient of 2 and 6, or $\frac{2}{6} - \frac{5}{3}$ is equal to the quotient of 2 and 2, or $\frac{2}{2}$. $\frac{2}{6} - \frac{5}{3} = \frac{1}{3} - \frac{5}{3} = -\frac{4}{3}$, and not $\frac{2}{2}$. Don't forget that the purpose of a guess is to better understand the problem.

2. ASSIGN. Let x = the unknown number.

3. ILLUSTRATE. No illustration is needed.

4. TRANSLATE:

In words:	The quotient of x and 6	minus	$\frac{5}{3}$	is	the quotient of x and 2.
Translate:	$\frac{x}{6}$	$-$	$\frac{5}{3}$	$=$	$\frac{x}{2}$

5. COMPLETE. Here, we solve the equation $\frac{x}{6} - \frac{5}{3} = \frac{x}{2}$. Begin solving this equation by multiplying both sides of the equation by the LCD 6.

$$6\left(\frac{x}{6} - \frac{5}{3}\right) = 6\left(\frac{x}{2}\right)$$

$$6\left(\frac{x}{6}\right) - 6\left(\frac{5}{3}\right) = 6\left(\frac{x}{2}\right) \quad \text{Apply the distributive property.}$$

$$x - 10 = 3x \quad \text{Simplify.}$$

$$-10 = 2x \quad \text{Subtract } x \text{ from both sides.}$$

$$-\frac{10}{2} = \frac{2x}{2} \quad \text{Divide both sides by 2.}$$

$$-5 = x \quad \text{Simplify.}$$

6. INTERPRET. *Check:* To check, verify that "the quotient of -5 and 6 minus $\frac{5}{3}$ is the quotient of -5 and 2, or $-\frac{5}{6} - \frac{5}{3} = -\frac{5}{2}$. *State:* The unknown number is -5.

2 The next example is often called a work problem. Work problems usually involve people or machines doing a certain task.

EXAMPLE 2 Sam Waterton and Frank Schaffer work in a plant that manufactures automobiles. Sam can complete a quality control tour of the plant in 3 hours while his assistant, Frank, needs 7 hours to complete the same job. The regional manager is coming to inspect the plant facilities, so both Sam and Frank are directed to complete a quality control tour together. How long will this take?

Solution: 1. **UNDERSTAND.** Read and reread the problem. The key idea here is the relationship between the **time** (hours) it takes to complete the job and the **part of the job** completed in 1 unit of time (hour). For example, if the **time** it takes Sam to complete the job is 3 hours, the **part of the job** he can complete in 1 hour is $\frac{1}{3}$. Similarly, Frank can complete $\frac{1}{7}$ of the job in 1 hour.

2. **ASSIGN.** Let x represent the **time** in hours it takes Sam and Frank to complete the job together. Then $\frac{1}{x}$ represents the **part of the job** they complete in 1 hour.

3. **ILLUSTRATE.** Here we summarize the information discussed above in a chart.

	HOURS TO COMPLETE TOTAL JOB	PART OF JOB COMPLETED IN 1 HOUR
Sam	3	$\frac{1}{3}$
Frank	7	$\frac{1}{7}$
Together	x	$\frac{1}{x}$

4. **TRANSLATE.**

In words:	part of job Sam completed in 1 hour	added to	part of job Frank completed in 1 hour	is equal to	part of job they completed together in 1 hour
Translate:	$\frac{1}{3}$	$+$	$\frac{1}{7}$	$=$	$\frac{1}{x}$

5. **COMPLETE.** Here, we solve the equation $\frac{1}{3} + \frac{1}{7} = \frac{1}{x}$. Begin solving the equation by multiplying both sides of the equation by the LCD $21x$.

$$21x\left(\frac{1}{3}\right) + 21x\left(\frac{1}{7}\right) = 21x\left(\frac{1}{x}\right)$$
$$7x + 3x = 21 \qquad \text{Simplify.}$$
$$10x = 21$$
$$x = \frac{21}{10} \quad \text{or} \quad 2\frac{1}{10} \text{ hours}$$

6. **INTERPRET.** *Check:* Our proposed solution is $2\frac{1}{10}$ hours. This proposed solution is reasonable since $2\frac{1}{10}$ hours is more than half of Sam's time and less than

half of Frank's time. Check this solution in the originally stated problem. *State:* Sam and Frank can complete the quality control tour in $2\frac{1}{10}$ hours.

 Next we look at a problem solved by the distance formula.

EXAMPLE 3 A car travels 180 miles in the same time that a semi truck travels 120 miles. If the car's speed is 20 miles per hour faster than the truck's, find the car's speed and the truck's speed.

Solution: **1. UNDERSTAND.** Read and reread the problem. Next, guess a solution. Suppose that the truck's speed is 45 miles per hour. Then the car's speed is 20 miles per hour more, or 65 miles per hour.

We are given that the car travels 180 miles in the same time that the truck travels 120 miles. To find the time it takes that car to travel 180 miles, remember that since $d = rt$, we know that $\frac{d}{r} = t$.

CAR'S TIME
$$t = \frac{d}{r} = \frac{180}{65} = 2\frac{50}{65} = 2\frac{10}{13} \text{ hours}$$

TRUCK'S TIME
$$t = \frac{d}{r} = \frac{120}{45} = 2\frac{30}{45} = 2\frac{2}{3} \text{ hours}$$

Since the times are not the same, we have not guessed the speeds correctly.

2. ASSIGN. Let

x = the speed of the truck.

Since the car's speed is 20 miles per hour faster than the truck's, then

$x + 20$ = the speed of the car

3. ILLUSTRATE. Use the formula $d = r \cdot t$ or **d**istance = **r**ate (speed) · **t**ime. Prepare a chart to organize the information in the problem. Recall that if $d = r \cdot t$, then $t = \frac{d}{r}$.

	distance	=	rate	·	time
Truck	120		x		$\frac{120}{x}\left(\frac{\text{distance}}{\text{rate}}\right)$
Car	180		$x + 20$		$\frac{180}{x + 20}\left(\frac{\text{distance}}{\text{rate}}\right)$

4. TRANSLATE. Since the car and the truck traveled the same amount of time, we have that

In words: car's time   =   truck's time

Translate: $$\frac{180}{x+20} = \frac{120}{x}$$

5. COMPLETE. Begin solving the equation by cross multiplying.

$$\frac{180}{x+20} = \frac{120}{x}$$

$180x = 120(x + 20)$	Cross multiply.
$180x = 120x + 2400$	Use the distributive property.
$60x = 2400$	Subtract $120x$ from both sides.
$x = 40$	Divide both sides by 60.

6. INTERPRET. The speed of the truck is 40 miles per hour. The speed of the car must then be $x + 20$ or 60 miles per hour. To *check*, find the time it takes the car to travel 180 miles and the time it takes the truck to travel 120 miles.

CAR'S TIME $\qquad$ TRUCK'S TIME

$$t = \frac{d}{r} = \frac{180}{60} = 3 \text{ hours} \qquad t = \frac{d}{r} = \frac{120}{40} = 3 \text{ hours}$$

Since both travel the same amount of time, the proposed solution is correct. *State:* The car's speed is 60 miles per hour and the truck's speed is 40 miles per hour.

Similar triangles have the same shape but not necessarily the same size. In similar triangles, the measures of corresponding angles are equal, and corresponding sides are in proportion.

If triangle ABC and triangle XYZ below are similar, then we know that the measure of angle A = the measure of angle X, the measure of angle B = the measure of angle Y, and the measure of angle C = the measure of angle Z, and we also know that corresponding sides are in proportion: $\frac{a}{x} = \frac{b}{y} = \frac{c}{z}$.

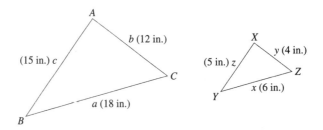

In this section, we will position similar triangles so that they have the same orientation.

To show that corresponding sides are in proportion for the triangles above, write the ratios of the corresponding sides.

$$\frac{a}{x} = \frac{18}{6} = 3 \qquad \frac{b}{y} = \frac{12}{4} = 3 \qquad \frac{c}{z} = \frac{15}{5} = 3$$

EXAMPLE 4 If the following two triangles are similar, find the missing length x.

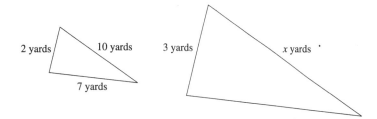

Solution: Since the triangles are similar, their corresponding sides are in proportion and we have

$$\frac{2}{3} = \frac{10}{x}$$

To solve, cross multiply.

$2x = 30$ Cross multiply.

$x = 15$ Divide both sides by 2.

The missing length is 15 yards.

EXERCISE SET 6.8

Solve the following. See Example 1.

1. Three times the reciprocal of a number equals 9 times the reciprocal of 6. Find the number.

2. Twelve divided by the sum of x and 2 equals the quotient of 4 and the difference of x and 2. Find x.

3. If twice a number added to 3 is divided by the number plus 1, the result is three halves. Find the number.

4. A number added to the product of 6 and the reciprocal of the number equals -5. Find the number.

See Example 2.

5. Smith Engineering found that an experienced surveyor surveys a roadbed in 4 hours. An apprentice surveyor needs 5 hours to survey the same stretch of road. If the two work together, find how long it takes them to complete the job.

6. An experienced bricklayer constructs a small wall in 3 hours. The apprentice completes the job in 6 hours. Find how long it takes if they work together.

7. In 2 minutes, a conveyor belt moves 300 pounds of recyclable aluminum from the delivery truck to a storage area. A smaller belt moves the same quantity of cans the same distance in 6 minutes. If both belts are used, find how long it takes to move the cans to the storage area.

8. Find how long it takes the conveyor belts described in Exercise 7 to move 1200 pounds of cans. (*Hint:* Think of 1200 pounds as four 300-pound jobs.)

See Example 3.

9. A jogger begins her workout by jogging to the park, a distance of 12 miles. She then jogs home at the same speed but along a different route. This return trip is 18 miles and her time is one hour longer. Find her jogging speed. Complete the accompanying chart and use it to find her jogging speed.

	distance	=	rate	·	time
Trip to park	12				x
Return trip	18				$x + 1$

10. A boat can travel 9 miles upstream in the same amount of time it takes to travel 11 miles downstream. If the current of the river is 3 miles per hour, complete the chart below and use it to find the speed of the boat in still water.

	distance	=	rate	·	time
Upstream	9		$r - 3$		
Downstream	11		$r + 3$		

11. A cyclist rode the first 20-mile portion of his workout at a constant speed. For the 16-mile cooldown portion of his workout, he reduced his speed by 2 miles per hour. Each portion of the workout took the same time. Find the cyclist's speed during the first portion and find his speed during the cooldown portion.

12. A semi truck travels 300 miles through the flatland in the same amount of time that it travels 180 miles through mountains. The rate of the truck is 20 miles per hour slower in the mountains than in the flatland. Find both the flatland rate and mountain rate.

Given that the following pairs of triangles are similar, find the missing lengths. See Example 4.

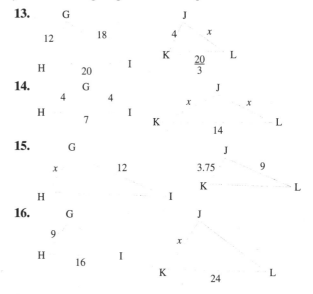

13.
14.
15.
16.

Solve the following.

17. One-fourth equals the quotient of a number and 8. Find the number.

18. Four times a number added to 5 is divided by 6. The result is $\frac{7}{2}$. Find the number.

19. Marcus and Tony work for Lombardo's Pipe and Concrete. Mr. Lombardo is preparing an estimate for a customer. He knows that Marcus lays a slab of concrete in 6 hours. Tony lays the same size slab in 4 hours. If both work on the job and the cost of labor is $45.00 per hour, decide what the labor estimate should be.

20. Mr. Dodson can paint his house by himself in 4 days. His son needs an additional day to complete the job if he works by himself. If they work together, find how long it takes to paint the house.

21. While road testing a new make of car, the editor of a consumer magazine finds that he can go 10 miles into a 3-mile-per-hour wind in the same amount of time he can go 11 miles with a 3-mile-per-hour wind behind him. Find the speed of the car in still air.

22. A fisherman on Pearl River rows 9 miles downstream in the same amount of time he rows 3 miles upstream. If the current is 6 miles per hour, find how long it takes him to cover the 12 miles.

Find the unknown length in the following pairs of similar triangles.

23.

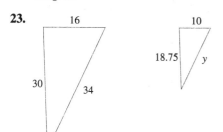

24.

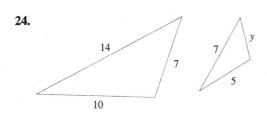

Solve the following.

25. Two divided by the difference of a number and 3 minus 4 divided by a number plus 3, equals 8 times the reciprocal of the difference of the number squared and 9. What is the number?

26. If 15 times the reciprocal of a number is added to the ratio of 9 times a number minus 7 and the number plus 2, the result is 9. What is the number?

27. A pilot flies 630 miles with a tail wind of 35 miles per hour. Against the wind, he flies only 455 miles. Find the rate of the plane in still air.

28. A marketing manager travels 1080 miles in a corporate jet and then an additional 240 miles by car. If the car ride takes one hour longer than the jet ride takes, and if the rate of the jet is 6 times the rate of the car, find the time the manager travels by jet and find the time the manager travels by car.

29. A cyclist rides 16 miles per hour on level ground on a still day. He finds that he rides 48 miles with the wind behind him in the same amount of time that he rides 16 miles into the wind. Find the rate of the wind.

30. The current on a portion of the Mississippi River is 3 miles per hour. A barge can go 6 miles upstream in the same amount of time it takes to go 10 miles downstream. Find the speed of the boat in still water.

31. One custodian cleans a suite of offices in 3 hours. When a second worker is asked to join the regular custodian, the job takes only $1\frac{1}{2}$ hours. How long does it take the second worker to do the same job alone?

32. One person proofreads copy for a small newspaper in 4 hours. If a second proofreader is also employed, the job can be done in $2\frac{1}{2}$ hours. How long does it take for the second proofreader to do the same job alone?

33. One pipe fills a storage pool in 20 hours. A second pipe fills the same pool in 15 hours. When a third pipe is added and all three are used to fill the pond, it takes only 6 hours. Find how long it takes the third pipe to do the job.

34. Mr. Jamison can do an audit in 400 hours. Mr. Ling can do the same audit in 300 hours with the help of a new computer program. How long will it take if both are assigned to the audit?

35. A toy maker wishes to make a triangular mainsail for a toy sailboat that will be the same shape as a regular-size sailboat's mainsail. Use the following diagram to find the missing dimensions.

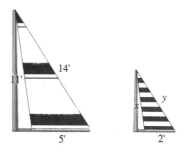

36. A seamstress wishes to make a doll's triangular diaper that will have the same shape as a full-size diaper. Use the following diagram to find the missing dimensions of the doll's diaper.

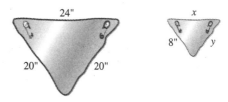

37. In 6 hours, an experienced cook prepares enough pies to supply a local restaurant's daily order. Another cook prepares the same number of pies in 7 hours. Together with a third cook, they prepare the pies in 2 hours. Find the work rate of the third cook.

38. Mrs. Smith balances the company books in 8 hours. It takes her assistant half again as long to do the same job. If they work together, find how long it takes them to balance the books.

39. One pump fills a tank 3 times as fast as another pump. If the pumps work together, they fill the tank in 21 minutes. How long does it take for each pump to fill the tank?

One of the great algebraists of ancient times was a man named Diophantus. Little is known of his life other than that he lived and worked in Alexandria. Some historians believe he lived during the first century of the Christian era, about the time of Nero. The only clue to his personal life is the following epigram found in a collection called the Palatine Anthology.

God granted him youth for a sixth of his life and added a twelfth part to this. He clothed his cheeks in down. He lit him the light of wedlock after a seventh part and five years after his marriage, He granted him a son. Alas, lateborn wretched child. After attaining the measure of half his father's life, cruel fate overtook him, thus leaving Diophantus during the last four years of his life only such consolation as the science of numbers. How old was Diophantus at his death?*

*From *The Nature and Growth of Modern Mathematics*, Edna Kramer, 1970, Fawcett Premier Books, Vol. 1, pages 107–108.

We are looking for Diophantus' age when he died, so let x represent that age. If we sum the parts of his life, we should get the total age.

Parts of his life
$$\begin{cases} \frac{1}{6} \cdot x + \frac{1}{12} \cdot x \text{ is the time of his youth.} \\ \frac{1}{7} \cdot x \text{ is the time between his youth and when he married.} \\ 5 \text{ years is the time between his marriage and the birth of his son.} \\ \frac{1}{2} \cdot x \text{ is the time Diophantus had with his son.} \\ 4 \text{ years is the time between his son's death and his own.} \end{cases}$$

The sum of these parts should equal Diophantus' age when he died.

$$\frac{1}{6} \cdot x + \frac{1}{12} \cdot x + \frac{1}{7} \cdot x + 5 + \frac{1}{2} \cdot x + 4 = x$$

40. Solve the epigram.

41. How old was Diophantus when his son was born? How old was the son when he died?

42. Solve the following epigram:

I was four when my mother packed my lunch and sent me off to school. Half my life was spent in school and another sixth was spent on a farm. Alas, hard times befell me. My crops and cattle fared poorly and my land was sold. I returned to school for 3 years and have spent one tenth of my life teaching. How old am I?

43. Write an epigram describing your life. Be sure that none of the time periods in your epigram overlap.

Review Exercises

Graph each linear equation by finding intercepts. See Section 3.3.

44. $5x + y = 10$ **45.** $-x + 3y = 6$
46. $x = -3y$ **47.** $y = 2x$
48. $x - y = -2$ **49.** $y - x = -5$

GROUP ACTIVITY

COMPARING FORMULAS FOR DOSES OF MEDICATION

MATERIALS:
• Calculator

Two dose formulas for prescribing medicines to children are well known among doctors. Unlike formulas for, say, area or distance, these dose formulas describe only an approximate relationship. Young's Rule and Cowling's Rule both relate a child's age A in years and an adult dose D of medication to the proper child's dose C. The formulas are most accurate when applied to children between the ages of 2 and 13.

$$\text{Young's Rule:} \quad C = \frac{DA}{A + 12}$$

$$\text{Cowling's Rule:} \quad C = \frac{D(A + 1)}{24}$$

1. Let the adult dose $D = 1000$ mg. Create a table comparing the doses predicted by both formulas for ages $A = 2, 3, \ldots, 13$.

2. Use the data from your table in Question 1 to form sets of ordered pairs for each formula. In your ordered pairs let x represent the age of the child. Graph the ordered pairs for each formula on the same graph. Describe the shapes of the graphed data.

3. Use your table, graph, or both to decide whether either formula will consistently predict a larger dose than the other. If so, which one? If not, is there an age at which the doses predicted by one becomes greater than the doses predicted by the other? If so, estimate that age.

4. Use your graph to estimate for what age the difference in the two predicted doses is greatest. Verify your answer by adding another column or row to your table giving the absolute value of the differences between Young's doses and Cowling's doses.

5. Does Cowling's Rule ever predict exactly the adult dose? If so, at what age? Explain. Does Young's Rule ever predict exactly the adult dose? If so, at what age? Explain.

6. Many doctors prefer to use formulas that relate doses to factors other than a child's age. Why is age not necessarily the most important factor when predicting a child's dose? What other factors might be used?

Chapter 6 Highlights

Definitions and Concepts	Examples
Section 6.1 Simplifying Rational Expressions	

A **rational expression** is an expression that can be written in the form $\frac{P}{Q}$, where P and Q are polynomials and Q does not equal 0.

Rational Expressions

$$\frac{7y^3}{4}, \quad \frac{x^2 + 6x + 1}{x - 3}, \quad \frac{-5}{s^3 + 8}$$

To find values for which a rational expression is undefined, find values for which the denominator is 0.

Find any values for which the expression $\frac{5y}{y^2 - 4y + 3}$ is undefined.

$$y^2 - 4y + 3 = 0 \quad \text{Set denominator equal to 0.}$$
$$(y - 3)(y - 1) = 0 \quad \text{Factor.}$$
$$y - 3 = 0 \text{ or } y - 1 = 0 \quad \text{Set each factor equal to 0.}$$
$$y = 3 \text{ or } y = 1 \quad \text{Solve.}$$

The expression is undefined when y is 3 and when y is 1.

Fundamental Principle of Rational Expressions

If P and Q are polynomials, and Q and R are not 0, then

$$\frac{PR}{QR} = \frac{P}{Q}$$

By the fundamental principle,

$$\frac{(x - 3)(x + 1)}{x(x + 1)} = \frac{x - 3}{x}$$

as long as $x \neq 0$ and $x \neq -1$.

To simplify a rational expression,

Step 1. Factor the numerator and denominator.

Step 2. Apply the fundamental principle to divide out common factors.

Simplify $\frac{4x + 20}{x^2 - 25}$.

$$\frac{4x + 20}{x^2 - 25} = \frac{4(x + 5)}{(x + 5)(x - 5)} = \frac{4}{x - 5}$$

| **Section 6.2 Multiplying and Dividing Rational Expressions** | |

To multiply rational expressions,

Step 1. Factor numerators and denominators.

Step 2. Multiply numerators and multiply denominators.

Step 3. Write the product in lowest terms.

$$\frac{P}{Q} \cdot \frac{R}{S} = \frac{PR}{QS}$$

Multiply: $\frac{4x + 4}{2x - 3} \cdot \frac{2x^2 + x - 6}{x^2 - 1}$

$$\frac{4x + 4}{2x - 3} \cdot \frac{2x^2 + x - 6}{x^2 - 1} = \frac{4(x + 1)}{2x - 3} \cdot \frac{(2x - 3)(x + 2)}{(x + 1)(x - 1)}$$

$$= \frac{4(x + 1)(2x - 3)(x + 2)}{(2x - 3)(x + 1)(x - 1)}$$

$$= \frac{4(x + 2)}{(x - 1)}$$

(continued)

DEFINITIONS AND CONCEPTS	EXAMPLES
SECTION 6.2 **MULTIPLYING AND DIVIDING RATIONAL EXPRESSIONS**	
To divide by a rational expression, multiply by the reciprocal. $$\frac{P}{Q} \div \frac{R}{S} = \frac{P}{Q} \cdot \frac{S}{R} = \frac{PS}{QR}$$	Divide: $\dfrac{15x + 5}{3x^2 - 14x - 5} \div \dfrac{15}{3x - 12}$ $\dfrac{15x + 5}{3x^2 - 14x - 5} \div \dfrac{15}{3x - 12} = \dfrac{5(3x + 1)}{(3x + 1)(x - 5)} \cdot \dfrac{3(x - 4)}{3 \cdot 5}$ $= \dfrac{x - 4}{x - 5}$
SECTION 6.3 **ADDING AND SUBTRACTING RATIONAL EXPRESSIONS WITH COMMON DENOMINATORS AND LEAST COMMON DENOMINATOR**	
To add or subtract rational expressions with the same denominator, add or subtract numerators, and place the sum or difference over a common denominator. $$\frac{P}{R} + \frac{Q}{R} = \frac{P + Q}{R}$$ $$\frac{P}{R} - \frac{Q}{R} = \frac{P - Q}{R}$$	Perform indicated operations. $\dfrac{5}{x + 1} + \dfrac{x}{x + 1} = \dfrac{5 + x}{x + 1}$ $\dfrac{2y + 7}{y^2 - 9} - \dfrac{y + 4}{y^2 - 9} = \dfrac{(2y + 7) - (y + 4)}{y^2 - 9}$ $= \dfrac{2y + 7 - y - 4}{y^2 - 9}$ $= \dfrac{y + 3}{(y + 3)(y - 3)}$ $= \dfrac{1}{y - 3}$
To find the least common denominator (LCD), *Step 1.* Factor the denominators. *Step 2.* The LCD is the product of all unique factors, each raised to a power equal to the greatest number of times that it appears in any one factored denominator.	Find the LCD for $\dfrac{7x}{x^2 + 10x + 25}$ and $\dfrac{11}{3x^2 + 15x}$ $x^2 + 10x + 25 = (x + 5)(x + 5)$ $3x^2 + 15x = 3x(x + 5)$ LCD $= 3x(x + 5)(x + 5)$ or $3x(x + 5)^2$
SECTION 6.4 **ADDING AND SUBTRACTING RATIONAL EXPRESSIONS WITH UNLIKE DENOMINATORS**	
To add or subtract rational expressions with unlike denominators. *Step 1.* Find the LCD. *Step 2.* Rewrite each rational expression as an equivalent expression whose denominator is the LCD. *Step 3.* Add or subtract numerators and place the sum or difference over the common denominator.	Perform the indicated operation. $\dfrac{9x + 3}{x^2 - 9} - \dfrac{5}{x - 3}$ $= \dfrac{9x + 3}{(x + 3)(x - 3)} - \dfrac{5}{x - 3}$ LCD is $(x + 3)(x - 3)$.

(continued)

DEFINITIONS AND CONCEPTS	EXAMPLES
SECTION 6.4 ADDING AND SUBTRACTING RATIONAL EXPRESSIONS WITH UNLIKE DENOMINATORS	

Step 4. Write the result in lowest terms.

$$= \frac{9x+3}{(x+3)(x-3)} - \frac{5(x+3)}{(x-3)(x+3)}$$

$$= \frac{9x+3-5(x+3)}{(x+3)(x-3)}$$

$$= \frac{9x+3-5x-15}{(x+3)(x-3)}$$

$$= \frac{4x-12}{(x+3)(x-3)}$$

$$= \frac{4(x-3)}{(x+3)(x-3)} = \frac{4}{x+3}$$

SECTION 6.5 COMPLEX FRACTIONS

Method 1: To Simplify a Complex Fraction

Step 1. Add or subtract fractions in the numerator and the denominator of the complex fraction.

Step 2. Perform the indicated division.

Step 3. Write the result in lowest terms.

Simplify.

$$\frac{\frac{1}{x}+2}{\frac{1}{x}-\frac{1}{y}} = \frac{\frac{1}{x}+\frac{2x}{x}}{\frac{y}{xy}-\frac{x}{xy}}$$

$$= \frac{\frac{1+2x}{x}}{\frac{y-x}{xy}}$$

$$= \frac{1+2x}{x} \cdot \frac{xy}{y-x}$$

$$= \frac{y(1+2x)}{y-x}$$

Method 2: To Simplify a Complex Fraction

Step 1. Find the LCD of all fractions in the complex fraction.

Step 2. Multiply the numerator and the denominator of the complex fraction by the LCD.

Step 3. Perform indicated operations and write in lowest terms.

$$\frac{\frac{1}{x}+2}{\frac{1}{x}-\frac{1}{y}} = \frac{xy\left(\frac{1}{x}+2\right)}{xy\left(\frac{1}{x}-\frac{1}{y}\right)}$$

$$= \frac{xy\left(\frac{1}{x}\right)+xy(2)}{xy\left(\frac{1}{x}\right)-xy\left(\frac{1}{y}\right)}$$

$$= \frac{y+2xy}{y-x} \quad \text{or} \quad \frac{y(1+2x)}{y-x}$$

DEFINITIONS AND CONCEPTS	EXAMPLES
SECTION 6.6 SOLVING EQUATIONS CONTAINING RATIONAL EXPRESSIONS	

To solve an equation containing rational expressions,

Step 1. Multiply both sides of the equation by the LCD of all rational expressions in the equation.

Step 2. Remove any grouping symbols and solve the resulting equation.

Step 3. Check the solution in the original equation.

Solve: $\dfrac{5x}{x+2} + 3 = \dfrac{4x-6}{x+2}$

$(x+2)\left(\dfrac{5x}{x+2} + 3\right) = (x+2)\left(\dfrac{4x-6}{x+2}\right)$

$(x+2)\left(\dfrac{5x}{x+2}\right) + (x+2)(3) = (x+2)\left(\dfrac{4x-6}{x+2}\right)$

$5x + 3x + 6 = 4x - 6$

$4x = -12$

$x = -3$

The solution checks and the solution set is $\{-3\}$.

SECTION 6.7 RATIO AND PROPORTION

A **ratio** is the quotient of two numbers or two quantities.

The ratio of *a* to *b* can be written as

Fractional Notation   Colon Notation

$\dfrac{a}{b}$ $a{:}b$

Write the ratio of 5 hours to 1 day using fractional notation.

$\dfrac{5 \text{ hours}}{1 \text{ day}} = \dfrac{5 \text{ hours}}{24 \text{ hours}} = \dfrac{5}{24}$

A **proportion** is a mathematical statement that two ratios are equal.

Proportions

$\dfrac{2}{3} = \dfrac{8}{12}$ $\dfrac{x}{7} = \dfrac{15}{35}$

In the proportion $\dfrac{2}{3} = \dfrac{8}{12}$, the 2 and the 12 are called the **extremes**; the 3 and the 8 are called the **means**.

extremes → $\dfrac{2}{3} = \dfrac{8}{12}$ ← means

In the proportion $\dfrac{a}{b} = \dfrac{c}{d}$, the products ad and bc are called **cross products.**

Cross Products

$\dfrac{2}{3} = \dfrac{8}{12}$ → $3 \cdot 8$ or 24
           → $2 \cdot 12$ or 24

In a true proportion, the product of the means equals the product of the extremes.

Cross multiplication:

If $\dfrac{a}{b} = \dfrac{c}{d}$, then $ad = bc$.

Solve: $\dfrac{3}{4} = \dfrac{x}{x-1}$

$\dfrac{3}{4} = \dfrac{x}{x-1}$

$3(x-1) = 4x$ Cross multiply.

$3x - 3 = 4x$

$-3 = x$

(continued)

DEFINITIONS AND CONCEPTS	EXAMPLES
SECTION 6.7 RATIO AND PROPORTION	
Problem-Solving Steps 1. UNDERSTAND. Read and reread the problem. 2. ASSIGN. 3. ILLUSTRATE. 4. TRANSLATE. 5. COMPLETE. 6. INTERPRET.	A sample of 200 size C batteries contained 3 defective batteries. How many defective batteries might we expect in a shipment of 25,000 size C batteries? Let x = number of defective batteries in the shipment. In words: $\dfrac{\text{total number in sample}}{\text{defective number in sample}} = \dfrac{\text{total number in shipment}}{\text{defective number in shipment}}$ Translate: $\dfrac{200}{3} = \dfrac{25{,}000}{x}$ Solve: $\dfrac{200}{3} = \dfrac{25{,}000}{x}$ $\quad 200x = 3(25{,}000)\quad$ Cross multiply. $\quad 200x = 75{,}000$ $\quad\quad\ x = 375$ **Check** the solution and **state** the conclusion. We expect the shipment of 25,000 batteries to contain 375 defective batteries.
SECTION 6.8 RATIONAL EQUATIONS AND PROBLEM SOLVING	
Problem-Solving Steps 1. UNDERSTAND. Read and reread the problem. 2. ASSIGN. 3. ILLUSTRATE.	A small plane and a car leave Kansas City, Missouri, and head for Minneapolis, Minnesota, a distance of 450 miles. The speed of the plane is 3 times the speed of the car, and the plane arrives 6 hours ahead of the car. Find the speed of the car. Let x = the speed of the car. Then $3x$ = the speed of the plane. |	DISTANCE	=	RATE	·	TIME | |---|---|---|---|---|---| | Car	450		x		$\dfrac{450}{x}\left(\dfrac{\text{distance}}{\text{rate}}\right)$ | | Plane	450		$3x$		$\dfrac{450}{3x}\left(\dfrac{\text{distance}}{\text{rate}}\right)$	

(continued)

DEFINITIONS AND CONCEPTS	EXAMPLES
SECTION 6.8	**RATIONAL EQUATIONS AND PROBLEM SOLVING**
4. TRANSLATE.	In words: Plane's time $+$ 6 hours $=$ car's time
	Translate: $\dfrac{450}{3x} + 6 = \dfrac{450}{x}$
5. COMPLETE.	Solve: $\dfrac{450}{3x} + 6 = \dfrac{450}{x}$
	$3x\left(\dfrac{450}{3x}\right) + 3x(6) = 3x\left(\dfrac{450}{x}\right)$
	$450 + 18x = 1350$
	$18x = 900$
	$x = 50$
6. INTERPRET.	**Check** the solution and **state** the conclusion. The speed of the car is 50 miles per hour.

CHAPTER 6 REVIEW

(1) *Find any real number for which each rational expression is undefined.*

1. $\dfrac{x+5}{x^2-4}$
2. $\dfrac{5x+9}{4x^2-4x-15}$

Find the value of each rational expression when $x = 5$, $y = 7$, and $z = -2$.

3. $\dfrac{z^2-z}{z+xy}$
4. $\dfrac{x^2+xy-z^2}{x+y+z}$

Simplify each rational expression.

5. $\dfrac{x+2}{x^2-3x-10}$
6. $\dfrac{x+4}{x^2+5x+4}$
7. $\dfrac{x^3-4x}{x^2+3x+2}$
8. $\dfrac{5x^2-125}{x^2+2x-15}$
9. $\dfrac{x^2-x-6}{x^2-3x-10}$
10. $\dfrac{x^2-2x}{x^2+2x-8}$
11. $\dfrac{x^2+6x+5}{2x^2+11x+5}$
12. $\dfrac{x^2+xa+xb+ab}{x^2-xc+bx-bc}$
13. $\dfrac{x^2+5x-2x-10}{x^2-3x-2x+6}$
14. $\dfrac{x^2-9}{9-x^2}$
15. $\dfrac{4-x}{x^3-64}$

Perform the indicated operations and simplify.

16. $\dfrac{15x^3y^2}{z} \cdot \dfrac{z}{5xy^3}$
17. $\dfrac{-y^3}{8} \cdot \dfrac{9x^2}{y^3}$
18. $\dfrac{x^2-9}{x^2-4} \cdot \dfrac{x-2}{x+3}$
19. $\dfrac{2x+5}{x-6} \cdot \dfrac{2x}{-x+6}$
20. $\dfrac{x^2-5x-24}{x^2-x-12} \div \dfrac{x^2-10x+16}{x^2+x-6}$
21. $\dfrac{4x+4y}{xy^2} \div \dfrac{3x+3y}{x^2y}$
22. $\dfrac{x^2+x-42}{x-3} \cdot \dfrac{(x-3)^2}{x+7}$
23. $\dfrac{2a+2b}{3} \cdot \dfrac{a-b}{a^2-b^2}$
24. $\dfrac{x^2-9x+14}{x^2-5x+6} \cdot \dfrac{x+2}{x^2-5x-14}$

25. $(x-3) \cdot \dfrac{x}{x^2+3x-18}$

26. $\dfrac{2x^2-9x+9}{8x-12} \div \dfrac{x^2-3x}{2x}$

27. $\dfrac{x^2-y^2}{x^2+xy} \div \dfrac{3x^2-2xy-y^2}{3x^2+6x}$

28. $\dfrac{x^2-y^2}{8x^2-16xy+8y^2} \div \dfrac{x+y}{4x-y}$

29. $\dfrac{x-y}{4} \div \dfrac{y^2-2y-xy+2x}{16x+24}$

30. $\dfrac{y-3}{4x+3} \div \dfrac{9-y^2}{4x^2-x-3}$

(6.3) *Perform the indicated operations and simplify.*

31. $\dfrac{5x-4}{3x-1} + \dfrac{6}{3x-1}$

32. $\dfrac{4x-5}{3x^2} - \dfrac{2x+5}{3x^2}$

33. $\dfrac{9x+7}{6x^2} - \dfrac{3x+4}{6x^2}$

Find the LCD of each pair of rational expressions.

34. $\dfrac{x+4}{2x}, \dfrac{3}{7x}$

35. $\dfrac{x-2}{x^2-5x-24}, \dfrac{3}{x^2+11x+24}$

Rewrite the following rational expressions as equivalent expressions whose denominator is the given polynomial.

36. $\dfrac{x+2}{x^2+11x+18}, (x+2)(x-5)(x+9)$

37. $\dfrac{3x-5}{x^2+4x+4}, (x+2)^2 \cdot (x+3)$

(6.4) *Perform the indicated operations and simplify.*

38. $\dfrac{4}{5x^2} - \dfrac{6}{y}$

39. $\dfrac{2}{x-3} - \dfrac{4}{x-1}$

40. $\dfrac{x+7}{x+3} - \dfrac{x-3}{x+7}$

41. $\dfrac{4}{x+3} - 2$

42. $\dfrac{3}{x^2+2x-8} + \dfrac{2}{x^2-3x+2}$

43. $\dfrac{2x-5}{6x+9} - \dfrac{4}{2x^2+3x}$

44. $\dfrac{x-1}{x^2-2x+1} - \dfrac{x+1}{x-1}$

45. $\dfrac{x-1}{x^2+4x+4} + \dfrac{x-1}{x+2}$

Find the perimeter and the area of each figure.

46.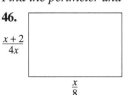

47.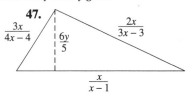

(6.5) *Simplify each complex fraction.*

48. $\dfrac{\dfrac{5x}{27}}{\dfrac{10xy}{21}}$

49. $\dfrac{\dfrac{8x}{x^2-9}}{\dfrac{4}{x+3}}$

50. $\dfrac{\dfrac{3}{5}+\dfrac{2}{7}}{\dfrac{1}{5}+\dfrac{5}{6}}$

51. $\dfrac{\dfrac{2}{a}+\dfrac{1}{2a}}{a+\dfrac{a}{2}}$

52. $\dfrac{3-\dfrac{1}{y}}{2-\dfrac{1}{y}}$

53. $\dfrac{2+\dfrac{1}{x^2}}{\dfrac{1}{x}+\dfrac{2}{x^2}}$

54. $\dfrac{\dfrac{1}{a}+\dfrac{1}{b}}{\dfrac{1}{ab}}$

55. $\dfrac{\dfrac{6}{x+2}+4}{\dfrac{8}{x+2}-4}$

(6.6) *Solve each equation for the variable or perform the indicated operation.*

56. $\dfrac{x+4}{9} = \dfrac{5}{9}$

57. $\dfrac{n}{10} = 9 - \dfrac{n}{5}$

58. $\dfrac{5y-3}{7} = \dfrac{15y-2}{28}$

59. $\dfrac{2}{x+1} - \dfrac{1}{x-2} = -\dfrac{1}{2}$

60. $\dfrac{1}{a+3} + \dfrac{1}{a-3} = -\dfrac{5}{a^2-9}$

61. $\dfrac{y}{2y+2} + \dfrac{2y-16}{4y+4} = \dfrac{y-3}{y+1}$

62. $\dfrac{4}{x+3} + \dfrac{8}{x^2-9} = 0$

63. $\dfrac{2}{x-3} - \dfrac{4}{x+3} = \dfrac{8}{x^2-9}$

64. $\dfrac{x-3}{x+1} - \dfrac{x-6}{x+5} = 0$

65. $x + 5 = \dfrac{6}{x}$

Solve the equation for the indicated variable.

66. $\dfrac{4A}{5b} = x^2$, for b

67. $\dfrac{x}{7} + \dfrac{y}{8} = 10$, for y

(6.7) *Write each phrase as a ratio in fractional notation.*

68. 20 cents to 1 dollar

69. four parts red to six parts white

Solve each proportion.

70. $\dfrac{x}{2} = \dfrac{12}{4}$

71. $\dfrac{20}{1} = \dfrac{x}{25}$

72. $\dfrac{32}{100} = \dfrac{100}{x}$

73. $\dfrac{20}{2} = \dfrac{c}{5}$

74. $\dfrac{2}{x-1} = \dfrac{3}{x+3}$

75. $\dfrac{4}{y-3} = \dfrac{2}{y-3}$

76. $\dfrac{y+2}{y} = \dfrac{5}{3}$

77. $\dfrac{x-3}{3x+2} = \dfrac{2}{6}$

Given the following prices charged for various sizes of an item, find the best buy.

78. Shampoo

    10 ounces for $1.29

    16 ounces for $2.15

79. Frozen green beans

    8 ounces for $0.89

    15 ounces for $1.63

    20 ounces for $2.36

Solve.

80. A machine can process 300 parts in 20 minutes. Find how many parts can be processed in 45 minutes.

81. As his consulting fee, Mr. Visconti charges $90.00 per day. Find how much he charges for 3 hours of consulting. Assume an 8-hour work day.

82. One fund raiser can address 100 letters in 35 minutes. Find how many he can address in 55 minutes.

(6.8) *Solve each problem.*

83. Five times the reciprocal of a number equals the sum of $\tfrac{3}{2}$ times the reciprocal of the number and $\tfrac{7}{6}$. What is the number?

84. The reciprocal of a number equals the reciprocal of the difference of 4 and the number. Find the number.

85. A car travels 90 miles in the same time that a car traveling 10 miles per hour slower travels 60 miles. Find the speed of each car.

86. The speed of a bayou near Lafayette, Louisiana, is 4 miles per hour. A paddle boat travels 48 miles upstream in the same amount of time it takes to travel 72 miles downstream. Find the speed of the boat in still water.

87. When Mark and Maria manicure Mr. Stergeon's lawn, it takes them 5 hours. If Mark works alone, it takes 7 hours. Find how long it takes Maria alone.

88. It takes pipe A 20 days to fill a fish pond. Pipe B takes 15 days. Find how long it takes both pipes together to fill the pond.

CHAPTER 6 TEST

1. Find any real numbers for which the following expression is undefined.

$$\dfrac{x+5}{x^2+4x+3}$$

2. For a certain computer desk, the manufacturing cost C per desk (in dollars) is

$$C = \dfrac{100x + 3000}{x}$$

where x is the number of desks manufactured.

a. Find the average cost per desk when manufacturing 200 computer desks.

b. Find the average cost per desk when manufacturing 1000 computer desks.

Simplify each rational expression.

3. $\dfrac{3x - 6}{5x - 10}$

4. $\dfrac{x + 10}{x^2 - 100}$

5. $\dfrac{x + 6}{x^2 + 12x + 36}$

6. $\dfrac{x + 3}{x^3 + 27}$

7. $\dfrac{2m^3 - 2m^2 - 12m}{m^2 - 5m + 6}$

8. $\dfrac{ay + 3a + 2y + 6}{ay + 3a + 5y + 15}$

9. $\dfrac{y - x}{x^2 - y^2}$

Perform the indicated operation and simplify if possible.

10. $\dfrac{x^2 - 13x + 42}{x^2 + 10x + 21} \div \dfrac{x^2 - 4}{x^2 + x - 6}$

11. $\dfrac{3}{x - 1} \cdot (5x - 5)$

12. $\dfrac{y^2 - 5y + 6}{2y + 4} \cdot \dfrac{y + 2}{2y - 6}$

13. $\dfrac{5}{2x + 5} - \dfrac{6}{2x + 5}$

14. $\dfrac{5a}{a^2 - a - 6} - \dfrac{2}{a - 3}$

15. $\dfrac{6}{x^2 - 1} + \dfrac{3}{x + 1}$

16. $\dfrac{x^2 - 9}{x^2 - 3x} \div \dfrac{xy + 5x + 3y + 15}{2x + 10}$

17. $\dfrac{x + 2}{x^2 + 11x + 18} + \dfrac{5}{x^2 - 3x - 10}$

18. $\dfrac{4y}{y^2 + 6y + 5} - \dfrac{3}{y^2 + 5y + 4}$

Solve each equation.

19. $\dfrac{4}{y} - \dfrac{5}{3} = \dfrac{-1}{5}$

20. $\dfrac{5}{y + 1} = \dfrac{4}{y + 2}$

21. $\dfrac{a}{a - 3} = \dfrac{3}{a - 3} - \dfrac{3}{2}$

22. $\dfrac{10}{x^2 - 25} = \dfrac{3}{x + 5} + \dfrac{1}{x - 5}$

Simplify each complex fraction.

23. $\dfrac{\dfrac{5x^2}{yz^2}}{\dfrac{10x}{z^3}}$

24. $\dfrac{\dfrac{b}{a} - \dfrac{a}{b}}{\dfrac{b}{a} + \dfrac{b}{a}}$

25. $\dfrac{5 - \dfrac{1}{y^2}}{\dfrac{1}{y} + \dfrac{2}{y^2}}$

26. In a sample of 85 fluorescent bulbs, 3 were found to be defective. At this rate, how many defective bulbs should be found in 510 bulbs?

27. One number plus five times its reciprocal is equal to six. Find the number.

28. A pleasure boat traveling down the Red River takes the same time to go 14 miles upstream as it takes to go 16 miles downstream. If the current of the river is 2 miles per hour, find the speed of the boat in still water.

29. An inlet pipe can fill a tank in 12 hours. A second pipe can fill the tank in 15 hours. If both pipes are used, find how long it takes to fill the tank.

30. Decide which is the best buy in crackers.

    6 ounces for $1.19

    10 ounces for $2.15

    16 ounces for $3.25

Chapter 6 Cumulative Review

1. Write each sentence as an equation. Let x represent the unknown number.

  a. The quotient of 15 and a number is 4.

  b. Three subtracted from 12 is a number.

  c. Four times a number added to 17 is 21.

2. Find each sum.

  a. $-3 + (-7)$ **b.** $5 + (+12)$

  c. $(-1) + (-20)$ **d.** $-2 + (-10)$

3. Name the property illustrated by each true statement.
  a. $2 \cdot 3 = 3 \cdot 2$
  b. $3(x + 5) = 3x + 15$
  c. $2 + (4 + 8) = (2 + 4) + 8$

4. Solve $3 - x = 7$ for x.

5. Twice a number added to seven is the same as three subtracted from the number. Find the number.

6. Solve $P = 2l + 2w$ for w.

7. Solve $x + 4 \leq -6$ for x. Graph the solution set.

8. Graph the linear equation $y = 3x$.

9. Simplify each quotient.
  a. $\dfrac{x^5}{x^2}$
  b. $\dfrac{4^7}{4^3}$
  c. $\dfrac{(-3)^5}{(-3)^2}$
  d. $\dfrac{2x^5y^2}{xy}$

10. Use the distributive property to find each product.
  a. $5x(2x^3 + 6)$
  b. $-3x^2(5x^2 + 6x - 1)$
  c. $(3n^2 - 5n + 4)(2n)$

11. Simplify each expression. Write answers with positive exponents.
  a. $\dfrac{y}{y^{-2}}$
  b. $\dfrac{p^{-4}}{q^{-9}}$
  c. $\dfrac{x^{-5}}{x^7}$

12. Divide: $\dfrac{2x^4 - x^3 + 3x^2 + x - 1}{x^2 + 1}$.

13. Factor $x^2 + 7x + 12$.

14. Factor $x^3 + 8$.

15. Solve $2x^3 - 4x^2 - 30x = 0$.

16. Multiply: $\dfrac{x^2 + x}{3x} \cdot \dfrac{6}{5x + 5}$.

17. Subtract: $\dfrac{3x^2 + 2x}{x - 1} - \dfrac{10x - 5}{x - 1}$.

18. Subtract: $\dfrac{6x}{x^2 - 4} - \dfrac{3}{x + 2}$.

19. Simplify $\dfrac{\dfrac{1}{z} - \dfrac{1}{2}}{\dfrac{1}{3} - \dfrac{z}{6}}$.

20. Solve $\dfrac{t - 4}{2} - \dfrac{t - 3}{9} = \dfrac{5}{18}$.

21. Sam Waterton and Frank Schaffer work in a plant that manufactures automobiles. Sam can complete a quality control tour of the plant in 3 hours while his assistant, Frank, needs 7 hours to complete the same job. The regional manager is coming to inspect the plant facilities, so both Sam and Frank are directed to complete a quality control tour together. How long will this take?

498 CHAPTER 9 ROOTS AND RADICALS

Having spent the last chapter studying equations, we return now to algebraic expressions. We expand on your skills of operating on expressions—adding, subtracting, multiplying, dividing, and raising to powers—to include finding roots. Just as subtraction is defined by addition and division by multiplication, finding roots is defined by raising to powers. This chapter also includes working with equations that contain roots and solving problems that can be modeled by such equations.

9.1 INTRODUCTION TO RADICALS

OBJECTIVES

 Find square roots of perfect squares.
 Find cube roots of perfect cubes.
 Find nth roots.
4 Identify rational and irrational numbers.

TAPE BA 9.1

In this section, we define finding the *root* of a number by its reverse operation, raising a number to a power. We begin with squares and square roots.

The square of 5 is $5^2 = 25$.
The square of -5 is $(-5)^2 = 25$.
The square of $\frac{1}{2}$ is $\left(\frac{1}{2}\right)^2 = \frac{1}{4}$.

The reverse operation of squaring a number is finding the *square root* of a number. For example,

A square root of 25 is 5, because $5^2 = 25$.
A square root of 25 is also -5, because $(-5)^2 = 25$.
A square root of $\frac{1}{4}$ is $\frac{1}{2}$, because $\left(\frac{1}{2}\right)^2 = \frac{1}{4}$.

In general, a number b is a square root of a number a if $b^2 = a$.

Notice that both 5 and -5 are square roots of 25. The symbol $\sqrt{}$ is used to denote the **positive** or **principal square root** of a number. For example,

$\sqrt{25} = 5$ since $5^2 = 25$ and 5 is positive.

The symbol $-\sqrt{}$ is used to denote the **negative square root**. For example,

$-\sqrt{25} = -5$

INTRODUCTION TO RADICALS   SECTION 9.1   **499**

> **SQUARE ROOT**
> The positive or principal square root of a positive number a is written as $\sqrt{a}$. The negative square root of a is written as $-\sqrt{a}$.
> $$\sqrt{a} = b \quad \text{only if } b^2 = a \text{ and } b > 0$$
> Also, the square root of 0, written as $\sqrt{0}$, is 0.

The symbol $\sqrt{}$ is called a **radical** or **radical sign.** The expression within or under a radical sign is called the **radicand.** An expression containing a radical is called a **radical expression.**

$$\sqrt{a} \quad \begin{matrix} \leftarrow \text{radical sign} \\ \leftarrow \text{radicand} \end{matrix}$$

EXAMPLE 1   Find each square root.

   a. $\sqrt{36}$ **b.** $\sqrt{64}$ **c.** $-\sqrt{25}$ **d.** $\sqrt{\dfrac{9}{100}}$ **e.** $\sqrt{0}$

Solution: **a.** $\sqrt{36} = 6$, because $6^2 = 36$ and 6 is positive.

b. $\sqrt{64} = 8$, because $8^2 = 64$ and 8 is positive.

c. $-\sqrt{25} = -5$. The negative sign in front of the radical indicates the negative square root of 25.

d. $\sqrt{\dfrac{9}{100}} = \dfrac{3}{10}$ because $\left(\dfrac{3}{10}\right)^2 = \dfrac{9}{100}$ and $\dfrac{3}{10}$ is positive.

e. $\sqrt{0} = 0$ because $0^2 = 0$.

Is the square root of a negative number a real number? For example, is $\sqrt{-4}$ a real number? To answer this question, we ask ourselves, is there a real number whose square is -4? Since there is no real number whose square is -4, we say that $\sqrt{-4}$ is not a real number. In general,

A square root of a negative number is not a real number.

2   Finding roots can be extended to other roots such as cube roots. For example, since $2^3 = 8$, we call 2 the **cube root** of 8. In symbols, we write
$$\sqrt[3]{8} = 2$$

> **CUBE ROOT**
> The **cube root** of a real number a is written as $\sqrt[3]{a}$, and
> $$\sqrt[3]{a} = b \quad \text{only if } b^3 = a$$

From the above definition, we have

$\sqrt[3]{27} = 3,$ since $3^3 = 27$
$\sqrt[3]{-64} = -4,$ since $(-4)^3 = -64$

Notice that unlike the square root of a negative number, the cube root of a negative number is a real number. This is so because while we cannot find a real number whose **square** is negative, we **can** find a real number whose **cube** is negative. In fact, the cube of a negative number is a negative number. Therefore, the cube root of a negative number is a negative number.

EXAMPLE 2 Find the cube roots.

  a. $\sqrt[3]{1}$ **b.** $\sqrt[3]{-27}$ **c.** $\sqrt[3]{\dfrac{1}{125}}$

Solution: **a.** $\sqrt[3]{1} = 1$ because $1^3 = 1$.
  b. $\sqrt[3]{-27} = -3$ because $(-3)^3 = -27$
  c. $\sqrt[3]{\dfrac{1}{125}} = \dfrac{1}{5}$ because $\left(\dfrac{1}{5}\right)^3 = \dfrac{1}{125}$.

Just as we can raise a real number to powers other than 2 or 3, we can find roots other than square roots and cube roots. In fact, we can take the nth root of a number where n is any natural number. An **nth root** of a number a is a number whose nth power is a. The natural number n is called the **index**.

In symbols, the nth root of a is written as $\sqrt[n]{a}$. The index 2 is usually omitted for square roots.

> **REMINDER** If the index is even, such as $\sqrt{}, \sqrt[4]{}, \sqrt[6]{}$, and so on, the radicand must be nonnegative for the root to be a real number. For example,
>
> $\sqrt[4]{16} = 2$ but $\sqrt[4]{-16}$ is not a real number
> $\sqrt[6]{64} = 2$ but $\sqrt[6]{-64}$ is not a real number

EXAMPLE 3 Simplify the following expressions.

  a. $\sqrt[4]{16}$ **b.** $\sqrt[5]{-32}$ **c.** $-\sqrt[3]{8}$ **d.** $\sqrt[4]{-81}$

Solution: a. $\sqrt[4]{16} = 2$ because $2^4 = 16$ and 2 is positive.
b. $\sqrt[5]{-32} = -2$ because $(-2)^5 = -32$.
c. $-\sqrt[3]{8} = -2$ since $\sqrt[3]{8} = 2$.
d. $\sqrt[4]{-81}$ is not a real number since the index 4 is even and the radicand -81 is negative.

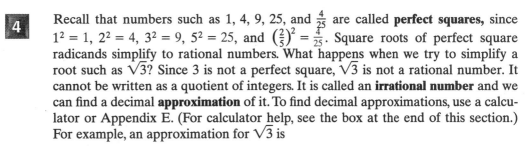

Recall that numbers such as 1, 4, 9, 25, and $\frac{4}{25}$ are called **perfect squares,** since $1^2 = 1$, $2^2 = 4$, $3^2 = 9$, $5^2 = 25$, and $\left(\frac{2}{5}\right)^2 = \frac{4}{25}$. Square roots of perfect square radicands simplify to rational numbers. What happens when we try to simplify a root such as $\sqrt{3}$? Since 3 is not a perfect square, $\sqrt{3}$ is not a rational number. It cannot be written as a quotient of integers. It is called an **irrational number** and we can find a decimal **approximation** of it. To find decimal approximations, use a calculator or Appendix E. (For calculator help, see the box at the end of this section.) For example, an approximation for $\sqrt{3}$ is

$$\sqrt{3} \approx 1.732$$
↑
approximation symbol

Radicands can also contain variables. Since the square root of a negative number is not a real number, we want to make sure variables in the radicand do not have replacement values that would make the radicand negative. To avoid negative radicands, assume for the rest of this chapter that **if a variable appears in the radicand of a radical expression, it represents positive numbers only.** Then

$$\sqrt{y^2} = y \quad \text{because } (y)^2 = y^2$$

Also,

$$\sqrt{x^8} = x^4 \quad \text{because } (x^4)^2 = x^8$$

Also,

$$\sqrt{9x^2} = 3x \quad \text{because } (3x)^2 = 9x^2$$

EXAMPLE 4 Simplify the following expressions. Assume that each variable represents a positive number.

a. $\sqrt{x^2}$ b. $\sqrt{x^6}$ c. $\sqrt[3]{27y^6}$ d. $\sqrt{16x^{16}}$

Solution: a. $\sqrt{x^2} = x$ because x times itself equals x^2.
b. $\sqrt{x^6} = x^3$ because $(x^3)^2 = x^6$.
c. $\sqrt[3]{27y^6} = 3y^2$ because $(3y^2)^3 = 27y^6$.
d. $\sqrt{16x^{16}} = 4x^8$ because $(4x^8)^2 = 16x^{16}$.

SCIENTIFIC CALCULATOR EXPLORATIONS

To simplify or approximate square roots using a calculator, locate the key marked $\boxed{\sqrt{}}$.
To simplify $\sqrt{25}$, press $\boxed{25}$ $\boxed{\sqrt{}}$. The display should read $\boxed{5}$.
To approximate $\sqrt{30}$, press $\boxed{30}$ $\boxed{\sqrt{}}$. The display should read $\boxed{5.4772256}$. This is an approximation for $\sqrt{30}$. Then a three-decimal-place approximation is

$$\sqrt{30} \approx 5.477$$

Is this answer reasonable? Since 30 is between perfect squares 25 and 36, $\sqrt{30}$ is between $\sqrt{25} = 5$ and $\sqrt{36} = 6$. The calculator result is then reasonable since 5.4772256 is between 5 and 6.

Use a calculator to approximate to three decimal places.

1. $\sqrt{7}$
2. $\sqrt{14}$
3. $\sqrt{10}$
4. $\sqrt{200}$
5. $\sqrt{82}$
6. $\sqrt{46}$

EXERCISE SET 9.1

Find each root. See Examples 1 through 3.

1. $\sqrt{16}$
2. $\sqrt{9}$
3. $\sqrt{81}$
4. $\sqrt{49}$
5. $\sqrt{\dfrac{1}{25}}$
6. $\sqrt{\dfrac{1}{64}}$
7. $-\sqrt{100}$
8. $-\sqrt{36}$
9. $\sqrt[3]{64}$
10. $\sqrt[3]{-1}$
11. $-\sqrt[3]{27}$
12. $-\sqrt[3]{8}$
13. $\sqrt[3]{\dfrac{1}{8}}$
14. $\sqrt[3]{\dfrac{1}{64}}$
15. $\sqrt[3]{-125}$
16. $\sqrt[3]{-27}$
17. $\sqrt[5]{32}$
18. $\sqrt[4]{-1}$
19. $\sqrt[4]{81}$
20. $\sqrt{121}$
21. $\sqrt{-4}$
22. $\sqrt[5]{\dfrac{1}{32}}$
23. $\sqrt[3]{\dfrac{1}{27}}$
24. $\sqrt[4]{256}$
25. $\sqrt{\dfrac{9}{25}}$
26. $\sqrt[3]{\dfrac{8}{27}}$
27. $-\sqrt{49}$
28. $-\sqrt[4]{625}$

29. Explain why the square root of a negative number is not a real number.

30. Explain why the cube root of a negative number is a real number.

Find each root. Assume that each variable represents a nonnegative real number. See Example 4.

31. $\sqrt{z^2}$
32. $\sqrt{y^{10}}$
33. $\sqrt{x^4}$
34. $\sqrt{z^6}$
35. $\sqrt{9x^8}$
36. $\sqrt{36x^{12}}$
37. $\sqrt{x^2y^6}$
38. $\sqrt{y^4z^{18}}$
39. $\sqrt[3]{x^{15}}$
40. $\sqrt[3]{y^{12}}$
41. $\sqrt{x^{12}}$
42. $\sqrt{z^{16}}$
43. $\sqrt{81x^2}$
44. $\sqrt{100z^4}$
45. $-\sqrt{144y^{14}}$
46. $-\sqrt{121z^{22}}$
47. $\sqrt{x^2y^2}$
48. $\sqrt{y^{20}z^{30}}$
49. $\sqrt{16x^{16}}$
50. $\sqrt{36y^{36}}$

Find each root that is a real number.

51. $\sqrt{0}$
52. $\sqrt[3]{0}$
53. $-\sqrt[5]{\dfrac{1}{32}}$
54. $-\sqrt[3]{\dfrac{27}{125}}$
55. $\sqrt{-64}$
56. $\sqrt[3]{-64}$
57. $-\sqrt{64}$
58. $\sqrt[6]{64}$
59. $-\sqrt{169}$

60. $\sqrt[4]{-16}$ **61.** $\sqrt{1}$ **62.** $\sqrt[3]{1}$

63. $\sqrt{\dfrac{25}{64}}$ **64.** $\sqrt{\dfrac{1}{100}}$ **65.** $-\sqrt[3]{-8}$

66. $-\sqrt[3]{-27}$

67. Simplify $\sqrt{\sqrt{81}}$. **68.** Simplify $\sqrt[3]{\sqrt[3]{1}}$.

*Determine whether each square root is rational or irrational. If it is rational, find its **exact value**. If it is irrational, use a calculator and write a three-decimal-place **approximation**.*

69. $\sqrt{9}$ **70.** $\sqrt{8}$ **71.** $\sqrt{37}$

72. $\sqrt{36}$ **73.** $\sqrt{169}$ **74.** $\sqrt{160}$

75. $\sqrt{4}$ **76.** $\sqrt{27}$

77. A fence is to be erected around a square garden with an area of 324 square feet. Each side of this garden has a length of $\sqrt{324}$ feet. Write a one-decimal-point approximation of this length.

78. The roof of the warehouse shown needs to be shingled. The total area of the roof is exactly $240\sqrt{41}$ square feet. Approximate this area to the nearest whole number.

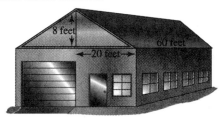

79. A standard baseball diamond is a square with 90-foot sides connecting the bases. The distance from first base to third base is $90 \cdot \sqrt{2}$ feet. Approximate $\sqrt{2}$ accurate to two decimal places and use it to approximate the distance $90 \cdot \sqrt{2}$ feet.

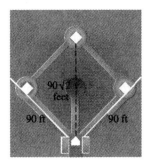

80. Graph $y = \sqrt{x}$. (*Hint:* Complete the table below, plot the ordered pair solutions, and draw a smooth curve through the points. Remember that since the radicand cannot be negative, this particular graph begins at the point with coordinates (0,0).)

x	y
0	0
1	
3	(approximate)
4	
9	

81. Graph $y = \sqrt[3]{x}$ (Complete the table below, plot the ordered pair solutions, and draw a smooth curve through the points.)

x	y
-8	
-2	(approximate)
-1	
0	
1	
2	(approximate)
8	

Use a grapher and graph each function. Observe the graph from left to right and give the ordered pair that corresponds to the "beginning" of the graph. Then tell why the graph starts at that point.

82. $y = \sqrt{x-2}$ **83.** $y = \sqrt{x+3}$

84. $y = \sqrt{x+4}$ **85.** $y = \sqrt{x-5}$

Review Exercises

Write each integer as a product of two integers such that one of the factors is a perfect square. For example, in $18 = 9 \cdot 2$, 9 is a perfect square.

86. 50 $25 \cdot 2$ **87.** 8 $4 \cdot 2$ **88.** 32

89. 75 $25 \cdot 3$ **90.** 28 $4 \cdot 7$ **91.** 44 $4 \cdot 11$

92. 27 $9 \cdot 3$ **93.** 90 $9 \cdot 10$

9.2 SIMPLIFYING RADICALS

OBJECTIVES

 Use the product rule to simplify radicals.
 Use the quotient rule to simplify radicals.

TAPE BA 9.2

Much of our work with expressions in this book has involved finding ways to write expressions in their simplest form. Writing radicals in simplest form requires recognizing several patterns, or rules, which we present here. Notice that

$$\sqrt{9 \cdot 16} = \sqrt{144} = 12$$

Also,

$$\sqrt{9} \cdot \sqrt{16} = 3 \cdot 4 = 12$$

Since both expressions simplify to 12, we can write

$$\sqrt{9 \cdot 16} = \sqrt{9} \cdot \sqrt{16}$$

This suggests the following product rule for square roots.

PRODUCT RULE FOR SQUARE ROOTS
If $\sqrt{a}$ and $\sqrt{b}$ are real numbers, then
$$\sqrt{a \cdot b} = \sqrt{a} \cdot \sqrt{b}$$

The product rule states that the square root of a product is equal to the product of the square roots. We use this rule to write radical expressions in **simplest form**. A radical is written in simplest form when the radicand has no perfect square factors other than 1. To simplify $\sqrt{20}$, for example, factor 20 so that one of its factors is a perfect square factor.

$$\sqrt{20} = \sqrt{4 \cdot 5} \qquad \text{Factor 20.}$$
$$\phantom{\sqrt{20}} = \sqrt{4} \cdot \sqrt{5} \qquad \text{Apply the product rule.}$$
$$\phantom{\sqrt{20}} = 2\sqrt{5} \qquad \text{Write } \sqrt{4} \text{ as 2.}$$

The notation $2\sqrt{5}$ means $2 \cdot \sqrt{5}$. Since the radicand 5 has no perfect square factor other than 1, $2\sqrt{5}$ is in simplest form.

When factoring a radicand, look for at least one factor that is a perfect square. Review the table of perfect squares in Appendix E to help locate perfect square factors more quickly.

When simplifying a radical, realize that a radical expression in simplest form does *not mean* a decimal approximation. The simplest form of a radical expression is an exact form and may still contain a radical.

> REMINDER When simplifying a radical, use **factors** of the radicand; **do not** write the radicand as a sum. For example, **do not** write $\sqrt{20}$ as $\sqrt{4+16}$ because $\sqrt{4+16} \neq \sqrt{4} + \sqrt{16}$. Correctly simplified, $\sqrt{20} = \sqrt{4 \cdot 5} = \sqrt{4} \cdot \sqrt{5} = 2\sqrt{5}$.

EXAMPLE 1 Simplify each expression.

a. $\sqrt{54}$ **b.** $\sqrt{12}$ **c.** $\sqrt{200}$ **d.** $\sqrt{35}$

Solution:

a. Try to factor 54 so that at least one of the factors is a perfect square. Since 9 is a perfect square and $54 = 9 \cdot 6$,

$$\begin{aligned}\sqrt{54} &= \sqrt{9 \cdot 6} && \text{Factor 54.}\\ &= \sqrt{9} \cdot \sqrt{6} && \text{Apply the product rule.}\\ &= 3\sqrt{6} && \text{Write } \sqrt{9} \text{ as 3.}\end{aligned}$$

b.
$$\begin{aligned}\sqrt{12} &= \sqrt{4 \cdot 3} && \text{Factor 12.}\\ &= \sqrt{4} \cdot \sqrt{3} && \text{Apply the product rule.}\\ &= 2\sqrt{3} && \text{Write } \sqrt{4} \text{ as 2.}\end{aligned}$$

c. The largest perfect square factor of 200 is 100.

$$\begin{aligned}\sqrt{200} &= \sqrt{100 \cdot 2} && \text{Factor 200.}\\ &= \sqrt{100} \cdot \sqrt{2} && \text{Apply the product rule.}\\ &= 10\sqrt{2} && \text{Write } \sqrt{100} \text{ as 10.}\end{aligned}$$

d. The radicand 35 contains no perfect square factors other than 1. Thus $\sqrt{35}$ is in simplest form.

In Example 1, part **(c)**, what happens if we don't use the largest perfect square factor of 200? Although using the largest perfect square factor saves time, the result is the same no matter what perfect square factor is used. For example, it is also true that $200 = 4 \cdot 50$. Then

$$\begin{aligned}\sqrt{200} &= \sqrt{4} \cdot \sqrt{50}\\ &= 2 \cdot \sqrt{50}\end{aligned}$$

Since $\sqrt{50}$ is not in simplest form, we continue.

$$\begin{aligned}\sqrt{200} &= 2 \cdot \sqrt{50}\\ &= 2 \cdot \sqrt{25} \cdot \sqrt{2}\\ &= 2 \cdot 5 \cdot \sqrt{2}\\ &= 10\sqrt{2}\end{aligned}$$

Next, let's examine the square root of a quotient.

$$\sqrt{\frac{16}{4}} = \sqrt{4} = 2$$

Also,

$$\frac{\sqrt{16}}{\sqrt{4}} = \frac{4}{2} = 2$$

Since both expressions equal 2, we can write

$$\sqrt{\frac{16}{4}} = \frac{\sqrt{16}}{\sqrt{4}}$$

This suggests the following quotient rule.

> **QUOTIENT RULE FOR SQUARE ROOTS**
> If $\sqrt{a}$ and $\sqrt{b}$ are real numbers and $b \neq 0$, then
> $$\sqrt{\frac{a}{b}} = \frac{\sqrt{a}}{\sqrt{b}}$$

The quotient rule states that the square root of a quotient is equal to the quotient of the square roots.

EXAMPLE 2 Simplify the following.

a. $\sqrt{\frac{25}{36}}$ b. $\sqrt{\frac{3}{64}}$ c. $\sqrt{\frac{4}{81}}$

Solution: Use the quotient rule.

a. $\sqrt{\frac{25}{36}} = \frac{\sqrt{25}}{\sqrt{36}} = \frac{5}{6}$ b. $\sqrt{\frac{3}{64}} = \frac{\sqrt{3}}{\sqrt{64}} = \frac{\sqrt{3}}{8}$

c. $\sqrt{\frac{40}{81}} = \frac{\sqrt{40}}{\sqrt{81}}$ Use the quotient rule.

$= \frac{\sqrt{4} \cdot \sqrt{10}}{9}$ Apply the product rule and write $\sqrt{81}$ as 9.

$= \frac{2\sqrt{10}}{9}$ Write $\sqrt{4}$ as 2.

EXAMPLE 3 Simplify each expression. Assume that variables represent positive numbers only.

a. $\sqrt{x^5}$ b. $\sqrt{8y^2}$ c. $\sqrt{\dfrac{45}{x^6}}$

Solution:
a. $\sqrt{x^5} = \sqrt{x^4 \cdot x} = \sqrt{x^4} \cdot \sqrt{x} = x^2\sqrt{x}$
b. $\sqrt{8y^2} = \sqrt{4 \cdot 2 \cdot y^2} = \sqrt{4y^2 \cdot 2} = \sqrt{4y^2} \cdot \sqrt{2} = 2y\sqrt{2}$
c. $\sqrt{\dfrac{45}{x^6}} = \dfrac{\sqrt{45}}{\sqrt{x^6}} = \dfrac{\sqrt{9 \cdot 5}}{x^3} = \dfrac{\sqrt{9} \cdot \sqrt{5}}{x^3} = \dfrac{3\sqrt{5}}{x^3}$

The product and quotient rules also apply to roots other than square roots. In general, we have the following product and quotient rules for radicals:

PRODUCT RULE FOR RADICALS
If $\sqrt[n]{a}$ and $\sqrt[n]{b}$ are real numbers, then
$$\sqrt[n]{a \cdot b} = \sqrt[n]{a} \cdot \sqrt[n]{b}$$

QUOTIENT RULE FOR RADICALS
If $\sqrt[n]{a}$ and $\sqrt[n]{b}$ are real numbers and $b \neq 0$, then
$$\sqrt[n]{\dfrac{a}{b}} = \dfrac{\sqrt[n]{a}}{\sqrt[n]{b}}$$

For example, to simplify cube roots, look for perfect cube factors of the radicand. For example, 8 is a perfect cube, since $2^3 = 8$.

To simplify $\sqrt[3]{48}$, factor 48 as $8 \cdot 6$.

$\sqrt[3]{48} = \sqrt[3]{8 \cdot 6}$ Factor 48.
$= \sqrt[3]{8} \cdot \sqrt[3]{6}$ Apply the product rule.
$= 2\sqrt[3]{6}$ Write $\sqrt[3]{8}$ as 2.

$2\sqrt[3]{6}$ is in simplest form since the radicand 6 contains no perfect cube factors other than 1.

EXAMPLE 4 Simplify each expression.

a. $\sqrt[3]{54}$ b. $\sqrt[3]{18}$ c. $\sqrt[3]{\dfrac{7}{8}}$ d. $\sqrt[3]{\dfrac{40}{27}}$

Solution: a. $\sqrt[3]{54} = \sqrt[3]{27 \cdot 2} = \sqrt[3]{27} \cdot \sqrt[3]{2} = 3\sqrt[3]{2}$

b. The number 18 contains no perfect cube factors, so $\sqrt[3]{18}$ cannot be simplified further.

c. $\sqrt[3]{\dfrac{7}{8}} = \dfrac{\sqrt[3]{7}}{\sqrt[3]{8}} = \dfrac{\sqrt[3]{7}}{2}$

d. $\sqrt[3]{\dfrac{40}{27}} = \dfrac{\sqrt[3]{40}}{\sqrt[3]{27}} = \dfrac{\sqrt[3]{8 \cdot 5}}{3} = \dfrac{\sqrt[3]{8} \cdot \sqrt[3]{5}}{3} = \dfrac{2\sqrt[3]{5}}{3}$

EXAMPLE 5 Simplify each expression. Assume variables represent positive numbers only.

a. $\sqrt[3]{x^5}$ **b.** $\sqrt[3]{40y^7}$ **c.** $\sqrt[3]{\dfrac{16}{x^6}}$

Solution:
a. $\sqrt[3]{x^5} = \sqrt[3]{x^3 \cdot x^2} = \sqrt[3]{x^3} \cdot \sqrt[3]{x^2} = x\sqrt[3]{x^2}$

b. $\sqrt[3]{40y^7} = \sqrt[3]{8 \cdot 5 \cdot y^6 \cdot y} = \sqrt[3]{8y^6 \cdot 5y} = \sqrt[3]{8y^6} \cdot \sqrt[3]{5y} = 2y^2\sqrt[3]{5y}$

c. $\sqrt[3]{\dfrac{16}{x^6}} = \dfrac{\sqrt[3]{16}}{\sqrt[3]{x^6}} = \dfrac{\sqrt[3]{8 \cdot 2}}{x^2} = \dfrac{\sqrt[3]{8} \cdot \sqrt[3]{2}}{x^2} = \dfrac{2\sqrt[3]{2}}{x^2}$

MENTAL MATH

Simplify each expression. Assume that all variables represent nonnegative real numbers.

1. $\sqrt{4 \cdot 9}$
2. $\sqrt{9 \cdot 36}$
3. $\sqrt{x^2}$
4. $\sqrt{y^4}$
5. $\sqrt{0}$
6. $\sqrt{1}$
7. $\sqrt{25x^4}$
8. $\sqrt{49x^2}$

EXERCISE SET 9.2

Use the product rule to simplify each expression. See Examples 1 and 4.

1. $\sqrt{20}$
2. $\sqrt{44}$
3. $\sqrt{18}$
4. $\sqrt{45}$
5. $\sqrt{50}$
6. $\sqrt{28}$
7. $\sqrt{33}$
8. $\sqrt{98}$
9. $\sqrt[3]{24}$
10. $\sqrt[3]{81}$
11. $\sqrt[3]{250}$
12. $\sqrt[3]{40}$

13. By using replacement values for a and b, show that $\sqrt{a^2 + b^2}$ does not equal $a + b$.

14. By using replacement values for a and b, show that $\sqrt{a + b}$ does not equal $\sqrt{a} + \sqrt{b}$.

Use the quotient rule and the product rule. Simplify each expression. See Examples 2 and 4.

15. $\sqrt{\dfrac{8}{25}}$
16. $\sqrt{\dfrac{63}{16}}$
17. $\sqrt{\dfrac{27}{121}}$
18. $\sqrt{\dfrac{24}{169}}$
19. $\sqrt{\dfrac{9}{4}}$
20. $\sqrt{\dfrac{100}{49}}$
21. $\sqrt{\dfrac{125}{9}}$
22. $\sqrt{\dfrac{27}{100}}$

23. $\sqrt[3]{\dfrac{5}{64}}$
24. $\sqrt[3]{\dfrac{32}{125}}$
25. $\sqrt[3]{\dfrac{7}{8}}$
26. $\sqrt[3]{\dfrac{10}{27}}$

Simplify each expression. Assume that all variables represent positive numbers only. See Examples 3 and 5.

27. $\sqrt{x^7}$
28. $\sqrt{y^3}$
29. $\sqrt{\dfrac{88}{x^4}}$
30. $\sqrt{\dfrac{x^{11}}{81}}$
31. $\sqrt[3]{x^{16}}$
32. $\sqrt[3]{y^{20}}$
33. $\sqrt[3]{\dfrac{2}{x^9}}$
34. $\sqrt[3]{\dfrac{48}{x^{12}}}$

Simplify each expression. Assume that all variables represent positive numbers only.

35. $\sqrt{60}$
36. $\sqrt{90}$
37. $\sqrt{180}$
38. $\sqrt{150}$
39. $\sqrt{52}$
40. $\sqrt{75}$
41. $\sqrt{\dfrac{11}{36}}$
42. $\sqrt{\dfrac{30}{49}}$
43. $-\sqrt{\dfrac{27}{144}}$
44. $-\sqrt{\dfrac{84}{121}}$
45. $\sqrt[3]{\dfrac{15}{64}}$
46. $\sqrt[3]{\dfrac{4}{27}}$
47. $\sqrt[3]{80}$
48. $\sqrt[3]{108}$
49. $\sqrt[4]{48}$
50. $\sqrt[4]{162}$
51. $\sqrt{x^{13}}$
52. $\sqrt{y^{17}}$
53. $\sqrt{75x^2}$
54. $\sqrt{72y^2}$
55. $\sqrt{96x^4y^2}$
56. $\sqrt{40x^8y^{10}}$
57. $\sqrt{\dfrac{12}{y^2}}$
58. $\sqrt{\dfrac{63}{x^4}}$
59. $\sqrt{\dfrac{9x}{y^2}}$
60. $\sqrt{\dfrac{6y^2}{x^4}}$
61. $\sqrt[3]{-8x^6}$
62. $\sqrt[3]{-54y^6}$

63. If a cube is to have a volume of 80 cubic inches, then each side must be $\sqrt[3]{80}$ inches long. Simplify the radical representing the side length.

64. Jeannie is swimming across a 40-foot-wide river, trying to head straight across to the opposite shore. However, the current is strong enough to move her downstream 100 feet by the time she reaches land. (See the figure.) Because of the current, the actual distance she swam is $\sqrt{11{,}600}$ feet. Simplify this radical.

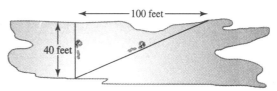

The cost C in dollars per day to operate a small delivery service is given by $C = 100\sqrt[3]{n} + 700$ where n is the number of deliveries per day.

65. Find the cost if the number of deliveries is 1000.
66. Approximate the cost if the number of deliveries is 500.

Review Exercises

Perform the following operations. See Sections 4.2 and 4.3.

67. $6x + 8x$
68. $(6x)(8x)$
69. $(2x + 3)(x - 5)$
70. $(2x + 3) + (x - 5)$
71. $9y^2 - 9y^2$
72. $(9y^2)(-8y^2)$

The following pairs of triangles are similar. Find the unknown lengths. See Section 6.7.

73.

 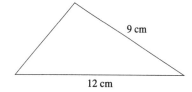

74.

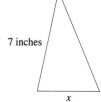

10.3 Solving Quadratic Equations by the Quadratic Formula

TAPE
BA 10.3

OBJECTIVE

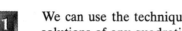 Use the quadratic formula to solve quadratic equations.

We can use the technique of completing the square to develop a formula to find solutions of any quadratic equation. We develop and use the **quadratic formula** in this section.

Recall that a quadratic equation in **standard form** is

$$ax^2 + bx + c = 0, \text{ providing } a \neq 0$$

To develop and use the quadratic formula, we need to practice identifying the values of a, b, and c in a quadratic equation.

Quadratic Equations in Standard Form	
$5x^2 - 6x + 2 = 0$	$a = 5, b = -6, c = 2$
$4y^2 - 9 = 0$	$a = 4, b = 0, c = -9$
$x^2 + x = 0$	$a = 1, b = 1, c = 0$
$\sqrt{2}x^2 + \sqrt{5}x + \sqrt{3} = 0$	$a = \sqrt{2}, b = \sqrt{5}, c = \sqrt{3}$

To derive the quadratic formula, we complete the square for the general quadratic equation in standard form.

$$ax^2 + bx + c = 0$$

First, divide both sides of the equation by the coefficient of x^2 and then isolate the variable terms.

$$x^2 + \frac{b}{a}x + \frac{c}{a} = 0 \qquad \text{Divide by } a; \text{ recall that } a \text{ cannot be 0.}$$

$$x^2 + \frac{b}{a}x = -\frac{c}{a} \qquad \text{Isolate the variable terms.}$$

The coefficient of x is $\frac{b}{a}$. Half of $\frac{b}{a}$ is $\frac{b}{2a}$ and $\left(\frac{b}{2a}\right)^2 = \frac{b^2}{4a^2}$. Add $\frac{b^2}{4a^2}$ to both sides of the equation.

$$x^2 + \frac{b}{a}x + \frac{b^2}{4a^2} = -\frac{c}{a} + \frac{b^2}{4a^2} \qquad \text{Add } \frac{b^2}{4a^2} \text{ to both sides.}$$

$$\left(x + \frac{b}{2a}\right)^2 = -\frac{c}{a} + \frac{b^2}{4a^2} \qquad \text{Factor the left side.}$$

$$\left(x + \frac{b}{2a}\right)^2 = -\frac{4ac}{4a^2} + \frac{b^2}{4a^2} \qquad \text{Multiply } -\frac{c}{a} \text{ by } \frac{4a}{4a} \text{ so that both terms on the right side have a common denominator.}$$

$$\left(x + \frac{b}{2a}\right)^2 = \frac{b^2 - 4ac}{4a^2}$$ Simplify the right side.

Now use the square root property.

$$x + \frac{b}{2a} = \pm\sqrt{\frac{b^2 - 4ac}{4a^2}}$$ Apply the square root property.

$$x + \frac{b}{2a} = \frac{\pm\sqrt{b^2 - 4ac}}{2a}$$ Simplify the radical.

$$x = -\frac{b}{2a} \pm \frac{\sqrt{b^2 - 4ac}}{2a}$$ Subtract $\frac{b}{2a}$ from both sides.

$$= \frac{-b \pm \sqrt{b^2 - 4ac}}{2a}$$ Simplify.

This final equation is called the **quadratic formula** and gives the solutions of any quadratic equation.

> **QUADRATIC FORMULA**
>
> If a, b, and c are real numbers and $a \neq 0$, a quadratic equation written in the form $ax^2 + bx + c = 0$ has solutions
>
> $$x = \frac{-b \pm \sqrt{b^2 - 4ac}}{2a}$$

EXAMPLE 1 Use the quadratic formula to solve $3x^2 + x - 3 = 0$.

Solution: This equation is in standard form with $a = 3$, $b = 1$, and $c = -3$. By the quadratic formula,

$$x = \frac{-b \pm \sqrt{b^2 - 4ac}}{2a}$$

$$x = \frac{-1 \pm \sqrt{1^2 - 4 \cdot 3 \cdot (-3)}}{2 \cdot 3}$$ Let $a = 3$, $b = 1$, and $c = -3$.

$$= \frac{-1 \pm \sqrt{1 + 36}}{6}$$ Simplify.

$$= \frac{-1 \pm \sqrt{37}}{6}$$

The solution set is $\left\{\frac{-1 + \sqrt{37}}{6}, \frac{-1 - \sqrt{37}}{6}\right\}$.

EXAMPLE 2 Use the quadratic formula to solve $2x^2 - 9x = 5$.

Solution: First, write the equation in standard form by subtracting 5 from both sides.

$$2x^2 - 9x = 5$$
$$2x^2 - 9x - 5 = 0$$

Next, $a = 2$, $b = -9$, and $c = -5$. Substitute these values into the quadratic formula.

$$x = \frac{-b \pm \sqrt{b^2 - 4ac}}{2a}$$

$$x = \frac{-(-9) \pm \sqrt{(-9)^2 - 4 \cdot 2 \cdot (-5)}}{2 \cdot 2} \quad \text{Substitute in the formula.}$$

$$= \frac{9 \pm \sqrt{81 + 40}}{4} \quad \text{Simplify.}$$

$$= \frac{9 \pm \sqrt{121}}{4} = \frac{9 \pm 11}{4}$$

Then, $x = \dfrac{9 - 11}{4} = -\dfrac{1}{2}$ or $x = \dfrac{9 + 11}{4} = 5$

Check by substituting $-\frac{1}{2}$ and 5 into the original equation. The solution set is $\left\{-\frac{1}{2}, 5\right\}$.

In the example above, the radicand of $\sqrt{121}$ is a perfect square, so the square root is rational.

TO SOLVE A QUADRATIC EQUATION BY THE QUADRATIC FORMULA

Step 1. Write the quadratic equation in standard form: $ax^2 + bx + c = 0$.

Step 2. If necessary, clear the equation of fractions to simplify calculations.

Step 3 Identify a, b, and c.

Step 4. Replace a, b, and c in the quadratic formula by known values, and simplify.

EXAMPLE 3 Use the quadratic formula to solve $7x^2 = 1$.

Solution: Write the equation in standard form by subtracting 1 from both sides.

$$7x^2 = 1$$
$$7x^2 - 1 = 0$$

Next, replace a, b, and c with values: $a = 7, b = 0, c = -1$.

$$x = \frac{0 \pm \sqrt{0^2 - 4 \cdot 7 \cdot (-1)}}{2 \cdot 7} \quad \text{Substitute in the formula.}$$

$$= \frac{\pm\sqrt{28}}{14} \quad \text{Simplify.}$$

$$= \frac{\pm 2\sqrt{7}}{14}$$

$$= \pm\frac{\sqrt{7}}{7}$$

The solution set is $\left\{-\frac{\sqrt{7}}{7}, \frac{\sqrt{7}}{7}\right\}$.

EXAMPLE 4 Use the quadratic formula to solve $x^2 = -x - 1$.

Solution: First, write the equation in standard form.

$$x^2 + x + 1 = 0$$

Next, replace a, b, and c in the quadratic formula by $a = 1, b = 1,$ and $c = 1$.

$$x = \frac{-1 \pm \sqrt{1^2 - 4 \cdot 1 \cdot 1}}{2 \cdot 1} \quad \text{Substitute in the formula.}$$

$$= \frac{-1 \pm \sqrt{-3}}{2} \quad \text{Simplify.}$$

There is no real number solution, because the radicand is negative.

EXAMPLE 5 Use the quadratic formula to solve $\frac{1}{2}x^2 - x = 2$.

Solution: Write the equation in standard form and then clear the equation of fractions by multiplying both sides by the LCD 2.

$$\frac{1}{2}x^2 - x = 2$$

$$\frac{1}{2}x^2 - x - 2 = 0 \quad \text{Write in standard form.}$$

$$x^2 - 2x - 4 = 0 \quad \text{Multiply both sides by 2.}$$

Here, $a = 1$, $b = -2$, and $c = -4$. Substitute these values into the quadratic formula.

$$x = \frac{-(-2) \pm \sqrt{(-2)^2 - 4 \cdot 1 \cdot (-4)}}{2 \cdot 1}$$

$$= \frac{2 \pm \sqrt{20}}{2} = \frac{2 \pm 2\sqrt{5}}{2} \quad \text{Simplify.}$$

$$= \frac{2(1 \pm \sqrt{5})}{2} = 1 \pm \sqrt{5} \quad \text{Factor and simplify.}$$

The solution set is $\{1 - \sqrt{5}, 1 + \sqrt{5}\}$.

> REMINDER When simplifying expressions such as
>
> $$\frac{3 \pm 6\sqrt{2}}{6}$$
>
> first factor out a common factor from the terms of the numerator and then simplify.
>
> $$\frac{3 \pm 6\sqrt{2}}{6} = \frac{3(1 \pm 2\sqrt{2})}{2 \cdot 3} = \frac{1 \pm 2\sqrt{2}}{2}$$

MENTAL MATH

Identify the value of a, b, and c in each quadratic equation.

1. $2x^2 + 5x + 3 = 0$
2. $5x^2 - 7x + 1 = 0$
3. $10x^2 - 13x - 2 = 0$
4. $x^2 + 3x - 7 = 0$
5. $x^2 - 6 = 0$
6. $9x^2 - 4 = 0$

EXERCISE SET 10.3

Simplify the following.

1. $\dfrac{-1 \pm \sqrt{1^2 - 4(1)(-2)}}{2(1)}$
2. $\dfrac{-(-5) \pm \sqrt{(-5)^2 - 4(2)(3)}}{2(2)}$
3. $\dfrac{-5 \pm \sqrt{5^2 - 4(1)(2)}}{2(1)}$
4. $\dfrac{-7 \pm \sqrt{7^2 - 4(2)(1)}}{2(2)}$
5. $\dfrac{-(-4) \pm \sqrt{(-4)^2 - 4(2)(1)}}{2(2)}$
6. $\dfrac{-6 \pm \sqrt{6^2 - 4(3)(1)}}{2(3)}$

Use the quadratic formula to solve each quadratic equation. See Examples 1 through 4.

7. $x^2 - 3x + 2 = 0$
8. $x^2 - 5x - 6 = 0$
9. $3k^2 + 7k + 1 = 0$
10. $7k^2 + 3k - 1 = 0$

11. $49x^2 - 4 = 0$
12. $25x^2 - 15 = 0$
13. $5z^2 - 4z + 3 = 0$
14. $3z^2 + 2x + 1 = 0$
15. $y^2 = 7y + 30$
16. $y^2 = 5y + 36$
17. $2x^2 = 10$
18. $5x^2 = 15$
19. $m^2 - 12 = m$
20. $m^2 - 14 = 5m$
21. $3 - x^2 = 4x$
22. $10 - x^2 = 2x$

Use the quadratic formula to solve each quadratic equation. See Example 5.

23. $3p^2 - \frac{2}{3}p + 1 = 0$
24. $\frac{5}{2}p^2 - p + \frac{1}{2} = 0$
25. $\frac{m^2}{2} = m + \frac{1}{2}$
26. $\frac{m^2}{2} = 3m - 1$
27. $4p^2 + \frac{3}{2} = -5p$
28. $4p^2 + \frac{3}{2} = 5p$

Use the quadratic formula to solve each quadratic equation.

29. $2a^2 - 7a + 3 = 0$
30. $3a^2 - 7a + 2 = 0$
31. $x^2 - 5x - 2 = 0$
32. $x^2 - 2x - 5 = 0$
33. $3x^2 - x - 14 = 0$
34. $5x^2 - 13x - 6 = 0$
35. $6x^2 + 9x = 2$
36. $3x^2 - 9x = 8$
37. $7p^2 + 2 = 8p$
38. $11p^2 + 2 = 10p$
39. $a^2 - 6a + 2 = 0$
40. $a^2 - 10a + 19 = 0$
41. $2x^2 - 6x + 3 = 0$
42. $5x^2 - 8x + 2 = 0$
43. $3x^2 = 1 - 2x$
44. $5y^2 = 4 - x$
45. $20y^2 = 3 - 11y$
46. $2z^2 = z + 3$
47. $x^2 + x + 1 = 0$
48. $k^2 + 2k + 5 = 0$
49. $4y^2 = 6y + 1$
50. $6z^2 + 3z + 2 = 0$
51. $5x^2 = \frac{7}{2}x + 1$
52. $2x^2 = \frac{5}{2}x + \frac{7}{2}$
53. $28x^2 + 5x + \frac{11}{4} = 0$
54. $\frac{2}{3}x^2 - 2x - \frac{2}{3} = 0$
55. $5z^2 - 2z = \frac{1}{5}$
56. $9z^2 + 12z = -1$
57. $x^2 + 3\sqrt{2}x - 5 = 0$
58. $y^2 - 2\sqrt{5}x - 1 = 0$

Use the quadratic formula and a calculator to solve each equation. Round solutions to the nearest tenth.

59. $x^2 + x = 15$
60. $y^2 - y = 11$

61. $1.2x^2 - 5.2x - 3.9 = 0$
62. $7.3z^2 + 5.4z - 1.1 = 0$

A rocket is launched from the top of an 80-foot cliff with an initial velocity of 120 feet per second. The height of the rocket h after t seconds is given by the equation

$$h = -16t^2 + 120t + 80$$

63. How long after the rocket is launched will it be 30 feet from the ground? Round to the nearest tenth of a second.

64. How long after the rocket is launched will it strike the ground? Round to the nearest tenth of a second. (*Hint:* The rocket will strike the ground when its height $h = 0$.)

65. Explain how the quadratic formula is derived and why it is useful.

Review Exercises

Solve the following linear equations. See Section 2.4.

66. $\frac{7x}{2} = 3$
67. $\frac{5x}{3} = 1$
68. $\frac{5}{7}x - \frac{2}{3} = 0$
69. $\frac{6}{11}x + \frac{1}{5} = 0$
70. $\frac{3}{4}z + 3 = 0$
71. $\frac{5}{2}z + 10 = 0$

10.4 Summary of Methods for Solving Quadratic Equations

OBJECTIVE

 Review methods for solving quadratic equations.

 An important skill in mathematics is learning when to use one technique in favor of another. We now practice this by deciding which method to use when solving quadratic equations. Although both the quadratic formula and completing the square can be used to solve any quadratic equation, the quadratic formula is usually less tedious and thus preferred. The following steps may be used to solve a quadratic equation.

TO SOLVE A QUADRATIC EQUATION

Step 1. If the equation is in the form $(ax + b)^2 = c$, use the square root property and solve. If not, go to *step 2*.

Step 2. Write the equation in standard form: $ax^2 + bx + c = 0$.

Step 3. Try to solve the equation by the factoring method. If not possible, go to *step 4*.

Step 4. Solve the equation by the quadratic formula.

EXAMPLE 1 Solve $m^2 - 2m - 7 = 0$.

Solution: The equation is in standard form, but the quadratic expression $m^2 - 2m - 7$ is not factorable, so use the quadratic formula with $a = 1$, $b = -2$, and $c = -7$.

$$m^2 - 2m - 7 = 0$$

$$m = \frac{-(-2) \pm \sqrt{(-2)^2 - 4 \cdot 1 \cdot (-7)}}{2 \cdot 1} = \frac{2 \pm \sqrt{32}}{2}$$

$$m = \frac{2 \pm 4\sqrt{2}}{2} = \frac{2(1 \pm 2\sqrt{2})}{2} = 1 \pm 2\sqrt{2}$$

The solution set is $\{1 - 2\sqrt{2}, 1 + 2\sqrt{2}\}$.

EXAMPLE 2 Solve $(3x + 1)^2 = 20$.

Solution: This equation is in a form that makes the square root property easy to apply.

$$(3x + 1)^2 = 20$$
$$3x + 1 = \pm\sqrt{20} \quad \text{Apply the square root property.}$$
$$3x + 1 = \pm 2\sqrt{5} \quad \text{Simplify } \sqrt{20}.$$
$$3x = -1 \pm 2\sqrt{5}$$
$$x = \frac{-1 \pm 2\sqrt{5}}{3}$$

The solution set is $\left\{\dfrac{-1 - 2\sqrt{5}}{3}, \dfrac{-1 + 2\sqrt{5}}{3}\right\}$.

EXAMPLE 3 Solve $x^2 - \dfrac{11}{2}x = -\dfrac{5}{2}$.

Solution: The fractions make factoring more difficult and also complicate the calculations for using the quadratic formula. Clear the equation of fractions by multiplying both sides of the equation by the LCD 2.

$$x^2 - \frac{11}{2}x = -\frac{5}{2}$$
$$x^2 - \frac{11}{2}x + \frac{5}{2} = 0 \quad \text{Write in standard form.}$$
$$2x^2 - 11x + 5 = 0 \quad \text{Multiply both sides by 2.}$$
$$(2x - 1)(x - 5) = 0 \quad \text{Factor.}$$
$$2x - 1 = 0 \quad \text{or} \quad x - 5 = 0 \quad \text{Apply the zero factor theorem.}$$
$$2x = 1 \quad \text{or} \quad x = 5$$
$$x = \frac{1}{2} \quad \text{or} \quad x = 5$$

The solution set is $\left\{\dfrac{1}{2}, 5\right\}$.

EXERCISE SET 10.4

Choose and use a method to solve each equation.

1. $5x^2 - 11x + 2 = 0$
2. $5x^2 + 13x - 6 = 0$
3. $x^2 - 1 = 2x$
4. $x^2 + 7 = 6x$
5. $a^2 = 20$
6. $a^2 = 72$
7. $x^2 - x + 4 = 0$
8. $x^2 - 2x + 7 = 0$
9. $3x^2 - 12x + 12 = 0$
10. $5x^2 - 30x + 45 = 0$
11. $9 - 6p + p^2 = 0$
12. $49 - 28p + 4p^2 = 0$
13. $4y^2 - 16 = 0$
14. $3y^2 - 27 = 0$
15. $x^4 - 3x^3 + 2x^2 = 0$
16. $x^3 + 7x^2 + 12x = 0$
17. $(2z + 5)^2 = 25$
18. $(3z - 4)^2 = 16$
19. $30x = 25x^2 + 2$
20. $12x = 4x^2 + 4$

21. $\frac{2}{3}m^2 - \frac{1}{3}m - 1 = 0$ 22. $\frac{5}{8}m^2 + m - \frac{1}{2} = 0$

23. $x^2 - \frac{1}{2}x - \frac{1}{5} = 0$ 24. $x^2 + \frac{1}{2}x - \frac{1}{8} = 0$

25. $4x^2 - 27x + 35 = 0$ 26. $9x^2 - 16x + 7 = 0$

27. $(7 - 5x)^2 = 18$ 28. $(5 - 4x)^2 = 75$

29. $3z^2 - 7z = 12$ 30. $6z^2 + 7z = 6$

31. $x = x^2 - 110$ 32. $x = 56 - x^2$

33. $\frac{3}{4}x^2 - \frac{5}{2}x - 2 = 0$ 34. $x^2 - \frac{6}{5}x - \frac{8}{5} = 0$

35. $x^2 - 0.6x + 0.05 = 0$ 36. $x^2 - 0.1x - 0.06 = 0$

37. $10x^2 - 11x + 2 = 0$ 38. $20x^2 - 11x + 1 = 0$

39. $\frac{1}{2}z^2 - 2z + \frac{3}{4} = 0$ 40. $\frac{1}{5}z^2 - \frac{1}{2}z - 2 = 0$

41. If a line segment AB is divided by a point C into two segments, AC and CB, such that the proportion $\frac{AB}{AC} = \frac{AC}{CB}$ is true, this ratio $\frac{AB}{AC}$ (or $\frac{AC}{CB}$) is called the golden ratio. If AC is 1 unit, find the length of AB. (*Hint:* Let x be the unknown length as shown and substitute into the given proportion.)

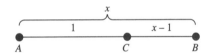

42. The formula $A = P(1 + r)^2$ is used to find the amount of money A in an account after P dollars have been invested in the account paying r annual interest rate for 2 years. Find the interest rate r if $1000 grows to $1690 in 2 years.

43. Explain how you will decide what method to use when solving quadratic equations.

Review Exercises

Simplify each expression. See Section 9.2.

44. $\sqrt{48}$ 45. $\sqrt{104}$

46. $\sqrt{50}$ 47. $\sqrt{80}$

Solve the following. See Section 2.6.

48. The height of a triangle is 4 times the length of the base. The area of the triangle is 18 square feet. Find the height and base of the triangle.

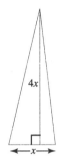

49. The length of a rectangle is 6 inches more than its width. The area of the rectangle is 391 square inches. Find the dimensions of the rectangle.

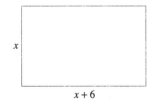

10.6 GRAPHING QUADRATIC EQUATIONS

OBJECTIVES

 Identify the graph of a quadratic equation as a parabola.
 Graph quadratic equations of the form $y = ax^2 + bx + c$.
3 Find the intercept points of a parabola.
4 Determine the vertex of a parabola.

 Recall from Section 3.2 that the graph of a linear equation in two variables $Ax + By = C$ is a straight line. Also recall from Section 7.3 that the graph of a quadratic equation in two variables $y = ax^2 + bx + c$ is a parabola. In this section, we further investigate the graph of a quadratic equation.

To graph the quadratic equation $y = x^2$, select a few values for x and find the corresponding y-values. Make a table of values to keep track. Then plot the points corresponding to these solutions.

If $x = 0$, then $y = 0^2 = 0$.
If $x = -2$, then $y = (-2)^2 = 4$. And so on.

$y = x^2$

x	y
0	0
1	1
2	4
3	9
−1	1
−2	4
−3	9

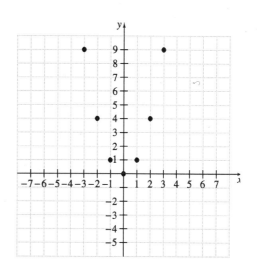

Clearly, these points are not on one straight line. As we saw in Chapter 7, the graph of $y = x^2$ is a smooth curve through the plotted points. This curve is called a **parabola.** The lowest point on a parabola opening upward is called the **vertex.** The vertex is $(0, 0)$ for the parabola $y = x^2$. If we fold the graph paper along the y-axis, the two pieces of the parabola match perfectly. For this reason, we say the graph is **symmetric about the y-axis,** and we call the y-axis the **axis of symmetry.**

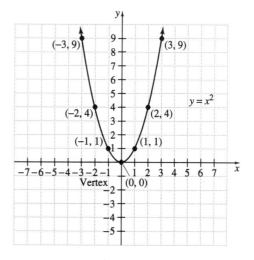

2 Notice that the parabola that corresponds to the equation $y = x^2$ opens upward. This happens when the coefficient of x^2 is positive. In the equation $y = x^2$, the coefficient of x^2 is 1. Example 1 shows the graph of a quadratic equation whose coefficient of x^2 is negative.

574 CHAPTER 10 SOLVING QUADRATIC EQUATIONS

EXAMPLE 1 Graph $y = -2x^2$.

Solution: Select x-values and calculate the corresponding y-values. Plot the ordered pairs found. Then draw a smooth curve through those points. When the coefficient of x^2 is negative, the corresponding parabola opens downward. When a parabola opens downward, the vertex is the highest point of the parabola. The vertex of this parabola is $(0, 0)$ and the axis of symmetry is again the y-axis.

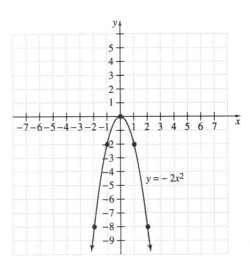

Just as for linear equations, we can use x- and y-intercepts to help graph quadratic equations. Recall from Chapter 3 that an x-intercept is the x-coordinate of the point where the graph intersects the x-axis. A y-intercept is the y-coordinate of the point where the graph intersects the y-axis.

> **REMINDER** Recall that:
> To find x-intercepts, let $y = 0$ and solve for x.
> To find y-intercepts, let $x = 0$ and solve for y.

EXAMPLE 2 Graph $y = x^2 - 4$.

Solution: First, find intercepts. To find the y-intercept, let $x = 0$. Then
$$y = 0^2 - 4 = -4$$

To find x-intercepts, let $y = 0$.

$$0 = x^2 - 4$$
$$0 = (x - 2)(x + 2)$$
$$x - 2 = 0 \quad \text{or} \quad x + 2 = 0$$
$$x = 2 \quad \text{or} \quad x = -2$$

Thus far, we have the y-intercept point $(0, -4)$ and the x-intercept points $(2, 0)$ and $(-2, 0)$. Next, select additional x-values, find y-values, plot the points, and draw a smooth curve through the points. The vertex is $(0, -4)$, and the axis of symmetry is the y-axis. This graph has the same shape as the graph of $y = x^2$. It is different from the graph of $y = x^2$ in that the vertex is 4 units lower on the y-axis.

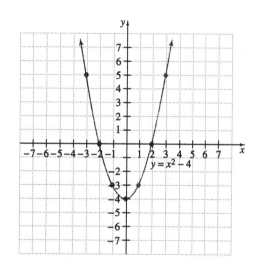

$y = x^2 - 4$

x	y
0	-4
1	-3
2	0
3	5
-1	-3
-2	0
-3	5

Notice that the graph above passes the vertical line test and is the graph of a function. Since the graph of $y = ax^2 + bx + c, a \neq 0$, is always a parabola opening upward or downward, by the vertical line test, it is always the graph of a function.

EXAMPLE 3 Graph $y = (x + 2)^2$.

Solution: Find the intercepts. To find x-intercepts, let $y = 0$.

$$0 = (x + 2)^2, \quad \text{so } x = -2$$

The x-intercept point is $(-2, 0)$. To find any y-intercepts, let $x = 0$.

$$y = (0 + 2)^2 = 4$$

The y-intercept point is $(0, 4)$. Plot the points $(-2, 0)$ and $(0, 4)$ and then select other values for x to obtain more ordered pairs.

576 CHAPTER 10 SOLVING QUADRATIC EQUATIONS

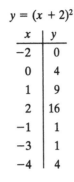

$y = (x + 2)^2$

x	y
-2	0
0	4
1	9
2	16
-1	1
-3	1
-4	4

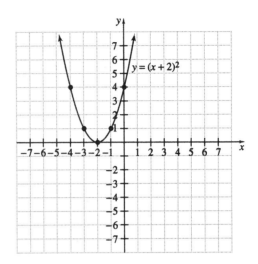

Notice that the graph of $y = (x + 2)^2$ is the same as the graph of $y = x^2$, except the vertex is shifted 2 units to the left, to $(-2, 0)$. The axis of symmetry is not the y-axis, but the vertical line through the vertex $(-2, 0)$. The axis of symmetry is the line $x = -2$.

4 So far we have located the vertex by graphing the equation. However, by writing the equation in a particular form, we can determine the vertex, the axis of symmetry, and whether the parabola opens upward or downward. This can be stated as follows.

> **THE GRAPH OF A QUADRATIC EQUATION IN THE FORM $y = a(x - h)^2 + k$**
>
> The graph of the quadratic equation $y = a(x - h)^2 + k$ is a parabola whose:
>
> - Vertex is (h, k).
> - Axis of symmetry is the line $x = h$.
> - Direction of opening is upward if $a > 0$ and downward if $a < 0$.

EXAMPLE 4 Graph $y = 3(x - 5)^2 + 2$.

Solution: The equation is written in the form $y = a(x - h)^2 + k$, with $a = 3, h = 5$, and $k = 2$. The vertex is then $(5, 2)$, the axis of symmetry is the line $x = 5$, and the graph opens upward since $a = 3$. By letting $x = 0$, we find that the y-intercept is 77, but our grid does not show coordinates as large as 77. Instead, take a point or two on each side of the vertex to determine the shape of the parabola. The x-coordinate of the vertex is 5, so we select $x = 4$ and $x = 6$, for example, to find two other points, as shown in the table.

$y = 3(x - 5)^2 + 2$

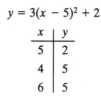

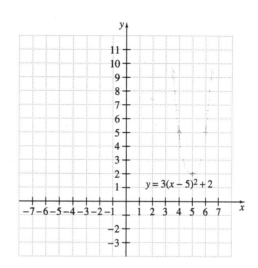

As this parabola demonstrates, some parabolas do not have an x-intercept. The vertex of this parabola is above the x-axis and the graph opens upward, so clearly the graph cannot intersect the x-axis.

If we try to find the x-intercept by replacing y with 0 in the equation, we solve $3(x - 5)^2 + 2 = 0$ or $3(x - 5)^2 = -2$ for x. But this equation has no real number solution. Since the x- and y-axes are real number lines, there is no x-intercept.

The figures below illustrate the possible x-intercepts of a parabola and solutions of the related quadratic equation.

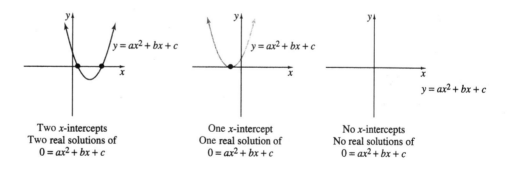

Two x-intercepts
Two real solutions of
$0 = ax^2 + bx + c$

One x-intercept
One real solution of
$0 = ax^2 + bx + c$

No x-intercepts
No real solutions of
$0 = ax^2 + bx + c$

EXAMPLE 5 Graph $y = -(x - 2)^2 + 1$.

Solution: The vertex is $(2, 1)$, the axis of symmetry is the line $x = 2$, and the parabola opens downward since $a = -1$. Again, selecting an x-value on each side of the vertex helps to determine the shape of the parabola. Finding the intercepts often yields a more accurate sketch.

To find x-intercepts, let $y = 0$.

$$0 = -(x-2)^2 + 1$$
$$(x-2)^2 = 1$$
$$x - 2 = \pm 1$$
$$x = 2 \pm 1$$
$$x = 3 \quad \text{or} \quad x = 1$$

There are two x-intercept points: $(3, 0)$ and $(1, 0)$. Plot $(3, 0)$ and $(1, 0)$.
To find y-intercepts, let $x = 0$.

$$y = -(0-2)^2 + 1 = -3$$

Plot the y-intercept point, $(0, -3)$

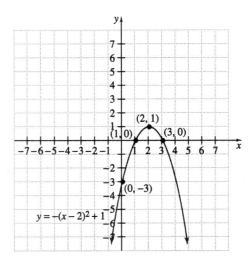

If a quadratic equation is not in the form $y = a(x - h)^2 + k$, then we cannot so easily identify the vertex. Notice that the x-coordinate of the parabola above is halfway between its x-intercepts. We can use this fact to find a formula for the vertex.

Recall that the x-intercepts of a parabola may be found by solving $0 = ax^2 + bx + c$. These solutions, by the quadratic formula, are

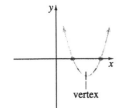

$$x = \frac{-b - \sqrt{b^2 - 4ac}}{2a}, \quad x = \frac{-b + \sqrt{b^2 - 4ac}}{2a}$$

The x-coordinate of the vertex of a parabola is halfway between its x-intercepts, so the x-value of the vertex may be found by computing the average, or $\frac{1}{2}$ of the sum of the intercepts.

$$x = \frac{1}{2}\left(\frac{-b - \sqrt{b^2 - 4ac}}{2a} + \frac{-b + \sqrt{b^2 - 4ac}}{2a}\right)$$

$$= \frac{1}{2}\left(\frac{-b - \sqrt{b^2 - 4ac} - b + \sqrt{b^2 - 4ac}}{2a}\right)$$

$$= \frac{1}{2}\left(\frac{-2b}{2a}\right)$$

$$= \frac{-b}{2a}$$

This formula may be used to find the x-value of the vertex of a parabola described by the equation $y = ax^2 + bx + c$. The corresponding y-value is found by substituting $\frac{-b}{2a}$ for x in the equation.

> **VERTEX FORMULA**
>
> The vertex of the parabola $y = ax^2 + bx + c$ has x-coordinate
>
> $$\frac{-b}{2a}$$

EXAMPLE 6 Graph $y = x^2 - 6x + 8$.

Solution: In the equation $y = x^2 - 6x + 8$, $a = 1$, $b = -6$, and $c = 8$.

The x-coordinate of the vertex is

$$\frac{-b}{2a} = \frac{-(-6)}{2 \cdot 1} = 3.$$

To find the corresponding y-coordinate, let $x = 3$.

$$y = 3^2 - 6 \cdot 3 + 8 = -1.$$

The vertex is $(3, -1)$ and the axis of symmetry is the line $x = 3$. The parabola opens upward since $a = 1$. We also plot intercepts.

To find x-intercepts, let $y = 0$.

$$0 = x^2 - 6x + 8$$

Factor the expression $x^2 - 6x + 8$ to find $(x - 4)(x - 2) = 0$. The x-intercepts are 4 and 2.

If we let $x = 0$ in the original equation, then $y = 8$, the y-intercept. Plot the vertex $(3, -1)$ and the intercept points $(4, 0)$, $(2, 0)$, and $(0, 8)$. Then sketch the parabola.

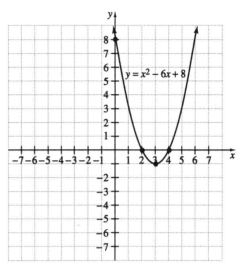

EXAMPLE 7   Graph $y = x^2 + 2x + 5$.

Solution: In the equation $y = x^2 + 2x + 5$, $a = 1$, $b = 2$, and $c = 5$.
The x-value of the vertex is

$$x = \frac{-b}{2a} = \frac{-2}{2 \cdot 1} = -1$$

The y-value is

$$y = (-1)^2 + 2(-1) + 5 = 4$$

Vertex: $(-1, 4)$
Axis of symmetry: $x = -1$
Opens: Upward

To find x-intercepts, let $y = 0$.

$$0 = (x + 1)^2 + 4$$
$$-4 = (x + 1)^2$$

The equation has no real number solution. Therefore, the graph has no x-intercepts.

To find y-intercepts, let $x = 0$ in the original equation and find that $y = 5$, and the resulting y-intercept point is $(0, 5)$.

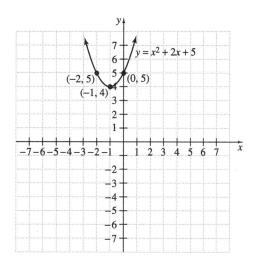

Graphing Calculator Explorations

Recall that a graphing calculator may be used to solve quadratic equations. The x-intercepts of the graph of $y = ax^2 + bx + c$ are solutions of $0 = ax^2 + bx + c$. To solve $x^2 - 7x - 3 = 0$, for example, graph $y_1 = x^2 - 7x - 3$. The x-intercepts of the graph are the solutions of the equation.

Use a grapher to solve each quadratic equation. Round solutions to two decimal places.

1. $x^2 - 7x - 3 = 0$
2. $2x^2 - 11x - 1 = 0$
3. $-1.7x^2 + 5.6x - 3.7 = 0$
4. $-5.8x^2 + 2.3x - 3.9 = 0$
5. $5.8x^2 - 2.6x - 1.9 = 0$
6. $7.5x^2 - 3.7x - 1.1 = 0$

Exercise Set 10.6

Graph each quadratic equation. Find the vertex and intercepts. See Examples 1 through 5.

1. $y = 2x^2$
2. $y = -2x^2$
3. $y = (x - 1)^2$
4. $y = (x + 2)^2$
5. $y = -x^2 + 4$
6. $y = x^2 - 4$
7. $y = \frac{1}{3}x^2$
8. $y = -\frac{1}{2}x^2$
9. $y = (x - 2)^2 + 1$
10. $y = -(x - 2)^2 - 1$
11. $y = -(x + 1)^2 + 4$
12. $y = (x - 1)^2 - 4$
13. $y = -4x^2 + 1$
14. $y = 4x^2 - 1$

Write the letter of the graph corresponding to each equation. See Examples 1 through 5.

◻ 15. $y = -x^2$
◻ 16. $y = x^2$
◻ 17. $y = (x-2)^2$
◻ 18. $y = -(x-1)^2$
◻ 19. $y = (x+3)^2 - 1$
◻ 20. $y = -(x-2)^2 + 3$
◻ 21. $y = 2(x+3)^2$
◻ 22. $y = -3(x+1)^2$
◻ 23. $y = -\frac{1}{2}x^2 + 1$

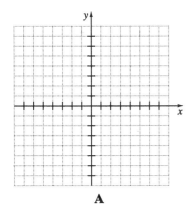
A

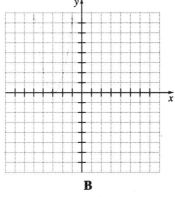
B

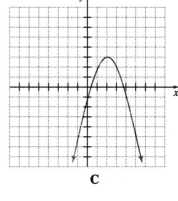
C

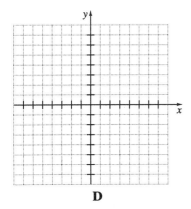
D

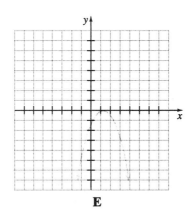
E

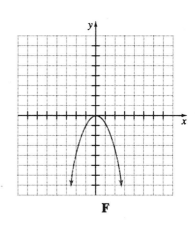
F

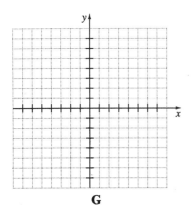
G

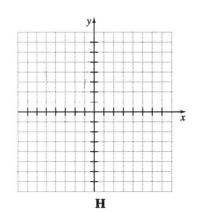
H

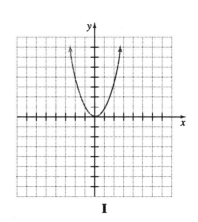
I

Sketch the graph of each equation. Identify the vertex and the intercepts. See Examples 6 and 7.

24. $y = x^2 + 6x$
25. $y = x^2 - 4x$
26. $y = x^2 + 2x - 8$
27. $y = x^2 - 2x - 3$
28. $y = x^2 - x - 2$
29. $y = x^2 + 2x + 1$
30. $y = x^2 + 5x + 4$
31. $y = x^2 + 7x + 10$
32. $y = x^2 - 4x + 3$
33. $y = x^2 - 6x + 8$

The graph of a quadratic equation that takes the form $y = ax^2 + bx + c$ is the graph of a function. Write the domain and the range of each of the functions graphed.

34.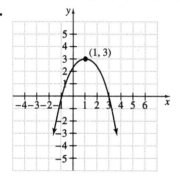

35.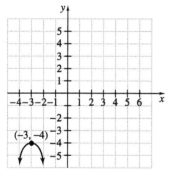

36.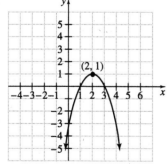

37.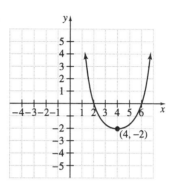

38. The height h of a fireball launched from a Roman candle with an initial velocity of 128 feet per second is given by the equation

$$h = -16t^2 + 128t$$

where t is time in seconds after launch.

Use the graph of this function to answer the questions.

a. Estimate the maximum height of the fireball.

b. Estimate the time when the fireball is at its maximum height.

c. Estimate the time when the fireball returns to the ground.

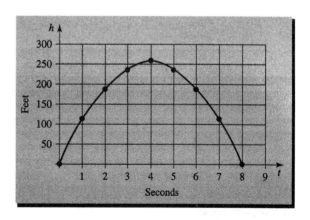

Match the graph of each quadratic equation of the form $y = a(x - h)^2 + k$ as shown on the next page.

39. $a > 0, h > 0, k > 0$
40. $a < 0, h > 0, k > 0$
41. $a > 0, h > 0, k < 0$
42. $a < 0, h > 0, k < 0$

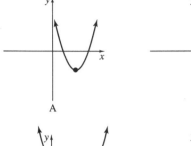
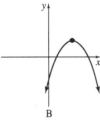
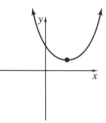
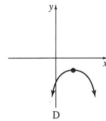

A B C D

Review Exercises

Simplify the following complex fractions. See Section 6.5.

43. $\dfrac{\frac{1}{7}}{\frac{2}{5}}$

44. $\dfrac{\frac{3}{8}}{\frac{1}{7}}$

45. $\dfrac{\frac{1}{x}}{\frac{2}{x^2}}$

46. $\dfrac{\frac{x}{5}}{\frac{2}{x}}$

47. $\dfrac{2x}{1-\frac{1}{x}}$

48. $\dfrac{x}{x-\frac{1}{x}}$

49. $\dfrac{\frac{a-b}{2b}}{\frac{b-a}{8b^2}}$

50. $\dfrac{\frac{2a^2}{a-3}}{\frac{a}{3-a}}$

GROUP ACTIVITY

MODELING A PHYSICAL SITUATION

MATERIALS:
- Metric ruler
- Grapher with curve-fitting capabilities (optional)

FIGURE 1

FIGURE 2

Model the physical situation of the parabolic path of water from a water fountain. For simplicity, use the given Figure 2 to investigate the following questions.

DATA TABLE

	x	y
Point A	0	0
Point B		
Point V		

1. Collect data for the x-intercepts of the parabolic path. Let points A and B in Figure 2 be on the x-axis and let the coordinates of point A be (0, 0). Use a ruler to measure the distance between points A and B **on Figure 2** to the nearest even one-tenth centimeter, and use this information to determine the coordinates of point B. Record this data in the data table. (*Hint:* If the distance from A to B measures 8 one-tenth centimeters, then the coordinates of point B are (8, 0).)

(continued)

APPENDIX

F ANSWERS TO SELECTED EXERCISES

CHAPTER 1
Review of Real Numbers

Exercise Set 1.1

1. <   **3.** >   **5.** =   **7.** <   **9.** $32 < 212$ **11.** True   **13.** False   **15.** False   **17.** True
19. $30 \leq 45$ **21.** $8 < 12$ **23.** $5 \geq 4$ **25.** $15 \neq -2$ **27.** 90   **29.** $70 \leq 90$
31. whole, integers, rational, real   **33.** integers, rational, real   **35.** natural, whole, integers, rational, real
37. rational, real   **39.** False   **41.** True   **43.** True   **45.** True   **47.** >   **49.** >   **51.** <   **53.** <
55. >   **57.** =   **59.** <   **61.** <   **63.** $0 > -26.7$ **65.** Sun   **67.** Sun   **69.** True   **71.** False
73. True   **75.** True   **77.** False   **79.** True   **81.** False   **83.** $20 \leq 25$ **85.** $6 > 0$ **87.** $-12 < -10$
89. Answers may vary

Exercise Set 1.2

1. $\frac{3}{8}$ **3.** $\frac{5}{7}$ **5.** $2 \cdot 2 \cdot 5$ **7.** $3 \cdot 5 \cdot 5$ **9.** $3 \cdot 3 \cdot 5$ **11.** $\frac{1}{2}$ **13.** $\frac{2}{3}$ **15.** $\frac{3}{7}$ **17.** $\frac{3}{5}$ **19.** $\frac{3}{8}$
21. $\frac{1}{2}$ **23.** $\frac{6}{7}$ **25.** 15   **27.** $\frac{1}{6}$ **29.** $\frac{25}{27}$ **31.** $\frac{11}{20}$ sq. miles   **33.** $\frac{3}{5}$ **35.** 1   **37.** $\frac{1}{3}$ **39.** $\frac{9}{35}$
41. $\frac{21}{30}$ **43.** $\frac{4}{18}$ **45.** $\frac{16}{20}$ **47.** $\frac{23}{21}$ **49.** $1\frac{2}{3}$ **51.** $\frac{5}{66}$ **53.** $\frac{7}{5}$ **55.** $\frac{2}{10}$ **57.** $\frac{3}{8}$ **59.** $\frac{5}{7}$ **61.** $\frac{65}{21}$
63. $\frac{2}{5}$ **65.** $\frac{9}{7}$ **67.** $\frac{3}{4}$ **69.** $\frac{17}{3}$ **71.** $\frac{7}{26}$ **73.** 1   **75.** $\frac{1}{5}$ **77.** $\frac{31}{6}$ **79.** $\frac{17}{18}$ **81.** $55\frac{1}{4}$ feet
83. $8\frac{1}{2}$ lbs.   **85.** Answers may vary   **87.** $3\frac{3}{8}$ miles   **89.** $\frac{3}{4}$ **91.** $\frac{49}{200}$

Scientific Calculator Explorations 1.3

1. 125   **3.** 59,049   **5.** 30   **7.** 9857   **9.** 2376

Exercise Set 1.3

1. 243   **3.** 27   **5.** 1   **7.** 5   **9.** $\frac{1}{125}$ **11.** $\frac{16}{81}$ **13.** 49   **15.** 32   **17.** 0   **19.** 16   **21.** 1.44
23. 5^2 square meters   **25.** 17   **27.** 20   **29.** 10   **31.** 21   **33.** 45   **35.** 0   **37.** $\frac{2}{7}$ **39.** 30
41. $\frac{27}{10}$ **43.** $\frac{7}{5}$ **45.** No; multiplication comes before addition in order of operations   **47.** $10 + 12 = 11 \cdot 2$
49. $5 + 6 > 10$ **51.** $3 \cdot 5 > 12$ **53.** $=$ **55.** $>$ **57.** $>$ **59.** $<$ **61.** Answers may vary   **63.** 2
65. $\frac{7}{18}$ **67.** $\frac{21}{8}$ **69.** 24.5   **71.** $(20 - 4) \cdot 4 \div 2$ **73.** $3^3 = 20 + 7$ **75.** $1 + 2 = 9 \div 3$
77. $9 \leq 11 \cdot 2$ **79.** $3 \neq 4 \div 2$

Exercise Set 1.4

1. 1   **3.** 11   **5.** 8   **7.** 45   **9.** 34   **11.** 16, 64, 144, 256   **13.** $x + 15$ **15.** $x - 5$ **17.** $3x + 22$
19. Solution   **21.** Not a solution   **23.** Not a solution   **25.** Not a solution   **27.** Solution
29. Not a solution   **31.** $5 + x = 20$ **33.** $13 - 3x = 13$ **35.** $\frac{12}{x} = \frac{1}{2}$ **37.** 54   **39.** $\frac{4}{9}$ **41.** 6
43. 18   **45.** 20   **47.** 27   **49.** 28   **51.** 132   **53.** $\frac{7}{5}$ **55.** $\frac{37}{18}$ **57.** Answers may vary   **59.** $\frac{x}{13}$
61. $2x - 10 = 18$ **63.** $20 - 30x$ **65.** $0.02x = 1.76$ **67.** $19 - x = 3x$ **69.** 28 meters   **71.** 12,000 sq. ft.
73. 6.5%   **75.** $340   **77.** 7   **79.** $\frac{29}{12}$ **81.** $\frac{13}{12}$

Exercise Set 1.5

1. 9   **3.** -14 **5.** -15 **7.** -16 **9.** 11   **11.** $2\frac{5}{8}$ **13.** -6 **15.** 2   **17.** 0   **19.** -6
21. Answers may vary   **23.** -2 **25.** 0   **27.** $-\frac{2}{3}$ **29.** -8 **31.** -12 **33.** 6   **35.** -4 **37.** 7
39. 12   **41.** -8 **43.** -59 **45.** -8 **47.** $-\frac{3}{16}$ **49.** $-\frac{13}{10}$ **51.** -19 **53.** 31   **55.** 59
57. -2.1 **59.** 38   **61.** -300 **63.** -9 **65.** -13.1 **67.** 5   **69.** -24 **71.** 19   **73.** Tues.
75. 7°   **77.** 1°   **79.** 45°   **81.** 146 ft.   **83.** $-\frac{3}{8}$ point   **85.** Solution   **87.** Not a solution
89. Answers may vary

Exercise Set 1.6

1. -10 **3.** -5 **5.** 19   **7.** $\frac{1}{6}$ **9.** 2   **11.** 13   **13.** -5 **15.** 3   **17.** -45 **19.** -4 **21.** 9
23. Sometimes positive and sometimes negative   **25.** -3 **27.** -16 **29.** 2   **31.** -10 **33.** $\frac{11}{6}$
35. -11 **37.** 11   **39.** 5   **41.** $5\frac{7}{10}$ **43.** $-\frac{31}{36}$ **45.** 4.1   **47.** 1   **49.** -7 **51.** 37   **53.** -17
55. $4\frac{1}{4}$ **57.** $\frac{31}{40}$ **59.** -6.4 **61.** 5.17   **63.** -36 **65.** -73 **67.** 24   **69.** -223 **71.** -18

73. −13 **75.** 25 **77.** 7 feet below sea level **79.** −9 **81.** 63 B.C. **83.** −308 ft. **85.** Monday
87. −9 **89.** −7 **91.** $\frac{7}{5}$ **93.** 21 **95.** $\frac{1}{4}$ **97.** Answers may vary **99.** Not a solution
101. Not a solution **103.** Solution

Scientific Calculator Explorations 1.7

1. 38 **3.** −441 **5.** $163\frac{1}{3}$ **7.** 54,499 **9.** 15,625

Exercise Set 1.7

1. −12 **3.** 42 **5.** 0 **7.** −18 **9.** $-\frac{2}{3}$ **11.** 2 **13.** −30 **15.** 90 **17.** 16 **19.** True
21. False **23.** $\frac{1}{9}$ **25.** $\frac{3}{2}$ **27.** $-\frac{1}{14}$ **29.** 1, −1 **31.** −9 **33.** 3 **35.** 5 **37.** 0 **39.** undefined
41. 16 **43.** −3 **45.** $-\frac{16}{7}$ **47.** −21 **49.** 41 **51.** −134 **53.** 3 **55.** −1
57. 12 **59.** −14 **61.** −6 **63.** undefined **65.** 5 **67.** 0 **69.** −36 **71.** −125 **73.** 16
75. −16 **77.** 2 **79.** $\frac{6}{5}$ **81.** −5 **83.** 18 **85.** −30 **87.** −24 **89.** $-\frac{1}{2}$ **91.** −8.372 **93.** 14
95. positive **97.** not possible to determine **99.** negative **101.** $-2 + \frac{-15}{3}; -7$
103. $2[-5 + (-3)]; -16$ **105.** Solution **107.** Solution **109.** Solution **111.** Sat., Oct. 2
113. Answers may vary

Exercise Set 1.8

1. commutative property of multiplication **3.** associative property of addition **5.** distributive property
7. associative property of multiplication **9.** identity property of addition **11.** distributive property
13. associative property of multiplication **15.** Answers may vary **17.** $18 + 3x$ **19.** $-2y + 2z$
21. $-21y + 35$ **23.** $5x + 20m + 10$ **25.** $-4 + 8m - 4n$ **27.** $-5x - 2$ **29.** $-r + 3 + 7p$ **31.** $4(1 + y)$
33. $11(x + y)$ **35.** $-1(5 + x)$ **37.** −16 **39.** 8 **41.** −9 **43.** $-\frac{2}{3}$ **45.** −1.2 **47.** 2 **49.** $\frac{3}{2}$
51. $-\frac{6}{5}$ **53.** $\frac{1}{6}$ **55.** $-\frac{1}{2}$ **57.** $-\frac{5}{3}$ **59.** $\frac{6}{23}$ **61.** $-x; \frac{1}{x}$ **63.** $3z; -\frac{1}{3z}$ **65.** $-a - b; \frac{1}{a + b}$
67. $\frac{2}{3} \cdot \frac{3}{2} = 1$ **69.** $(-4)(-3) = (-3)(-4)$ **71.** $3 + (8 + 9) = (3 + 8) + 9$ **73.** $y + 0 = y$
75. $x(a + b) = xa + xb$ **77.** $a(b + c) = (b + c)a$ **79. (a)** distributive property
(b) commutative property of addition **81. (a)** commutative property of addition
(b) commutative property of addition **(c)** associative property of addition **83.** Answers may vary

Exercise Set 1.9

1. 7¢ **3.** Alaska Village Electric Coop. **5.** Answers may vary **7.** $130 million **9.** $66 million
11. *The Lion King* **13.** $125 **15.** 50 miles **17.** 85 heartbeats per minute **19.** 95 heartbeats per minute
21. Philadelphia Flyers **23.** 8 years **25.** 6 years **27.** 1988; $633 **29.** $533
31. latitude 30° North, longitude 90° West **33.** Answers may vary

Chapter 1 Review

1. < **3.** > **5.** < **7.** = **9.** > **11.** $4 \geq -3$ **13.** $0.03 < 0.3$ **15. (a)** $\{1, 3\}$ **(b)** $\{0, 1, 3\}$
(c) $\{-6, 0, 1, 3\}$ **(d)** $\{-6, 0, 1, 1\frac{1}{2}, 3, 9.62\}$ **(e)** $\{\pi\}$ **(f)** $\{-6, 0, 1, 1\frac{1}{2}, 3, 9.62, \pi\}$ **17.** Friday **19.** $2 \cdot 2 \cdot 3 \cdot 3$
21. $\frac{12}{25}$ **23.** $\frac{13}{10}$ **25.** $9\frac{3}{8}$ **27.** 15 **29.** $\frac{7}{12}$ **31.** $A = \frac{34}{121}$ sq. in.; $P = 2\frac{4}{11}$ in. **33.** 70 **35.** 37
37. $\frac{18}{7}$ **39.** $20 - 12 = 2 \cdot 4$ **41.** 18 **43.** 5 **45.** 63° **47.** Not a solution **49.** $-\frac{2}{3}$ **51.** 7
53. -17 **55.** -5 **57.** 3.9 **59.** -11.5 **61.** -11 **63.** 4 **65.** 1 **67.** $-\frac{1}{6}$ **69.** -48 **71.** 3
73. -36 **75.** undefined **77.** commutative property of addition **79.** distributive property
81. associative property of addition **83.** distributive property **85.** multiplicative inverse
87. commutative property of addition **89.** $400 million **91.** revenue is increasing

Chapter 1 Test

1. $|-7| > 5$ **2.** $(9 + 5) \geq 4$ **3.** -5 **4.** -11 **5.** -14 **6.** -39 **7.** 12 **8.** -2 **9.** undefined
10. -8 **11.** $-\frac{1}{3}$ **12.** $4\frac{5}{8}$ **13.** $\frac{51}{40}$ **14.** -32 **15.** -48 **16.** 3 **17.** 0 **18.** > **19.** >
20. > **21.** = **22. (a)** $\{1, 7\}$ **(b)** $\{0, 1, 7\}$ **(c)** $\{-5, -1, 0, 1, 7\}$ **(d)** $\{-5, -1, \frac{1}{4}, 0, 1, 7, 11.6\}$
(e) $\{\sqrt{7}, 3\pi\}$ **(f)** $\{-5, -1, \frac{1}{4}, 0, 1, 7, 11.6, \sqrt{7}, 3\pi\}$ **23.** 40 **24.** 12 **25.** 22 **26.** -1
27. associative property of addition **28.** commutative property of multiplication **29.** distributive property
30. multiplicative inverse **31.** 9 **32.** -3 **33.** second down **34.** yes **35.** 17° **36.** loss of $420
37. $8 billion **38.** $3 billion **39.** $5.5 billion **40.** 1994

CHAPTER 2
Equations, Inequalities, and Problem Solving

Mental Math, Sec. 2.1

1. -7 **3.** 1 **5.** 17 **7.** like **9.** unlike **11.** like

Exercise Set 2.1

1. $15y$ **3.** $13w$ **5.** $-7b - 9$ **7.** $-m - 6$ **9.** $5y - 20$ **11.** $7d - 11$ **13.** $-3x + 2y - 1$
15. $2x + 14$ **17.** Answers may vary **19.** $10x - 3$ **21.** $-4x - 9$ **23.** $2x - 4$ **25.** $\frac{3}{4}x + 12$
27. $-2 + 12x$ **29.** $5x^2$ **31.** $4x - 3$ **33.** $8x - 53$ **35.** -8 **37.** $7.2x - 5.2$ **39.** $k - 6$
41. $0.9m + 1$ **43.** $-12y + 16$ **45.** $x + 5$ **47.** -11 **49.** $1.3x + 3.5$ **51.** $x + 2$ **53.** $-15x + 18$
55. $2k + 10$ **57.** $-3x + 5$ **59.** $(18x - 2)$ ft. **61.** $-4m - 3$ **63.** $8(x + 6)$ **65.** $x - 10$ **67.** $\frac{7x}{6}$
69. balanced **71.** balanced **73.** $(15x + 23)$ in. **75.** 2 **77.** -23 **79.** -25 **81.** $5b^2c^3 + b^3c^2$
83. $5x^2 + 9x$ **85.** $-7x^2y$

Mental Math Sec. 2.2

1. $x = 2$ **3.** $n = 12$ **5.** $b = 17$

Exercise Set 2.2

1. $x = -13$ **3.** $y = -14$ **5.** $x = -8$ **7.** $x = -2$ **9.** $f = \frac{1}{4}$ **11.** $x = 11$ **13.** $t = 0$ **15.** $x = -3$
17. $y = -10$ **19.** $y = 8.9$ **21.** $b = -0.7$ **23.** $x = -1$ **25.** $t = 12$ **27.** $y = 0.2$
29. Answers may vary   **31.** $x = 11$ **33.** $w = -30$ **35.** $x = -7$ **37.** $n = 2$ **39.** $x = -12$
41. $n = 21$ **43.** $t = -6$ **45.** $y = -25$ **47.** $m = 0$ **49.** $t = 1.83$ **51.** $20 - p$ **53.** $(10 - x)$ ft.
55. $(180 - x)°$ **57.** $(173 - 3x)°$ **59.** $(n + 284)$ votes   **61.** Solution   **63.** Not a solution   **65.** $\frac{8}{5}$
67. $\frac{1}{2}$ **69.** -9 **71.** 5 hours   **73.** 2 hours   **75.** 6 hours and 7 hours after arrival

Mental Math Sec. 2.3

1. $a = 9$ **3.** $b = 2$ **5.** $x = -5$

Exercise Set 2.3

1. $x = -4$ **3.** $x = 0$ **5.** $x = 12$ **7.** $x = -12$ **9.** $d = 3$ **11.** $a = -2$ **13.** $k = 0$ **15.** $x = 10$
17. $x = -4$ **19.** $x = -5$ **21.** $y = -6$ **23.** $x = -3$ **25.** $x = 5$ **27.** $x = 2$ **29.** $a = -2$
31. Answers may vary   **33.** Answers may vary   **35.** $w = -6$ **37.** $z = 4$ **39.** $h = \frac{3}{4}$ **41.** $a = 0$
43. $k = 0$ **45.** $x = \frac{3}{2}$ **47.** $x = 21$ **49.** $x = \frac{11}{2}$ **51.** $x = -\frac{1}{4}$ **53.** $x = -\frac{5}{6}$ **55.** $n = 1$
57. $z = \frac{9}{10}$ **59.** $x = -2$ **61.** $y = -7$ **63.** $z = 1$ **65.** $n = 2$ **67.** $x = -30$ **69.** $x + 2$
71. $2x + 2$ **73.** $2x + 2$ **75.** $3x + 6$ **77.** $x = -2.95$ **79.** $x = 0.02$ **81.** $7x - 12$ **83.** $12z + 44$
85. 1   **87.** >   **89.** =   **91.** <

Scientific Calculator Explorations 2.4

1. solution   **3.** not a solution   **5.** solution

Exercise Set 2.4

1. $x = 1$ **3.** $n = \frac{9}{2}$ **5.** $x = \frac{9}{4}$ **7.** $x = 0$ **9.** $x = 2$ **11.** $x = -5$ **13.** $x = 10$ **15.** $z = 18$
17. $x = 1$ **19.** $x = 50$ **21.** $y = 0.2$ **23.** all real numbers   **25.** no solution   **27.** no solution
29. Answers may vary   **31.** Answers may vary   **33.** $x = 4$ **35.** $y = -4$ **37.** $x = 3$ **39.** $y = -2$
41. $c = 4$ **43.** $x = \frac{7}{3}$ **45.** no solution   **47.** $z = \frac{9}{5}$ **49.** $y = \frac{4}{19}$ **51.** $a = 1$ **53.** no solution
55. $x = \frac{7}{2}$ **57.** $v = -17$ **59.** $t = \frac{19}{6}$ **61.** $t = -\frac{2}{7}$ **63.** $x = 3$ **65.** $x = 13$ **67.** $x = 15.3$
69. $x = -0.2$ **71.** $2x + \frac{1}{5} = 3x - \frac{4}{5}; x = 1$ **73.** $2x + 7 = x + 6; x = -1$ **75.** $3x - 6 = 2x + 8; x = 14$

77. $\frac{1}{3}x = \frac{5}{6}$; $x = \frac{5}{2}$ **79.** $x - 4 = 2x$; $x = -4$ **81.** $\frac{x}{4} + \frac{1}{2} = \frac{3}{4}$; $x = 1$ **83.** $x = 4$ cm, $2x = 8$ cm   **85.** $\frac{5}{4}$
87. Midway   **89.** 145   **91.** -1 **93.** $\frac{1}{5}$ **95.** $(6x - 8)$ meters   **97.** $x = -\frac{7}{8}$ **99.** no solution
101. $n = 0$

Exercise Set 2.5

1. Nebraska: \$65,000; New York: \$130,000   **3.** 1st: 5 in; 2nd: 10 in; 3rd: 25 in.   **5.** $-\frac{3}{4}$ **7.** -16
9. 172 miles   **11.** shorter piece: 5 ft.; longer piece: 12 ft.   **13.** $45°$ and $135°$ **15.** Bulls: 99; Suns: 98
17. There are none.   **19.** -25 **21.** $-11, -9$ **23.** height 34 in.; diameter: 49 in.
25. son: \$5000; husband: \$10,000   **27.** $65°, 115°$ **29.** $58°, 60°, 62°$ **31.** $30°, 60°, 90°$
33. Answers may vary   **35.** Answers may vary   **37.** -4 **39.** -6 **41.** -4 **43.** $\frac{1}{2}(x - 1) = 37$
45. $\frac{3(x + 2)}{5} = 0$

Exercise Set 2.6

1. $h = 3$ **3.** $h = 3$ **5.** $r = 2.5$ **7.** $T = 3$ **9.** $h = 20$ **11.** $c = 12$ **13.** $h = 15$ **15.** $h = \frac{f}{5g}$
17. $W = \frac{V}{LH}$ **19.** $y = 7 - 3x$ **21.** $R = \frac{A - P}{PT}$ **23.** $A = \frac{3V}{h}$ **25.** $a = P - b - c$ **27.** $h = \frac{S - 2\pi r^2}{2\pi r}$
29. It multiplies the volume by 8.   **31.** 3000 miles   **33.** 500 sec. or $8\frac{1}{3}$ min.   **35.** 25,120 miles
37. $-109.3°$F   **39.** $-10°$C   **41.** 2000 mph   **43.** 800 cubic ft.   **45.** 96 piranhas   **47.** 2.25 hours
49. $-40°$ **51.** 10.8   **53.** 33,493,333,333 cubic miles   **55.** 565.2 cubic in.   **57.** $25\frac{5}{9}°$C   **59.** 6.25 hours
61. 8 ft.   **63.** $\frac{1}{2}(5x)$ **65.** $\frac{1}{3}\left(\frac{x}{6}\right)$ **67.** $\frac{2x}{3x}$ **69.** $x - (x + 6)$

Exercise Set 2.7

1. 1.2   **3.** 0.225   **5.** 0.0012   **7.** 75%   **9.** 200%   **11.** 12.5%   **13.** 38%   **15.** 54%   **17.** 136.8°
19. Answers may vary   **21.** 11.2   **23.** 55%   **25.** 180   **27.** 4.6   **29.** 50   **31.** 30%
33. \$39 decrease; \$117 sale price   **35.** 15.2% increase   **37.** 647.5 ft.   **39.** 55.40%   **41.** 54 people
43. No, many people use several medications.   **45.** 75% increase   **47.** 31 men
51. 13%; Yes   **53.** 0.48% increase   **55.** 39%
49.

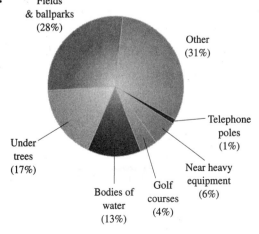

57. 5%   **59.** 9.6%   **61.** 26.9%; Yes   **63.** 17.1%   **65.** 6   **67.** 208   **69.** −55   **71.** $240
73. 199.5 miles

Exercise Set 2.8

1. length: 78 ft.; width: 52 ft.   **3.** 18 ft., 36 ft., 48 ft.   **5.** $666\frac{2}{3}$ miles   **7.** 160 miles   **9.** 2 gal.   **11.** $6\frac{2}{3}$ lbs.
13. No   **15.** $11,500 @ 8%; $13,500 @ 9%   **17.** $700 @ 11% profit; $3000 @ 4% loss
19. square's side length: 8 in.; triangle's side length: 13 in.   **21.** $30,000 @ 8%; $24,000 @ 10%
23. $5000 @ 12%; $15,000 @ 4%   **25.** $4500   **27.** $4500 @ 9%; $9000 @ 10%; $13,500 @ 11%
29. 3 adult tickets   **31.** 2.2 mph, 3.3 mph   **33.** 27.5 miles   **35.** $30,000   **37.** 25 monitors   **39.** −4
41. $\frac{9}{16}$ **43.** −4   **45.** >   **47.** =

Mental Math, Sec. 2.9

1. $x > 2$ **3.** $x \geq 8$

Exercise Set 2.9

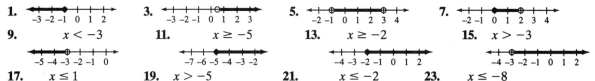

9. $x < -3$ **11.** $x \geq -5$ **13.** $x \geq -2$ **15.** $x > -3$
17. $x \leq 1$ **19.** $x > -5$ **21.** $x \leq -2$ **23.** $x \leq -8$
25. $x > 4$ **27.** $-1 < x < 2$ **29.** $4 \leq x \leq 5$
31. $1 < x < 5$ **33.** Answers may vary   **35.** $x \geq 20$ **37.** $x > 16$
39. $x > -3$ **41.** $x \geq -\frac{2}{3}$ **43.** $x > \frac{8}{3}$ **45.** $x > -13$
47. $x > 0$ **49.** $x \geq 0$ **51.** $1 < x < 4$ **53.** $x > 3$
55. $x \leq 0$ **57.** $0 < x < \frac{14}{3}$ **59.** $x > -10$ **61.** 35 cm   **63.** $-38.2° \leq F \leq 113°$
65. $0.924 \leq d \leq 0.987$ **67.** $-3 < x < 3$ **69.** 10%   **71.** 193   **73.** 8   **75.** 1   **77.** $\frac{16}{49}$
79. 32 million   **81.** 1992   **83.** $x > 1$ **85.** $x < \frac{5}{8}$ **87.** $x \leq 0$

Chapter 2 Review

1. $6x$ **3.** $4x - 2$ **5.** $3n - 18$ **7.** $-6x + 7$ **9.** $3x - 7$ **11.** $x = 4$ **13.** $x = -6$ **15.** $x = -9$
17. $10 - x$ **19.** $(175 - x)°$ **21.** $x = 4$ **23.** $x = -1$ **25.** $y = -1$ **27.** $x = 6$ **29.** $x = 2$
31. no solution   **33.** $z = \frac{3}{4}$ **35.** $n = 20$ **37.** $z = 0$ **39.** $n = \frac{20}{7}$ **41.** $c = \frac{23}{7}$ **43.** $x = 102$
45. 3   **47.** 1052 ft.   **49.** 307; 955   **51.** $w = 9$ **53.** $m = \frac{y - b}{x}$ **55.** $x = \frac{2y - 7}{5}$ **57.** $\pi = \frac{C}{D}$
59. 15 meters   **61.** 1 hr. and 20 min.   **63.** 93.5   **65.** 70%   **67.** 1280   **69.** 6%   **71.** 120 travelers
73. 14.3%   **75.** 80 nickels   **77.** 48 miles   **79.** $x > 0$ **81.** $0.5 \leq y < 1.5$
83. $x < -4$ **85.** $x \leq 4$ **87.** $-\frac{1}{2} < x < \frac{3}{4}$
89. $x \leq \frac{19}{3}$ **91.** score must be less than 83

Chapter 2 Test

1. $y - 10$ **2.** $5.9x + 1.2$ **3.** $-2x + 10$ **4.** $-15y + 1$ **5.** $x = -5$ **6.** $n = 8$ **7.** $y = \frac{7}{10}$
8. $z = 0$ **9.** $x = 27$ **10.** $y = -\frac{19}{6}$ **11.** $x = 3$ **12.** $y = \frac{3}{11}$ **13.** $x = 0.25$ **14.** $a = \frac{25}{7}$ **15.** 21
16. 7 gal.   **17.** $8500 @ 10%; $17,000 @ 12%   **18.** $2\frac{1}{2}$ hours   **19.** $x = 6$ **20.** $h = \frac{V}{\pi r^2}$
21. $y = \frac{3x - 10}{4}$ **22.** $x < -2$ **23.** $x < 4$ **24.** $-1 < x < \frac{7}{3}$
25. $\frac{7}{4} < x < 4$ **26.** $x > \frac{2}{5}$ **27.** 81.3%   **28.** $5.9314 billion   **29.** 24.12°

Chapter 2 Cumulative Review

1. (a) True   **(b)** True   **(c)** False   **(d)** True; *Sec. 1.1, Ex. 2*   **2. (a)** <   **(b)** =   **(c)** >   **(d)** <
(e) >; *Sec. 1.1, Ex. 7*   **3.** $\frac{2}{39}$; *Sec. 1.2, Ex. 3*   **4.** $\frac{8}{3}$; *Sec. 1.3, Ex. 3*   **5. (a)** -4 **(b)** 8
(c) -0.3; *Sec. 1.5, Ex. 2*   **6. (a)** -12 **(b)** -3; *Sec. 1.6, Ex. 3*   **7. (a)** -0.06 **(b)** $-\frac{7}{15}$; *Sec. 1.7, Ex. 3*
8. (a) 3   **(b)** -5 **(c)** 0   **(d)** -2; *Sec. 1.8, Ex. 5*   **9. (a)** $70   **(b)** 278 miles; *Sec. 1.9, Ex. 3*
10. (a) unlike   **(b)** like   **(c)** like   **(d)** like; *Sec. 2.1, Ex. 2*   **11.** $-2x - 1$; *Sec. 2.1, Ex. 7*
12. $x = 17$; *Sec. 2.2, Ex. 2*   **13.** $x = 4$; *Sec. 2.3, Ex. 6*   **14.** $a = 0$; *Sec. 2.4, Ex. 4*
15. 54 Republicans; 46 Democrats; *Sec. 2.5, Ex. 2*   **16.** 79.2 years; *Sec. 2.6, Ex. 1*
17. (a) 0.35   **(b)** 0.895   **(c)** 1.5; *Sec. 2.7, Ex. 1*   **18.** 87.5%; *Sec. 2.7, Ex. 5*
19. 7 hours; *Sec. 2.8, Ex. 2*   **20.** $2 < x \leq 4$; *Sec. 2.9, Ex. 2*   **21.** $x \geq 1$; *Sec. 2.9, Ex. 9*

CHAPTER 3
Graphing

Mental Math, Sec. 3.1

1. Answers may vary; Ex.: (5, 5), (7, 3)   **3.** Answers may vary; Ex.: (3, 5), (3, 0)

Exercise Set 3.1

1. quadrant I   **3.** no quadrant, *x*-axis   **5.** quadrant IV

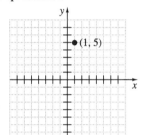 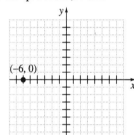 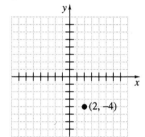

7. no quadrant, *x*-axis   **9.** no quadrant, origin   **11.** no quadrant, *y*-axis

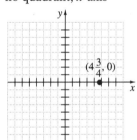 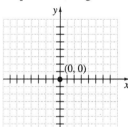 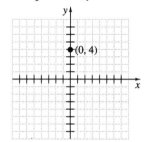

13. $a = b$ **15.** 26 units   **17.** yes; no; yes   **19.** no; yes; yes   **21.** no; yes; yes   **23.** yes; no
25. no; no   **27.** yes; yes   **29.** yes; yes
31. A, (0, 0); B, $\left(3\frac{1}{2}, 0\right)$; C, (3, 2); D, (−1, 3); E, (−2, −2); F, (0, −1); G, (2, −1)   **33.** quadrant IV
35. quadrants II or III   **37.** (−4, −2); (4, 0)   **39.** (0, 9); (3, 0)   **41.** (11, −7); answers may vary, Ex.: (2, −7)
43. (0, 2), (6, 0), (3, 1)   **45.** (0, −12), (5, −2), (−3, −18)   **47.** $\left(0, \frac{5}{7}\right), \left(\frac{5}{2}, 0\right), (-1, 1)$

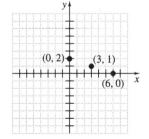 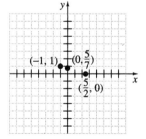

49. $(3, 0), (3, -0.5), \left(3, \dfrac{1}{4}\right)$ **51.** $(0, 0), (-5, 1), (10, -2)$

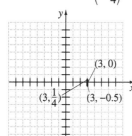
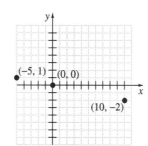

53. Answers may vary   **55. (a)** 13,000; 21,000; 29,000   **(b)** 45 desks

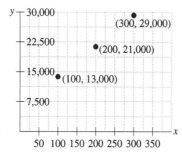

57. year 1: $500 million; year 2: $1500 million; year 3: $1000 million; year 4: $1500 million   **59.** $y = 5 - x$
61. $y = -\dfrac{1}{2}x + \dfrac{5}{4}$ **63.** $y = -2x$ **65.** $y = \dfrac{1}{3}x - 2$

Graphing Calculator Explorations 3.2

1. 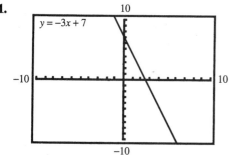   **3.**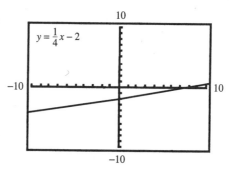

5. 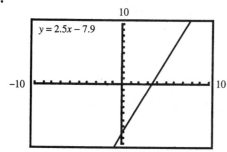   **7.**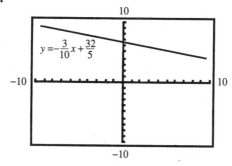

Exercise Set 3.2

1. yes **3.** yes **5.** no **7.** yes **9.** 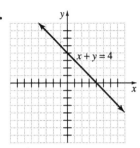 **11.**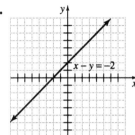

13. 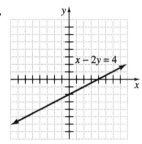 **15.** 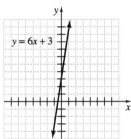 **17.** $y = x + 5$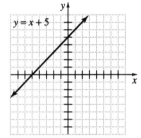

19. $2x + 3y = 6$ 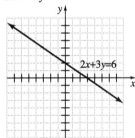 **21.** 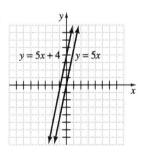 **23.**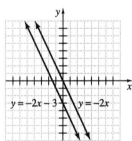

25. 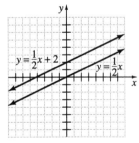 **27.** 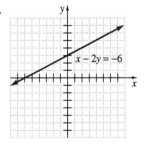 **29.**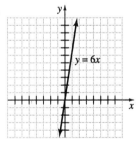

31. 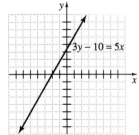 **33.** 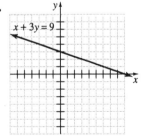 **35.**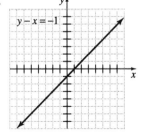

37. 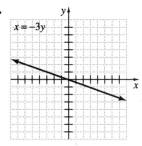 **39.** 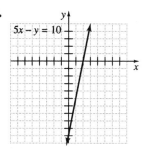 **41.**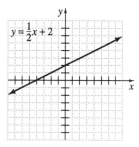

43. 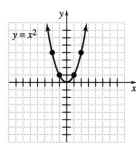 **45.** C **47.** D **49.** D **51.** A **53.** B **55.** Answers may vary

57. Yes; Answers may vary **59.** $(4, -1)$ **61.** $x = -5$ **63.** $x = -\dfrac{1}{10}$ **65.** $(0, 3), (-3, 0)$ **67.** $(0, 0), (0, 0)$

Graphing Calculator Explorations 3.3

1. 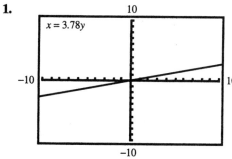 **3.**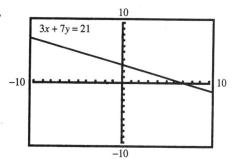

5.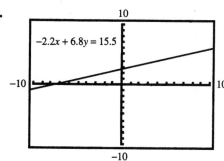

Exercise Set 3.3

1. $x = -1$; $y = 1$; $(-1, 0)$; $(0, 1)$ **3.** $x = -2$; $(-2, 0)$
5. $x = -1$; $x = 1$; $y = 1$; $y = -2$; $(-1, 0)$; $(1, 0)$; $(0, 1)$; $(0, -2)$ **7.** infinite   **9.** 0

11. 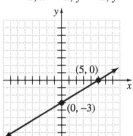   **13.** 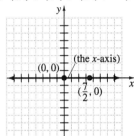   **15.**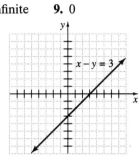

17. 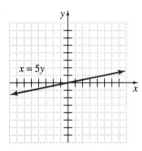   **19.** 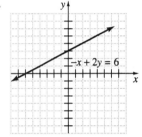   **21.**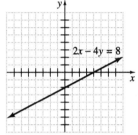

23. 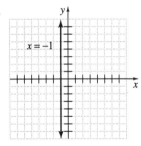   **25.** 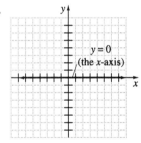   **27.**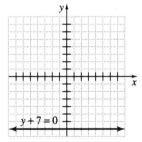

29. $x = 1$
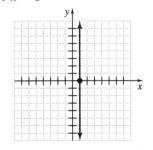   **31.** 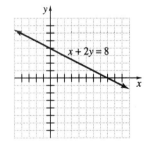   **33.**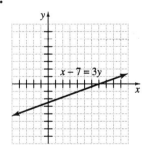

35.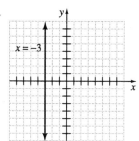
37.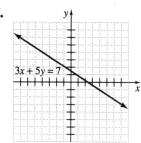
39.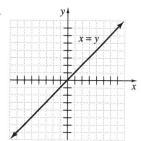
41.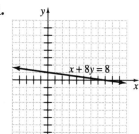
43.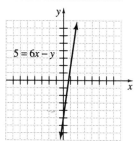
45.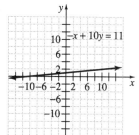
47.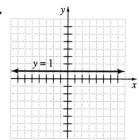
49.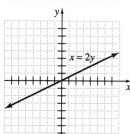
51.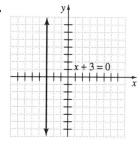
53. 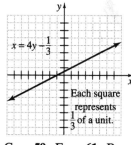 Each square represents $\frac{1}{3}$ of a unit.
55.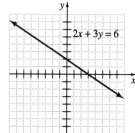

57. C **59.** E **61.** B **63.** 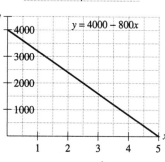 **65.** Answers may vary

67. Answers may vary **69.** $\frac{3}{2}$ **71.** 6 **73.** $-\frac{6}{5}$

Graphing Calculator Explorations 3.4

1. 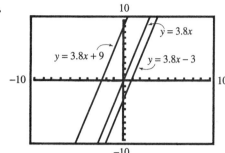   **3.**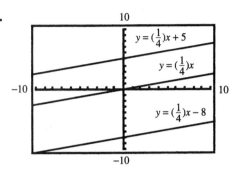

Mental Math, Sec. 3.4

1. upward **3.** horizontal

Exercise Set 3.4

1. $-\dfrac{4}{3}$ **3.** undefined   **5.** $\dfrac{5}{2}$ **7.** $\dfrac{8}{7}$ **9.** -1 **11.** $-\dfrac{1}{4}$ **13.** undefined   **15.** $-\dfrac{2}{3}$ **17.** undefined
19. 0   **21.** line 1   **23.** line 2   **25.** D   **27.** B   **29.** E

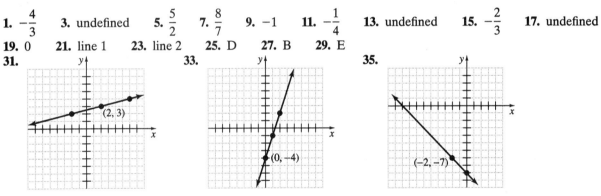

39. 5   **41.** -2 **43.** undefined   **45.** 0   **47.** $\dfrac{2}{3}$ **49.** $\dfrac{1}{2}$ **51.** $\dfrac{3}{5}$

37.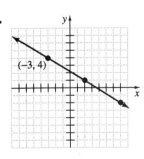

53. 6   **55.** 16%   **57.** (a) 1 (b) -1 **59.** (a) $\dfrac{9}{11}$ (b) $-\dfrac{11}{9}$ **61.** parallel   **63.** perpendicular
65. parallel   **67.** neither   **69.** perpendicular   **71.** Answers may vary   **73.** 20 mpg   **75.** 3.4 mpg
77. between 1986 and 1987   **79.** 1987   **81.** $-\dfrac{1}{5}$ **83.** $\dfrac{1}{3}$ **85.** 1   **87.** -0.25 **89.** 0.875

91. $m = -\dfrac{1}{3}$; $b = 2$; answers may vary   **93.** The line becomes steeper   **95.** $x \geq 3$ **97.** $x < 12$

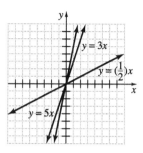

99. 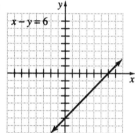   **101.** 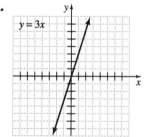   **103.**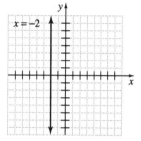

Mental Math, Sec. 3.5

1. yes   **3.** yes   **5.** yes   **7.** no

Exercise Set 3.5

1. no; no; yes   **3.** yes; no; no   **5.** no; no; yes   **7.**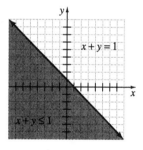

9. 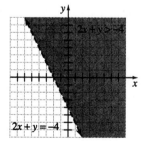   **11.** 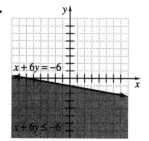   **13.**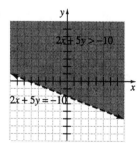

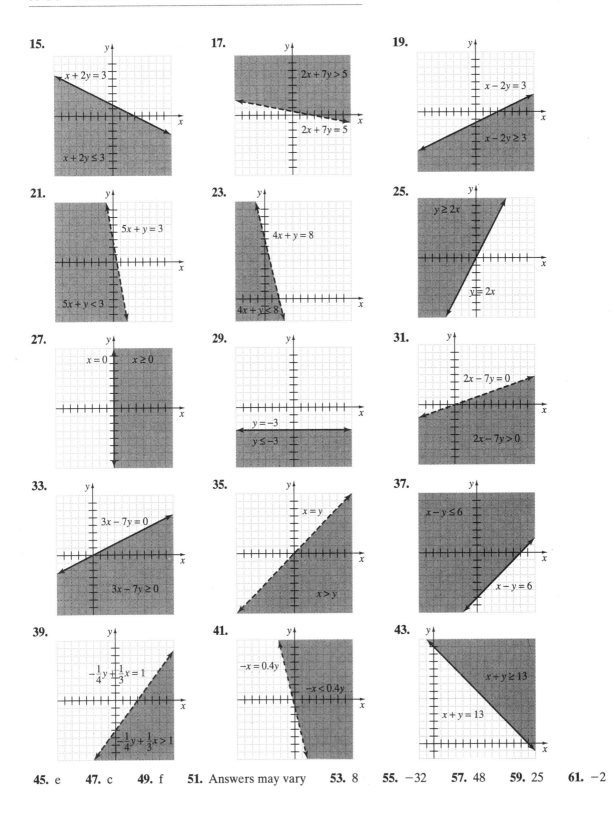

45. e **47.** c **49.** f **51.** Answers may vary **53.** 8 **55.** −32 **57.** 48 **59.** 25 **61.** −2

Chapter 3 Review

1.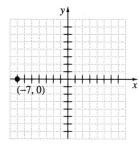

3.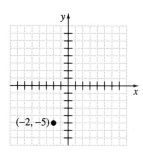

5.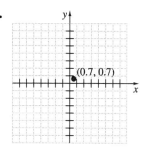

7. no; yes

9. yes; yes **11.** (7, 44) **13.** (−3, 0), (1, 3), (9, 9) **15.** (0, 0), (10, 5), (−10, −5)

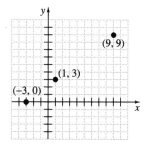

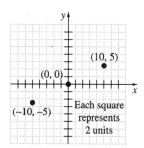

17.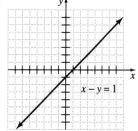

19.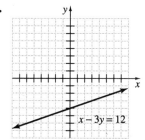

21.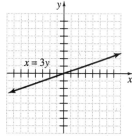

23.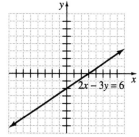

25. $x = 4$; $y = -2$; $(4, 0)$; $(0, -2)$

27. $x = -2$; $x = 2$; $y = 2$; $y = -2$; $(-2, 0)$; $(2, 0)$; $(0, 2)$; $(0, -2)$

29. 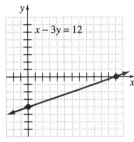   **31.** 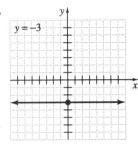   **33.**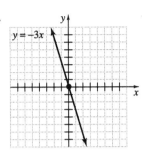

35. 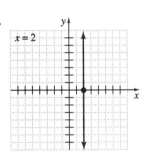   **37.** $-\dfrac{3}{4}$ **39.** d   **41.** c   **43.** e   **45.** $\dfrac{5}{3}$ **47.** -1 **49.** $\dfrac{1}{2}$

51. undefined   **53.** 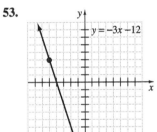   **55.** neither   **57.** perpendicular

59. 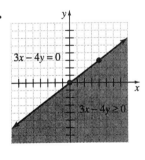   **61.** 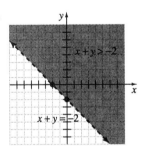   **63.**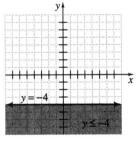

65.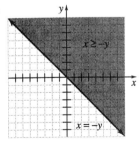

Chapter 3 Test

1. no **2.** no **3.** (1, 1) **4.** (−4, 17) **5.** $\frac{2}{5}$ **6.** 0 **7.** −1 **8.** −7 **9.** 3 **10.** undefined
11. parallel **12.** neither **13.** **14.**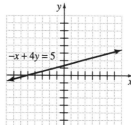

15. 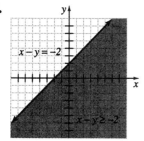 **16.** 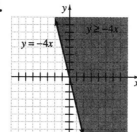 **17.**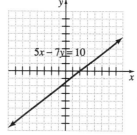

18. 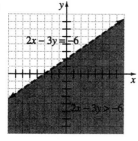 **19.** 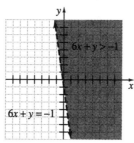 **20.**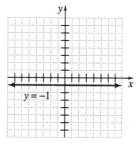

21. 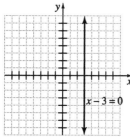 **22.** 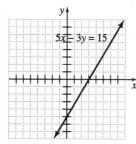 **23.**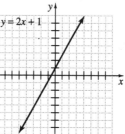

24. 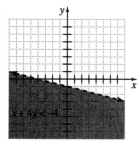 **25.** $x + 2y = 21$; $x = 5$ meters

Chapter 3 Cumulative Review

1. (a) < (b) = (c) >; *Sec. 1.1, Ex. 5* **2.** (a) $\frac{6}{7}$ (b) $\frac{11}{27}$ (c) $\frac{22}{5}$; *Sec. 1.2, Ex. 2* **3.** $\frac{14}{3}$; *Sec. 1.3, Ex. 5*

4. (a) $x + 3$ (b) $3x$ (c) $2x$ (d) $10 - x$ (e) $5x + 7$; *Sec. 1.4, Ex. 2* **5.** (a) $\frac{1}{2}$ (b) 9; *Sec. 1.6, Ex. 4*

6. (a) 6 (b) -12 (c) $-\frac{8}{15}$; *Sec. 1.7, Ex. 6* **7.** (a) $5x + 10$ (b) $-2y - 0.6z + 2$
(c) $-x - y + 2z - 6$; *Sec. 2.1, Ex. 5* **8.** $a = 8$; *Sec. 2.2, Ex. 5* **9.** $y = 140$; *Sec. 2.3, Ex. 3*
10. $x = 4$; *Sec. 2.4, Ex. 5* **11.** 10; *Sec. 2.5, Ex. 3* **12.** 40 ft; *Sec. 2.6, Ex. 2* **13.** 144; *Sec. 2.7, Ex. 3*

14. $x \leq 4$

; *Sec. 2.9, Ex. 6* **15.** (a) quadrant I (b) quadrant III (c) quadrant IV
(d) quadrant II (e) no quadrant; origin (f) no quadrant; y-axis (g) no quadrant; x-axis

(h) no quadrant; y-axis   ; *Sec. 3.1, Ex. 1*

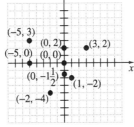

16. 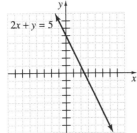 *Sec. 3.2, Ex. 2* **17.** $(-1, -3), (0, 0), (-3, -9)$; *Sec. 3.1, Ex. 4*

18. 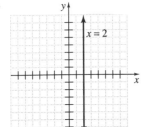 *Sec. 3.3, Ex. 6* **19.** $-\frac{8}{3}$; *Sec. 3.4, Ex. 1*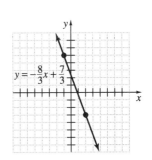

20. Sec. 3.4, Ex. 3   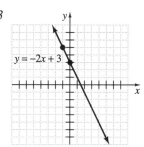   **21.** 0; Sec. 3.4, Ex. 5   **22.** Sec. 3.5, Ex. 2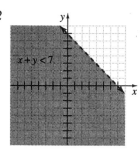

CHAPTER 4
Exponents and Polynomials

Mental Math, Sec. 4.1

1. base: 3; exponent: 2   **3.** base: -3; exponent: 6   **5.** base: 4; exponent: 2
7. base 5 has an understood exponent of 1; base 3 has an exponent of 4
9. base 5 has an understood exponent of 1; base x has an exponent of 2

Exercise Set 4.1

1. 49   **3.** -5 **5.** -16 **7.** 16   **9.** $\frac{1}{27}$ **11.** 112   **13.** Answers may vary   **15.** 4   **17.** 135
19. 150   **21.** $\frac{32}{5}$ **23.** 343 cu. meters   **25.** volume   **27.** x^7 **29.** $(-3)^{12}$ **31.** $15y^5$ **33.** $-24z^{20}$
35. $20x^5$ sq. ft.   **37.** p^7q^7 **39.** $\frac{m^9}{n^9}$ **41.** $x^{10}y^{15}$ **43.** $\frac{4x^2z^2}{y^{10}}$ **45.** $64z^{10}$ sq. decimeters   **47.** $27y^{12}$ cu. ft.
49. x^2 **51.** 4   **53.** p^6q^5 **55.** $\frac{y^3}{2}$ **57.** 1   **59.** -2 **61.** 2   **63.** Answers may vary   **65.** $\frac{-27a^6}{b^9}$
67. x^{39} **69.** $\frac{z^{14}}{625}$ **71.** $7776m^4n^3$ **73.** -25 **75.** $\frac{1}{64}$ **77.** $81x^2y^2$ **79.** 1   **81.** 40   **83.** b^6
85. a^9 **87.** $-16x^7$ **89.** $64a^3$ **91.** $36x^2y^2z^6$ **93.** $\frac{y^{15}}{8x^{12}}$ **95.** x **97.** $2x^2y$ **99.** $243x^7y^{24}$
101. $-2x + 7$ **103.** $2y - 10$ **105.** $-x - 4$ **107.** $-x + 5$ **109.** x^{9a} **111.** a^{5b} **113.** x^{5a}
115. $x^{5a^2}y^{5ab}z^{5ac}$

Mental Math, Sec. 4.2

1. $-14y$ **3.** $7y^3$ **5.** $7x$

Exercise Set 4.2

1. 1; binomial   **3.** 3; none of these   **5.** 6; trinomial   **7.** 2; binomial   **9.** Answers may vary
11. Answers may vary   **13.** (a) 6; (b) 5   **15.** (a) -2; (b) 4   **17.** (a) -15; (b) -16
19. -45.24 ft.; the object has reached the ground.   **21.** $23x^2$ **23.** $12x^2 - y$ **25.** $7s$ **27.** $-1.1y^2 + 4.8$
29. $12x + 12$ **31.** $-3x^2 + 10$ **33.** $-x + 14$ **35.** $-2x + 9$ **37.** $-3x^2 + 4$ **39.** $2x^2 + 7x - 16$
41. $(x^2 + 7x + 4)$ ft.   **43.** $(3y^2 + 4y + 11)$ meters   **45.** $8t^2 - 4$ **47.** $-2z^2 - 16z + 6$
49. $2x^3 - 2x^2 + 7x + 2$ **51.** $15a^3 + a^2 + 16$ **53.** $62x^2 + 5$ **55.** $12x + 2$ **57.** $9xy^2 - x - 27$
59. $-11x$ **61.** $7x - 13$ **63.** $-y^2 - 3y - 1$ **65.** $y^2 - 7$ **67.** $2x^2 + 11x$ **69.** $-16x^4 + 8x + 9$
71. $7x^2 + 14x + 18$ **73.** $3x - 3$ **75.** $7x^2 - 2x + 2$ **77.** $4y^2 + 12y + 19$ **79.** $6x^2 - 5x + 21$

81. $-6.6x^2 - 1.8x - 1.8$ **83. (a)** 184 ft.   **(b)** 600 ft.   **(c)** 595.84 ft.   **(d)** 362.56 ft.   **85.** 12.5 sec.   **87.** $6x^2$
89. $-12x^8$ **91.** $200x^3y^2$
93. 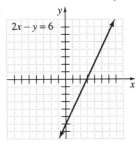   **95.**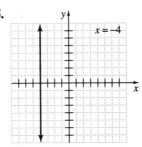

Mental Math, Sec. 4.3

1. $10xy$ **3.** x^7 **5.** $18x^3$

Exercise Set 4.3

1. $4a^2 - 8a$ **3.** $7x^3 + 14x^2 - 7x$ **5.** $6x^4 - 3x^3$ **7.** $x^2 + 3x$ **9.** $a^2 + 5a - 14$ **11.** $4y^2 - 16y + 16$
13. $30x^2 - 79xy + 45y^2$ **15.** $4x^4 - 20x^2 + 25$ **17.** $x^2 + 5x + 6$ **19.** $x^3 - 5x^2 + 13x - 14$
21. $x^4 + 5x^3 - 3x^2 - 11x + 20$ **23.** $10a^3 - 27a^2 + 26a - 12$ **25.** $x^3 + 6x^2 + 12x + 8$
27. $8y^3 - 36y^2 + 54y - 27$ **29.** $2x^3 + 10x^2 + 11x - 3$ **31.** $x^4 - 2x^3 - 51x^2 + 4x + 63$ **33 (a)** 25; 13
(b) 324; 164; No; Answers may vary   **35.** $2a^2 + 8a$ **37.** $6x^3 - 9x^2 + 12x$ **39.** $15x^2 + 37xy + 18y^2$
41. $x^3 + 7x^2 + 16x + 12$ **43.** $49x^2 + 56x + 16$ **45.** $-6a^4 + 4a^3 - 6a^2$ **47.** $x^3 + 10x^2 + 33x + 36$
49. $a^3 + 3a^2 + 3a + 1$ **51.** $x^2 + 2xy + y^2$ **53.** $x^2 - 13x + 42$ **55.** $3a^3 + 6a$ **57.** $-4y^3 - 12y^2 + 44y$
59. $25x^2 - 1$ **61.** $5x^3 - x^2 + 16x + 16$ **63.** $8x^3 - 60x^2 + 150x - 125$ **65.** $32x^3 + 48x^2 - 6x - 20$
67. $49x^2y^2 - 14xy + y^2$ **69.** $5y^4 - 16y^3 - 4y^2 - 7y - 6$ **71.** $6x^4 - 8x^3 - 7x^2 + 22x - 12$
73. $(4x^2 - 25)$ sq. yds.   **75.** $(6x^2 - 4x)$ sq. in.   **77. (a)** $a^2 - b^2$ **(b)** $4x^2 - 9y^2$
(c) $16x^2 - 49$; Answers may vary   **79.** $16p^2$ **81.** $49m^4$ **83.** $3500   **85.** $500
87. There is a loss in value each year.

Scientific Calculator Explorations 4.5

1. 5.31 EE 03 **3.** 6.6 EE −09 **5.** 1.5×10^{13} **7.** 8.15×10^{19}

Mental Math, Sec. 4.5

1. $\dfrac{5}{x^2}$ **3.** y^6 **5.** $4y^3$

Exercise Set 4.5

1. $\dfrac{1}{64}$ **3.** $\dfrac{7}{x^3}$ **5.** -64 **7.** $\dfrac{5}{6}$ **9.** p^3 **11.** $\dfrac{q^4}{p^5}$ **13.** $\dfrac{1}{x^3}$ **15.** z^3 **17.** Answers may vary
19. $\dfrac{a^{30}}{b^{12}}$ **21.** $\dfrac{1}{x^{10}y^6}$ **23.** $\dfrac{z^2}{4}$ **25.** Answers may vary **27.** $\dfrac{1}{9}$ **29.** $-p^4$ **31.** -2 **33.** r^6 **35.** $\dfrac{1}{x^{15}y^9}$
37. $\dfrac{4}{3}$ **39.** $\dfrac{1}{32x^5}$ **41.** $\dfrac{49a^4}{b^6}$ **43.** $a^{24}b^8$ **45.** x^9y^{19} **47.** $-\dfrac{y^8}{8x^2}$ **49.** $\dfrac{27}{x^6z^3}$ cu. in. **51.** $2x^7$
53. 7.8×10^4 **55.** 1.67×10^{-6} **57.** 6.35×10^{-3} **59.** 1.16×10^6 **61.** 2.0×10^7 **63.** 9.3×10^7
65. 786,000,000 **67.** 0.0000000008673 **69.** 0.033 **71.** 20,320 **73.** 6,250,000,000,000,000,000
75. 9,460,000,000,000 km **77.** 0.000036 **79.** 0.00000000000000000028 **81.** 0.0000005 **83.** 200,000
85. 1.512×10^{10} cu. ft. **87.** 1.248×10^6 **89.** -394.5 **91.** 1.3 sec. **93.** $\dfrac{5x^3}{3}$ **95.** $\dfrac{5z^3y^2}{7}$
97. $5y - 6 + \dfrac{5}{y}$ **99.** $20x^3 - 12x^2 + 2x$ **101.** a^m **103.** $27y^{6z}$ **105.** y^{5a} **107.** $\dfrac{1}{z^{6a+4}}$

Mental Math, Sec. 4.6

1. a^2 **3.** a^2 **5.** k^3

Exercise Set 4.6

1. $4k^3$ **3.** $3m$ **5.** $-\dfrac{4a^5}{b}$ **7.** $-\dfrac{6x}{7y^2}$ **9.** $5p^2 + 6p$ **11.** $-\dfrac{3}{2x} + 3$ **13.** $-3x^2 + x - \dfrac{4}{x^3}$
15. $-1 + \dfrac{3}{2x} - \dfrac{7}{4x^4}$ **17.** $5x^3 - 3x + \dfrac{1}{x^2}$ **19.** $(3x^3 + x - 4)$ ft. **21.** $x + 1$ **23.** $2x + 3$
25. $2x + 1 + \dfrac{7}{x-4}$ **27.** $4x + 9$ **29.** $3a^2 - 3a + 1 + \dfrac{2}{3a+2}$ **31.** $2b^2 + b + 2 - \dfrac{12}{b+4}$
33. Answers may vary **35.** $(2x + 5)$ meters **37.** $\dfrac{4}{x} + \dfrac{1}{x^2} + \dfrac{9}{5x^3}$ **39.** $5x - 2 + \dfrac{2}{x+6}$ **41.** $x^2 - \dfrac{12x}{5} - 1$
43. $6x - 1 - \dfrac{1}{x+3}$ **45.** $4x^2 + 1$ **47.** $2x^2 + 6x - 5 - \dfrac{2}{x-2}$ **49.** $6x - 1$ **51.** $-x^3 + 3x^2 - \dfrac{4}{x}$
53. $4x + 3 - \dfrac{2}{2x+1}$ **55.** $2x + 9$ **57.** $2x^2 + 3x - 4$ **59.** $x^2 + 3x + 9$ **61.** $x^2 - x + 1$
63. $-3x + 6 - \dfrac{11}{x+2}$ **65.** $2b - 1 - \dfrac{6}{2b-1}$ **67.** $2a^3 + 2a$ **69.** $2x^3 + 14x^2 - 10x$

71. $-3x^2y^3 - 21x^3y^2 - 24xy$ **73.** $9a^2b^3c + 36ab^2c - 72ab$ **75.** Thriller
77. Born in the U.S.A.; Eagles Greatest Hits

Chapter 4 Review

1. base: 3; exponent: 2   **3.** base: 5; exponent: 4   **5.** 36   **7.** -65 **9.** 1   **11.** 8   **13.** $-10x^5$ **15.** $\dfrac{b^4}{16}$
17. $\dfrac{x^6y^6}{4}$ **19.** $40a^{19}$ **21.** $\dfrac{3}{64}$ **23.** $\dfrac{1}{x}$ **25.** 5   **27.** 1   **29.** $6a^6b^9$ **31.** 7   **33.** 8   **35.** 5   **37.** 6
39. 22; 78; 154.02; 400   **41.** $22x^2y^3 + 3xy + 6$ **43.** cannot be combined   **45.** $2s^5 + 3s^4 + 4s^3 + s^2 - 7s - 6$
47. $-2x^2 - x + 20$ **49.** $-56xz^2$ **51.** $-12x^2a^2y^4$ **53.** $9x - 63$ **55.** $54a - 27$ **57.** $-32y^3 + 48y$
59. $-3a^3b - 3a^2b - 3ab^2$ **61.** $42b^4 - 28b^2 + 14b$ **63.** $6x^2 - 11x - 10$ **65.** $42a^2 + 11a - 3$
67. $x^6 + 2x^5 + x^2 + 3x + 2$ **69.** $x^6 + 8x^4 + 16x^2 - 16$ **71.** $8x^3 - 60x^2 + 150x - 125$ **73.** $x^2 - 10x + 25$
75. $16x^2 + 16x + 4$ **77.** $25x^2 - 1$ **79.** $a^2 - 4b^2$ **81.** $16a^4 - 4b^2$ **83.** $-\dfrac{1}{49}$ **85.** $\dfrac{1}{16x^4}$ **87.** $\dfrac{9}{4}$
89. $\dfrac{1}{42}$ **91.** $-q^3r^3$ **93.** $\dfrac{r^5}{s^3}$ **95.** $-\dfrac{s^4}{4}$ **97.** $\dfrac{y^6}{3}$ **99.** $\dfrac{x^6y^9}{64z^3}$ **101.** $\dfrac{y^8}{49}$ **103.** $\dfrac{x^3}{y^3}$ **105.** $\dfrac{a^{10}}{b^{10}}$
107. $-\dfrac{4}{27}$ **109.** x^{10+3h} **111.** a^{2m+5} **113.** 8.868×10^{-1} **115.** -8.68×10^5 **117.** 4.0×10^3
119. 0.00386   **121.** 893,600   **123.** 0.00000000000000000000000003   **125.** 400,000,000,000   **127.** $\dfrac{y}{8z}$
129. $-a^2 + 3b - 4$ **131.** $4x + \dfrac{7}{x+5}$ **133.** $3b^2 - 4b - \dfrac{1}{3b-2}$ **135.** $-x^2 - 16x - 117 - \dfrac{684}{x-6}$

Chapter 4 Test

1. 32   **2.** 81   **3.** -81 **4.** $\dfrac{1}{64}$ **5.** $\dfrac{y^4}{49x^2}$ **6.** $7x^2y^2$ **7.** $\dfrac{4}{x^6y^9}$ **8.** $\dfrac{x^2}{y^{14}}$ **9.** $\dfrac{1}{6xy^8}$ **10.** 5.63×10^5
11. 8.63×10^{-5} **12.** 0.0015   **13.** 62,300   **14.** 0.036   **15.** 5   **16.** $3xyz + 18x^2y$
17. $16x^3 + 7x^2 - 3x - 13$ **18.** $-3x^3 + 5x^2 + 4x + 5$ **19.** $x^3 + 8x^2 + 3x - 5$ **20.** $3x^3 + 22x^2 + 41x + 14$
21. $2x^5 - 5x^4 + 12x^3 - 8x^2 + 4x + 7$ **22.** $3x^2 + 16x - 35$ **23.** $9x^2 - 49$ **24.** $16x^2 - 16x + 4$
25. $64x^2 + 48x + 9$ **26.** $x^4 - 81b^2$ **27.** 1001 ft.; 985 ft.; 857 ft.; 601 ft.   **28.** $\dfrac{2}{x^2yz}$ **29.** $\dfrac{x}{2y} + \dfrac{1}{4} - \dfrac{7}{8y}$
30. $x + 2$ **31.** $9x^2 - 6x + 4 - \dfrac{16}{3x+2}$

Chapter 4 Cumulative Review

1. (a) $9 \le 11$ **(b)** $8 > 1$ **(c)** $3 \ne 4$; Sec. 1.1, Ex. 3   **2. (a)** $2 \cdot 2 \cdot 2 \cdot 5$ **(b)** $3 \cdot 3 \cdot 7$; Sec. 1.2, Ex. 1
3. (a) $\dfrac{6}{7}$ **(b)** $\dfrac{1}{2}$ **(c)** 1   **(d)** $\dfrac{4}{3}$; Sec. 1.2, Ex. 5   **4.** Solution; Sec. 1.4, Ex. 3   **5.** -12; Sec. 1.6, Ex. 2
6. (a) -6 **(b)** -24 **(c)** $\dfrac{3}{4}$; Sec. 1.7, Ex. 9
7. (a) $5x + 7$ **(b)** $-4a - 1$ **(c)** $4y - 3y^2$ **(d)** $7.3x - 6$; Sec. 2.1, Ex. 4   **8.** $x = -11$; Sec. 2.3, Ex. 2
9. all real numbers; Sec. 2.4, Ex. 7   **10.** $l = \dfrac{V}{wh}$; Sec. 2.6, Ex. 4   **11.** 21.2%; Sec. 2.7, Ex. 7

12. $12,500 @ 7\%; \$7,500 @ 9\%$; Sec. 2.8, Ex. 4   **13.** Sec. 3.2, Ex. 3

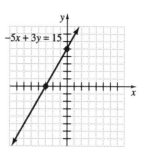

14. Sec. 3.3, Ex. 2

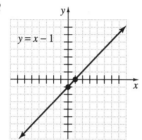

15. $m = 2$; Sec. 3.4, Ex. 2

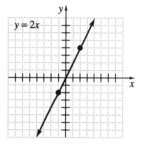

16. (a) x^{10} (b) y^{16} (c) $(-5)^{12}$; Sec. 4.1, Ex. 5   **17.** (a) $4x$ (b) $13x^2 - 2$; Sec. 4.2, Ex. 5
18. $4x^2 - 4xy + y^2$; Sec. 4.3, Ex. 3
19. (a) $t^2 + 4t + 4$ (b) $p^2 - 2pq + q^2$ (c) $4x^2 + 12xy + 9y^2$ (d) $25r^2 - 70rs + 49s^2$; Sec. 4.4, Ex. 5
20. (a) $\dfrac{2x}{25}$ (b) $\dfrac{b^3}{27a^6}$ (c) $\dfrac{16y^7}{x^5}$ (d) $y^{18}z^{36}$ (e) $-8x^6y^6$; Sec. 4.5, Ex. 4   **21.** $2xy - 4 + \dfrac{1}{2y}$; Sec. 4.6, Ex. 3

CHAPTER 5

Factoring Polynomials

Mental Math, Sec. 5.1

1. $2 \cdot 7$ **3.** $2 \cdot 5$ **5.** 3 **7.** 3

Exercise Set 5.1

1. 4   **3.** 6   **5.** y^2 **7.** xy^2 **9.** 4   **11.** $4y^3$ **13.** $3x^3$ **15.** $9x^2y$ **17.** $3(a + 2)$ **19.** $15(2x - 1)$
21. $6cd(4d^2 - 3c)$ **23.** $-6a^3x(4a - 3)$ **25.** $4x(3x^2 + 4x - 2)$ **27.** $5xy(x^2 - 3x + 2)$
29. Answers may vary   **31.** $(x + 2)(y + 3)$ **33.** $(y - 3)(x - 4)$ **35.** $(x + y)(2x - 1)$
37. $(x + 3)(5 + y)$ **39.** $(y - 4)(2 + x)$ **41.** $(y - 2)(3x + 8)$ **43.** $(y + 3)(y^2 + 1)$
45. $12x^3 - 2x; 2x(6x^2 - 1)$ **47.** $200x + 25\pi; 25(8x + \pi)$ **49.** $3(x - 2)$ **51.** $-2(4x + 9)$
53. $2x(16y - 9x)$ **55.** $4(x - 2y + 1)$ **57.** $(x + 2)(8 - y)$ **59.** $-8x^8y^5(5y + 2x)$ **61.** $5(x + 2)$
63. $-3(x - 4)$ **65.** $6x^3y^2(3y - 2 + x^2)$ **67.** $-2a^2(a + 3b)$ **69.** $(x - 2)(y^2 + 1)$ **71.** $(y + 3)(5x + 6)$
73. $(x - 2y)(4x - 3)$ **75.** $42yz(3x^3 + 5y^3z^2)$ **77.** $4(y - 3)(y + z)$ **79.** $(3 - x)(5 + y)$
81. $3(6x + 5)(2 + y)$ **83.** $2(3x^2 - 1)(2y - 7)$ **85.** Answers may vary   **87.** factored   **89.** not factored
91. $(n^3 - 6)$ units   **93.** $x^2 + 7x + 10$ **95.** $a^2 - 15a + 56$ **97.** $b^2 - 3b - 4$ **99.** $y^2 - 7y - 18$

Mental Math, Sec. 5.2

1. $(x + 5)$ **3.** $(x - 3)$ **5.** $(x + 2)$

Exercise Set 5.2

1. $(x + 6)(x + 1)$ **3.** $(x + 5)(x + 4)$ **5.** $(x - 5)(x - 3)$ **7.** $(x - 9)(x - 1)$ **9.** prime
11. $(x - 6)(x + 3)$ **13.** prime **15.** $(x + 5y)(x + 3y)$ **17.** $(x - y)(x - y)$ **19.** $(x - 4y)(x + y)$
21. $2(z + 8)(z + 2)$ **23.** $2x(x - 5)(x - 4)$ **25.** $7(x + 3y)(x - y)$ **27.** $(x + 12)(x + 3)$
29. $(x - 2)(x + 1)$ **31.** $(r - 12)(r - 4)$ **33.** $(x - 7)(x + 3)$ **35.** $(x + 5y)(x + 2y)$ **37.** prime
39. $2(t + 8)(t + 4)$ **41.** $x(x - 6)(x + 4)$ **43.** $(x - 9)(x - 7)$ **45.** $(x + 2y)(x - y)$ **47.** $3(x + 5)(x - 2)$
49. $3(x - 18)(x - 2)$ **51.** $(x - 24)(x + 6)$ **53.** $6x(x + 4)(x + 5)$ **55.** $2t^3(t - 4)(t - 3)$
57. $5xy(x - 8y)(x + 3y)$ **59.** $4y(x^2 + x - 3)$ **61.** $2b(a - 7b)(a - 3b)$ **63.** 8; 16 **65.** 6; 26
67. 5; 8; 9 **69.** 3; 4 **71.** Answers may vary **73.** $2x^2 + 11x + 5$ **75.** $15y^2 - 17y + 4$
77. $9a^2 + 23a - 12$
79. 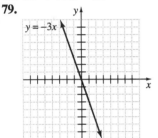 **81.** 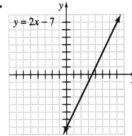 **83.** $2y(x + 5)(x + 10)$

85. $-12y^3(x^2 + 2x + 3)$ **87.** $(x + 1)(y - 5)(y + 3)$

Mental Math, Sec. 5.4

1. 1^2 **3.** 9^2 **5.** 3^2 **7.** 1^3 **9.** 2^3

Exercise Set 5.4

1. $(5y - 3)(5y + 3)$ **3.** $(11 - 10x)(11 + 10x)$ **5.** $3(2x - 3)(2x + 3)$ **7.** $(13a - 7b)(13a + 7b)$
9. $(xy - 1)(xy + 1)$ **11.** $(x + 6)$ **13.** $(a + 3)(a^2 - 3a + 9)$ **15.** $(2a + 1)(4a^2 - 2a + 1)$
17. $5(k + 2)(k^2 - 2k + 4)$ **19.** $(xy - 4)(x^2y^2 + 4xy + 16)$ **21.** $(x + 5)(x^2 - 5x + 25)$

23. $3x(2x - 3y)(4x^2 + 6xy + 9y^2)$ **25.** $(2x + y)$ **27.** $(x - 2)(x + 2)$ **29.** $(9 - p)(9 + p)$
31. $(2r - 1)(2r + 1)$ **33.** $(3x - 4)(3x + 4)$ **35.** prime **37.** $(3 - t)(9 + 3t + t^2)$
39. $8(r - 2)(r^2 + 2r + 4)$ **41.** $(t - 7)(t^2 + 7t + 49)$ **43.** $(x - 13y)(x + 13y)$ **45.** $(xy - z)(xy + z)$
47. $(xy + 1)(x^2y^2 - xy + 1)$ **49.** $(s - 4t)(s^2 + 4st + 16t^2)$ **51.** $2(3r - 2)(3r + 2)$ **53.** $x(3y - 2)(3y + 2)$
55. $25y^2(y - 2)(y + 2)$ **57.** $xy(x - 2y)(x + 2y)$ **59.** $4s^3t^3(2s^3 + 25t^3)$ **61.** $xy^2(27xy - 1)$
63. $4x^3 + 2x^2 - 1 + \dfrac{3}{x}$ **65.** $2x + 1$ **67.** $3x + 4 - \dfrac{2}{x + 3}$ **69.** $(x - 2)(x + 2)(x^2 + 4)$
71. $(a - 2 - b)(a + 2 + b)$ **73.** $(x - 2)^2(x + 1)(x + 3)$

Exercise Set 5.5

1. $(a + b)^2$ **3.** $(a - 3)(a + 4)$ **5.** $(a + 2)(a - 3)$ **7.** $(x + 1)^2$ **9.** $(x + 1)(x + 3)$ **11.** $(x + 3)(x + 4)$
13. $(x + 4)(x - 1)$ **15.** $(x + 5)(x - 3)$ **17.** $(x - 6)(x + 5)$ **19.** $2(x - 7)(x + 7)$ **21.** $(x + 3)(x + y)$
23. $(x + 8)(x - 2)$ **25.** $4x(x + 7)(x - 2)$ **27.** $2(3x + 4)(2x + 3)$ **29.** $(2a - b)(2a + b)$
31. $(5 - 2x)(4 + x)$ **33.** prime **35.** $(4x - 5)(x + 1)$ **37.** $4(t^2 + 9)$ **39.** $(x + 1)(a + 2)$
41. $4a(3a^2 - 6a + 1)$ **43.** prime **45.** $(5p - 7q)^2$ **47.** $(5 - 2y)(25 + 10y + 4y^2)$ **49.** $(5 - x)(6 + x)$
51. $(7 - x)(2 + x)$ **53.** $3x^2y(x + 6)(x - 4)$ **55.** $5xy^2(x - 7y)(x - y)$ **57.** $3xy(4x^2 + 81)$
59. $(x - y - z)(x - y + z)$ **61.** $(s + 4)(3r - 1)$ **63.** $(4x - 3)(x - 2y)$ **65.** $6(x + 2y)(x + y)$
67. $(x + 3)(y - 2)(y + 2)$ **69.** $(5 + x)(x + y)$ **71.** $(7t - 1)(2t - 1)$ **73.** $(3x + 5)(x - 1)$
75. $(x + 12y)(x - 3y)$ **77.** $(1 - 10ab)(1 + 2ab)$ **79.** $(x - 3)(x + 3)(x - 1)(x + 1)$
81. $(x - 4)(x + 4)(x^2 + 2)$ **83.** $(x - 15)(x - 8)$ **85.** $2x(3x - 2)(x - 4)$ **87.** $(3x - 5y)(9x^2 + 15xy + 25y^2)$
89. $(xy + 2z)(x^2y^2 - 2xyz + 4z^2)$ **91.** $2xy(1 - 6x)(1 + 6x)$ **93.** $(x - 2)(x + 2)(x + 6)$ **95.** $2a^2(3a + 5)$
97. $(a^2 + 2)(a + 2)$ **99.** $(x - 2)(x + 2)(x + 7)$ **101.** Answers may vary **103.** $x = 6$ **105.** $m = -2$
107. $z = \dfrac{1}{5}$ **109.** 8 in. **111.** $(-2, 0), (4, 0), (0, 2), (0, -2)$ **113.** $(-1, 0), (3, 0), (0, -3)$

Graphing Calculator Explorations 5.6

1. $x = -0.9, x = 2.2$ **3.** No real number solution **5.** $x = -1.8, x = 2.8$

Mental Math, Sec. 5.6

1. $a = 3, a = 7$ **3.** $x = -8, x = -6$ **5.** $x = -1, x = 3$

Exercise Set 5.6

1. $x = 2, x = -1$ **3.** $x = 0, x = -6$ **5.** $x = -\dfrac{3}{2}, x = \dfrac{5}{4}$ **7.** $x = \dfrac{7}{2}, x = -\dfrac{2}{7}$ **9.** $(x - 6)(x + 1) = 0$
11. $x = 9, x = 4$ **13.** $x = -4, x = 2$ **15.** $x = 8, x = -4$ **17.** $x = \dfrac{7}{3}, x = -2$ **19.** $x = \dfrac{8}{3}, x = -9$
21. $x^2 - 12x + 35 = 0$ **23.** $x = 0, x = 8, x = 4$ **25.** $x = \dfrac{3}{4}$ **27.** $x = 0, x = \dfrac{1}{2}, x = -\dfrac{1}{2}$
29. $x = 0, x = \dfrac{1}{2}, x = -\dfrac{3}{8}$ **31.** $-\dfrac{4}{3}, 1$ **33.** $-2, 5$ **35.** $-6, \dfrac{1}{2}$ **37.** E **39.** B **41.** C
43. $x = 0, x = -7$ **45.** $x = -5, x = 4$ **47.** $x = -5, x = 6$ **49.** $y = -\dfrac{4}{3}, y = 5$
51. $x = -\dfrac{3}{2}, x = 3, x = -\dfrac{1}{2}$ **53.** $x = -5, x = 3$ **55.** $x = 0, x = 16$ **57.** $y = -\dfrac{9}{2}, y = \dfrac{8}{3}$
59. $x = \dfrac{5}{4}, x = \dfrac{11}{3}$ **61.** $x = -\dfrac{2}{3}, x = \dfrac{1}{6}$ **63.** $x = -\dfrac{5}{3}, x = \dfrac{1}{2}$ **65.** $x = \dfrac{17}{2}$ **67.** $x = 2, x = -\dfrac{4}{5}$

69. $y = -4, y = 3$ **71.** $y = \frac{1}{2}, y = -\frac{1}{2}$ **73.** $t = -2, t = -11$ **75.** $t = 3$ **77.** $x = -3$
79. $x = \frac{3}{4}, x = -\frac{4}{3}$ **81.** $t = 0, t = -1, t = \frac{5}{2}$
83. (a) 300; 304; 276; 216; 124; 0; −156 (b) 5 sec. (c) 304 ft.

(d) Answers may vary

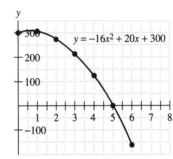

85. $\frac{47}{45}$ **87.** $\frac{17}{60}$ **89.** $\frac{15}{8}$ **91.** $\frac{7}{10}$

93. $x = 0, x = \frac{1}{2}$ **95.** $x = 0, x = -15$ **97.** $x = 0, x = 13$

Exercise Set 5.7

1. x and $36 - x$ **3.** x and $x + 2$ if x is an odd integer **5.** width: x; length: $3x - 4$
7. age now: x; age 10 years ago: $x - 10$ **9.** $x, x + 1, x + 2$ if x is an integer
11. 1st side: $2x - 2$; 2nd side: x; 3rd side: $x + 10$ **13.** 5 sec. **15.** width: 6 cm; length: 5 cm
17. base: 8 cm; altitude: 24 cm **19.** 5 in. **21.** hypotenuse: 25 cm; legs: 15 cm, 20 cm **23.** 36 ft.
25. −12 or 11 **27.** 13 and 7 **29.** 5 sides **31.** length: 13 miles; width: 8 miles **33.** 4 and 5; −1 and 0
35. width: 29 m; length: 35 m **37.** −26 and −24; 24 and 26 **39.** 12 mm, 16 mm, 20 mm **41.** 10 km
43. −11 and −10 **45.** width: 5 yd.; length: 12 yd. **47.** length: 15 miles; width: 8 miles **49.** 20%
51. 105 units **53.** 8 mph and 15 mph **55.** length: 8 yd.; width: 13 yd. **57.** Answers may vary
59. 467 acres **61.** 2.1 million **63.** Answers may vary **65.** $\frac{4}{7}$ **67.** $\frac{3}{2}$ **69.** $\frac{1}{3}$

Chapter 5 Review

1. $2x - 5$ **3.** $4x(5x + 3)$ **5.** $-2x^2y(4x - 3y)$ **7.** $(x + 1)(5x - 1)$ **9.** $(2x - 1)(3x + 5)$
11. $(x + 4)(x + 2)$ **13.** prime **15.** $(x + 4)(x - 2)$ **17.** $(x + 5y)(x + 3y)$ **19.** $2(3 - x)(12 + x)$
21. $(2x - 1)(x + 6)$ **23.** $(2x + 3)(2x - 1)$ **25.** $(6x - y)(x - 4y)$ **27.** $(2x + 3y)(x - 13y)$
29. $(6x + 5y)(3x - 4y)$ **31.** $(2x - 3)(2x + 3)$ **33.** prime **35.** $(2x + 3)(4x^2 - 6x + 9)$
37. $2(3 - xy)(9 + 3xy + x^2y^2)$ **39.** $(2x - 1)(2x + 1)(4x^2 + 1)$ **41.** $(2x - 3)(x + 4)$ **43.** $(x - 1)(x + 3)$
45. $2xy(2x - 3y)$ **47.** $(5x + 3)(25x^2 - 15x + 9)$ **49.** $(x + 7 - y)(x + 7 + y)$ **51.** $x = -6, x = 2$
53. $x = -\frac{1}{5}, x = -3$ **55.** $x = -4, x = 6$ **57.** $x = 2, x = 8$ **59.** $x = -\frac{2}{7}, x = \frac{3}{8}$ **61.** $x = -\frac{2}{5}$
63. $t = 3$ **65.** $t = 0, t = -\frac{7}{4}, t = 3$ **67.** 36 yd

69. (a) 17.5 sec. and 10 sec.; the rocket reaches a height of 2800 ft. on its way up and on its way back down.
(b) 27.5 sec.

Chapter 5 Test

1. $3x(3x + 1)(x + 4)$ 2. prime 3. prime 4. $(y - 12)(y + 4)$ 5. $(3a - 7)(a + b)$
6. $(3x - 2)(x - 1)$ 7. prime 8. $(x + 12y)(x + 2y)$ 9. $x^4(26x^2 - 1)$ 10. $5x(10x^2 + 2x - 7)$
11. $5(6 - x)(6 + x)$ 12. $(4x - 1)(16x^2 + 4x + 1)$ 13. $(6t + 5)(t - 1)$ 14. $(y - 2)(y + 2)(x - 7)$
15. $x(1 - x)(1 + x)(1 + x^2)$ 16. $-xy(y^2 + x^2)$ 17. $x = -7, x = 2$ 18. $x = -7, x = 1$
19. $x = 0, x = \dfrac{3}{2}, x = -\dfrac{4}{3}$ 20. $t = 0, t = 3, t = -3$ 21. $x = 0, x = -4$ 22. $t = -3, t = 5$
23. $x = -3, x = 8$ 24. $x = 0, x = \dfrac{5}{2}$ 25. width: 6 ft.; length: 11 ft. 26. 17 ft. 27. 8 and 9 28. 7 sec.

Chapter 5 Cumulative Review

1. **(a)** 4 **(b)** 5 **(c)** 0; *Sec. 1.1, Ex. 6* 2. **(a)** 4 **(b)** $\dfrac{9}{4}$ **(c)** $\dfrac{5}{2}$ **(d)** 5; *Sec. 1.4, Ex. 1*
3. 14,776 ft.; *Sec. 1.6, Ex. 5* 4. $t = -7$; *Sec. 2.2, Ex. 3* 5. $x = 2$; *Sec. 2.4, Ex. 1* 6. 138 min.; *Sec. 2.5, Ex. 4*
7. $x \geq 2$ ⟵—+—+—+—●—+—+—+—+—+—⟶ ; *Sec. 2.9, Ex. 4* 8. **(a)** solution **(b)** not a solution
     −1 0 1 2 3 4 5 6
(c) solution; *Sec. 3.1, Ex. 2* 9. *Sec. 3.2, Ex. 5*

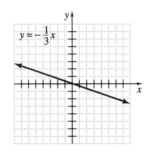

10. **(a)** $x = -3; y = 2; (-3, 0); (0, 2)$ **(b)** $x = -4; x = -1; y = 1; (-4, 0); (-1, 0); (0, 1)$ **(c)** $x = 0; y = 0; (0, 0)$
(d) $x = 2; (2, 0)$ **(e)** $x = -1; x = 3; y = -1; y = 2; (-1, 0); (3, 0); (0, -1); (0, 2)$; *Sec. 3.3, Ex. 1*
11. $\dfrac{2}{3}$; *Sec. 3.4, Ex. 4* 12. *Sec. 3.5, Ex. 3*

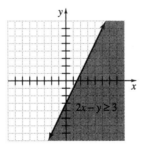

13. **(a)** s^4t^4 **(b)** $\dfrac{m^7}{n^7}$ **(c)** $8a^3$ **(d)** $25x^4y^6z^2$ **(e)** $\dfrac{16x^{16}}{81y^{20}}$; *Sec. 4.1, Ex. 6* 14. $4x^3 - x^2 + 7$; *Sec. 4.2, Ex. 7*
15. $x + 4$; *Sec. 4.6, Ex. 5* 16. **(a)** $6(t + 3)$ **(b)** $y^5(1 + y)(1 - y)$; *Sec. 5.1, Ex. 4*
17. $(x - 5)(x - 3)$; *Sec. 5.2, Ex. 2* 18. $(3x + 2)(x + 3)$; *Sec. 5.3, Ex. 1*
19. $9(x + 2)(x - 2)$; *Sec. 5.4, Ex. 3* 20. $x = 5, x = -\dfrac{7}{2}$; *Sec. 5.6, Ex. 1*
21. base: 6 m; height: 10 m; *Sec. 5.7, Ex. 3*

CHAPTER 6
Rational Expressions

Mental Math, Sec. 6.1

1. $x = 0$ 3. $x = 0, x = 1$

Exercise Set 6.1

1. $\dfrac{7}{4}$ 3. $\dfrac{13}{3}$ 5. $-\dfrac{11}{2}$ 7. $\dfrac{7}{4}$ 9. $-\dfrac{8}{3}$ 11. (a) $37.5 million   (b) $85.7 million   (c) $48.2 million
13. $x = -2$ 15. $x = 4$ 17. $x = -2$ 19. none   21. Answers may vary   23. $\dfrac{2}{x^4}$ 25. $\dfrac{5}{x+1}$
27. -5 29. $\dfrac{1}{x-9}$ 31. $5x + 1$ 33. $\dfrac{x+4}{2x+1}$ 35. Answers may vary   37. -1 39. $-\dfrac{y}{2}$
41. $\dfrac{2-x}{x+2}$ 43. $x + y$ 45. $\dfrac{5-y}{2}$ 47. $-\dfrac{3y^5}{x^4}$ 49. $\dfrac{x-2}{5}$ 51. -6 53. $\dfrac{x+2}{2}$ 55. $\dfrac{11x}{6}$
57. $\dfrac{1}{x-2}$ 59. $x + 2$ 61. $\dfrac{x+1}{x-1}$ 63. $\dfrac{m-3}{m+3}$ 65. $\dfrac{2a-6}{a+3}$ 67. -1 69. $-x - 1$ 71. $\dfrac{x+5}{x-5}$
73. $\dfrac{x+2}{x+4}$ 75. $\dfrac{x+3}{x}$ 77. $x^2 - 2x + 4$ 79. $\dfrac{x+5}{3}$ 81. $-x^2 - x - 1$
83. $y = \dfrac{x^2 - 25}{x + 5}$ 85. $y = \dfrac{x^2 + x - 12}{x + 4}$ 87. $\dfrac{3}{11}$ 89. $\dfrac{4}{3}$ 91. $\dfrac{50}{99}$ 93. $\dfrac{117}{40}$

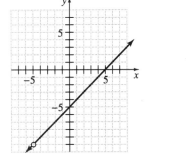
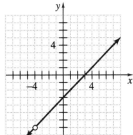

Mental Math, Sec. 6.2

1. $\dfrac{2x}{3y}$ 3. $\dfrac{5y^2}{7x^2}$ 5. $\dfrac{9}{5}$

Exercise Set 6.2

1. $\dfrac{21}{4y}$ 3. x^4 5. $-\dfrac{b^2}{6}$ 7. $\dfrac{x^2}{10}$ 9. $\dfrac{1}{3}$ 11. 1 13. $\dfrac{x+5}{x}$ 15. $\dfrac{2}{9x^2(x-5)}$ sq. ft.   17. x^4 19. $\dfrac{12}{y^6}$
21. $x(x + 4)$ 23. $\dfrac{3(x+1)}{x^3(x-1)}$ 25. $m^2 - n^2$ 27. $-\dfrac{x+2}{x-3}$ 29. $-\dfrac{x+2}{x-3}$ 31. Answers may vary
33. $\dfrac{1}{6b^4}$ 35. $\dfrac{9}{7x^2y^7}$ 37. $\dfrac{5}{6}$ 39. $\dfrac{3x}{8}$ 41. $\dfrac{3}{2}$ 43. $\dfrac{3x+4y}{2(x+2y)}$ 45. $-2(x + 3)$ 47. $\dfrac{2(x+2)}{x-2}$

49. $\dfrac{(a+5)(a+3)}{(a+2)(a+1)}$ **51.** $-\dfrac{1}{x}$ **53.** $\dfrac{2(x+3)}{x-4}$ **55.** $-(x^2+2)$ **57.** -1 **59.** $\dfrac{(a+b)^2}{a-b}$ **61.** $\dfrac{3x+5}{x^2+4}$

63. $4x^3(x-3)$ **65.** $\dfrac{4}{x-2}$ **67.** $\dfrac{a-b}{6(a^2+ab+b^2)}$ **69.** 1 **71.** $-\dfrac{10}{9}$ **73.** $-\dfrac{1}{5}$

75. 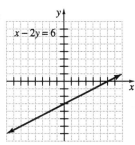 **77.** $\dfrac{x}{2}$ **79.** $\dfrac{5a(2a+b)(3a-2b)}{b^2(a-b)(a+2b)}$

Mental Math, Sec. 6.3

1. 1 **3.** $\dfrac{7x}{9}$ **5.** $\dfrac{1}{9}$ **7.** $\dfrac{7-10y}{5}$

Exercise Set 6.3

1. $\dfrac{a+9}{13}$ **3.** $\dfrac{y+10}{3+y}$ **5.** $\dfrac{3m}{n}$ **7.** $\dfrac{5x+7}{x-3}$ **9.** $\dfrac{1}{2}$ **11.** 4 **13.** $x+5$ **15.** $x+4$ **17.** 1 **19.** $\dfrac{12}{x^3}$

21. $-\dfrac{5}{x+4}$ **23.** $\dfrac{x-2}{x+y}$ **25.** 4 **27.** 3 **29.** $\dfrac{1}{a+5}$ **31.** $\dfrac{1}{x-6}$ **33.** $\dfrac{20}{x-2}m$

35. Answers may vary **37.** 33 **39.** $4x^3$ **41.** $8x(x+2)$ **43.** $6(x+1)^2$ **45.** $8-x$ or $x-8$

47. $40x^3(x-1)^2$ **49.** $(2x+1)(2x-1)$ **51.** $(2x-1)(x+4)(x+3)$ **53.** Answers may vary **55.** $\dfrac{6x}{4x^2}$

57. $\dfrac{24b^2}{12ab^2}$ **59.** $\dfrac{18}{2(x+3)}$ **61.** $\dfrac{9ab+2b}{5b(a+2)}$ **63.** $\dfrac{x(x+1)}{x(x+4)(x+2)(x+1)}$ **65.** $\dfrac{15x(x-7)}{3x(2x+1)(x-7)(x-5)}$

67. $\dfrac{18y-2}{30x^2-60}$ **69.** $\dfrac{x(x-4)}{x(x-4)^2(x+4)}$ **71.** $-\dfrac{5}{x-2}$ **73.** $\dfrac{7+x}{x-2}$ **75.** $x=0, x=3$ **77.** $x=-5, x=-1$

79. $\dfrac{29}{21}$ **81.** $-\dfrac{5}{12}$

Exercise Set 6.4

1. $\dfrac{5}{x}$ **3.** $\dfrac{75a+6b^2}{5b}$ **5.** $\dfrac{6x+5}{2x^2}$ **7.** $\dfrac{21}{2(x+1)}$ **9.** $\dfrac{17x+30}{2(x-2)(x+2)}$ **11.** $\dfrac{35x-6}{4x(x-2)}$ **13.** $\dfrac{5+10y-y^3}{y^2(2y+1)}$

15. Answers may vary **17.** $-\dfrac{2}{x-3}$ **19.** $-\dfrac{1}{x^2-1}$ **21.** $\dfrac{1}{x-2}$ **23.** $\dfrac{5+2x}{x}$ **25.** $\dfrac{6x-7}{x-2}$

27. $-\dfrac{y+4}{y+3}$ **29.** $\left(\dfrac{90x-40}{x}\right)^\circ$ **31.** 2 **33.** $3x^3-4$ **35.** $\dfrac{x+2}{(x+3)^2}$ **37.** $\dfrac{9b-4}{5b(b-1)}$ **39.** $\dfrac{2+m}{m}$

41. $\dfrac{10}{1-2x}$ **43.** $\dfrac{15x-1}{(x+1)^2(x-1)}$ **45.** $\dfrac{x^2-3x-2}{(x-1)^2(x+1)}$ **47.** $\dfrac{a+2}{2(a+3)}$ **49.** $\dfrac{x-10}{2(x-2)}$

51. $\dfrac{-2y-3}{(y-1)(y-2)}$ **53.** $\dfrac{-5x+23}{(x-3)(x-2)}$ **55.** $\dfrac{12x-32}{(x+2)(x-2)(x-3)}$ **57.** $\dfrac{6x+9}{(3x-2)(3x+2)}$
59. $\dfrac{2x^2-2x-46}{(x+1)(x-6)(x-5)}$ **61.** $\dfrac{2x-16}{(x-4)(x+4)}$ in. **63.** 10 **65.** 2 **67.** $\dfrac{25a}{9(a-2)}$ **69.** $\dfrac{x+4}{(x-2)(x-1)}$
71. Answers may vary **73.** $(x-1)(x^2+x+1)$ **75.** $(5z+2)(25z^2-10z+4)$ **77.** $(x+3)(y+2)$
79. 2 **81.** $-\dfrac{1}{3}$ **83.** $\dfrac{4x^2-15x+6}{(x-2)^2(x+2)(x-3)}$ **85.** $\dfrac{-3x^2+7x+55}{(x+2)(x+7)(x+3)}$ **87.** $\dfrac{2(x+1)}{(x-3)(x^2+3x+9)}$

Mental Math, Sec. 6.6

1. $x=10$ **3.** $z=36$

Exercise Set 6.6

1. $x=30$ **3.** $x=0$ **5.** $x=-5, x=2$ **7.** $a=5$ **9.** $x=3$ **11.** 100°, 80° **13.** 22.5°, 67.5°
15. $a=\dfrac{1}{4}$ **17.** extraneous solution **19.** $x=6, x=-4$ **21.** $y=5$ **23.** $x=-\dfrac{10}{9}$ **25.** $x=\dfrac{11}{14}$
27. extraneous solution **29.** Expression; $\dfrac{3+2x}{3x}$ **31.** Equation; $x=3$ **33.** Expression; $\dfrac{x-1}{x(x+1)}$
35. Equation; no solution **37.** Answers may vary **39.** $x=-\dfrac{21}{11}$ **41.** $y=1$ **43.** $x=\dfrac{9}{2}$
45. extraneous solution **47.** $x=-\dfrac{9}{4}$ **49.** $a=-\dfrac{3}{2}, a=4$ **51.** $x=-2$ **53.** $r=\dfrac{12}{5}$ **55.** $t=-2, t=8$
57. $x=\dfrac{1}{5}$ **59.** $R=\dfrac{D}{T}$ **61.** $y=\dfrac{3x+6}{5x}$ **63.** $b=-\dfrac{3a^2+2a-4}{6}$ **65.** $B=\dfrac{2A}{H}$ **67.** $r=\dfrac{C}{2\pi}$
69. $a=\dfrac{bc}{c+b}$ **71.** $n=\dfrac{m^2-3p}{2}$ **73.** $a=\dfrac{17}{4}$ **79.** $x=2; y=-2; (2,0); (0,-2)$

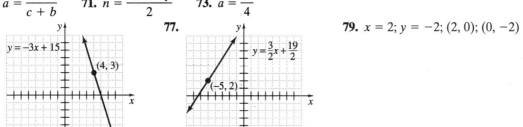

81. $x=-4; x=-2; x=3; y=4; (-4,0); (-2,0); (3,0); (0,4)$

Exercise Set 6.7

1. $\dfrac{2}{15}$ **3.** $\dfrac{5}{6}$ **5.** $\dfrac{5}{12}$ **7.** $\dfrac{1}{10}$ **9.** $\dfrac{7}{20}$ **11.** $\dfrac{19}{18}$ **13.** Answers may vary **15.** $x = 4$ **17.** $x = \dfrac{50}{9}$
19. $x = \dfrac{21}{4}$ **21.** $a = 30$ **23.** $x = 7$ **25.** $x = -\dfrac{1}{3}$ **27.** $x = -3$ **29.** $x = \dfrac{14}{9}$ **31.** $x = 5$
33. $x = -\dfrac{2}{3}$ **35.** Yes; answers may vary **37.** 110 oz. for $5.79 **39.** 8 oz. for $0.90 **41.** 123 lb.
43. 73 brown m&m's **45.** 165 calories **47.** 472.5 miles **49.** 337 yds./game **51.** 360 sq. ft.
53. 26 students **55.** 245 miles **57.** 9 gal. **59.** $19.13 per hr. **61.** $1348.58 **63.** 1408 megawatts
65. 1,666,560 people **67.** Their ratio is 1. **69.** $m = 3$; upward **71.** $m = -\dfrac{9}{5}$; downward
73. $m = 0$; horizontal

Exercise Set 6.8

1. 2 **3.** -3 **5.** $2\dfrac{2}{9}$ hr. **7.** $1\dfrac{1}{2}$ min.
9. trip to park rate: r; return trip rate: r; to park time: $12/r$; return time: $18/r$; $r = 6$ mph
11. 1st portion: 10 mph; cool down: 8 mph **13.** $x = 6$ **15.** $x = 5$ **17.** 2 **19.** $108.00 **21.** 63 mph
23. $y = 21.25$ **25.** 5 **27.** 217 mph **29.** 8 mph **31.** 3 hr **33.** 20 hr **35.** $x = 4.4$ ft; $y = 5.6$ ft
37. $5\dfrac{1}{4}$ hr **39.** 1st pump: 28 min.; 2nd pump: 84 min. **41.** 38 yr; 42 yr **43.** Answers may vary

45. 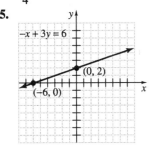 **47.** 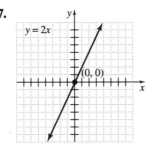 **49.**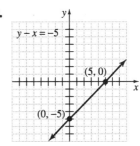

Chapter 6 Review

1. $x = 2, x = -2$ **3.** $\dfrac{2}{11}$ **5.** $\dfrac{1}{x-5}$ **7.** $\dfrac{x(x-2)}{x+1}$ **9.** $\dfrac{x-3}{x-5}$ **11.** $\dfrac{x+1}{2x+1}$ **13.** $\dfrac{x+5}{x-3}$
15. $-\dfrac{1}{x^2+4x+16}$ **17.** $-\dfrac{9x^2}{8}$ **19.** $-\dfrac{2x(2x+5)}{(x-6)^2}$ **21.** $\dfrac{4x}{3y}$ **23.** $\dfrac{2}{3}$ **25.** $\dfrac{x}{x+6}$ **27.** $\dfrac{3(x+2)}{3x+y}$
29. $-\dfrac{2(2x+3)}{y-2}$ **31.** $\dfrac{5x+2}{3x-1}$ **33.** $\dfrac{2x+1}{2x^2}$ **35.** $(x-8)(x+8)(x+3)$ **37.** $\dfrac{3x^2+4x-15}{(x+2)^2(x+3)}$
39. $\dfrac{-2x+10}{(x-3)(x-1)}$ **41.** $\dfrac{-2x-2}{x+3}$ **43.** $\dfrac{x-4}{3x}$ **45.** $\dfrac{x^2+2x-3}{(x+2)^2}$ **47.** $\dfrac{29x}{12(x-1)}; \dfrac{3xy}{5(x-1)}$ **49.** $\dfrac{2x}{x-3}$
51. $\dfrac{5}{3a^2}$ **53.** $\dfrac{2x^2+1}{x+2}$ **55.** $-\dfrac{7+2x}{2x}$ **57.** $n = 30$ **59.** $x = 3, x = -4$ **61.** extraneous solution
63. $x = 5$ **65.** $x = -6, x = 1$ **67.** $y = \dfrac{560-8x}{7}$ **69.** $\dfrac{2}{3}$ **71.** $x = 500$ **73.** $c = 50$

75. extraneous solution   **77.** extraneous solution   **79.** 15 oz. for $1.63   **81.** $33.75   **83.** 3
85. 30 mph; 20 mph   **87.** $17\frac{1}{2}$ hr

Chapter 6 Test

1. $x = -1, x = -3$ **2. (a)** $115 **(b)** $103   **3.** $\frac{3}{5}$ **4.** $\frac{1}{x - 10}$ **5.** $\frac{1}{x + 6}$ **6.** $\frac{1}{x^2 - 3x + 9}$
7. $\frac{2m(m + 2)}{m - 2}$ **8.** $\frac{a + 2}{a + 5}$ **9.** $-\frac{1}{x + y}$ **10.** $\frac{(x - 6)(x - 7)}{(x + 7)(x + 2)}$ **11.** 15   **12.** $\frac{y - 2}{4}$ **13.** $-\frac{1}{2x + 5}$
14. $\frac{3a - 4}{(a - 3)(a + 2)}$ **15.** $\frac{3}{x - 1}$ **16.** $\frac{2(x + 5)}{x(y + 5)}$ **17.** $\frac{x^2 + 2x + 35}{(x + 9)(x + 2)(x - 5)}$ **18.** $\frac{4y^2 + 13y - 15}{(y + 4)(y + 5)(y + 1)}$
19. $y = \frac{30}{11}$ **20.** $y = -6$ **21.** extraneous solution   **22.** extraneous solution   **23.** $\frac{xz}{2y}$ **24.** $\frac{b^2 - a^2}{2b^2}$
25. $\frac{5y^2 - 1}{y + 2}$ **26.** 18 bulbs   **27.** 5 or 1   **28.** 30 mph   **29.** $6\frac{2}{3}$ hr   **30.** 6 oz. for $1.19

Chapter 6 Cumulative Review

1. (a) $\frac{15}{x} = 4$ **(b)** $12 - 3 = x$ **(c)** $4x + 17 = 21$; Sec. 1.4, Ex. 4   **2. (a)** -10 **(b)** 17   **(c)** -21
(d) -12; Sec. 1.5, Ex. 1   **3. (a)** commutative property for multiplication   **(b)** distributive property
(c) associative property for addition; Sec. 1.8, Ex. 7   **4.** $x = -4$; Sec. 2.2, Ex. 6   **5.** -10; Sec. 2.4, Ex. 8
6. $w = \frac{P - 2l}{2}$; Sec. 2.6, Ex. 6   **7.** $x \leq -10$; [number line from −13 to −9]; Sec. 2.9, Ex. 3
8. Sec. 3.2, Ex. 4

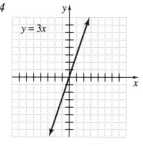

9. (a) x^3 **(b)** 256   **(c)** -27 **(d)** $2x^4y$; Sec. 4.1, Ex. 7   **10. (a)** $10x^4 + 30x$ **(b)** $-15x^4 - 18x^3 + 3x^2$
(c) $6n^3 - 10n^2 + 8n$; Sec. 4.3, Ex. 1   **11. (a)** y^3 **(b)** $\frac{q^9}{p^4}$ **(c)** $\frac{1}{x^{12}}$; Sec. 4.5, Ex. 3
12. $2x^2 - x + 1 + \frac{2x - 2}{x^2 + 1}$; Sec. 4.6, Ex. 8   **13.** $(x + 3)(x + 4)$; Sec. 5.1, Ex. 1
14. $(x + 2)(x^2 - 2x + 4)$; Sec. 5.4, Ex. 5   **15.** $x = 0, x = -3, x = 5$; Sec. 5.6, Ex. 7   **16.** $\frac{2}{5}$; Sec. 6.2, Ex. 2
17. $3x - 5$; Sec. 6.3, Ex. 3   **18.** $\frac{3}{x - 2}$; Sec. 6.4, Ex. 2   **19.** $\frac{3}{z}$; Sec. 6.5, Ex. 4   **20.** $t = 5$; Sec. 6.6, Ex. 1
21. $2\frac{1}{10}$ hr; Sec. 6.8, Ex. 2

CHAPTER 7
Further Graphing

Graphing Calculator Explorations 7.1

1. 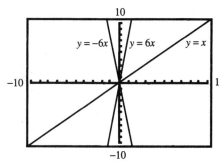   **3.**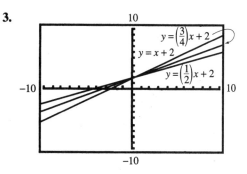

5.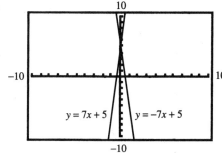

Mental Math, Sec. 7.1

1. $m = 2$; $(0, -1)$ **3.** $m = 1$; $\left(0, \dfrac{1}{3}\right)$ **5.** $m = \dfrac{5}{7}$; $(0, -4)$

Exercise Set 7.1

1. $m = -2$; $(0, 4)$ **3.** $m = -\dfrac{1}{9}$; $\left(0, \dfrac{1}{9}\right)$ **5.** $m = \dfrac{4}{3}$; $(0, -4)$ **7.** $m = -1$; $(0, 0)$ **9.** $m = 0$; $(0, -3)$
11. $m = \dfrac{1}{5}$; $(0, 4)$ **13.** B   **15.** D   **17.** neither   **19.** neither   **21.** perpendicular   **23.** parallel
25. Answers may vary   **27.** $y = -x + 1$ **29.** $y = 2x + \dfrac{3}{4}$ **31.** $y = \dfrac{2}{7}x$ **33.**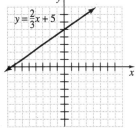

35.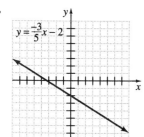
37.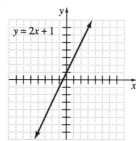
39.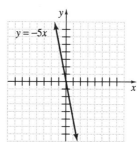
41.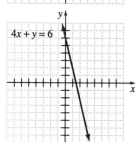
43.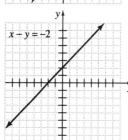
45.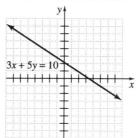
47. 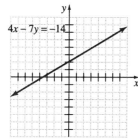 **49.** Answers may vary **51.** $K = C + 273$
53.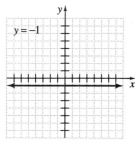
55. 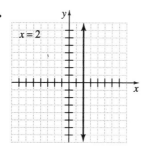 **57.** $3x - y = 16$ **59.** $6x + y = -10$

Mental Math, Sec. 7.2

1. $m = 3$; Answers may vary, Ex. $(4, 8)$ **3.** $m = -2$; Answers may vary, Ex. $(10, -3)$
5. $m = \dfrac{2}{5}$; Answers may vary, Ex. $(-1, 0)$

Exercise Set 7.2

1. $6x - y = 10$ **3.** $8x + y = -13$ **5.** $x - 2y = 17$ **7.** $2x - y = 4$ **9.** $8x - y = -11$
11. $4x - 3y = -1$ **13.** $x = 0$ **15.** $y = 3$ **17.** $x = -\dfrac{7}{3}$ **19.** $y = 2$ **21.** $y = 5$ **23.** $x = 6$
25. (a) $s = 32t$ (b) 128 ft/sec **27.** $3x + 6y = 10$ **29.** $x - y = -16$ **31.** $x + y = 17$ **33.** $y = 7$
35. $4x + 7y = -18$ **37.** $x + 8y = 0$ **39.** $3x - y = 0$ **41.** $x - y = 0$ **43.** $5x + y = 7$

45. $11x + y = -6$ **47.** $x = -\dfrac{3}{4}$ **49.** $y = -3$ **51.** $7x - y = 4$ **53.** $31x - 5y = -5$
55. (a) $v = -200t + 3000$ **(b)** \$1000   **57.** Answers may vary   **59. (a)** $3x - y = -5$ **(b)** $x + 3y = 5$
61. (a) $3x + 2y = -1$ **(b)** $2x - 3y = 21$
63. 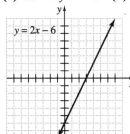 **65.** 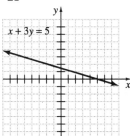 **67.**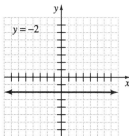

Graphing Calculator Explorations 7.3

1. $(0.56, 0), (-3.56, 0)$ **3.** $(-0.87, 0), (2.78, 0)$ **5.** $(-0.65, 0), (0.65, 0)$ **7.** $(-1.51, 0)$

Exercise Set 7.3

1. 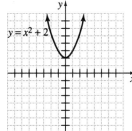   **3.** 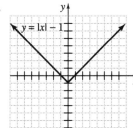   **5.**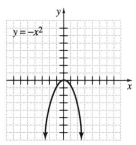

7. 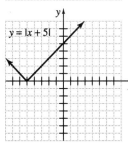   **9.** C   **11.** D   **13.** A   **15.** $(0, 1), (0, -1), (-2, 0), (2, 0)$

17. $(2, 0), \left(\dfrac{2}{3}, -2\right)$ **19.** (2, any real number)   **21.** There is no such point.   **23.** $(2, -1)$
25. linear   **27.** linear   **29.** linear

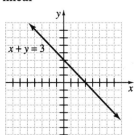   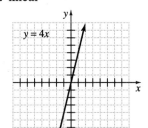   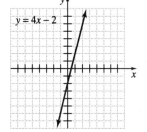

31. not linear 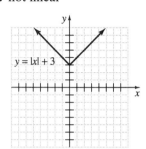 **33.** linear 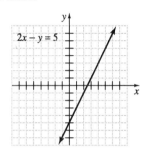 **35.** not linear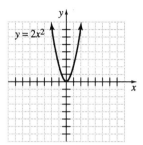

37. not linear 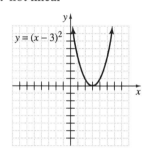 **39.** linear 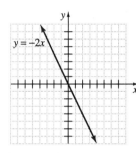 **41.** linear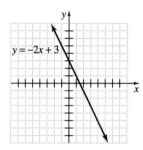

43. not linear 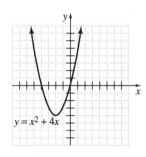 **45.** $(-3, -27), (-2, -8), (-1, -1), (0, 0), (1, 1), (2, 8), (3, 27)$

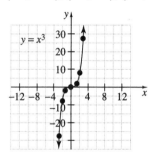

47. -1 **49.** 5   **51.** no   **53.** yes

Exercise Set 7.4

1. domain: $\{-7, 0, 2, 10\}$; range: $\{-7, 0, 4, 10\}$ **3.** domain: $\{0, 1, 5\}$; range: $\{-2\}$ **5.** yes   **7.** no   **9.** yes
11. no   **13.** yes   **15.** yes   **17.** no   **19.** yes   **21.** no   **23.** yes   **25.** yes   **27.** no   **29.** no
31. yes   **33.** $-9; -5; 1$ **35.** $6; 2; 11$ **37.** $-8; 0; 27$ **39.** $2; 0; 3$ **41.** $-5; 0; 20$ **43.** $5; 3; 35$
45. $4; 3; -21$ **47.** $6; 6; 6$ **49.** 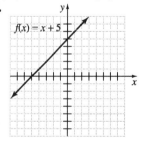   **51.**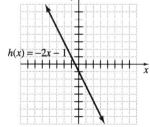

53. 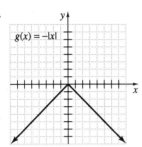 **55.** 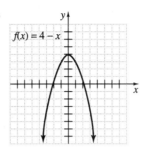 **57.**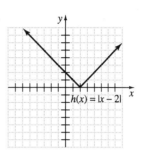

59. 5:20 A.M.   **61.** Answers may vary   **63.** all real numbers   **65.** all real numbers except -5
67. all real numbers   **69.** domain: all real numbers; range: $y \geq -4$
71. domain: all real numbers; range: all real numbers   **73.** domain: all real numbers; range: $y = 2$ **75.** 9 P.M.
77. January 1; December 1   **79.** Yes; it passes the vertical line test   **81.** Answers may vary
83. (a) 166.38 cm   (b) 148.25 cm   **85.** $\dfrac{19}{2x}$ meters   **87.** $(-2, 1)$ **89.** $(-3, -1)$ **91.** (a) 11   (b) $2a + 7$
(c) $2a + 11$ **93.** (a) 16   (b) $a^2 + 7$ (c) $a^2 - 6a + 16$

Chapter 7 Review

1. $m = -3; (0, 7)$ **3.** $m = 0; (0, 2)$ **5.** perpendicular   **7.** neither   **9.** $y = \dfrac{2}{3}x + 6$
11. 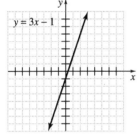   **13.** 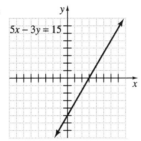   **15.** C   **17.** B   **19.** $3x + y = -5$

21. $y = -3$ **23.** $6x + y = 11$ **25.** $x + y = 6$ **27.** $x = 5$ **29.** $x = 6$ **31.** (a) $3x + y = 15$
(b) $x - 3y = 5$ **33.** 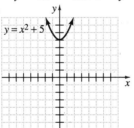   **35.** 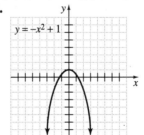   **37.**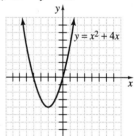

39. $(-1, -3)$ **41.** $(6, -1)$ **43.** yes   **45.** yes   **47.** yes   **49.** no   **51.** (a) 6   (b) 10   (c) 5   **53.** (a) 45
(b) -35 (c) 0   **55.** all real numbers   **57.** domain: $-3 \leq x \leq 5$; range: $-4 \leq y \leq 2$
59. domain: $x = 3$; range: all real numbers

Chapter 7 Test

1. $m = \frac{7}{3}$; $\left(0, -\frac{2}{3}\right)$ **2.** neither   **3.** $x + 4y = 10$ **4.** $7x + 6y = 0$ **5.** $8x + y = 11$ **6.** $x = -5$
7. $x - 8y = -96$ **8.** 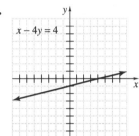   **9.**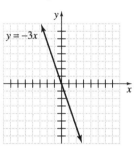

10. 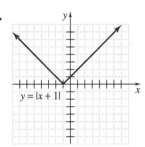   **11.** 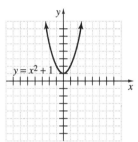   **12.** yes   **13.** yes   **14.** yes   **15.** no

16. no   **17.** (a) -8 (b) -3.6 (c) -4 **18.** (a) 0 (b) 0 (c) 60   **19.** (a) 6 (b) 6 (c) 6
20. all real numbers except -1 **21.** domain: all real numbers; range: $y \leq 4$
22. domain: all real numbers; range: all real numbers

Chapter 7 Cumulative Review

1. (a) -12 (b) -9; *Sec. 1.5, Ex. 3*   **2.** (a) *The Lion King*; $313 million (b) $72 million; *Sec. 1.9, Ex. 2*
3. (a) $2x + 6$ (b) $\frac{x - 4}{7}$ (c) $3x + 8$; *Sec. 2.1, Ex. 8*   **4.** $x = 6$; *Sec. 2.3, Ex. 1*
5. $x < -2$; 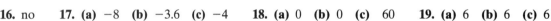 ; *Sec. 2.9, Ex. 5*   **6.** *Sec. 3.2, Ex. 6*

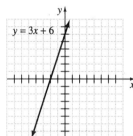

7. undefined; *Sec. 3.4, Ex. 6*   **8.** (a) 2; trinomial (b) 1; binomial (c) 3; none of these
(d) 3; binomial; *Sec. 4.2, Ex. 2*   **9.** $(x + 6)(x - 2)$; *Sec. 5.2, Ex. 3*   **10.** $x = 4, x = 5$; *Sec. 5.6, Ex. 2*
11. 1; *Sec. 6.2, Ex. 7*   **12.** $\frac{12ab^2}{27a^2b}$; *Sec. 6.3, Ex. 9*   **13.** $\frac{2m + 1}{m + 1}$; *Sec. 6.4, Ex. 5*   **14.** $\frac{x + y}{x + 2y}$; *Sec. 6.5, Ex. 3*
15. $x = -3, x = -2$; *Sec. 6.6, Ex. 2*   **16.** $m = -5$; $(0, 2)$; *Sec. 7.1, Ex. 2*   **17.** $x = -1$; *Sec. 7.2, Ex. 3*

18. *Sec. 7.3, Ex. 1*   **19.** (a) $(-1, 2)$ (b) $(-2, 0)$ (c) $\left(-2\frac{2}{3}, -1\right), (1, -1), (3, -1)$; *Sec. 7.3, Ex. 4*

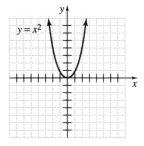

20. (a) 1   (b) 1   (c) -3; *Sec. 7.4, Ex. 7*

CHAPTER 8
Solving Systems of Linear Equations

Graphing Calculator Explorations

1. $(0.37, 0.23)$ **3.** $(0.03, -1.89)$

Mental Math, Sec. 8.1

1. consistent, independent   **3.** consistent, dependent   **5.** inconsistent, independent
7. consistent, independent

Exercise Set 8.1

1. (a) no   (b) yes   (c) no   **3.** (a) no   (b) yes   (c) no   **5.** (a) yes   (b) yes   (c) yes   **7.** Answers may vary
9. $(2, 3)$; consistent; independent   **11.** $(1, -2)$; consistent; independent

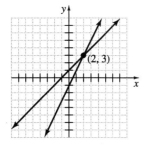   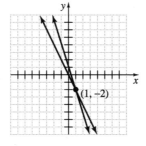

13. $(-2, 1)$; consistent; independent   **15.** $(4, 2)$; consistent; independent

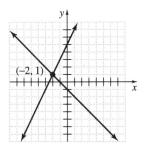 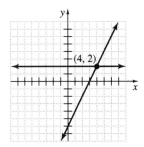

17. no solution; inconsistent; independent   **19.** infinite number of solutions; consistent; dependent

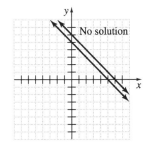 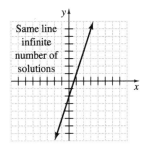

21. $(0, -1)$; consistent; independent   **23.** $(4, -3)$; consistent; independent

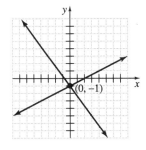 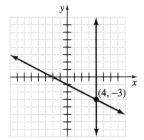

25. $(-5, -7)$; consistent; independent   **27.** $(5, 2)$; consistent; independent

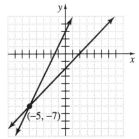 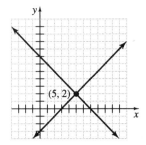

29. Answers may vary   **31.** intersecting; one solution   **33.** parallel; no solution
35. identical lines; infinite number of solutions   **37.** intersecting; one solution   **39.** intersecting; one solution
41. identical lines; infinite number of solutions   **43.** parallel; no solution   **45.** Answers may vary
47. 1984, 1988   **49. (a)** $(4, 9)$ **(b)** See graph; yes   **51.** $x = -1$ **53.** $y = 3$ **55.** $z = -7$

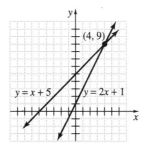

Exercise Set 8.2

1. $(2, 1)$ **3.** $(-3, 9)$ **5.** $(4, 2)$ **7.** $(-2, 4)$ **9.** $(-2, -1)$ **11.** no solution
13. infinite number of solutions   **15.** Answers may vary   **17.** $(10, 5)$ **19.** $(2, 7)$ **21.** $(3, -1)$
23. $(3, 5)$ **25.** $\left(\frac{2}{3}, -\frac{1}{3}\right)$ **27.** $(-1, -4)$ **29.** $(-6, 2)$ **31.** $(2, 1)$ **33.** no solution   **35.** $\left(-\frac{1}{5}, \frac{43}{5}\right)$
37. $(1, -3)$ **39.** $(-2.6, 1.3)$ **41.** $(3.28, 2.11)$ **43.** $-6x - 4y = -12$ **45.** $-12x + 3y = 9$
47. $5n$ **49.** $-15b$

Exercise Set 8.3

1. $(1, 2)$ **3.** $(2, -3)$ **5.** $(5, -2)$ **7.** $(-2, -5)$ **9.** $(-7, 5)$ **11.** $\left(\frac{12}{11}, -\frac{4}{11}\right)$ **13.** no solution
15. $\left(\frac{3}{2}, 3\right)$ **17.** $(1, 6)$ **19.** Answers may vary   **21.** $(6, 0)$ **23.** no solution   **25.** $\left(2, -\frac{1}{2}\right)$
27. $(6, -2)$ **29.** infinite number of solutions   **31.** infinite number of solutions   **33.** $(-2, 0)$
35. (a) $b = 15$ **(b)** any real number except 15   **37.** $(2, 5)$ **39.** $(-3, 2)$ **41.** $(0, 3)$ **43.** $(5, 7)$
45. $\left(\frac{1}{3}, 1\right)$ **47.** infinite number of solutions   **49.** $(-8.9, 10.6)$ **51.** $2x + 6 = x - 3$ **53.** $20 - 3x = 2$
55. $4(n + 6) = 2n$

Exercise Set 8.4

1. $x + y = 15; x - y = 7$ **3.** larger: x; smaller: y; $x + y = 6500; x = y + 800$ **5.** c   **7.** a   **9.** a
11. 33 and 50   **13.** adult's: \$29; children's: \$18   **15.** 27 nickels; 53 quarters
17. rowing: $6\frac{1}{2}$ mph; current: $2\frac{1}{2}$ mph   **19.** plane: 455 mph; wind: 65 mph
21. $4\frac{1}{2}$ oz. of 4% solution; $7\frac{1}{2}$ oz. of 12% solution   **23.** 113 lb high quality; 87 lb cheaper quality
25. 14 and -3 **27.** 23   **29.** $x = 9$ ft.; $y = 15$ ft.   **31.** $4\frac{1}{2}$ hr   **33.** \$23 daily fee; \$0.21 per mile

35.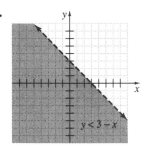

37.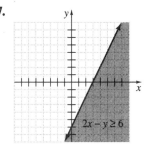

Exercise Set 8.5

1.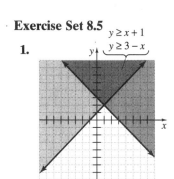

3.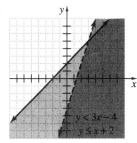

5.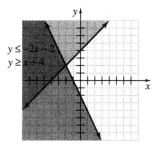

7.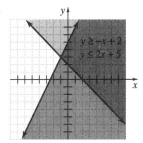

9.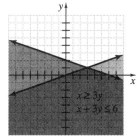

11.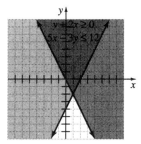

13.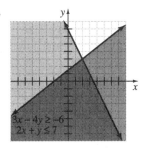

15.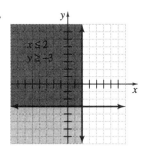

17.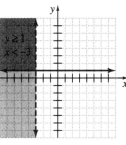

19.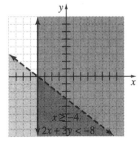

21.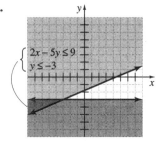

23.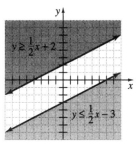

25. C **27.** D **29. (a)** $x + y \leq 8, x < 3, x \geq 0, y \geq 0$ **(b)** See graph. **31.** Answers may vary **33.** 9

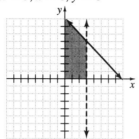

35. $\dfrac{4}{9}$ **37.** 1 **39.** -5

Chapter 8 Review

1. no; yes; no **3.** no; yes; yes **5.** $(3, -1)$ **7.** $(-3, -2)$

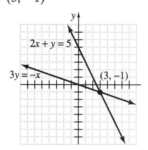   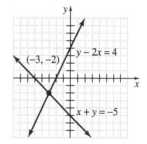

9. $\left(\dfrac{1}{2}, \dfrac{1}{2}\right)$ **11.** intersecting lines, one solution **13.** identical lines, infinite number of solutions

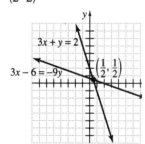

15. $(-1, 4)$ **17.** $(3, -2)$ **19.** no solutions, inconsistent **21.** $(3, 1)$ **23.** $(8, -6)$ **25.** $(-6, 2)$
27. $(3, 7)$ **29.** infinite number of solutions, dependent **31.** $\left(2, -2\dfrac{1}{2}\right)$ **33.** $(-6, 15)$ **35.** $(-3, 1)$
37. -6 and 22 **39.** ship: 21.1 mph; current: 3.2 mph **41.** width: 1.15 ft.; length: 1.85 ft.
43. one egg: $0.40; one strip of bacon: $0.65 **45.**

47. 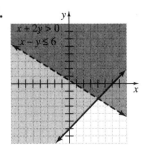 **49.** 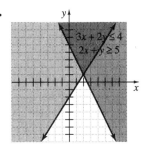 **51.**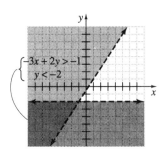

Chapter 8 Test

1. no **2.** yes **3.** $(-4, 2)$ **4.** $(-4, 1)$ **5.** $\left(\dfrac{1}{2}, -2\right)$ **6.** $(4, -2)$ **7.** $\left(2, \dfrac{1}{2}\right)$

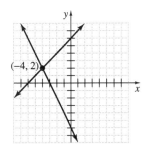

8. $(4, -5)$ **9.** $(7, 2)$ **10.** $(5, -2)$ **11.** 20 $1 bills; 42 $5 bills **12.** $1225 at 5%; $2775 at 9%
13. 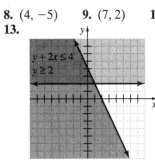 **14.**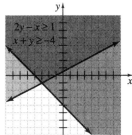

Chapter 8 Cumulative Review

1. (a) $\dfrac{64}{25}$ **(b)** $\dfrac{1}{20}$ **(c)** $\dfrac{5}{4}$; Sec. 1.2, Ex. 4 **2.** $2 loss; Sec. 1.5, Ex. 4 **3. (a)** $\dfrac{1}{22}$ **(b)** $\dfrac{16}{3}$ **(c)** $-\dfrac{1}{10}$
(d) $-\dfrac{13}{9}$; Sec. 1.7, Ex. 4 **4. (a)** 5 **(b)** $8 - x$; Sec. 2.2, Ex. 7 **5.** no solution; Sec. 2.4, Ex. 6 **6. (a)** 73%
(b) 139% **(c)** 25%; Sec. 2.7, Ex. 2 **7.** $x \leq -18$; ; Sec. 2.9, Ex. 7

8. *Sec. 3.3, Ex. 3*   **9.** *Sec. 3.5, Ex. 4*   **10.** (a) $\dfrac{25x^4}{y^6}$ (b) x^6 (c) $32x^2$

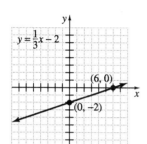
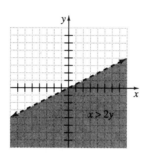

(d) a^3b; *Sec. 4.1, Ex. 9*   **11.** $-4x^2 + 6x + 2$; *Sec. 4.2, Ex. 6*   **12.** $4x^2 - 4x + 6 - \dfrac{11}{2x+3}$; *Sec. 4.6, Ex. 7*
13. $x = -\dfrac{1}{2}, x = 4$; *Sec. 5.6, Ex. 3*   **14.** 6, 8, 10; *Sec. 5.7, Ex. 4*   **15.** 1; *Sec. 6.3, Ex. 2*
16. $\dfrac{3}{4}$; *Sec. 7.1, Ex. 1*   **17.** yes; no; *Sec. 8.1, Ex. 1*   **18.** no solution; *Sec. 8.1, Ex. 5*   **19.** $\left(6, \dfrac{1}{2}\right)$; *Sec. 8.2, Ex. 2*
20. $\left(-\dfrac{5}{4}, -\dfrac{5}{2}\right)$; *Sec. 8.3, Ex. 5*   **21.** 29 and 8; *Sec. 8.4, Ex. 3*   **22.** *Sec. 8.5, Ex. 3*

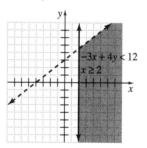

CHAPTER 9
Roots and Radicals

Scientific Calculator Explorations 9.1
1. 2.646   **3.** 3.162   **5.** 9.055

Exercise Set 9.1
1. 4   **3.** 9   **5.** $\dfrac{1}{5}$ **7.** -10 **9.** 4   **11.** -3 **13.** $\dfrac{1}{2}$ **15.** -5 **17.** 2   **19.** 3
21. not a real number   **23.** $\dfrac{1}{3}$ **25.** $\dfrac{3}{5}$ **27.** -7 **29.** Answers may vary   **31.** z **33.** x^2 **35.** $3x^4$
37. xy^3 **39.** x^5 **41.** x^6 **43.** $9x$ **45.** $-12y^7$ **47.** xy **49.** $4x^8$ **51.** 0   **53.** $-\dfrac{1}{2}$
55. not a real number   **57.** -8 **59.** -13 **61.** 1   **63.** $\dfrac{5}{8}$ **65.** 2   **67.** 3   **69.** rational, 3

71. irrational, 6.083   **73.** rational, 13   **75.** rational, 2   **77.** 18.0 ft.   **79.** $\sqrt{2} \approx 1.41$; 126.90 ft.
81. $(-8, -2), (-2, -1.3), (-1, -1), (0, 0), (1, 1), (2, 1.3), (8, 2)$

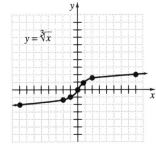

83. See graph; $(-3, 0)$

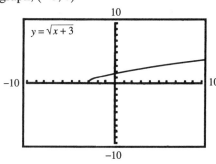

85. See graph; $(5, 0)$

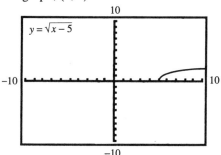

87. $4 \cdot 2$ **89.** $25 \cdot 3$ **91.** $4 \cdot 11$ **93.** $9 \cdot 10$

Mental Math, Sec. 9.2

1. 6   **3.** x **5.** 0   **7.** $5x^2$

Exercise Set 9.2

1. $2\sqrt{5}$ **3.** $3\sqrt{2}$ **5.** $5\sqrt{2}$ **7.** $\sqrt{33}$ **9.** $2\sqrt[3]{3}$ **11.** $5\sqrt[3]{2}$ **13.** Answers may vary   **15.** $\dfrac{2\sqrt{2}}{5}$
17. $\dfrac{3\sqrt{3}}{11}$ **19.** $\dfrac{3}{2}$ **21.** $\dfrac{5\sqrt{5}}{3}$ **23.** $\dfrac{\sqrt[3]{5}}{4}$ **25.** $\dfrac{\sqrt[3]{7}}{2}$ **27.** $x^3\sqrt{x}$ **29.** $\dfrac{2\sqrt{22}}{x^2}$ **31.** $x^5\sqrt[3]{x}$
33. $\dfrac{\sqrt[3]{2}}{x^3}$ **35.** $2\sqrt{15}$ **37.** $6\sqrt{5}$ **39.** $2\sqrt{13}$ **41.** $\dfrac{\sqrt{11}}{6}$ **43.** $-\dfrac{\sqrt{3}}{4}$ **45.** $\dfrac{\sqrt[3]{15}}{4}$ **47.** $2\sqrt[3]{10}$
49. $2\sqrt[4]{3}$ **51.** $x^6\sqrt{x}$ **53.** $5x\sqrt{3}$ **55.** $4x^2y\sqrt{6}$ **57.** $\dfrac{2\sqrt{3}}{y}$ **59.** $\dfrac{3\sqrt{x}}{y}$ **61.** $-2x^2$ **63.** $2\sqrt[3]{10}$ in.
65. $1700   **67.** $14x$ **69.** $2x^2 - 7x - 15$ **71.** 0   **73.** 8 cm

Mental Math, Sec. 9.3

1. $8\sqrt{2}$ **3.** $7\sqrt{x}$ **5.** $3\sqrt{7}$

Exercise Set 9.3

1. $-4\sqrt{3}$ **3.** $9\sqrt{6} - 5$ **5.** $\sqrt{5} + \sqrt{2}$ **7.** $7\sqrt[3]{3} - \sqrt{3}$ **9.** $-5\sqrt[3]{2} - 6$ **11.** $8\sqrt{5}$ in.
13. Answers may vary   **15.** $5\sqrt{3}$ **17.** $9\sqrt{5}$ **19.** $-\sqrt{2} + 4\sqrt{3}$ **21.** $\sqrt[3]{4}$ **23.** $x + \sqrt{x}$ **25.** 0
27. $4\sqrt{x} - x\sqrt{x}$ **29.** $7\sqrt{5}$ **31.** $2\sqrt{5}$ **33.** $6 - 3\sqrt{3}$ **35.** $9\sqrt{3}$ **37.** $5\sqrt{5}$ **39.** $3\sqrt{6} + 9$

41. $\dfrac{3\sqrt{3}}{8}$ **43.** $4 + 4\sqrt[3]{2}$ **45.** $-3 + 3\sqrt[3]{2}$ **47.** $23\sqrt{2}$ **49.** $12z\sqrt{2x}$ **51.** $13\sqrt{x}$ **53.** $x\sqrt{3x} + 3x\sqrt{x}$
55. $4x\sqrt{2} + 2\sqrt[3]{4x^2} + 2x$ **57.** $x\sqrt[3]{5x}$ **59.** $\left(48 + \dfrac{9\sqrt{3}}{2}\right)$ sq. ft.   **61.** $x^2 + 12x + 36$ **63.** $4x^2 - 4x + 1$
65. $(4, 2)$

Mental Math, Sec. 9.4

1. $\sqrt{6}$ **3.** $\sqrt{6}$ **5.** $\sqrt{10y}$

Exercise Set 9.4

1. 4   **3.** $5\sqrt{2}$ **5.** $2\sqrt[3]{6}$ **7.** $2\sqrt{5} + 5\sqrt{2}$ **9.** $15 - 12\sqrt{15} - 5\sqrt{2} + 4\sqrt{30}$ **11.** $x - 36$
13. $67 + 16\sqrt{3}$ **15.** $130\sqrt{3}$ sq. meters   **17.** 4   **19.** $3\sqrt{2}$ **21.** $5y^2$ **23.** $\dfrac{\sqrt{15}}{5}$ **25.** $\dfrac{\sqrt{6y}}{6y}$
27. $\dfrac{\sqrt{10}}{6}$ **29.** $3\sqrt[3]{4}$ **31.** $\dfrac{\sqrt[3]{3}}{3}$ **33.** $\dfrac{\sqrt[3]{6x}}{3x}$ **35.** $\dfrac{\sqrt{A\pi}}{\pi}$ **37.** Answers may vary
39. $3\sqrt{2} - 3$ **41.** $2\sqrt{10} + 6$ **43.** $\sqrt{30} + 5 + \sqrt{6} + \sqrt{5}$ **45.** $3 + \sqrt{3}$ **47.** $3 - 2\sqrt{5}$ **49.** $3\sqrt{3} + 1$
51. $24\sqrt{5}$ **53.** 20   **55.** $36x$ **57.** $\sqrt{30} + \sqrt{42}$ **59.** $4x\sqrt{5} - 60\sqrt{x}$ **61.** $\sqrt{6} - \sqrt{15} + \sqrt{10} - 5$
63. -5 **65.** $x - 9$ **67.** $15 + 6\sqrt{6}$ **69.** $9x - 30\sqrt{x} + 25$ **71.** $5\sqrt{3}$ **73.** $2y\sqrt{6}$ **75.** $2xy\sqrt{3y}$
77. $12\sqrt[3]{10}$ **79.** $5\sqrt[3]{3}$ **81.** $3\sqrt[3]{x^2}$ **83.** $\dfrac{\sqrt{30}}{15}$ **85.** $\dfrac{\sqrt{15}}{10}$ **87.** $\dfrac{3\sqrt{2x}}{2}$ **89.** $\dfrac{\sqrt[3]{10}}{2}$ **91.** $-8 - 4\sqrt{5}$
93. $5\sqrt{10} - 15$ **95.** $\dfrac{6\sqrt{5} - 4\sqrt{3}}{11}$ **97.** $\sqrt{6} + \sqrt{3} + \sqrt{2} + 1$ **99.** $\dfrac{2}{\sqrt{6} - \sqrt{2} - \sqrt{3} + 1}$
101. $x + 4$ **103.** $2x - 1$ **105.** $x = 44$ **107.** $z = 2$

Exercise Set 9.5

1. $x = 81$ **3.** $x = -1$ **5.** $x = 16$ **7.** $x = 2$ **9.** $x = -3$ **11.** $x = 2$ **13.** no solution   **15.** $x = 5$
17. $x = 49$ **19.** no solution   **21.** $x = -2$ **23.** $x = \dfrac{3}{2}$ **25.** $x = 4$ **27.** $x = 3$ **29.** $x = 9$
31. $x = 12$ **33.** $x = 3, x = 1$ **35.** $x = -1$ **37.** $x = 2$ **39.** $x = 2$ **41.** no solution
43. $x = 0, x = -3$ **45.** 9   **47.** (a) 3.2; 10; 31.6 (b) no   **49.** Answers may vary   **51.** 2.43   **53.** 0.48
55. $2x - (x + 3) = 11; x = 14$ **57.** $2x + 2(x + 2) = 24$; length: 7 in.   **59.** $x - y = 58; y = 3x$; 29 and 87

Exercise Set 9.6

1. $\sqrt{13}$ **3.** $3\sqrt{3}$ **5.** 25   **7.** $\sqrt{22}$ **9.** $3\sqrt{17}$ **11.** $\sqrt{41}$ **13.** $4\sqrt{2}$ **15.** $3\sqrt{10}$
17. $x = -4 + 2\sqrt{10}$ **19.** $\sqrt{29}$ **21.** $\sqrt{73}$ **23.** $2\sqrt{10}$ **25.** $\dfrac{3\sqrt{5}}{2}$ **27.** $\sqrt{85}$ **29.** 20.6 ft.
31. 24 cubic ft.   **33.** 51.2 ft.   **35.** 11.7 ft.   **37.** $\dfrac{3\sqrt{2}}{2}$ in.   **39.** 126 ft.   **41.** 360 ft.   **43.** 130.6 m
45. Answers may vary   **47.** -27 **49.** $\dfrac{8}{343}$ **51.** x^6 **53.** x^8

Exercise Set 9.7

1. 2   **3.** 3   **5.** 8   **7.** 4   **9.** $-\dfrac{1}{2}$ **11.** $\dfrac{1}{64}$ **13.** $\dfrac{1}{729}$ **15.** $\dfrac{5}{2}$ **17.** Answers may vary
19. 2   **21.** 2   **23.** $\dfrac{1}{x^{2/3}}$ **25.** x^3 **27.** Answers may vary   **29.** 9   **31.** -2 **33.** -3 **35.** $\dfrac{1}{9}$

37. $\dfrac{3}{4}$ **39.** 27 **41.** 512 **43.** -4 **45.** 32 **47.** $\dfrac{8}{27}$ **49.** $\dfrac{1}{27}$ **51.** $\dfrac{1}{2}$ **53.** $\dfrac{1}{5}$ **55.** $\dfrac{1}{125}$ **57.** 9 **59.** $6^{1/3}$ **61.** x^6 **63.** 36 **65.** $\dfrac{1}{3}$ **67.** $\dfrac{x^{2/3}}{y^{3/2}}$ **69.** $\dfrac{x^{16/5}}{y^6}$ **71.** 11,224 people **73.** 3.344 **75.** 5.665 **77.** 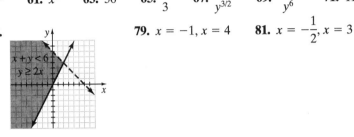 **79.** $x=-1, x=4$ **81.** $x=-\dfrac{1}{2}, x=3$

Chapter 9 Review

1. 9 **3.** 3 **5.** $-\dfrac{3}{8}$ **7.** not a real number **9.** irrational, 8.718 **11.** x^6 **13.** $3x^3y$ **15.** $-2x^2$ **17.** $3\sqrt{6}$ **19.** $5xy^3\sqrt{6x}$ **21.** $3\sqrt[3]{2}$ **23.** $2y\sqrt[4]{3x^3y^2}$ **25.** $\dfrac{3\sqrt{2}}{5}$ **27.** $\dfrac{3y\sqrt{5}}{2x^2}$ **29.** $\dfrac{\sqrt[4]{9}}{2}$ **31.** $\dfrac{y^2\sqrt[3]{3}}{2x}$ **33.** $2\sqrt[3]{3}-\sqrt[3]{2}$ **35.** $6\sqrt{6}-2\sqrt[3]{6}$ **37.** $5\sqrt{7}+2\sqrt[3]{7}$ **39.** $\dfrac{\sqrt{5}}{6}$ **41.** 0 **43.** $30\sqrt{2}$ **45.** $6\sqrt{2}-18$ **47.** $3\sqrt{2}-5\sqrt{3}+2\sqrt{6}-10$ **49.** $4\sqrt{2}$ **51.** $\dfrac{x\sqrt{5x}}{2y^4}$ **53.** $\dfrac{\sqrt{30}}{6}$ **55.** $\dfrac{\sqrt{6x}}{2x}$ **57.** $\dfrac{\sqrt{35}}{10y}$ **59.** $\dfrac{\sqrt[3]{21}}{3}$ **61.** $\dfrac{\sqrt[3]{12x}}{2x}$ **63.** $3\sqrt{5}+6$ **65.** $4\sqrt{6}-8$ **67.** $\dfrac{2\sqrt{2}-1}{7}$ **69.** $-\dfrac{3+5\sqrt{3}}{11}$ **71.** $x=18$ **73.** $x=25$ **75.** $x=12$ **77.** $x=6$ **79.** $2\sqrt{14}$ **81.** $4\sqrt{34}$ ft. **83.** $\sqrt{130}$ **85.** 2.4 in. **87.** $a^{5/2}$ **89.** $x^{5/2}$ **91.** 4 **93.** -2 **95.** -512 **97.** $\dfrac{8}{27}$ **99.** $\dfrac{1}{5}$ **101.** 32 **103.** $\dfrac{1}{3^{2/3}}$ **105.** $\dfrac{1}{x^2}$

Chapter 9 Test

1. 4 **2.** 5 **3.** 8 **4.** $\dfrac{3}{4}$ **5.** not a real number **6.** $\dfrac{1}{9}$ **7.** $3\sqrt{6}$ **8.** $2\sqrt{23}$ **9.** $x^3\sqrt{3}$ **10.** $2x^2y^3\sqrt{2y}$ **11.** $3x^4\sqrt{x}$ **12.** $2\sqrt[3]{5}$ **13.** $2x^2y^3\sqrt[3]{y}$ **14.** $-8\sqrt{3}$ **15.** $3\sqrt[3]{2}-2x\sqrt{2}$ **16.** $\dfrac{\sqrt{5}}{4}$ **17.** $\dfrac{x\sqrt[3]{2}}{3}$ **18.** $6\sqrt{2x}$ **19.** $\dfrac{\sqrt{6}}{3}$ **20.** $\dfrac{\sqrt[3]{15}}{3}$ **21.** $\dfrac{\sqrt[3]{6x}}{2x}$ **22.** $4\sqrt{6}-8$ **23.** $-1-\sqrt{3}$ **24.** $x=9$ **25.** $x=5$ **26.** $x=9$ **27.** $4\sqrt{5}$ in. **28.** $\sqrt{5}$ **29.** $\dfrac{1}{16}$ **30.** $x^{10/3}$

Chapter 9 Cumulative Review

1. 54; Sec. 1.3, Ex. 4 **2.** $x=-3$; Sec. 2.2, Ex. 4 **3.** (a) 45% (b) 83% (c) 9900 homeowners (d) 57.6°; Sec. 2.7, Ex. 4 **4.** $x>\dfrac{13}{7}$; Sec. 2.9, Ex. 8

5. Sec. 3.3, Ex. 7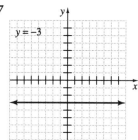
6. Sec. 3.5, Ex. 5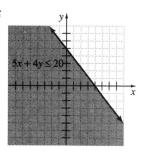

7. $6x^2 - 11x - 10$; Sec. 4.3, Ex. 2 **8. (a)** 102,000 **(b)** 0.007358 **(c)** 84,000,000 **(d)** 0.00003007; Sec. 4.5, Ex. 6
9. $-3a(3a^4 - 6a + 1)$; Sec. 5.1, Ex. 5 **10. (a)** $x = 3$ **(b)** $x = 1, x = 2$ **(c)** none **(d)** none; Sec. 6.1, Ex. 2
11. $-\dfrac{2 + x}{3x + 1}$; Sec. 6.1, Ex. 8 **12. (a)** 0 **(b)** $\dfrac{15 + 14x}{50x^2}$; Sec. 6.4, Ex. 1 **13.** $x = -\dfrac{3}{2}, x = 1$; Sec. 6.6, Ex. 3
14. $2x + y = 3$; Sec. 7.2, Ex. 1 **15.** $(3, 1)$; Sec. 8.3, Ex. 4 **16.** Sec. 8.5, Ex. 2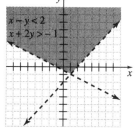

17. (a) 1 **(b)** -3 **(c)** $\dfrac{\sqrt[3]{5}}{5}$; Sec. 9.1, Ex. 2 **18. (a)** $3\sqrt{6}$, **(b)** $2\sqrt{3}$ **(c)** $10\sqrt{2}$ **(d)** $\sqrt{35}$; Sec. 9.2, Ex. 1
19. (a) $2x - 4\sqrt{x}$ **(b)** $19x\sqrt[3]{2x}$; Sec. 9.3, Ex. 3 **20. (a)** $\dfrac{2\sqrt{7}}{7}$ **(b)** $\dfrac{\sqrt{15}}{6}$ **(c)** $\dfrac{\sqrt{2x}}{6x}$; Sec. 9.4, Ex. 5
21. $x = \dfrac{1}{2}$; Sec. 9.5, Ex. 4 **22.** $8\sqrt{5}$ ft. per sec.; Sec. 9.6, Ex. 4 **23. (a)** 5 **(b)** 2 **(c)** -2 **(d)** -3
(e) $\dfrac{1}{3}$; Sec. 9.7, Ex. 1

CHAPTER 10
Solving Quadratic Equations

Exercise Set 10.1

1. $k = \pm 3$ **3.** $m = -5, 3$ **5.** $x = \pm\dfrac{9\sqrt{2}}{2}$ **7.** $a = \pm 3$ **9.** $x = \pm 8$ **11.** $p = \pm\dfrac{1}{7}$
13. no real solution **15.** $x = \pm 5$ **17.** $x = \pm\dfrac{2\sqrt{3}}{3}$ **19.** 15 ft. by 15 ft. **21.** $x = -2, 12$ **23.** $x = -2 \pm \sqrt{7}$
25. $m = 0, 1$ **27.** $p = -2 \pm \sqrt{10}$ **29.** $y = -4, \dfrac{8}{3}$ **31.** no real solution **33.** $x = \dfrac{11 \pm 5\sqrt{2}}{2}$
35. $x = 10\sqrt{2}$ cm **37.** $q = \pm 10$ **39.** $x = 9, 17$ **41.** $z = \pm 2\sqrt{3}$ **43.** $x = -5 \pm \sqrt{10}$

45. no real solution **47.** $y = \pm\dfrac{\sqrt{22}}{2}$ **49.** $p = -3, 8$ **51.** $x = \dfrac{1 \pm \sqrt{7}}{3}$ **53.** $x = \dfrac{7 \pm 4\sqrt{2}}{3}$
55. $x = \pm 1.33$ **57.** $x = -1.02, 3.76$ **59.** $y = -3.09, 1.51$ **61.** $x = -6, 2$ **63.** $y = 5 \pm \sqrt{11}$
65. $r = 6$ in. **67.** $(x + 3)^2$ **69.** $(x - 2)^2$ **71.** 35 million **73.** 44.8 million

Mental Math, Sec. 10.2

1. 16 **3.** 100 **5.** 49

Exercise Set 10.2

1. $(x + 2)^2$ **3.** $(k - 6)^2$ **5.** $\left(x - \dfrac{3}{2}\right)^2$ **7.** $\left(m - \dfrac{1}{2}\right)^2$ **9.** $x = 0, 6$ **11.** $x = -6, -2$
13. $x = -1 \pm \sqrt{6}$ **15.** $k = \pm 8$ **17.** $x = -\dfrac{1}{2}, \dfrac{13}{2}$ **19.** no real solution **21.** $x = \dfrac{3 \pm \sqrt{19}}{2}$
23. $x = -3 \pm \sqrt{34}$ **25.** $z = \dfrac{-5 \pm \sqrt{53}}{2}$ **27.** $x = 1 \pm \sqrt{2}$ **29.** $y = -4, -1$ **31.** $x = -2, 4$
33. $y = \dfrac{-4 \pm \sqrt{6}}{2}$ **35.** $y = \dfrac{1}{2}, 1$ **37.** $y = \dfrac{1 \pm \sqrt{13}}{3}$ **39.** $y = -2, 7$ **41.** $x = -6, 3$
43. Answers may vary **45.** $x = -6, -2$ **47.** $x \approx -0.68, 3.68$ **49.** $-\dfrac{1}{2}$ **51.** -1 **53.** $3 + 2\sqrt{5}$
55. $\dfrac{1 - 3\sqrt{2}}{2}$

Mental Math, Sec. 10.3

1. $a = 2, b = 5, c = 3$ **3.** $a = 10, b = -13, c = -2$ **5.** $a = 1, b = 0, c = -6$

Exercise Set 10.3

1. $-2, 1$ **3.** $\dfrac{-5 \pm \sqrt{17}}{2}$ **5.** $\dfrac{2 + \sqrt{2}}{2}$ **7.** $x = 1, 2$ **9.** $k = \dfrac{-7 \pm \sqrt{37}}{6}$ **11.** $x = \pm\dfrac{2}{7}$
13. no real solution **15.** $y = -3, 10$ **17.** $x = \pm\sqrt{5}$ **19.** $m = -3, 4$ **21.** $x = -2 \pm \sqrt{7}$
23. no real solution **25.** $m = 1 \pm \sqrt{2}$ **27.** $p = -\dfrac{3}{4}, -\dfrac{1}{2}$ **29.** $a = \dfrac{1}{2}, 3$ **31.** $x = \dfrac{5 \pm \sqrt{33}}{2}$
33. $x = -2, \dfrac{7}{3}$ **35.** $x = \dfrac{-9 \pm \sqrt{129}}{12}$ **37.** $p = \dfrac{4 \pm \sqrt{2}}{7}$ **39.** $a = 3 \pm \sqrt{7}$ **41.** $x = \dfrac{3 \pm \sqrt{3}}{2}$
43. $x = -1, \dfrac{1}{3}$ **45.** $y = -\dfrac{3}{4}, \dfrac{1}{5}$ **47.** no real solution **49.** $y = \dfrac{3 \pm \sqrt{13}}{4}$ **51.** $x = \dfrac{7 \pm \sqrt{129}}{20}$
53. no real solution **55.** $z = \dfrac{1 \pm \sqrt{2}}{5}$ **57.** $x = \dfrac{-3\sqrt{2} \pm \sqrt{38}}{2}$ **59.** $x = -4.4, 3.4$ **61.** $x = -0.7, 5.0$
63. 7.9 sec. **65.** Answers may vary **67.** $x = \dfrac{3}{5}$ **69.** $x = -\dfrac{11}{30}$ **71.** $z = -4$

Exercise Set 10.4

1. $x = \dfrac{1}{5}, 2$ **3.** $x = 1 \pm \sqrt{2}$ **5.** $a = \pm 2\sqrt{5}$ **7.** no real solution **9.** $x = 2$ **11.** $p = 3$ **13.** $y = \pm 2$

15. $x = 0, 1, 2$ **17.** $z = -5, 0$ **19.** $x = \dfrac{3 \pm \sqrt{7}}{5}$ **21.** $m = -1, \dfrac{3}{2}$ **23.** $x = \dfrac{5 \pm \sqrt{105}}{20}$ **25.** $x = \dfrac{7}{4}, 5$

27. $x = \dfrac{7 \pm 3\sqrt{2}}{5}$ **29.** $z = \dfrac{7 \pm \sqrt{193}}{6}$ **31.** $x = -10, 11$ **33.** $x = -\dfrac{2}{3}, 4$ **35.** $x = 0.1, 0.5$

37. $x = \dfrac{11 \pm \sqrt{41}}{20}$ **39.** $z = \dfrac{4 \pm \sqrt{10}}{2}$ **41.** $\dfrac{1 + \sqrt{5}}{2}$ **43.** Answers may vary **45.** $2\sqrt{26}$ **47.** $4\sqrt{5}$

49. width: 17 in.; length: 23 in.

Graphing Calculator Explorations 10.6

1. $x = -0.41, 7.41$ **3.** $x = 0.91, 2.38$ **5.** $x = -0.39, 0.84$

Exercise Set 10.6

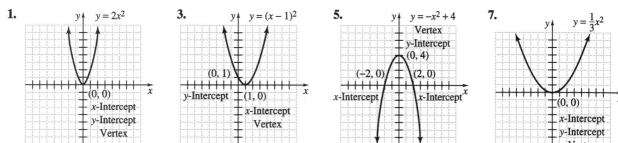

9. 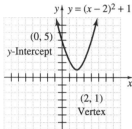   **11.** 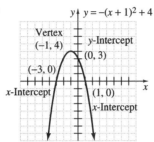   **13.** 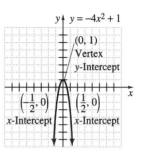   **15.** F

17. A   **19.** H   **21.** B   **23.** D   **25.** 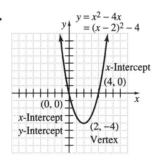   **27.**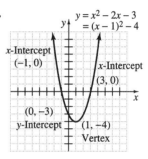

29. 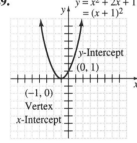   **31.** 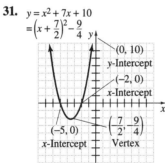   **33.**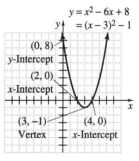

35. domain: all real numbers; range: $y \leq -4$ **37.** domain: all real numbers; range: $y \geq -2$ **39.** C   **41.** A

43. $\dfrac{5}{14}$ **45.** $\dfrac{x}{2}$ **47.** $\dfrac{2x^2}{x-1}$ **49.** $-4b$

Chapter 10 Review

1. $x = -\dfrac{3}{5}, 4$ **3.** $m = -\dfrac{1}{3}, 2$ **5.** $k = \pm 5\sqrt{2}$ **7.** $x = -1, 7$ **9.** $x = 4, 18$ **11.** $x = 0, \pm 3$

13. $p = -3, 2$ **15.** $x^2 - 10x + 25 = (x-5)^2$ **17.** $a^2 + 4a + 4 = (a+2)^2$ **19.** $m^2 - 3m + \dfrac{9}{4} = \left(m - \dfrac{3}{2}\right)^2$

21. $x = 3 \pm \sqrt{2}$ **23.** $y = -1, \dfrac{1}{2}$ **25.** $x = 5 \pm 3\sqrt{2}$ **27.** $x = -1, \dfrac{1}{2}$ **29.** $x = -\dfrac{5}{3}$ **31.** $x = \dfrac{2}{5}, \dfrac{1}{3}$

33. $x = \dfrac{-1 \pm i\sqrt{39}}{4}$ **35.** $z = \dfrac{-1 \pm \sqrt{21}}{10}$ **37.** $x = 0, \pm \dfrac{1}{2}$ **39.** $x = \dfrac{1}{2}, 7$ **41.** $x = \dfrac{1}{3}$ **43.** $x = 3$

45. $x = -10, 22$ **47.** $x = \dfrac{3 \pm \sqrt{5}}{20}$ **49.** $x = -5 \pm \sqrt{30}$ **51.** $12i$ **53.** $6i\sqrt{3}$ **55.** $21 - 10i$

57. $-8 - 2i$ **59.** $-13i$ **61.** 25 **63.** $-\frac{3}{2} - \frac{1}{2}i$ **65.** $\frac{2}{5} - \frac{9}{5}i$ **67.** $x = \pm 4i$ **69.** $x = 2 \pm 3i$
71. vertex: $(0, 0)$; axis of symmetry: $x = 0$; opens downward
73. vertex: $(3, 0)$; axis of symmetry: $x = 3$; opens upward
75. vertex: $(0, -7)$; axis of symmetry: $x = 0$; opens upward
77. vertex: $(72, 14)$; axis of symmetry: $x = 72$; opens downward
79. 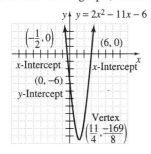 **81.** 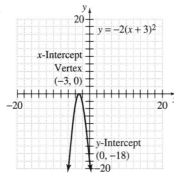 **83.**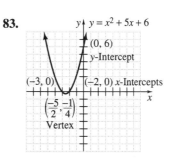

85. Note: In the graph, each square represents 2 units. **87.** 1 solution: $x = -2$ **89.** no real solution

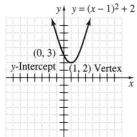

Chapter 10 Test

1. $x = -\frac{3}{2}, 7$ **2.** $x = -2, 0, 1$ **3.** $k = \pm 4$ **4.** $m = \frac{5 \pm 2\sqrt{2}}{3}$ **5.** $x = 10, 16$ **6.** $x = -2, \frac{1}{5}$
7. $x = -2, 5$ **8.** $p = \frac{5 \pm \sqrt{37}}{6}$ **9.** $x = -\frac{4}{3}, 1$ **10.** $x = -1, \frac{5}{3}$ **11.** $x = \frac{7 \pm \sqrt{73}}{6}$ **12.** $x = 2 \pm i$
13. $x = \frac{1}{3}, 2$ **14.** $x = \frac{3 \pm \sqrt{7}}{2}$ **15.** $x = -3, 0, \frac{1}{2}$ **16.** $x = 0, \pm\frac{1}{3}$ **17.** $5i$ **18.** $10i\sqrt{2}$ **19.** $8 + i$
20. $7 - 6i$ **21.** $4i$ **22.** 6 **23.** 13 **24.** $\frac{1}{5} - \frac{7}{5}i$

25.   **26.**   **27.**

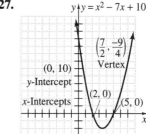

Chapter 10 Cumulative Review

1. $y = -1.6$; Sec. 2.2, Ex. 2   **2.** $t = \dfrac{16}{3}$; Sec. 2.4, Ex. 2   **3.** 8 liters @ 40%; 4 liters @ 70%; Sec. 2.8, Ex. 3
4. (a) $(0, 12)$ **(b)** $(2, 6)$ **(c)** $(-1, 15)$; Sec. 3.1, Ex. 3   **5.** $\dfrac{3}{5}$; Sec. 3.4, Ex. 8   **6. (a)** 1 **(b)** 1 **(c)** 1
(d) -1; Sec. 4.1, Ex. 8   **7.** $27a^3 + 27a^2b + 9ab^2 + b^3$; Sec. 4.3, Ex. 5   **8.** $2x + 4 - \dfrac{1}{3x - 1}$; Sec. 4.6, Ex. 6
9. $(r - 7)(r + 6)$; Sec. 5.2, Ex. 4   **10.** $(x + 6)^2$; Sec. 5.3, Ex. 7   **11.** $(y - 3)(y^2 + 3y + 9)$; Sec. 5.4, Ex. 6
12. $(4x + 1)(16x^2 - 4x + 1)$; Sec. 5.4, Ex. 7   **13.** $x = \dfrac{1}{5}, -\dfrac{3}{2}, -6$; Sec. 5.6, Ex. 6   **14.** $x - y$; Sec. 6.1, Ex. 9
15. 14 oz. for $3.79; Sec. 6.7, Ex. 6   **16. (a)** all real numbers except 0   **(b)** all real numbers; Sec. 7.4, Ex. 9
17. $(4, 2)$; Sec. 8.2, Ex. 1   **18. (a)** 6 **(b)** 8 **(c)** -5 **(d)** $\dfrac{3}{10}$ **(e)** 0; Sec. 9.1, Ex. 1   **19. (a)** -44
(b) $7x + 4\sqrt{7x} + 4$; Sec. 9.4, Ex. 3   **20. (a)** $-1 + \sqrt{3}$ **(b)** $\dfrac{9 + 5\sqrt{5}}{4}$; Sec. 9.4, Ex. 7
21. $x = -1, 7$; Sec. 10.1, Ex. 4   **22.** $x = \dfrac{-3 \pm i\sqrt{5}}{2}$; Sec. 10.2, Ex. 5   **23.** $m = 2 \pm i$; Sec. 10.5, Ex. 8

INDEX

$a^{-m/n}$, definition of, 535
a^n, definition of, 228, 234, 274
$a^{1/n}$, definition of, 534
Absolute value, 21, 31, 32
  bars, 21
  of real numbers, 2, 8, 63
Absolute value equations, graphs of, 427
Acute angle, A-3
Addition:
  associative property of, 50, 53, 66
  commutative property of, 49, 348
  of complex numbers, 566, 587
  of decimals, A-1
  distributive property of multiplication over, 50, 51
  of exponential expressions, 230, 235
  of fractions, 11, 267
  of fractions with same denominator, 14, 64
  of fractions with unlike denominators, 358
  identity element for, 52
  of polynomials, 227, 238, 240, 242, 243, 274–75, 568, 587
  property of equality, 81–87, 88, 95, 117, 151, 468
  property of inequality, 141, 143, 155
  of radicals, 497, 510–12, 539
  of rational expressions with common denominators, 343, 356–57, 401
  of rational expressions with least common denominators, 343, 356, 358–59, 401
  of rational expressions with unlike denominators, 363–69, 401–2
  of real numbers, 1, 30–35, 36, 49, 65
  of real numbers with unlike signs, 31–32, 65
  solving systems of linear equations by, 453, 468–74, 476, 480
  of two numbers with same sign, 30–31
Additive identity(ies), identifying of, 48
Additive inverse property(ies), identifying of, 48
Additive inverses (or opposites), 33, 34, 52, 53, 65
Adjacent angles, A-5
Algebra, 2, 25, 548
Algebraic equations, 498
Algebraic expressions:
  definition of, 25, 65
  evaluating with real numbers, 36
  evaluation of, 25, 65
  finding value of, 41
  multiplication of, 515
  simplifying of, 73, 74–80, 150
  translating phrases into, 25–26
  writing word phrases as, 78, 81, 85, 92, 104
Alternate interior angles, A-5–A-6
Angles, A-3–A-11
Arithmetic, language of, 2
Associative property, 88
  of addition, 50, 53, 66
  identifying of, 48, 50
  of multiplication, 50, 66
  and multiplying monomials, 246
Axis of symmetry, 573, 574, 575, 576, 579, 588

Bar graphs:
  reading of, 55–58
  three-dimensional, 103
Base, 64
  of exponential expression, 228, 231
Binomials, 238 (*See also* Polynomials)
  definition of, 239, 274
  distributive property in multiplying two, 247
  factoring of, 281, 289, 290, 291, 304–9, 311, 335
  multiplying using FOIL method, 252, 253
  special products for squaring, 299
  squaring of, 252, 253, 254, 275
Boundary lines, 210, 485
  graphing of, 211, 212, 213, 214, 215, 221
Braces, 20
Brackets, 20, 22
Broken line graph, 59, 67
Building options, choosing among, 281, 333

Calculators (*See* Graphing calculators; Scientific calculators)
Centers of circles, A-13
Circle graphs, 55
  percent in, 124
Circles, A-13
Closure properties, 49
Coefficients, 239
  of slopes, 413
  of squared variables, 554
  of term, 274
Common base, 230
Common denominator(s), 14, 15
  adding and subtracting rational expressions with, 343, 356–57, 401
Common factors:

Common factors *(cont.)*
 and fundamental principle, 347
 fundamental principle for dividing, 400
Common terms, 347
Commutative properties:
 of addition, 49, 66, 348
 identifying of, 48, 49
 of multiplication, 53, 66
 and multiplying monomials, 246
Complementary angles, A-3–A-4
Completing the square, 564
 solving quadratic equations by, 547, 552–57, 585
Complex conjugates, 569, 587
Complex expressions (or complex fractions), simplifying of, 369–375
Complex fractions:
 simplifying of, 343, 369–75, 402
 steps in simplifying, 370
Complex numbers:
 addition of, 566, 587
 definition of, 587
 division of, 566
 multiplication of, 566
 subtraction of, 566, 587
 system of, 567
 writing using *i* notation, 566
Complex solutions of quadratic equations, 566–72
Compound inequalities, 140
 definition of, 155
 solving of, 139, 145, 156
Computers, graphing software for, 181
Congruent triangles, A-7
Conjugates:
 of denominators, 540
 rationalizing of, 514, 517, 518
Consecutive integers, 328
Consistent system of equations, 454, 456, 489
Constant, degree of, 240
Constant term, 239, 412
Coplanar figures, A-4
Corner points, coordinates of, 484
Corresponding angles, A-7–A-8
Cowling's Rule, 399
Cross multiplication:
 definition of, 384
 in rational expressions, 394
Cross products, definition of, 384, 403
Cube roots, 499, 500

of negative numbers, 500
of perfect cubes, 498
of real numbers, 539
simplifying of, 507
Cubes, A-13
 factoring and difference of, 335
 factoring sum or difference of two, 304, 306–7, 311
Cylinders, investigating dimensions of, 497, 537–38

Decimal approximations, of irrational numbers, 501
Decimal numbers, 6
Decimal points in scientific notation, 261
Decimals:
 operations on, A-1–A-2
 solving equations containing, 94, 98
 writing as percents, 122, 123, 153
Degree, 238
Degree of a polynomial, definition of, 275
Degree of a term, 274
Denominators, 11, 12, 63
 adding and subtracting fractions with same, 14
 factoring in rational expressions, 346, 347, 352, 360, 365, 366, 400
 multiplying of, 351
 prime factorization of, 359
 rationalizing of, 514, 516–18, 540
 simplifying of, 20, 21, 45, 46, 64
 of slope, 199
Dependent equations, 454, 457, 466, 489
Descending powers of *x*, 239
Difference of squares, 254
 and factoring, 335
Diophantus, 398
Distance formula, 114, 529
 to solve problems, 526, 528–29
Distance, solving problems involving, 131, 133, 390, 393–94
Distributive property, 53, 67, 96, 132
 and combining like radicals, 439
 and combining like terms, 75, 510
 and complex fractions, 373
 in equations containing rational expressions, 376, 377, 378, 380
 and factoring, 285
 identifying of, 48

of multiplication over addition, 50, 51
 in multiplying complex numbers, 568, 587
 in multiplying polynomials, 246–47
 in multipling radical expressions, 515
 and point-slope form, 419, 420, 421
 and rational equations, 391, 394
 and rational expressions, 358, 364, 365
 and removing parentheses, 74, 76–77, 84, 95, 97, 117, 143, 150
 and solving systems of equations by substitution, 464
Dividend polynomials, 268, 269, 270
Dividends, A-2
Division:
 of both sides of an equation, 88, 90, 91
 of complex numbers, 566
 of decimals, A-2
 of exponential expressions, 230
 of fractions, 11, 13, 64
 and inequalities, 142
 of polynomials, 227, 266–72, 276
 of radicals, 497, 514–20, 540
 of rational expressions, 353–54, 400–401
 of real numbers, 66
Divisor polynomial, 268, 269, 270
Divisors, A-2
Domains, 433
 of functions, 439, 440
 of relations, 434, 447

Elements, 63
 of sets, 2
Elimination method, 468
Equality:
 addition property of, 73, 81–87, 88, 95, 117, 468
 multiplication property of, 73, 88–94, 95, 117
Equal measures, vertical angles with, A-5
Equal symbol, 3, 26
Equations, 1 *(See also* Formulas; Linear equations; Quadratic equations)
 checking solutions of, 37
 clearing of fractions in, 561, 565
 consistent systems of, 454, 489

containing roots, 498
definition of, 26
dependent, 454, 457, 466, 489
inconsistent system of, 454, 467, 489
independent, 457, 489
matching description of linear data to, 444–45
methods for solving of, 548
modeling a problem, 104
with no solution, 94, 98, 99
solution of system in two variables, 454, 455, 489
solving of with decimals, 94
solving of with fractions, 44
solving of with more than one property, 90
solving of with radicals, 497, 521–36, 540
system of, 454
translating of, 114
translating sentences into, 27
in two variables, 412
and variable expressions, 25–29, 65
writing sentences as percent, 125–26
writing in slope-intercept form, 413, 456, 457, 458
Equilateral triangles, A-12
Equivalent equalities, 142
Equivalent equations, 81, 82, 151
Equivalent fractions, 14, 64
Even integers, consecutive, 92
Exponential expressions:
 base of, 228, 231
 definition of, 228
 evaluating of, 21
 evaluating on scientific calculators, 23
 form $a^{m/n}$ and evaluation of, 533, 534
 form $a^{-m/n}$ and evaluation of, 533, 535
 form $a_{1/n}$ andnote evaluation of, 533
 multiplying with common base, 246
 simplifying of, 233, 235, 257, 258, 259, 260
Exponential notation, 19
Exponents, 227, 228–38
 definitions for, 274
 negative, 259, 533
 and order of operations, 19–24, 64

positive, 535, 536
power of a power rule for, 231, 235, 236, 259, 274, 533
power of a product or quotient, 259, 274, 533
power rule for, 228, 230, 231, 259, 274, 533
product rule for, 228, 230, 231, 259, 274, 533
quotient rule for, 228, 233, 234, 235, 236, 258, 259, 533
rules for simplifying expressions containing fractional, 533, 536
summary of rules for, 259
zero, 259, 274, 533
Exterior angles, A-5
Extraneous solution, 377
Extremes, 383, 403

Factoring, 334
 of binomials, 281, 304–9, 335
 difference of two squares, 304
 by grouping, 281, 285–87, 334
 of perfect squares, 505
 of perfect trinomials, 553
 of radicands, 504, 505
 solving quadratic equations by, 314–19, 336, 548
 strategies of polynomials with, 281, 284–85, 289, 292, 293, 298, 335–36
 and sum or difference of two cubes, 304
 of trinomials, 281, 289–304, 311, 334–335
Factors, 11, 12, 63
 zero as, 42
Finance and financial analysis:
 applications of radicals in, 527
 evaluating data for, 163, 217
FOIL method, 252
 finding a product by, 254
 of multiplication, 515
 in multiplying complex numbers, 568
 in multiplying two binomials, 275
 squaring a binomial using, 253
Formulas:
 comparing for doses of medication, 343, 399
 for converting degrees Celsius to degrees Fahrenheit, 115, 116
 definition of, 112
 for distance, 114, 133

examples of common, 112
for perimeter of rectangle, 115, 132
and problem solving, 73, 112–22, 152–53
Fractional exponents, 533
 using rules for exponents to simplify expressions containing, 533, 536
Fractions, 1, 11–18, 63–64
 adding and subtracting with same denominator, 14
 addition of, 267
 division by, 13, 353
 fundamental principle of, 12, 233, 234
 multiplying of, 13, 42
 simplifying of, 11, 12, 346
 solving equations containing, 94, 95–96
 with unlike denominators, 358
 written in lowest or simplest form, 346
Function notation, 433, 437, 447
Functions, 411, 412, 447, 433–43
 definition of, 434, 447
 domain of, 439, 440
 graphing of, 575
 range of, 439, 440
Fundamental principle:
 applied in multiplying rational expressions, 352, 360
 of rational expressions, 400
 for simplifying rational expressions, 364

GCF (*See* Greatest common factor)
Geometric figures, review of, A-12–A-14
Geometry, 176, 199
 application of radicals in, 527
 meaning of, A-3
 solving problems involving, 131, 132
Golden rectangle, 110, 111
Graphing:
 of equations, 25
 of equations using slope-intercept form, 415
 of equations with two variables, 454
 and functions, 435, 436, 438, 439
 of inequalities, 140

Graphing (cont.)
  of inequalities in two variables, 412
  of linear equations, 163, 175–80, 436
  of linear equations in two variables, 412, 572
  of linear inequalities, 163, 210–15
  of linear inequalities in two variables, 210–15
  of nonlinear equations, 411, 425–29, 447
  of ordered pairs, 166–67, 218
  of parabolas, 437–38
  of quadratic equations, 547, 572–81, 588–89
  of quadratic equations of form $y = ax^2 + bx + c$, 572, 576
  and solving linear equations in one variable, 164, 169
  and solving linear equations in two variables, 164, 168, 170, 171, 176–80
  and solving linear inequalities in two variables, 164
  and solving systems of linear equations by, 453, 454–59, 467, 48–89
  systems of linear inequalities, 483, 492
Graphing calculators:
  creating and interpreting graphs with, 62
  finding solutions of quadratic equations on, 320
  and linear equations, 181, 192
  and quadratic equations, 430–31
  and slope, 204
  slope and graphing equations using, 416
  and solutions of systems of equations, 459
  used in solving quadratic equations, 581
Graphs:
  creating and interpreting, 1, 62
  finding and plotting intercept points with, 187–90
  of functions, 447
  idenfifying intercepts of, 185, 186
  of inequalities, 140
  intercept points of, 219
  matching descriptions of linear data to, 444–45
  reading and interpreting percent in, 122, 124

  reading of, 55–61, 67
  slope, 412
Greatest common factor (GCF):
  and factoring by grouping, 281, 282–89, 311, 334
  factoring out, 305, 317, 318
Greatest y-value, 429
Grouping, 285
  factoring by, 334
  factoring four-term polynomial by, 286, 300–301
Grouping symbols, 19, 37
  examples of, 20, 22
  in exponential expressions, 21
  removal of, 95, 117, 143

Half planes, 210
  graphing of, 212, 213, 214, 215, 221
Historical data, making predictions based on, 227, 273
Horizontal axis, 164, 165 (See also x-axis)
  finding value on, 67
  slopes, 195, 200–201
Horizontal lines, 185, 191
  finding equation of, 418, 421
  and linear equations, 422
Horizontal number line, 218
Hypotenuse, of right triangle, 327, 328, 527, 528, A-9–A-10

Identity, 99
Identity element for addition, 52
Identity element for multiplication, 52
Identity properties, 52
Imaginary numbers, 567, 587
Imaginary unit, 567, 586
Inconsistent system of equations, 454, 456, 467, 489
Independent equations, 457, 489
Index, 539
  and natural numbers, 500
  in radicals, 534
Inequalities:
  addition property of, 141
  multiplication property of, 142
  solution of, 140
Inequality applications, solving of, 139
Inequality signs, 143

Inequality symbols, 3, 7, 139, 140, 141, 142, 143
Integers, 63
  consecutive, 328
  finding greatest common factor of, list of, 282, 334
  identifying of, 2
  multiplying by, 42
  quotient of, 344
  set of, 5
Intercepts and intercept points, 163, 185–91, 219–20
  finding on parabola, 572, 574–76, 588
Interest:
  formula for simple, 152
  solving problems involving, 131, 135–36
Interior angles, A-5
Intersecting lines, A-4
Irrational numbers, 5, 63, 501
  identifying of, 2, 498
  set of, 6
Isoceles trapezoids, A-13
Isoceles triangles, A-12

LCD (See Lowest common denominator)
Least common denominator, 15, 95, 96
  adding and subtracting rational expressions with, 343, 356, 358–59, 401
Least y-value, 429
Legs, of right angles, 327, 328, 527, A-9
Like radicals, 539
  addition of, 512
  definition of, 510
Like terms, 74, 75, 150
  combining in rational expressions, 358, 365
  combining of, 75, 76, 77, 91, 95, 150, 238, 242, 247, 248, 252, 510
  partial products and combining, 249
  in problem solving, 116, 117, 118, 143
Linear equations, 314 (See also Point-slope forms; Quadratic equations)
  addition property of equality in solving, 81
  forms of, 422
  graphing of, 163

and graphing in two variables, 412
and horizontal lines, 422
identifying of, 175, 176
intercepts and graphs of, 185
and linear inequalities, 139
matching descriptions of situations to, 411
in one variable, 548
point-slope form of, 422
problem solving and systems of, 453, 474–82
slope-intercept form and graphing of, 414–15
slope-intercept form of, 422
solving and addition property of equality, 88
solving and graphing in one variable, 164, 169
solving and graphing in two variables, 164, 168, 170, 171, 176, 190
solving and multiplication property of equality, 88
solving in one variable, 81, 85, 112
solving of, 73, 94–100, 151–52, 154–56
solving systems of, 453
solving systems of, by addition, 453, 468–74, 476, 480, 490–91
solving systems of, by graphing, 454–59, 488–89
solving systems of, by substitution, 453, 462–68, 490
standard form of, 422
in two variables, 176
and vertical lines, 422
x-intercept of graph of, 318
and zero factor theorem to solve, 315–16
Linear inequalities:
defining in one variable, 139, 140
graphing of, 208–217
in one variable, 154
solutions of, 73, 139, 141–49, 155
solutions of systems of, 483–85
solving and graphing in one variable, 143, 164
solving and graphing in two variables, 164
systems of, 453, 483–87, 492
Linear programming, 483, 484
Linear systems, 454
Line graphs, reading of, 55, 58–60
Lines, A-4, 412 (*See also* Slopes)
finding x-and y-intercepts on, 187
graphs of intercepts of, 185
on graphs of linear equations in two variables, 176–80
shapes of parallel, 194
shapes of perpendicular, 195
slopes of, 197–203
Long division (*See also* Division)
of polynomials, 268–70, 276
Lowest common denominators (LCDs), 117, 118
with addition method of solving systems of linear equations, 472
in linear inequalities, 143
Lowest terms, fractions in, 12, 64

Market research, graphs and, 1
Mathematical statements, translating sentences into, 2, 4
Mathematics, 2, 74, 412, 564
Means, 383, 403, A-15–A-17
Measures of central tendency, A-15
Medians, A-15–A-17
Members of sets, 2
Mixed numbers, rewriting of, 16
Mixtures, solving problems involving, 131, 134–35
Modes, A-15–A-17
Monomials, 238, 240 (*See also* Polynomials)
definition of, 239, 274
multiplication of, 246
polynomial divided by, 266, 267, 276
Multiplication (*See also* Factoring)
associative property of, 50, 66
commutative property of, 49, 53
of complex numbers, 566
of decimals, A-1
distributive property of, over addition, 50, 51
of exponential expressions, 230, 235
FOIL method of, 515
of fractions, 11, 13, 64
identity element for, 52
of polynomials, 227, 275
property of equality, 88–94, 95, 117, 151
property of inequality, 142, 143, 155
of radicals, 497, 514–20, 540
of rational expressions, 400–401
of real numbers, 1, 16, 30, 41–46, 49
sign rules for, 41
of sum and differences of two terms, 252, 255
Multiplicative identity(ies), identifying of, 48
Multiplicative inverse properties, identifying of, 48, 52
Multiplicative inverses (or reciprocals), 43, 44, 52, 53, 66

Natural numbers, 2, 63
Negative exponents, 259
definition of, 258
and scientific notation, 227, 257–66, 276, 533
Negative fractions, 345
Negative integers, 5, 276
Negative numbers, 6, 32, 41, 155
entering on scientific calculators, 46
on number line, 30
square roots of, 499, 501
Negative radicands, 567
Negative rational numbers, 535
Negative square root, 498, 538
Nonlinear equations, graphing of, 411
Number line, 2–3, 63
addition of real numbers on, 30
graphing solution sets on, 139, 140
numbers less than zero on, 5
Numbers (*See also* Irrational numbers; Natural numbers; Rational numbers; Real numbers; Whole numbers)
common sets of, 7
finding opposites of, 30, 33
finding percents of, 122
on number line, 5
ordered pairs of, 164, 166
symbols and sets of, 1, 2–9, 63
symbols to compare, 4
written in scientific notation, 257
Numerators, 11, 12, 63
factoring in rational expressions, 346, 347, 348, 352, 400
multiplication of, 351
in rational exponents, 534
simplifying of, 20, 21, 45, 46, 64
of slope, 199
Numerical coefficients, 74, 75
positive, 295, 296, 298
negative, 297, 310
of terms, 150, 239, 274
of trinomials, 295

Obtuse angle, A-3
Odd integers, consecutive, 92
Operation symbols, 19
Opposites (or additive inverses), 33, 34, 52, 53, 65
Ordered pairs, 164, 427, 434
  finding and plotting solutions for, 175, 177–80
  and graphing calculators, 181
  graphing of, 166, 167, 169, 188–91, 218
  and graphing quadratic equations, 589
  and linear inequalities in two variables, 208, 209
  reading from a graph, 428
  and solutions of equations, 454, 463
  and solutions of systems of equations
    in two variables, 454, 455
  and solutions of systems of linear inequalities, 483, 484
  written from graphs of nonlinear equations, 425
Ordered pair solutions, 412
Order of operations:
  exponents and, 1, 19–24, 45, 64
  with scientific calculators, 23
Order property, for real numbers, 7, 63
Origin, definition of, 164–65
Original inequality, 212, 213

Parabolas, 319, 426, 437, 447, 573, 588
  identifying graph of quadratic equation as, 572, 573, 574
  locating intercept points of, 572, 574–76
  and modeling a physical situation, 547, 584–85
  vertex of, 572, 576–80
Parallel lines, 422, 456, 459, A-4, A-6
  cut by transversal, A-5
  nonvertical, 414, 422
  shape of, 195
  slope of, 202
  slope of nonvertical, 220
Parallelogram, A-13
Parentheses, 20, 22
  in rational expressions, 345
  using distributive property to remove, 74, 76–77, 84, 95, 97, 117, 143, 150
Partial product, 249
Percent:
  decrease, 126
  increase, 126–27
  and problem solving, 73, 122–31, 153–54
  writing as decimals, 122, 123, 153
Perfect cube, cube root of, 498, 517
Perfect squares, 501
  factoring of, 505
  square roots of, 498
  table of, 504
Perfect square trinomials, 299–300
  definition of, 335
  factoring of, 335
  finding of, 552, 553
Perpendicular lines, A-4
  nonvertical, 414, 422
  shapes of, 195, 203
  slope of, 221
Physical situations: modeling by using quadratic equations, 547, 584–85
Pie charts, 55
  percent in, 124
Plane figures, A-4
Planes, description of, A-4
Point-slope form, 411, 418–24, 446
  of equations of a line, 419
  of linear equations, 422
  using to find equation of a line given point of line, 418, 419
  using to find equation of line given two points of line, 418, 420
  using to solve problems, 418
Polygons, definition of, A-6, A-12
Polynomials, 228
  addition of, 227, 238, 242, 274–75, 568, 587
  definition of, 274
  degree of, 240, 275
  divided by monomials, 267, 276
  divided by polynomials, 266, 268–70, 276
  division of, 227, 266–72, 276
  factoring by grouping, 282, 286
  factoring greatest common factor from, 282, 284–85, 305
  factoring of, 281, 289
  and like terms, 242
  multiplication of, 227, 246–51, 275
  multiplication of two, 247
  multiplying vertically, 246, 248
  prime, 305
  quotient of, 344
  subtraction of, 227, 238, 243, 274–75, 568, 587
  values of, 240–41
  in x, 239
Positive integers, 5, 258, 276
Positive numbers, 6, 32, 41, 155, 324
  on number line, 30
  products of, 567
Positive square root, 498, 499, 538
Powers of a product or quotient rule, 232, 233
  raising product to, 233
Power of a power rule for exponents, 259, 260, 274, 533
Power of a product rule, 259
  and exponents, 274, 533
Power of a quotient rule and exponents, 274, 533
Powers, real numbers raised to, 30
Price per unit, calculating of, 73, 149
Prime numbers, 12, 283
Prime polynomials, 292
Primes, product of, 334
Principal square root, 498, 499, 538
Problem solving:
  addition of real numbers and, 30
  and distance formulas, 528–29
  formulas and, 73, 77, 112–22, 152–53
  further, 154
  introduction to, 73, 104–11, 152
  involving distance, 131, 133
  involving geometry concepts, 131, 132
  involving interest, 131, 135–36
  involving mixtures, 131, 134–35
  and linear inequalities, 139–49, 154–56
  percent and, 73, 122–31, 153
  proportions and, 383
  and quadratic equations, 281, 324–33
  and radical equations, 497, 526–32, 540–41
  and rational equations, 343, 390–98, 404–5
  steps for, 104, 113, 131, 152
  of systems of linear equations and, 455, 474–82, 491
Product, 11, 12
  factors of, 282
Product rule:

for exponents, 259, 274, 533
for radicals, 507, 514, 515, 540
to simplify radicals, 504, 505, 511, 512
for square roots, 504
Products, 63
  of positive numbers, 567
  power rules for, 228, 232, 233, 235
  of real numbers, 567
  written as sums, 50, 52
  and zero, 66
Proportion:
  definition of, 383, 403
  and ratio, 383–90, 403–4
Pythagorean formula, 327, 328, 329, 527, A-9, A-10, A-11
  to solve problems, 426, 528, 529

Quadrants, 165, 167
Quadratic equations, 426, 437, 586–87
  complex solutions of, 547, 566–72
  definition of, 314, 336
  graphing of, 547, 572–81
  problem solving, 281, 324–33, 336–37
  solving by completing the square, 547, 552–57, 585
  solving in one variable, 548
  solving by the quadratic formula, 547, 558–63, 586
  solving by square root property, 547, 548–50
  in standard form, 315, 558, 559, 560, 561, 564, 565, 570, 571
  steps in solving of, 564
  summary of methods for solving, 547, 564–66, 586
  in two variables, 426, 447
  written in standard form, 378
  x-intercepts of graph in two variables of, 319
Quadratic formula, 559, 564
  solving quadratic equations by, 547, 558–63, 586
Quadrilaterals, A-12
Quotient:
  decimal points in, A-2
  of two polynomials, 267
Quotient polynomials, 269
Quotient rule:
  for exponents, 258, 259, 274
  for radicals, 507, 514, 515, 516, 540

to simplify radicals, 504
for square roots, 506
Quotients:
  of integers, 5
  power rules for, 228, 233, 235
  simplifying of, 234
  square roots of, 506
  of two real numbers, 43, 44, 66
  written as complex numbers, 569
  of zero and nonzero numbers, 44, 45, 66

Radical equations:
  and problem solving, 497, 526–32, 540–41
  solving those containing square roots, 523
Radical expressions, 499
  simplifying of, 510
  with variables, 511
  variables in radicands of, 501
Radical notation, 534
Radicals, 567
  addition of, 497, 510–12, 539
  introduction to, 497, 498–501, 538–39
  multiplication and division of, 497, 514–20, 540
  product rule for, 507, 539
  product rule to simplify, 504
  quotient rule for, 507, 539
  quotient rule to simplify, 504
  simplifying of, 497, 504–9, 512, 539
  solving equations containing, 497, 521–26, 540
  solving problems using formulas containing, 526
  subtraction of, 497, 510, 539
  symbols for, 499
Radical sign, symbol for, 499
Radicands, 499, 500, 501, 539
  factoring of, 504, 505
  in quadratic equations, 561
Range, 433
  of functions, 440
  of relation, 434, 447
Ratio:
  definition of, 383, 403
  and proportion, 343, 383, 390, 403–4
Rational equations, and problem solving, 343, 390–98, 404–5

Rational exponents, 497, 533–37, 541
Rational expressions, 267
  adding and subtracting with common denominators and least common denominators, 356–63, 401
  adding and subtracting with unlike denominators, 363–69
  definition of, 344, 400
  dividing of, 353–54, 400
  as equivalent rational expressions, 360, 364, 366
  factoring of, 282
  fundamental principle of, 346, 347, 360, 400–401
  multiplying of, 351–53, 400–401
  performing operations on, 379
  simplifying of, 344–50, 400
  solving equations containing, 376–82, 403
Rational numbers, 6, 63, 344
  identifying of, 2, 498
  set of, 5
Real numbers, 63
  absolute value of, 8
  addition of, 1, 30–35, 65
  evaluating on scientific calculators, 46
  identifying of, 2
  multiplying and dividing, 41–46, 66
  operations on scientific calculators, 46
  order property for, 7, 63
  properties of, 1, 48–55, 67
  quotients of two, 43, 66
  set of, 6
  subtraction of, 1, 33, 36–41, 66
Reciprocals (or multiplicative inverses), 43, 44, 52, 53, 66
  defined, 13
  fractions and, 64
Rectangles:
  area of, 16, 152, A-13
  formula for perimeter of, 115, 132
Rectangular coordinate system, 163, 164–74, 218
  on graphing calculator, 181
Rectangular solids, A-13
Relations, 433, 434, 435
  domains, 447
Remainder polynomials, 269
Replacement values, evaluating expresions for given, 37

Rhombus, A-13
Right angle, A-3
Right circular cones, A-14
Right circular cylinders, A-14
Right triangles, 327, A-9, A-10, A-12
Rise, between points, 196
ROOT feature on graphers, 320
Roots:
   of equations, 25, 26, 65
   finding nth of, 498, 500, 514, 539
   and raising to powers, 498
Run, between points, 196

Scalene triangles, A-12
Science, applications of radicals in, 527
Scientific calculators:
   checking equations on, 100–101
   exponential expressions on, 23
   numbers in scientific notation on, 263
   for simplifying or approximating square roots, 502
Scientific notation:
   defined, 260
   negative exponents and, 227, 257–66, 276
   writing number in, 261
Sets, 63
   definition of, 2
   of numbers, 63
   symbols and number, 1, 2–9
Ships, analyzing courses of, 453, 487–88
Signed decimals, multiplication of, 42
Signed numbers, 6, 38
   addition of real numbers with same, 30–31, 65
   addition of real numbers with unlike, 31–32, 65
   and zero, 42
Similar triangles (*See also* Triangles)
   solving problems involving, 390, 394–95, A-8, A-9
Simple inequalities, 141
Simplest form, radical expressions written in, 504
Slope, 163, 195–221
   denominator of, 199
   of horizontal lines, 201
   of a line, 197–203
   negative, 198, 202

numerator of, 199
   of parallel lines, 202
   of perpendicular lines, 203
   positive, 198, 202
   undefined, 201, 202, 220
   of vertical lines, 201
   of zero, 201, 202, 220
Slope-intercept form, 411, 412–18, 446
   definition of, 413
   determining parallel or perpendicular property of lines, 412, 414
   equations written in, 457, 458
   finding slope and y-intercept of line, 412, 413
   in graphing linear equations, 412, 414
   of linear equations, 422
   and writing an equation of line, 412, 414
Solutions:
   of equations, 25, 26, 65
   of system of two equations in two variables, 454, 455, 489
   systems without, 467
Solution sets: graphing on number line, 139
Special products, 227, 252–57, 275
   for squaring binomials, 299
   used to multiply expressions containing radicals, 515
Spheres, A-14
Square root property, 565
   of negative numbers, 501
   of perfect squares, 498
   solving quadratic equations by, 547, 548–50, 559, 570, 585
   table of, A-18, A-19
Squares, A-13
   factoring and difference of, 335
   factoring difference of two, 304–5
   table of, A-18, A-19
Squaring property of equality, 521, 522, 523
Standard form, 176, 219
   complex numbers written in, 567, 568
   equations written in, 419, 420, 524
   of linear equations, 422
   numbers converted from scientific notation to, 257, 262
   quadratic equations written in, 315, 336, 378, 558, 559, 560, 561, 564, 565, 570, 571
Standard order of operations, 344

Standard windows on graphing calculators, 181, 430
Statistics, percents used in, 122, 124
Straight angle, A-3
Substitution method, 463, 471
   solving systems of linear equations by, 453, 462–68, 490
Subtraction:
   from both sides of equations, 82–83
   of complex numbers, 566, 587
   of decimals, A-1
   of exponential expressions, 230
   of fractions, 11
   of fractions with same denominators, 14, 64
   of fractions with unlike denominators, 358
   of polynomials, 227, 238, 242, 243, 274–75, 568, 587
   of radicals, 497, 510–12, 539
   rational expressions with common denominators, 343, 356–57, 401
   of rational expressions with least common denominators, 343, 356, 358–59, 401
   of rational expressions with unlike denominators, 363–69, 401–2
   of real numbers, 1, 30, 33, 36–41, 49, 66
   of values, 141
Sum of opposites, 33
Sums, written as products, 50, 52
Supplementary angles, A-3, A-6
Surveys, graphs and, 1
Symbols:
   for inequality, 139, 140, 141, 142, 143, 155
   for negative square root, 498
   for percent, 123, 153
   for principal square root, 498
   for radicals (or radical sign), 499, 539
   and sets of numbers, 1, 2–9, 63
   translating words and sentences to, 22
System of equations, 454

Table of values, 170, 171
Technology, applications of radicals in, 527
Terms, 239
   combining like, 242, 243, 244
   coefficient of, 274

definition of, 74, 274
degree of, 240, 274
finding greatest common factor of list of, 282, 283, 284
multiplying the sum and difference of two, 252, 275
numerical coefficient of, 150
in polynomials, 309–12
Trace feature on graphing calculator, 320, 430, 431
Transversal:
  definition of, A-5
  parallel lines cut by, A-5
Trapezoids, A-13
Triangles, A-12
  areas of, 16
  congruent, A-7
  definition of, A-6
Trinomials, 238, 240 (*See also* Polynomials)
  definition of, 239, 274
  factoring of, 281, 289–304, 305, 310, 334–35
  perfect square, 299–300, 335

Undefined slope, 201, 202, 220
Unit pricing, determining, 383
Unknown number, finding, 390, 391
Unlike denominators:
  adding and subtracting fractions with, 358, 401–2
  adding and subtracting rational expressions with, 343, 363–69
Unlike terms, 74

Values:
  with denominator of zero, 400
  and equations representing different, 454
  with rational expression undefined, 345, 400
  of rational numbers, 344
Variable expressions, 1
  and equations, 25–29, 65
  formula used to determine, 529–30
Variables, 2, 25, 65, 74
  factors containing, 317
  of formulas, 112
  graphing linear equations in two, 318
  graphing linear inequalities in two, 210–15

greatest common factor of list of, 283
isolating of, 82, 83, 88, 89, 90, 91, 95, 141, 143, 145
and linear equations in one, 81, 95, 112, 151
and linear equations in two, 176
radical expressions containing, 511
radicands containing, 501
replacement values of, 344
solving equations containing rational expressions for specified, 376, 379–380
solving equations for, 26
solving equations for specified, 116–18
solving formula for specified, 153
value of polynomial with replacement for, 238, 240
Vertex, 573, 576–80, 588
  formula for, 579
Vertical angles, A-5, A-6
Vertical axis, 164, 165 (*See also* y-axis)
  finding value on, 67
Vertical lines:
  finding equations of, 418, 420
  and linear equations, 422
  slopes of, 195, 200–201, 220
Vertical line test, 433, 435, 436, 437, 438, 447, 575
Vertical number line, 218

Whole numbers, 2, 4, 30, 63, 283
  multiplied by zero, 42
  on number line, 3
Window(s) on graphing calculators, 181
Word phrases, written as algebraic expressions, 74, 78, 81, 104
Work, solving problems involving, 390, 391–93

x-axis, 165, 218, 219, 428
  on graphing calculator, 181
  y-coordinate on, 168, 187
x-coordinate, 165
  and domain of a relation, 434
  on graphers, 320
  horizontal change in, 196
  on y-axis, 168

x-intercept points, 185, 186, 219, 220, 429
  finding, 187, 189, 412
  on graphs of linear equations, 318
  on graphs of quadratic equations, 574, 575, 588
  on graphs of quadratic equations in two variables, 319
x-value, 426, 427, 434
  on graphs of quadratic equations, 572, 574
  of vertex of parabola, 588
  y-coordinates corresponding to, 428

y-axis, 165, 218, 219, 428
  on graphing calculator, 181
  graph symmetric about, 573
  x-coordinate on, 168, 187
y-coordinate, 165, 428
  on x-axis, 168
  on graphers, 320
  and range of relation, 434
  vertical change in, 196
y-intercept points, 185, 186, 219, 220, 412, 413, 429, 456
  finding of, 187, 189
  to graph quadratic equations, 574, 575, 588
  and slope–intercept form, 446
Young's Rule, 399
Y-value, 426, 427, 428, 429, 431, 434
  on graphs of quadratic equations, 572, 574
  of vertex of parabola, 588

Zero:
  as divisor or dividend, 45
  and lack of multiplicative inverse, 43
  products and quotients involving, 66
  quotient of, 44
  and signed numbers, 42
  slope of, 201, 202, 220
Zero exponent, 235, 259, 274, 533
Zero factor theorem, 315–16, 317, 548
  applied to quadratic equations, 565
ZOOM features:
  on graphing calculators, 320, 430

Geometric Formulas

Rectangle

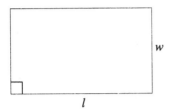

Perimeter: $P = 2l + 2w$
Area: $A = lw$

Square

Perimeter: $P = 4s$
Area: $A = s^2$

Triangle

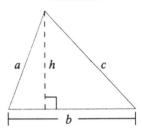

Perimeter $P = a + b + c$
Area: $A = \frac{1}{2}bh$

Sum of Angles of Triangle

$A + B + C = 180°$
The sum of the measures of the three angles is 180°.

Right Triangle

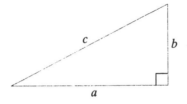

Perimeter: $P = a + b + c$
Area: $A = \frac{1}{2}ab$
One 90° (right) angle

Pythagorean Theorem (for Right Triangles)

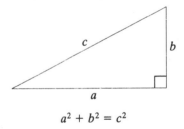

$a^2 + b^2 = c^2$

Isosceles Triangle

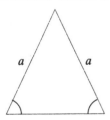

Triangle has:
two equal sides and
two equal angles.

Equilateral Triangle

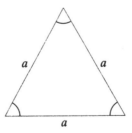

Triangle has:
three equal sides and
three equal angles.
Measure of each angle is 60°.

Trapezoid

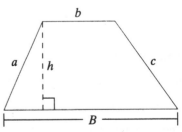

Perimeter: $P = a + b + c + B$
Area: $A = \frac{1}{2}h(B + b)$

PARALLELOGRAM

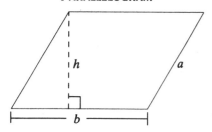

Perimeter: $P = 2a + 2b$
Area: $A = bh$

CIRCLE

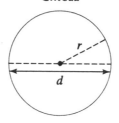

Circumference: $C = \pi d$
$C = 2\pi r$
Area: $A = \pi r^2$

RECTANGULAR SOLID

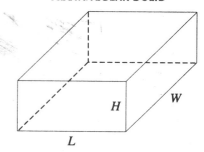

Volume: $V = LWH$
Surface Area: $A = 2HW + 2LW + 2LH$

CUBE

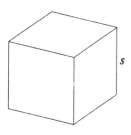

Volume: $V = s^3$
Surface Area: $A = 6s^2$

CONE

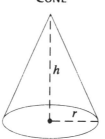

Volume: $V = \frac{1}{3}\pi r^2 h$
Surface Area: $A = \pi r \sqrt{r^2 + h^2}$

RIGHT CIRCULAR CYLINDER

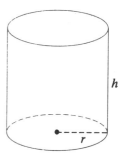

Volume: $V = \pi r^2 h$
Surface Area:
$A = 2\pi rh + 2\pi r^2$

OTHER FORMULAS

Distance: $d = rt$ (r = rate, t = time)

Temperature: $F = \frac{9}{5}C + 32 \quad C = \frac{5}{9}(F - 32)$

Simple Interest: $I = Prt$
  (P = principal, r = annual interest rate, t = time in years)

Compound Interest: $A = P\left(1 + \frac{r}{n}\right)^{nt}$
  (P = principal, r = annual interest rate, t = time in years, n = number of compoundings per year)

Common Graphs

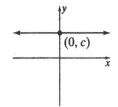
Horizontal Line;
Zero Slope
$y = c$

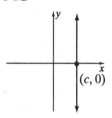
Vertical Line;
Undefined Slope
$x = c$

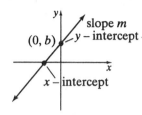
Linear Equation;
Positive Slope
$y = mx + b; m > 0$

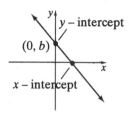
Linear Equation;
Negative Slope
$y = mx + b; m < 0$

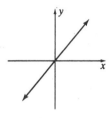
$y = x$

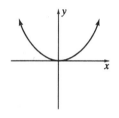
$y = x^2$

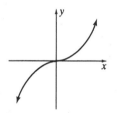
$y = x^3$

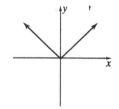
$y = |x|$

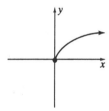
$y = \sqrt{x}; x \geq 0$

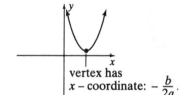
Quadratic Equation
$y = ax^2 + bx + c; a \neq 0$
Parabola opens upward if $a > 0$
Parabola opens downward if $a < 0$

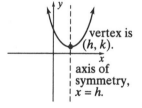
Quadratic Equation
$y = a(x - h)^2 + k; a \neq 0$
Parabola opens upward if $a > 0$
Parabola opens downward if $a < 0$

Systems of Linear Equations

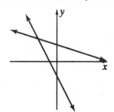
Independent and
consistent; one solution

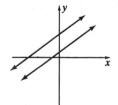
Independent and
inconsistent; no solution

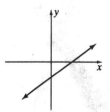
Dependent and
consistent; infinitely many solutions

Algebraic Formulas

Slope of a Line Containing Points (x_1, y_1) and (x_2, y_2)

$$m = \frac{y_2 - y_1}{x_2 - x_1}, \text{ if } x_1 \neq x_2$$

The slope is m.

Slope-Intercept Form of the Equation of a Line

$$y = mx + b$$

The slope is m and y-intercept is b.

Point-Slope Form of the Equation of a Line

$$y - y_1 = m(x - x_1)$$

The slope is m and point (x_1, y_1) is on the line.

Quadratic Formula

$$x = \frac{-b \pm \sqrt{b^2 - 4ac}}{2a}$$

Yields solutions of an equation of the form $ax^2 + bx + c = 0$, where $a \neq 0$.

Reference to Selected Keys on Scientific and Graphing Calculators

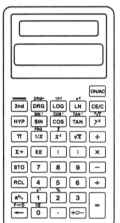

Scientific Calculator

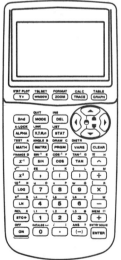

Graphing Calculator (Optional)

$=$ or ENTER	Gives the result of the current operation.
y^x or a^b or $\wedge$	Raises any base to a power. For example, 4^5 as 1024.
EXP or EE	Use to enter a number in scientific notation. For example, 6.1×10^8.
$+/-$ or ⌐ or $(-)$	Changes the sign of the displayed number. Also, use to enter a negative value. (This is not the subtraction key.)
$1/x$ or x^{-1}	Finds the reciprocal of a displayed number.
STO or STO→	Stores a number in memory.
RCL or RCL/STO→	Recalls a number stored in memory. May need to press 2nd STO
x^2	Squares the displayed number.
DEL	On a graphing calculator, deletes previously entered characters.
X, T or X, T, θ or X, T, θ, n	On a graphing calculator, use to write the variable x.
$y=$	On a graphing calculator, use to enter equation(s) to be graphed.
WINDOW or RANGE	On a graphing calculator, use to set the viewing window to choose the part of the coordinate plane to be displayed.
ZOOM	On a graphing calculator, use to "zoom-in" or "zoom-out" to magnify or reduce a part of the curve. Can also be used to obtain a "square" setting, providing equal spacing between tic marks on the x- and y-axes to obtain a true geometric perspective.
TRACE	On a graphing calculator, use to find the x- and y-coordinates of points on a curve.
GRAPH	On a graphing calculator, use to graph a previously entered equation(s).
TABLE GRAPH or 2nd GRAPH	On a graphing calculator, use for a table of values relating to an entered equation. The table can be used to estimate an answer to a problem.
OFF or 2nd OFF	Turns calculator off.